THE BASAL FOREBRAIN
Anatomy to Function

ADVANCES IN EXPERIMENTAL MEDICINE AND BIOLOGY

Recent Volumes in this Series

THE BASAL FOREBRAIN

Anatomy to Function

Edited by

T. Celeste Napier

Loyola University of Chicago
Stritch School of Medicine
Maywood, Illinois

Peter W. Kalivas

College of Veterinary Medicine
Washington State University
Pullman, Washington

and

Israel Hanin

Loyola University of Chicago
Stritch School of Medicine
Maywood, Illinois

PLENUM PRESS • NEW YORK AND LONDON

Library of Congress Cataloging-in-Publication Data

The Basal forebrain : anatomy to function / edited by T. Celeste
Napier, Peter W. Kalivas, and Israel Hanin.
    p.   cm. -- (Advances in experimental medicine and biology ; v.
295)
  The proceedings of a conference held May 24-27, 1990 in Chicago,
Ill.
  Includes bibliographical references and index.
   ISBN 978-1-4757-0147-0   ISBN 978-1-4757-0145-6 (eBook)
   DOI 10.1007/978-1-4757-0145-6
  1. Prosencephalon--Congresses. I. Napier, T. Celeste.
II. Kalivas, Peter W., 1952- . III. Hanin, Israel. IV. Series.
  [DNLM: 1. Basal Ganglia--anatomy & histology--congresses.
2. Electrophysiology--congresses. 3. Neurons--physiology-
-congresses. 4. Neuropharmacology--congresses. W1 AD559 v. 295 /
WL 307 B2965 1990]
QP382.F7B36 1991
599'.0188--dc20
DNLM/DLC
for Library of Congress   91-23984
                                          CIP

Proceedings of a conference on the Basal Forebrain: Anatomy to Function,
held May 24-27, 1990, in Chicago, Illinois

ISBN 978-1-4757-0147-0

© 1991 Plenum Press, New York
Softcover reprint of the hardcover 1st edition 1991
A Division of Plenum Publishing Corporation
233 Spring Street, New York, N.Y. 10013

PREFACE

The basal forebrain has received considerable attention in recent years. This emphasis resulted from observations that the cortically projecting cholinergic neurons found in this region are critical for normal information processing. However, to achieve a complete understanding of such a complex function as "information processing" it is necessary to consider the basal forebrain not as an autonomous structure with a solitary task, but one that plays an <u>integrative role</u>; a structure that is connected intimately with many brain regions. This view evolved from the realization that the basal forebrain interfaces cognitive and reward functions with motor outputs. It is from this integrative and functional perspective that the present book was organized.

The book is a unique collection of reports pertaining to the basal forebrain that encompasses a diversity of research approaches and techniques. It provides the reader with a progression of information that begins with anatomical descriptions of the afferent and efferent systems, stressing the integrative nature of various neurotransmitters located within the basal forebrain. The chapters focusing on anatomy are complemented by electrophysiologic studies that merge anatomical concepts with synaptic pharmacology and behavior. *In vitro* experiments demonstrate physiologic variations in anatomically identified neuronal subtypes and, together with *in vivo* techniques, provide pharmacologic descriptions of neuronal consequences to various neurotransmitter influences. Additional *in vivo* reports correlate changes in neuronal activity with specific motivational states and motor behaviors. These functional approaches culminate with behavioral studies that overview current understanding of basal forebrain involvement in mnemonic, reward, and motor processes. The pharmacology of neurotransmitter systems involved in basal forebrain regulation of these functions is discussed. Finally, descriptions of normal basal forebrain anatomy and physiology are augmented by behavioral, *in vitro*, and clinical approaches to analyzing and treating basal forebrain dysfunction.

The research diversity of the chapters within this book illustrates the enthusiasm that has been generated with regard to the basal forebrain. The novelty of findings in a particular subject underscores the pioneering nature of the topic. Thus, this book provides the first collection of interdisciplinary subjects related to the basal forebrain, serving as a guide for future research. Much remains to be learned before a mature understanding of the physiology and pathology of this complex brain region can be realized.

Each of the chapters in the book resulted from a presentation at The Basal Forebrain: Anatomy to Function conference held in Chicago, May 23-27, 1990. Gratitude is extended to the following Chairpersons of the conference sessions: Umberto Cornelli, Suzanne Haber, Steven Henriksen, and Derek Van Der Kooy. These scientists guided the lively discussions which ultimately influenced the conclusions summarized in the chapters of this book. The editors also are grateful to the following agencies which provided funds to support the conference:

National Institute on Aging, Washington, D.C.
National Science Foundation, Washington, D.C.

Abbott Laboratories, Abbott Park, IL
Alzheimer's Association, Chicago, IL
Berlex Laboratories, Wayne, NJ
Burroughs Wellcome Co., Research Triangle Park, NC
Fidia SPA, Padvoa, Italy
Hoechst-Roussel Pharmaceuticals, Somerville, NJ
Institut De Recherches Internationales Servier, Courbevoie,
   France
Miles, Inc., West Haven, CT
Neuroscience and Aging Institute, Loyola Univ. Chicago,
   Stritch School of Medicine, Maywood, IL
Office of the Associate Dean for Research, Loyola Univ.
   Chicago, Stritch School of Medicine, Maywood, IL
Schering-Plough Research, Bloomfield, NJ
Syntex Research, Palo Alto, CA
Wyeth-Ayerst Research, Princeton, NJ

T. Celeste Napier   Peter W. Kalivas   Israel Hanin
Maywood, IL   Pullman, WA   Maywood, IL

CONTENTS

ANATOMY OF THE BASAL FOREBRAIN AND RELATED STRUCTURES

ELECTROPHYSIOLOGY AND PHARMACOLOGY OF BASAL FOREBRAIN NEURONS

CLINICAL MANIFESTATION OF BASAL FOREBRAIN DYSFUNCTION, AND MODELS OF THIS DYSFUNCTION

HISTORY OF ACETYLCHOLINE

PIECING TOGETHER THE PUZZLE OF BASAL FOREBRAIN ANATOMY

Lennart Heimer[1] and George F. Alheid[2]

[1]Departments of Otolaryngology and Neurosurgery
[2]Department of Behavioral Medicine and Psychiatry
University of Virginia Health Sciences Center
Charlottesville, Virginia 22908

INTRODUCTION

The basal forebrain contains a seemingly heterogeneous collection of structures including nucleus accumbens, olfactory tubercle, septum, diagonal band nuclei, bed nucleus of stria terminalis, substantia innominata, olfactory cortex, hippocampus formation and amygdaloid body. It is also traversed by a number of large fiber tracts, e.g. fornix, stria terminalis, diagonal band of Broca, medial forebrain bundle, inferior thalamic peduncle, and ventral amygdalofugal pathway, to which the various basal forebrain structures contribute axons in order to establish connections between themselves and with other parts of the brain. Hypothalamus, the main diencephalic component of the basal forebrain, is one such region closely related to many of the telencephalic basal forebrain structures and fiber tracts. These intimate relations to the hypothalamus provided much of the anatomical rationale to bring the above-mentioned basal forebrain structures together as integral parts of the "limbic system". This has contributed to the popular view of forebrain organization in which the neocortex is related to the basal ganglia or the "extrapyramidal motor system" through the well-known cortico-striato-pallidal pathways, while so-called limbic structures, e.g. septum, nucleus accumbens, amygdaloid body and allocortical areas like hippocampus and olfactory cortex are characterized foremost by their relation to the hypothalamus, a major regulator of autonomic and endocrine functions.

Attractive as this viewpoint may have been, modern anatomical studies during the last 20 years have challenged some of the long-standing assumptions that provided the impetus for the dichotomy between an extrapyramidal and limbic system. For instance, it had long been assumed, primarily on the basis of normal-anatomical studies during the first half of this century, that the large fiber systems like the deep olfactory radiation (Dejerine, 1895; Herrick, 1910; Ramón y Cajal, 1911) and ventral amygdalofugal pathway (Gloor, 1955; de Olmos, 1972) served primarily as the anatomical substrate for olfacto-hypothalamic and amygdalo-hypothalamic interactions. However, the notion that the preoptic-anterior hypothalamic continuum constitutes a formidable gateway for olfactory impulses to the subcortical structures has not survived the test by modern experimental tract-tracing studies. Instead, a significant part of the fibers that converge on the rostral mediobasal forebrain from structures such as the olfactory cortex, as well as from the cortical-like basolateral amygdala, were shown to terminate more prominently in extrahypothalamic areas than within the hypothalamus (de Olmos, 1972; Heimer, 1972; Krettek and Price, 1978). Furthermore, some of these extrahypothalamic forebrain areas, including the accumbens and medium-celled part of the olfactory tubercle, were shown to project massively to a ventral subcommissural extension of the globus pallidus (Heimer and Wilson, 1975; Swanson and Cowan 1975), which occupies a large part of the so-called subcommissural substantia innominata (see below). These more than 15 year old studies, aided by sensitive silver impregnation techniques and autoradiographic methods, paved the way for the notion of the <u>ventral striatopallidal system</u> (Heimer and Wilson, 1975; Heimer, 1978; Heimer et al., 1982; Switzer et al., 1982). This concept has been supported by many subsequent studies, and it is now generally accepted that the ventral parts of the basal ganglia, i.e. the ventral striatum and ventral pallidum, extend in a continuous fashion to the ventral surface of the brain in the region of the olfactory tubercle or the anterior perforated space in every mammal that has been examined, including the human (see reviews by Heimer et al., 1985; Alheid and Heimer, 1988; Alheid et al., 1990; Heimer et al., 1991a).

As indicated by Price et al. (1987), the ventral amygdalofugal pathway is represented by a multitude of different components, and the medial sweep of this diversified system from various parts of the amygdaloid complex is so broad as to defy the term "pathway". Later in this discussion, we will focus the attention on a prominent component of the ventral amygdalofugal pathway which terminates densely in another extrahypothalamic continuum composed of bed nucleus of stria terminalis and continuous cell columns in the sublenticular substantia innominata. Together with centromedial amygdala, and we suggest portions of the medial accumbens, these extrahypothalamic areas form a large forebrain continuum, i.e. the <u>extended amygdala</u>. Although the term "extended amygdala" is a recent contrivance (Alheid and Heimer, 1988), the concept of an extension of the amygdala into the mediobasal forebrain traces its origin to J.B. Johnston, who already in 1923 argued on comparative and developmental grounds that the centromedial amygdala, bed nucleus of stria terminalis and nucleus accumbens should be recognized as an anatomical entity.

VENTRAL STRIATOPALLIDAL SYSTEM

The ventral striatopallidal system occupies an extensive region in the mediobasal forebrain, including the nucleus accumbens, olfactory tubercle, and large territories of the general area referred to as subcommissural substantia innominata. Based on the pattern of afferents from allocortical (hippocampal formation and primary olfactory cortex), periallocortical (e.g. entorhinal area) and frontal and temporal proisocortical areas (e.g. insular, anterior cingulate, orbitofrontal, anterior temporal and perirhinal cortex), the accumbens, medium-celled parts of the olfactory tubercle and the most ventromedial parts of the caudate-putamen were referred to as ventral striatum (Heimer and Wilson, 1975). In typical striatal fashion, the ventral striatum projects to an adjoining pallidal area referred to as ventral pallidum, which occupies an extensive region underneath the temporal limb of the anterior commissure (see below). The ventral striatopallidal system of the rat and the primate has been reviewed in detail in several recent publications (Heimer et al., 1985; Alheid and Heimer, 1988; Alheid et al., 1989b, 1990; Groenewegen et al., 1989; 1991; Heimer et al., 1991a), and it will be sufficient to emphasize only its most salient anatomical features.

Olfactory Tubercle

That the nucleus accumbens or fundus striatum reaches the ventral surface of the human brain in the region of the anterior perforated space was
well appreciated by earlier anatomists (e.g., Beccari, 1910, 1911; von
Economo and Koskinas, 1925; Macchi, 1951). However, by the use of striatal
markers, e.g. acetylcholinesterase, it can be appreciated that the striatal complex reaches the ventral surface of the brain not only in primates
including the human, but also in macrosmatic mammals with a well developed
olfactory tubercle (Fig. 1). The concept of the ventral striatopallidal
system includes a realignment of the olfactory tubercle. Since the tubercle of macrosmatic mammals has some laminar characteristics and is a recipient of direct olfactory bulb input, it was traditionally considered an
integral, albeit modified part of the olfactory cortex (prepiriform cortex). But when evidence is considered from data obtained by a number of cytoarchitectonic, histochemical and hodological techniques, it becomes apparent that the many striatal features of the olfactory tubercle by far
outnumber its purported similarities with neighboring olfactory cortex. The
medium-sized cell territories of the tubercle have cellular and histochemical features reminiscent of striatum (Heimer, 1978; Krieger, 1981; Millhouse and Heimer, 1984; Mugnaini and Oertel, 1985; Phelps and Vaughn,
1986), and its pattern of extrinsic connections are generally typical of
more dorsal striatal areas (Heimer and Wilson, 1975; Heimer, 1978; Newman
and Winans, 1980; Fallon, 1983; Young et al., 1984; Heimer et al., 1987;
Zahm and Heimer, 1987; Zahm et al., 1987). Furthermore, a closer analysis
of the suggested similarities between the olfactory cortex and olfactory
tubercle, i.e. lamination and direct relations to the olfactory bulb, suggests that even those few similarities are rather tenuous. Contrary to the
situation in the neighboring olfactory cortex, the olfactory tubercle is
not characterized by a radial and tangential organization of its structural elements (Blackstad, 1967). Nor does the olfactory cortex, as often
was contended, show a gradual transition into the olfactory tubercle. The
border between these two structures is sharply defined not only in Nissl
preparations (see Figs. 27 and 28 in Heimer, 1978), but also in sections
prepared by histochemical methods that stain preferentially in either the
striatum or cortex (Fig. 1).

Although both olfactory tubercle and neighboring olfactory cortex receive direct input from the olfactory bulb, the olfactory bulb projection
to the tubercle is of much smaller magnitude than that to the olfactory
cortex. Furthermore, the olfactory bulb input to the tubercle is not
matched by a massive reciprocal pathway to the bulb as is the case for the
olfactory cortex. Likewise, the projection from olfactory cortex to the tubercle is not reciprocated (Haberly and Price, 1978; Luskin and Price,
1983) as might be expected if the tubercle were an integral part of olfactory cortex. We could consider the olfactory tubercle a rather exceptional
part of the striatal complex, in that it is reached by direct sensory, i.e.
olfactory input, in addition to receiving such information indirectly from
cortex. However, a growing number of experiments also have shown direct
projections to the dorsal striatum from sensory relays in the thalamus.
This information appears to converge in the striatal zones receiving fibers
from the related sensory areas of cortex, so that the rather direct olfactory projections to a striatal territory receiving afferents from olfactory
cortex appear to be a rather typical pattern for sensory afferents to striatum (see LeDoux et al., 1985; Matelli et al., 1988; Alheid et al., 1990,
p. 558 for review). As with other striatal areas, the efferents of the
striatal parts of the olfactory tubercle proceed via a relay in pallidal
areas (i.e. ventral pallidum, see below). This includes an extensive layer

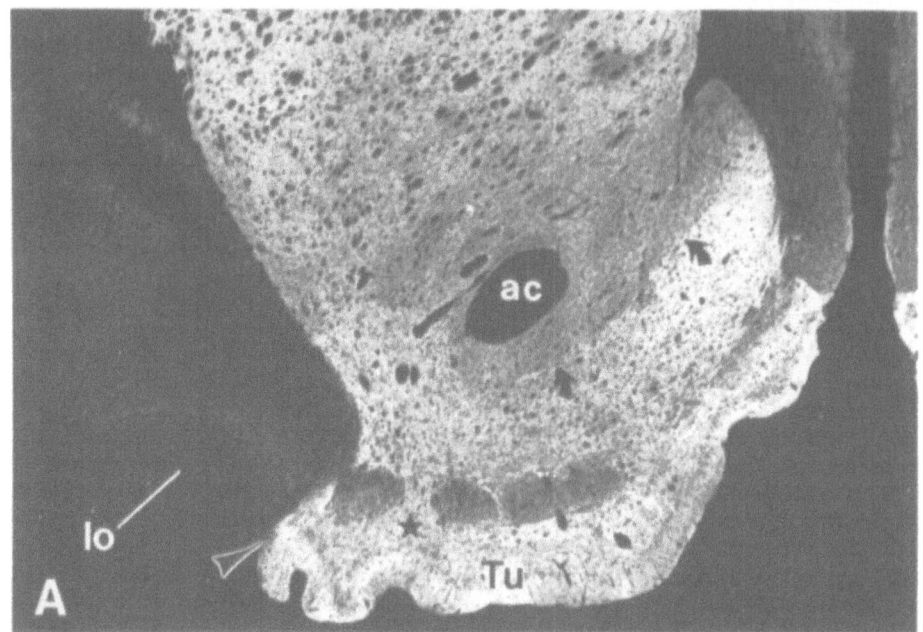

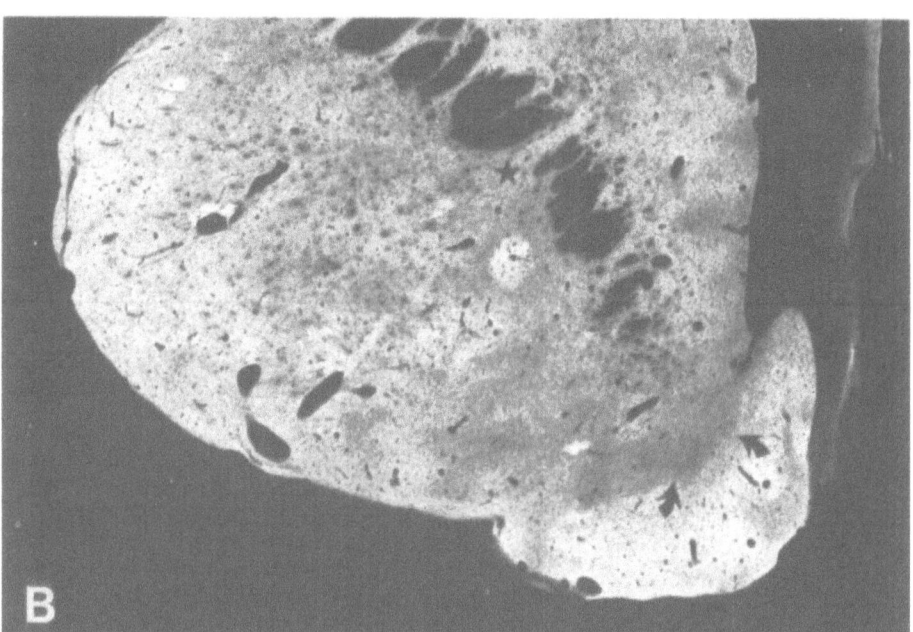

Fig. 1.   Acetylcholinesterase staining in the nucleus accumbens of the rat
(A) and squirrel monkey (B). The black curved arrows indicate the
boundary of the shell of the accumbens with the adjacent core
area. The solid stars mark cell bridges between the accumbens and
olfactory tubercle (Tu) in (A) and between the caudate and
putamen in (B). Note the sharp border (arrowhead in A) between
the olfactory tubercle and olfactory cortex. ac = anterior
commissure, lo = lateral olfactory tract.

of pallidal cells in the deep portions of the tubercle. The tubercle-related pallidal cells share with the remainder of ventral pallidum efferent axons targeting the mediodorsal thalamus as a major output (e.g., Young et al., 1984; Zahm and Heimer, 1987; Zahm et al., 1987; Groenewegen, 1988).

Cells in the general region of the olfactory tubercle and olfactory cortex previously have also been identified as representing the origin for a remarkable pathway to the nuclei gemini in the rat posterolateral hypothalamus. The pathway, which is located in the heavily myelinated ventro-lateral part of the medial forebrain bundle, apparently carries olfactory-related information (Scott and Pfaffman, 1967, 1972). This would seem to fit the suggested location of gemini-projecting cells in olfactory areas of the basal forebrain, including the olfactory tubercle (e.g., Scott and Leonard, 1971; Scott and Chafin, 1975; Fallon, 1983; Price et al., 1991). However, the presence of a projection from the olfactory tubercle to the hypothalamus appears anomalous in the view of the tubercle as a ventral extension of the basal ganglia. In a recent study we have proposed an alternative explanation. Anterograde and retrograde tracing indicated that most of the large gemini-projecting cells are diffusely scattered among the fibers of the medial forebrain bundle (Heimer et al., 1990; Price et al., 1991), and we have suggested that they might be considered a subpopulation of the cells of the nucleus of the horizontal limb of the diagonal band. On close analysis, gemini-projecting cells are generally not found within the striatal or pallidal parts of the olfactory tubercle. However, where the ventral pallidum interdigitates with the longitudinal fascicles of the medial fore-brain bundle, many gemini-projecting cells encroach on ventral pallidal areas, apparently in a similar fashion as the large corticopetal cells of basal forebrain traverse certain pallidal areas (see section on the Magno-cellular Corticopetal Complex below). Therefore, it remains to be seen whether gemini-projecting cells receive olfactory-related information through pathways originating in the olfactory tubercle, or by other routes, for example, those originating in the olfactory cortex.

Nucleus Accumbens

There is little doubt about the striatal characteristics of nucleus accumbens; like the rest of the striatum it receives prominent afferents from cerebral cortex, e.g. from prefrontal cortex, hippocampus, entorhinal cortex, and from the cortical-like basolateral amygdala. As with the dorsal striatum it also receives input from mesencephalic dopamine neurons and "non-specific" thalamic nuclei. Its compartmentalized infrastructure (Graybiel et al., 1979; Gerfen, 1984; Herkenham et al., 1984; Gerfen, 1985; Groenewegen et al., 1989; Voorn et al., 1989; Graybiel, 1990) is typical of striatum, and it projects prominently to the ventral pallidum and substantia nigra-ventral tegmental area (e.g., Heimer and Wilson, 1975; Swanson and Cowan, 1975; Williams et al., 1977; Nauta and Domesick, 1978; Nauta et al., 1978; Haber and Nauta, 1983; Groenewegen and Russchen, 1984; Alheid et al., 1989a; 1990; Haber et al., 1990; Heimer et al., 1991b). Unlike the main dorsal expanse of the striatum, however, nucleus accumbens has direct connections also with the hypothalamus (Koikegami et al., 1967; Nauta et al., 1978; Groenewegen and Russchen, 1984; Heimer et al., 1991b). In other words, it encompasses essentially all of the features of the basal ganglia, but also includes several anomalous features from the viewpoint of basal ganglia circuitry, most prominently direct relations with hypothalamus. The hypothalamic relations of the nucleus accumbens projections, together with

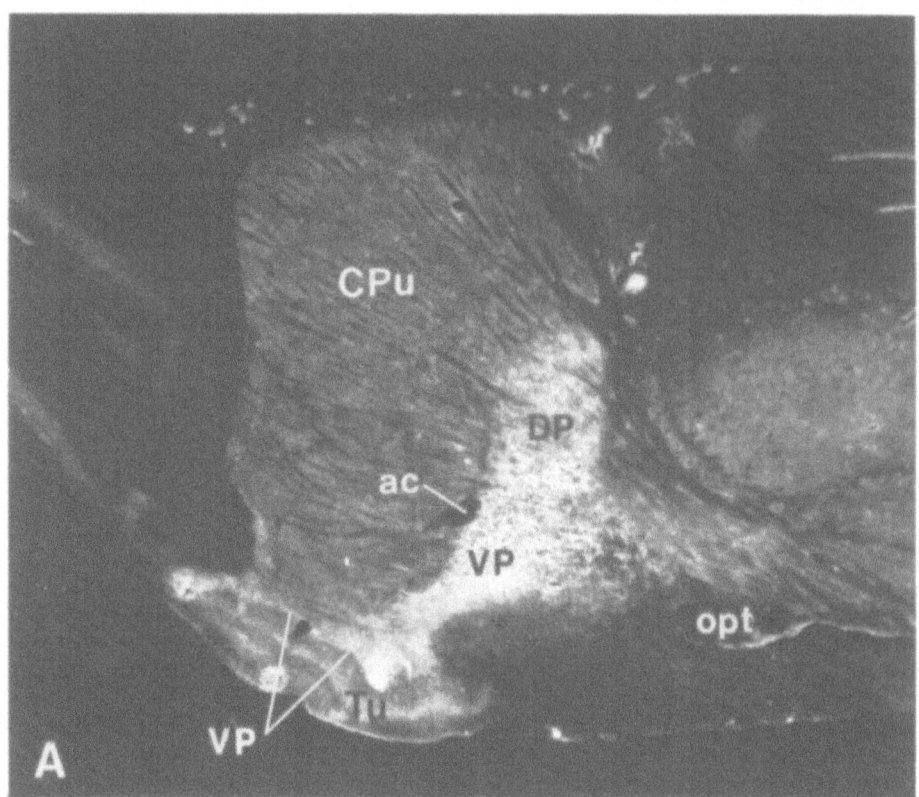

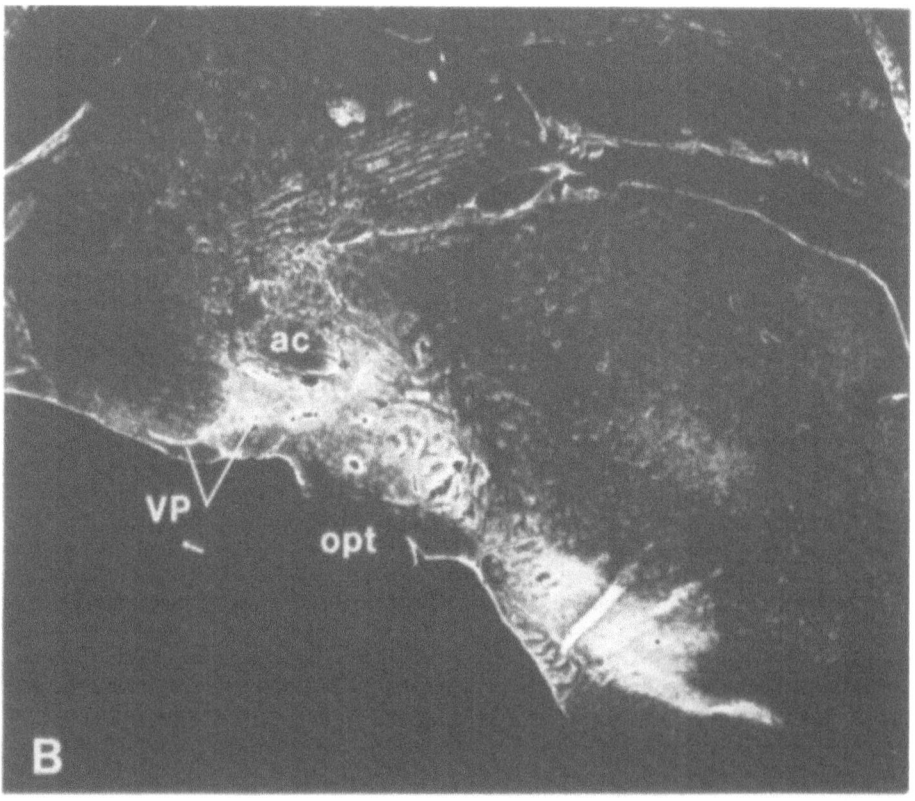

its close relations with allocortical areas and amygdaloid body have resulted in its classification as "limbic striatum" (Nauta and Domesick, 1984), or as an "interface" between the limbic system and the motor system (Mogenson, 1987).

Recent experimental studies have shed new light on this apparent dual nature of nucleus accumbens. For example, it was noted occasionally that the nucleus accumbens can be divided into a central core surrounded on its medial and ventral aspects by a peripheral shell (Zaborszky et al., 1985; see also Fig. 1), and this distinction has subsequently been made by the aid of a number of histochemical methods (e.g., Groenewegen et al., 1989; Voorn et al., 1989; Heimer et al., 1991b). This dichotomy, which is most clearly appreciated in the middle and caudal portions of the nucleus, is reinforced by a similar medial-to-lateral dichotomy in the efferent projections of the nucleus accumbens (Groenewegen and Russchen, 1984; Zahm, 1989; Zahm and Heimer, 1990; Heimer et al., 1991b). For example, although both the core and the shell of the accumbens project topographically and in typical striatal fashion to ventral pallidum, entopeduncular nucleus, and the substantia nigra-ventral tegmental area, the shell projects additionally to the hypothalamus and to the extended amygdala. As indicated above, we have suggested that the more unusual features of the accumbens might be better understood in a systematic anatomical framework if the shell is considered, not only as part of the striatum, but also as a part of a larger cellular continuum which stretches from the bed nucleus of stria terminalis medially to the centromedial amygdala in the temporal lobe (de Olmos et al., 1985; Heimer et al., 1985; Alheid and Heimer, 1988; Alheid et al., 1989b; Alheid et al., 1990; Heimer et al., 1991ab). It appears that this continuum, i.e. the extended amygdala, includes at least in part the shell in the caudomedial and ventral aspects of the accumbens, and possibly contiguous parts of the medial olfactory tubercle (Zahm and Heimer, 1988, 1990). In other words, while the core of the accumbens parallels in almost every aspect the rest of the striatum, the shell invariably stands out as a mixed area or as an area of transition. It has features that are reminiscent both of the striatal complex and the extended amygdala. Such mixed areas characterize most of the fringes of the striatal complex where it adjoins the extended amygdala. These difficult zones will be discussed following a description of the extended amygdala.

Ventral Pallidum

The extent to which the pallidal complex accompanies the ventral extension of the striatum towards the surface of the brain, not only in the rat (Fig. 2A) and the monkey (Fig. 2B), but also in human (e.g., Haber and

Fig. 2.   Perl's stain for iron, intensified with diaminobenzidine in sagittal brain sections from the rat (A) and Macaque (B) (see Fig. 19.28 in Alheid et al., 1990, for similar sections in the human). The continuity between the dorsal pallidum (DP) and ventral pallidum (VP) is readily apparent, as is its extension into the deep polymorph layers of the olfactory tubercle (Tu) of the rat. In the primate, the ventral pallidum sends finger-like extensions rostrally into the anterior perforated space. This latter structure is not entirely homologous to the rat olfactory tubercle which is in the primate confined to the rostral portions of the anterior perforated space. Abbrev. ac = anterior commissure, CPu = caudate-putamen; NR = nucleus ruber, pars reticulata. (Rat section, courtesy of J. Hill, NIMH).

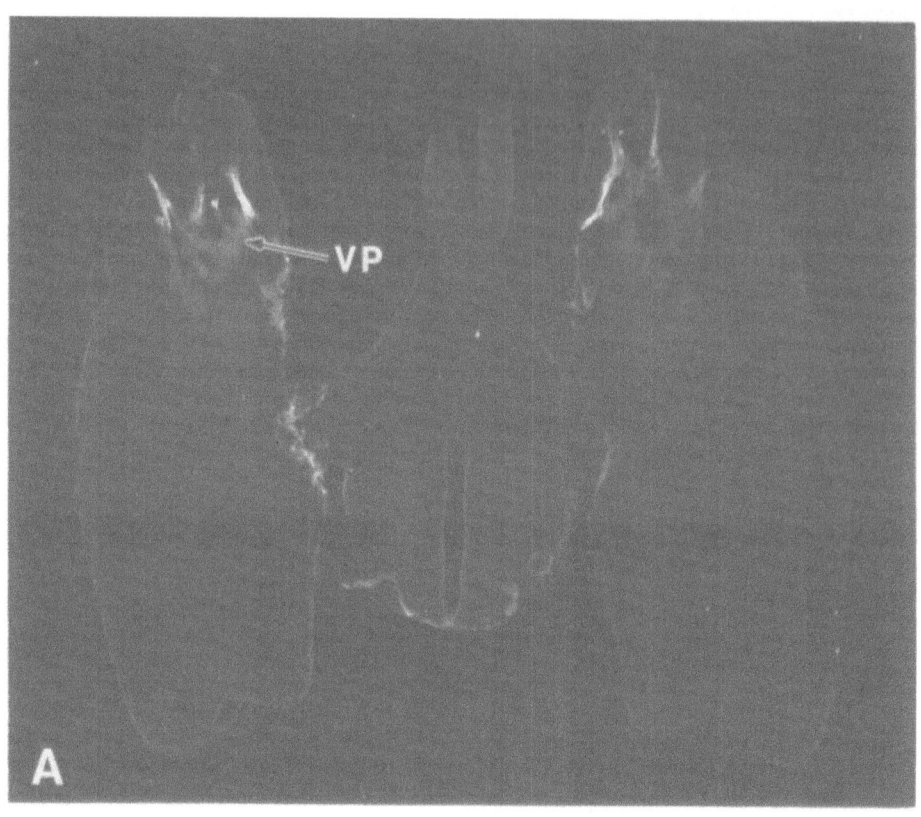

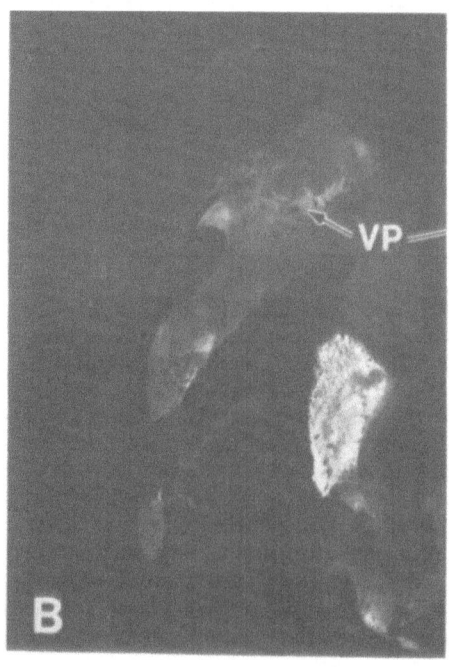

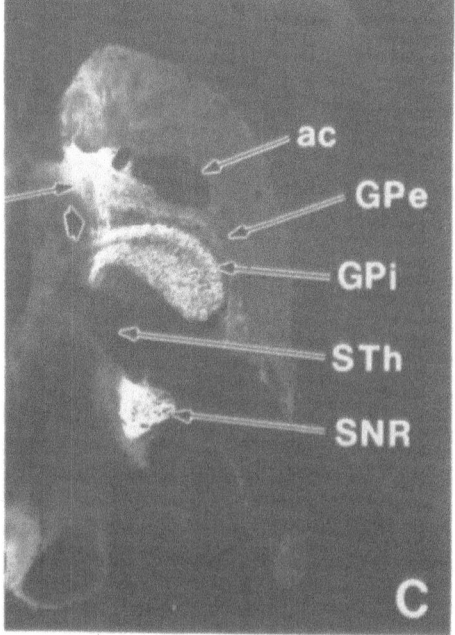

Elde, 1981; Haber and Watson, 1985; Mai et al., 1986; Alheid and Heimer, 1988; Alheid et al., 1989b, 1990; Pioro et al., 1990) has been acknowledged (Fig. 3). Although the forward sweep of the ventral pallidum is more pronounced in the rat and the monkey than in the human, the ventral extension of the pallidal complex close to the surface of the brain is apparent in all mammals studied. Therefore, the ventral pallidum occupies an extensive area underneath the temporal limb of the anterior commissure. This region of the basal forebrain has in the past been referred to as the subcommissural substantia innominata (Miodonski, 1967). In contrast, the more caudal sublenticular substantia innominata is occupied mainly by the sublenticular portions of the extended amygdala (see below).

Connections of the Ventral Striatopallidal System

The introduction of the concept of the ventral striatopallidal system demonstrated the existence of an additional channel in the striatopallidal circuitry, which is related to allocortical, periallocortical, and proisocortical areas, and to the basolateral amygdala, in the same way as the rest of the neocortex is related to the dorsal part of the basal ganglia. We and others have argued for the basic similarity of its pallidothalamic output with the dorsal pallidum targeting the ventral thalamus and subsequent projections of the latter to premotor areas, while the ventral pallidum sends terminals to the mediodorsal thalamus that subsequently projects to prefrontal cortex and anterior cingulate gyrus (Heimer et al., 1982; Young et al., 1984; Haber et al., 1985; Alexander et al., 1986; Groenewegen, 1988; Alheid et al., 1990). Like its dorsal counterpart, the ventral pallidum also has reciprocal relations with the subthalamic nucleus and projects to the territory of the substantia nigra/ventral tegmental area (Heimer et al., 1983; Haber et al., 1985; Zahm, 1989; Groenewegen and Berendse, 1990).

Fig. 3.   Horizontal sections close to the ventral surface of the rat (A) and Cebus monkey (B and C) brain. A. Rat section immunostained for epidermal growth factor which stains primarily in pallidal areas of forebrain. Note the large interconnected network of the ventral pallidum (VP) within the polymorph layer of the olfactory tubercle; the rostral densely stained areas are located within the hilus of the Islands of Calleja. B and C. Sections from the Cebus monkey immunostained for Substance P, which is particularly dense within the internal pallidal segment (GPi) and the closely related pars reticulata substantia nigra (SNR), while staining only lightly within the external pallidal segment (GPe). Note that the dense substance P immunoreactivity also extends into the ventral pallidum suggesting that elements of both the internal and external pallidal segment may be found here. In B, finger-like extensions of the ventral pallidum protrude into the anterior perforated space, especially medially and laterally. Slightly more dorsal (C), the ventral pallidum can be seen as an almost solid wall at the caudal end of the accumbens. At the caudal surface of the ventral pallidum, bands of substance P-rich axons can be seen to cross the sublenticular portion of the extended amygdala (wide arrow) in order to reach terminal sites in the internal pallidal segment. Abbrev. ac = anterior commissure, STh = subthalamic nucleus.

It only recently has been realized that the ventral striatum also sends efferents to the medial pallidal segment, either to the most medial elements of the entopeduncular nucleus in the rat (Heimer et al., 1991b) or to the internal pallidal segment of the monkey (Hedreen and DeLong, 1986, unpublished observation cited in Alexander et al., 1986; Lynd et al., 1988; Alheid et al., 1989a; Alheid et al., 1990; Haber et al., 1990). This presents a new perspective for current interpretations of basal ganglia circuitry. For instance, it is reasonable to assume that ventral thalamus is a potential target of ventral striatal processing. Moreover, it suggests that the brainstem output of the ventral striatum may proceed via the well-known efferents of the internal pallidal segment, rather than directly from the ventral pallidum. In fact, very few neurons within the ventral pallidum are retrogradely labeled even after large brainstem tracer injections caudal to the substantia nigra (Jackson and Crossman, 1981; Swanson et al., 1984; Alheid, unpublished observations), although many neurons are labeled in the parts of the extended amygdala that are immediately adjacent to ventral pallidum (e.g. Alheid and Heimer, 1988, Fig. 10).

But the connections of the ventral striatopallidum also show special features, which in part reflect the dichotomy between the core and the shell of the accumbens. Whereas the accumbens, like the rest of striatum, projects heavily to the ventral pallidum, Groenewegen and Russchen (1984) noted that in the cat the medial part of the accumbens in addition projects to the lateral hypothalamus, mesencephalic reticular formation and central gray. This dichotomy in accumbal projections has recently been extended to the rat (Heimer et al., 1991b) and appears to be related to the core-shell dichotomy discussed earlier. It also has been shown that the shell of the accumbens projects to a neurotensin-rich ventromedial district of the ventral pallidum, whereas the core projects to a dorsolateral region of the ventral pallidum normally devoid of neurotensin immunoreactivity (Zahm and Heimer, 1988; Zahm, 1989). The evidence so far suggests that this dorsolateral portion of the ventral pallidum differs from the shell-related ventromedial portion of the ventral pallidum by the occurrence of subthalamopetal neurons, which are sparse or lacking in the ventromedial pallidal zone (Zahm, 1989; Zahm and Heimer, 1990). The ventromedial pallidal zone, on the other hand, is more closely related to the mediodorsal thalamus. Following a discussion of the extended amygdala the "anomalous" projections of the shell-related parts of the ventral striatopallidal system will be reconsidered (see below).

Significance of the ventral striatopallidal system.

The concept of the ventral striatopallidal system has facilitated the structural and functional analysis of the basal forebrain. For instance, as predicted from analogy with the dorsal parts of the basal ganglia, the output channels for the ventral striatopallidal system are directed towards the thalamus (mediodorsal nucleus and probably ventrolateral thalamus), subthalamic nucleus and substantia nigra, and it seems, to some extent, midbrain reticular formation (e.g., Nauta et al., 1978; Groenewegen and Russchen, 1984; Haber et al., 1985; Alheid et al., 1990; Haber et al.,

1990; Heimer et al., 1991b). Since the ventral parts of the basal ganglia receive cortical input from fronto-temporal association areas, insular cortices, allocortical and periallocortical regions, as well as the basolateral amygdala, the notion of the ventral striatopallidal system also acknowledges the fact that not only neocortex, but the whole cortical mantle including the cortical-like basolateral amygdala, is being subserved by striatopallidal mechanisms.

Although theories about basal ganglia functions abound (see books and reviews by Cools et al., 1977; Divac and Öberg, 1979ab; Penney and Young, 1983; Evered and O'Conner, 1984; McKenzie et al., 1984; Mishkin et al., 1984; Alexander et al., 1986; Schneider and Lidsky, 1986; Phillips and Carr, 1987; Sandler et al., 1987), it seems clear that they must participate in the execution and modulation of motor responses resulting from various sensory and cognitive activities in the cerebral cortex. It is a fair guess that whatever functions are being carried out by the dorsal striatopallidal system, comparable functions characterize the ventral parts of the basal ganglia in response to stimuli emanating from allocortex, periallocortex, frontal and temporal association areas, including basolateral amygdala, i.e. areas to which the ventral striatum is closely connected.

Despite the progress in recent years in understanding the anatomy of the olfactory tubercle, little can be said about the function of this area. Since in macrosmatic mammals the olfactory tubercle-related parts of the ventral striatopallidal system receive significant olfactory input, both directly from the olfactory bulb and indirectly from the olfactory cortex and nucleus of the lateral olfactory tract (Luskin and Price, 1983), this part of the ventral striatopallidal system appears specialized for conveying this sensation to the effector mechanisms of the basal ganglia. At least in developmental experiments, this type of interaction has been demonstrated (Small and Leonard, 1983). However, higher order processing of olfactory stimuli also appears to be routed through the olfactory tubercle via its connections with the mediodorsal thalamus (e.g., Slotnick, 1990).

Considerably more attention has been given to the nucleus accumbens as the most prominent part of the ventral striatal complex, especially in theories of motivation, limbic-motor integration, and as a possible substrate for the effects of psychoactive drugs (e.g., Matthyse, 1973; Stevens, 1973; Fibiger and Phillips, 1986; Nielsen and Scheel-Krüger, 1986; Mogenson, 1987; Swerdlow and Koob, 1987; Wood and Emmett-Oglesby, 1989). In spite of much effort, however, the functional role of accumbens remains elusive. It is clear that its designation as a "limbic striatum" together with its input from the so-called "limbic" and "mesolimbic" areas has dominated behavioral and physiological studies directed at this structure (e.g., Mogenson, 1984; Everitt et al., 1989). Only in very few instances (see chapter by Churchill et al. in this book) has the dual nature of the accumbens been appreciated, a view that may help to explain the sometimes contradictory results of manipulations of this structure (Albert et al., 1989; Russell et al., 1989; Vaccarino and Rankin, 1989; Yoshikawa et al., 1989; see below under the "Significance of the Extended Amygdala").

Johnston Revisited

J.B. Johnston (1923) observed that a continuity between two large mediobasal forebrain structures, i.e. bed nucleus of stria terminalis and accumbens, and the centromedial amygdala in the temporal lobe, were indicated by cell columns along the dorsal course of the stria terminalis (see
also Strenge et al., 1978). The re-emergence of this idea in recent years
has provided a new perspective on basal forebrain organization that should
be of immediate relevance for theories of motivation, cognition, and memory. The anatomical organization of this forebrain continuum has been studied primarily by de Olmos (1972), who realized that the continuity between
the centromedial amygdala and bed nucleus of stria terminalis is reflected
not only by cells along the stria terminalis, but also by cell columns in
the sublenticular substantia innominata (Fig. 4A). While the details of the
extended amygdala have been best described for the rat (see especially de
Olmos et al., 1985; but also Alheid and Heimer, 1988; Grove, 1988ab; Heimer
et al., 1991a), this continuum has also been recognized in other species,
including the rabbit (Schwaber et al., 1982) and the primate including the
human (Price et al., 1987; Alheid and Heimer, 1988; Martin et al., 1988;
Alheid et al., 1989ab, 1990; de Olmos, 1990; Heimer et al., 1991a). For
convenience we have termed this large forebrain continuum the "extended
amygdala", since in many respects its histochemistry (Figs. 4 and 5), connections and functions most resemble those ascribed to the cells of the
centromedial amygdala. It should be emphasized, however, that the term "extended amygdala" does not include the large basolateral amygdaloid complex.
This latter complex is better understood as a modified portion of cortex
(Crosby and Humphrey, 1941; Lauer, 1945; Hall, 1972; Millhouse and de Olmos, 1983; McDonald, 1984; de Olmos et al., 1985; Carlsen and Heimer, 1988;
McDonald, 1991).

The Two Subdivisions of the Extended Amygdala

The existence of the extended amygdala is reinforced by studies that
demonstrate parallels between the various parts of the extended amygdala in
regard to afferent and efferent connections, as well as in their transmitters and peptide content (e.g. Schwaber et al., 1982; de Olmos et al.,
1985; Holstege et al., 1985; Gray, 1987; Gray and Magnuson, 1987; Price et
al., 1987; Alheid and Heimer, 1988; Grove, 1988ab; Martin et al., 1988;
Amaral et al., 1989; Chang, 1989; Chang and Kuo, 1989; Lesur et al., 1989;
de Olmos, 1990; Holstege, 1990; Heimer et al., 1991a). As defined, the
extended amygdala represents a macrostructure whose internal complexity
rivals or even exceeds that of the striatopallidal system. Two major subdivisions of the extended amygdala can be recognized; one consists of the
central amygdaloid nucleus, cell columns in the anterodorsal part of the
sublenticular substantia innominata and the lateral bed nucleus of stria
terminalis, whereas the other is formed by the medial amygdaloid nucleus,
cell columns in posteroventral sublenticular substantia innominata and the
medial bed nucleus of stria terminalis (de Olmos et al., 1985; Grove,
1988ab; de Olmos, 1990). These two major divisions of the extended amygdala
may be further divided into component nuclei, in most cases with comparable

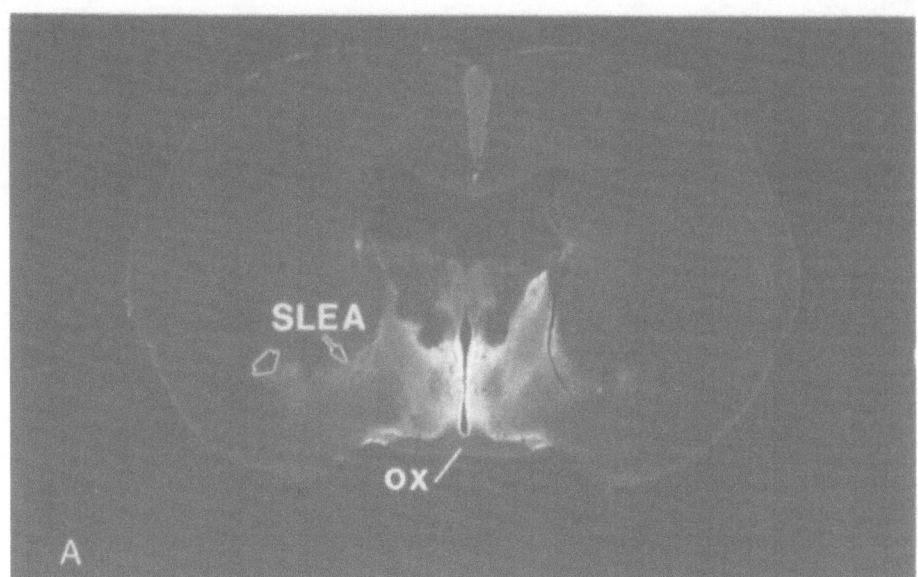

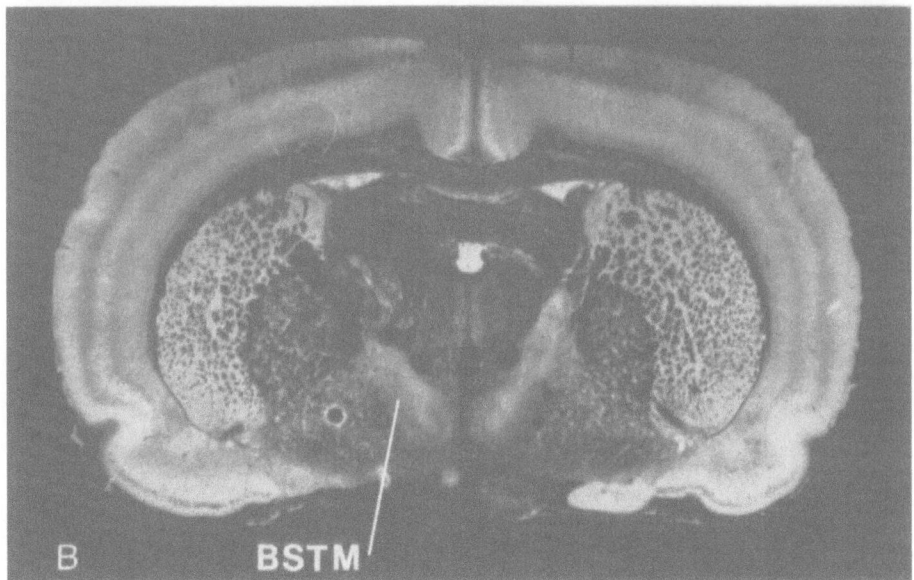

Fig. 4.   <u>A</u>. Angiotensin-II immunoreactivity within the posterior part of
         bed nucleus of the stria terminalis and sublenticular portion of
         the extended amygdala (SLEA) in the rat. The wide arrowhead
         indicates angiotensin II immunoreactivity in the interstitial
         nucleus of the posterior limb of the anterior commissure (see
         text and Fig. 5B). <u>B</u>. Timm's stain for zinc from a section at
         about the same level as in A. Note that the zinc staining in the
         posterior portion of the medial bed nucleus of the stria
         terminalis (BSTM) helps to define its border with the
         hypothalamus, which is not the case with angiotensin-II
         immunoreactivity. The continuation of the bed nucleus into the
         sublenticular extended amygdala is also apparent in the Timm's
         stain. ox = optic chiasm.

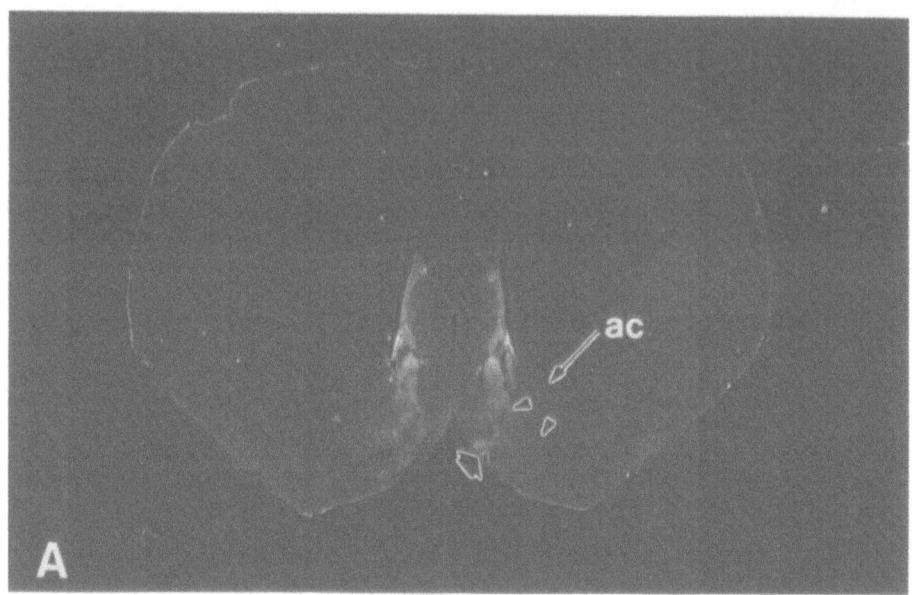

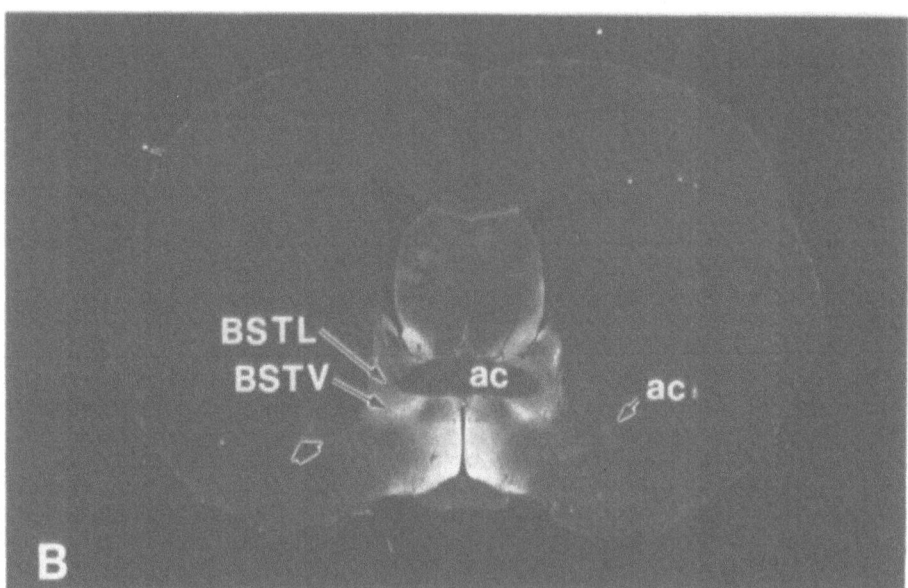

Fig. 5.   Angiotensin-II immunoreactivity in the shell area of the nucleus accumbens (A) and in adjacent bed nucleus of the stria terminalis (B). The small arrowheads in A mark the border of the accumbens shell with the core area of the accumbens. Note also the continuity of this immunoreactivity into the medial portion of the olfactory tubercle (wide arrowhead in A). The wide arrow in B indicates angiotensin-II immunoreactivity within the interstitial nucleus of the posterior limb of the anterior commissure, which is contiguous with the accumbens rostrally and with the central nucleus of the amygdala caudally (see text and Fig. 11 in Alheid and Heimer, 1988). <u>Abbrev</u>. ac = anterior commissure, BSTL = lateral division of the bed nucleus of the stria terminalis, BSTV = ventral division of the bed nucleus of the stria terminalis.

components recognizable in both the mediobasal forebrain and temporal lobe (Fig. 6).

Connections of the extended amygdala

Cortical projections to the extended amygdala mirror in part those to the ventral striatum inasmuch as they originate in allocortical, periallo-cortical and proisocortical areas (see reviews by de Olmos et al., 1985; Price et al., 1987; de Olmos 1990). These zones, which include the frontal cortex, hippocampus, entorhinal cortex, olfactory cortex, perirhinal and insular cortex, as well as the cortical-like basolateral amygdala, taken together represent some of the more complex association areas of the brain, many of which have been frequently implicated in memory, learning, and motivation.

The descending projections from the extended amygdala are complicated and target a variety of areas in the hypothalamus and brainstem. In general, the central division of the extended amygdala is more closely related to the lateral hypothalamus and the brainstem, whereas the medial division of the extended amygdala projects more prominently to the medial hypothal-amus. These main trends in the efferent organization are also reflected in the fact that the extensive intrinsic association pathways appear to ter-minate preferentially within, rather than across the two main divisions of extended amygdala (de Olmos et al., 1985; Price et al., 1987). While these are useful generalizations, it should be kept in mind that there are pro-jections from either division that deviate from these rules. For example, sparse but readily demonstrated terminals from the central division of the extended amygdala terminate in the medial hypothalamus (de Olmos et al., 1985; Gray et al., 1989), and notably, there are efferents from the medial subdivision of the extended amygdala that reach the central division. These latter projections implicate the central division as a main corridor for the effects of the extended amygdala on autonomic and somatomotor func-tions.

Ascending projections of the extended amygdala are another potential corridor of action. In general, little direct reciprocity is seen from the extended amygdala to the cortical areas that provide inputs to this struc-ture. However, light microscopic evidence suggests that both the medial and central divisions of the extended amygdala provide afferents not only to the topographically organized cholinergic and non-cholinergic corticopetal cell systems of the basal forebrain (Price and Amaral, 1981; Grove and Nau-ta, 1984), but presumably also to the diffusely projecting corticopetal sy-stems in the hypothalamus and brainstem. While the exact role of most of these systems remain controversial, the range of functions suggested for the non-thalamic corticopetal systems range from mediation of rewarding stimuli and learning, alteration of behavioral state (waking, sleeping, at-tention) to specific effects on the size and organization of somatosensory fields and on the regulation of cerebral microcirculation,(Bartus et al., 1982; Vanderwolf, 1983; Saper, C.B., 1987; Biesold et al., 1989; Ma et al., 1989; Szymusiak and McGinty, 1989; Delacour et al., 1990; Hallström et al.,

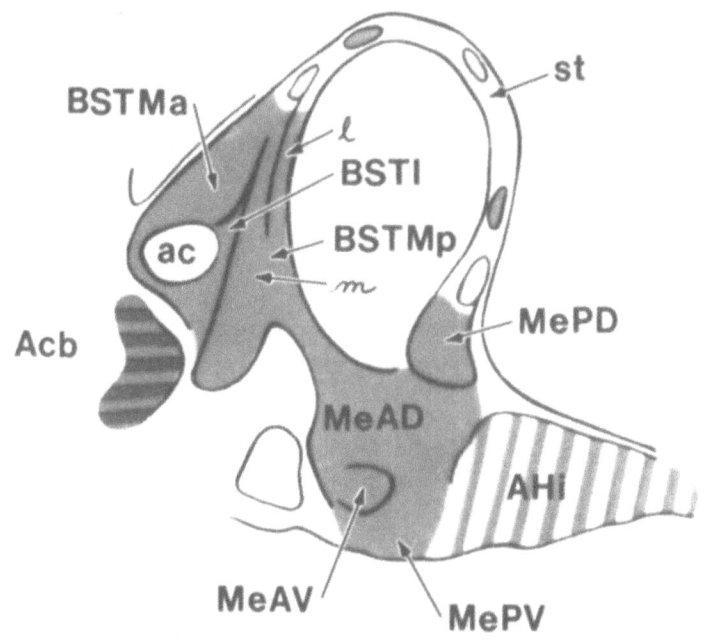

A

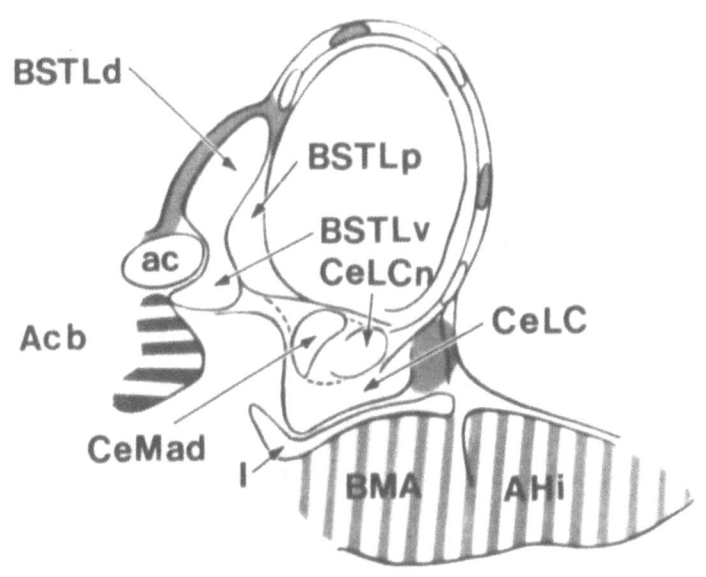

B

1990; Maeda et al, 1990; Richardson and DeLong, 1990; see also Záborzsky et al., this volume).

Special features of the striatopallidal system and the extended amygdala

Core and shell zones of the nucleus accumbens. The strong similarities between the nucleus accumbens and the rest of the striatum is apparent in almost every histochemical or hodological approach to this problem. As discussed above, however, a growing body of data also suggests the special nature of the shell of the accumbens. While we first noted this in relation to immunohistochemistry for cholecystokinin in the rat (Záborzsky et al., 1985), earlier experiments in the cat (Groenewegen and Russchen, 1984) had already suggested that the medial and lateral portions of the accumbens possessed divergent projections. These observations based on modern anatomical techniques echo superficially similar subdivisions recognized in early cytoarchitectonic studies. Specifically, the accumbens was often subdivided into a medial or lateral part (see review by Koikegami et al., 1967), or into an outer and internal division (Ziehen, 1909). It appears as if the shell portion of the accumbens is particularly rich in a variety of neuro-

---

Fig. 6.   Schematic diagram of the medial (A) and central division (B) of the extended amygdala depicting the various component subnuclei in a sagittal view. The striped areas indicate zones that are closely allied with either the medial or central division of the extended amygdala. For the latter, this includes the caudomedial portions of the nucleus accumbens (Acb) in B, and the interstitial nucleus of the posterior limb of the stria terminalis (not shown). For the medial division these associated areas include the amygdalo-hippocampal transition area (AHi) and the basomedial nucleus of the amygdala (BMA), and potentially the caudomedial accumbens. Interrupted cell columns accompanying the dorsal course of the stria terminalis are indicated as clear or grey ovals that are related to the central or medial divisions of the extended amygdala respectively (adapted from a drawing by de Olmos in Heimer et al., 1991a). Note that the ventral bed nucleus of the stria terminalis (BSTV) is not shown. Abbrev. ac = anterior commissure, BST = bed nucleus of the stria terminalis, BSTI = intermediate division of the BST, BSTLd, BSTLv, BSTLp = dorsal, ventral, and posterior portions of the lateral division of the BST, respectively; BSTMa, BSTMp = anterior and posterior divisions of the medial BST, respectively; CeL, CeM = lateral and medial divisions of the central nucleus of the amygdala, respectively; CeLC = lateral capsular subdivision of the CeL, CeLCn = central subdivision of the CeL, CeMad = anterodorsal subdivision of the CeL, I = intercalated cell groups of the amygdala, l = lateral, large-celled column of BSTMp, m = medial, medium-celled column of BSTMp, Me = medial amygdala, MeAD = dorsal anterior subdivision of Me, MeAV = ventral anterior subdivision of Me, MePD = dorsal posterior subdivision of Me, MePV = ventral posterior subdivision of Me, st = stria terminalis.

transmitters or their enzymes, including many that are found at only much lower levels, if at all, in the core of the accumbens or in the overlying striatum. As a rule of thumb, it appears that many of the transmitters, as well as their enzymes and receptors found in unusual concentrations within the shell of the accumbens and its extension into the medial olfactory tubercle also are found in the immediately adjacent bed nucleus of the stria terminalis as well as in the rest of the extended amygdala (Figs. 4 and 5). Roberts et al. (1980) noted that it was difficult, if not impossible, to discriminate immunohistochemical borders between the rostral portions of the bed nucleus of the stria terminalis and the nucleus accumbens. Our interest in the amygdala and its relation to the sublenticular zone led us to suggest that one explanation for the anomalous histochemistry and connections of the shell region of the accumbens, was that portions of the "extended amygdala" invade this territory of the striatum (de Olmos et al., 1985; Heimer et al., 1985; Alheid and Heimer, 1988; Alheid et al., 1989b, 1990).

This partially revives the earlier appraisal by Johnston (1923), who, based on developmental and comparative studies, included the nucleus accumbens and bed nucleus of the stria terminalis within the gray matter accompanying the longitudinal association bundle that extends from the central and medial amygdala into the forebrain in lower vertebrates and in fetal mammals, including humans. A difference between Johnston's assessment and ours is that we cannot assign the core of the accumbens to the "extended amygdala", since the core aligns itself very closely with the rest of the caudate-putamen. Nor can we assign the shell entirely to the extended amygdala, since it also projects densely to the ventral pallidum, a projection that does not characterize any other part of the extended amygdala.

It should be appreciated that this position was taken only after we had struggled for some time with reconciling the unusual projections of the nucleus accumbens with regular histochemical and hodological relations seen within the rest of the the striatopallidal system. For example, our initial notion was that the well documented projections of the nucleus accumbens to the lateral hypothalamus could represent axons terminating on ectopic neurons of the entopeduncular nucleus (Heimer et al., 1985) and, in fact, this does seem to be the case. However, the shell of the accumbens also sends a concentrated column of axons that terminate within the entire rostrocaudal extent of the lateral hypothalamic zone of the rat (Nauta et al., 1978; Groenewegen and Russchen, 1984; Haber et al., 1990; Heimer et al, 1991b). This type of hypothalamic projection is typical, particularly of the central division of the extended amygdala but is not observed for any part of the striatum, including the core of the accumbens.

The interstitial nucleus of the posterior limb of the anterior commissure. The caudal face of the accumbens that abuts the rostral surface of the ventral pallidum is also unusual with respect to possessing a significant number of medium-sized neurons that project to the lower brainstem particularly to the parabrachial area. This field of neurons is contiguous with similarly labeled cells in the bed nucleus of the stria terminalis and further lateral with similar cells that accompany the posterior limb of the anterior commissure (Fig. 10 in Alheid and Heimer, 1988) all along its course into the amygdala, where they join similarly labeled neurons in the lateral part of the central nucleus of the amygdala. This zone has previously been termed the "interstitial nucleus of the posterior limb of the anterior commissure" (see especially Figs. 21 and 31 in de Olmos, 1972). This area, which is generally considered part of the caudate-putamen, has several features in common with the extended amygdala (de Olmos et al., 1985; Alheid and Heimer, 1988). This includes similarities in histochemistry and receptor binding that are not found in striatum. For example, one can appreciate this continuity in sections immunostained for cholecystokinin (Záborszky et al., 1985) or for angiotensin II (Alheid and Heimer, 1988), and in spirodecanone (Fig. 16 in Alheid and Heimer, 1988), quisqualate (Fig. 6 in Insel et al., 1990), and vasopressin receptor binding (Fig. 4 in Freund-Mercier et al., 1988; Fig. 2 in Phillips et al., 1988). As indicated earlier, this histochemical similarity is reinforced by the fact that cells along this entire course project to the region of the parabrachial nuclei (Jackson and Crossman, 1981; Alheid and Heimer, 1988; Grove, 1988b), an area that is not targeted by the striatum, and that cells within this "lateral column" of the extended amygdala appear to send association fibers into the sublenticular zone, particularly to the anterodorsal column of the extended amygdala, potentially including terminals directed at the corticopetal cells that are intermingled with this part of the extended amygdala (Grove, 1988b). In these connections the interstitial nucleus most resembles an extension of the lateral part of the central amygdaloid nucleus with which it is contiguous caudally, and the lateral part of the bed nucleus of the stria terminalis, with which it is contiguous rostrally. The different parts of the extended amygdala and its relations to the shell of the accumbens can be appreciated best in horizontal sections as indicated in the schematic drawing in Figure 7.

The marginal division of the striatum. Recently, Shu and colleagues (1988, 1990) called attention to an unusual portion of the striatum of the rat lying adjacent to its border with the dorsal pallidum. They have termed this the "marginal division" of the striatum. Basically, they have described a narrow zone populated by fusiform neurons that apparently project to cholinergic neurons located in the caudomedial portion of the globus pallidus rather than to the remainder of the pallidum. These cells are further characterized by extensive collaterals that spread dorsoventrally within the marginal division. This corridor is characterized by a lighter activity of acetylcholinesterase than that found in the remainder of the striatum, dense immunoreactivity for substance P, dynorphin B, and met-enkephalin, as well as by a particularly dense staining for zinc (Fig. 7; Shu et al., 1988, 1990), and for calcitonin gene-related peptide (Chang and Kuo, 1989). Serotonin is also denser in this zone (Mori et al., 1985), and angiotensin II fibers appear to follow portions of this corridor (Lind et al., 1984; Alheid, unpublished observations).

Although Shu and her colleagues do not specify the ventral extent of this marginal division, it appears that it includes a rather broad band along the caudal portion of the nucleus accumbens, just adjacent to the rostral face of the ventral pallidum (Fig. 8). This caudomedial portion of the accumbens corresponds to the shell, which we have earlier noted to be similar to the remainder of the extended amygdala in several respects. This

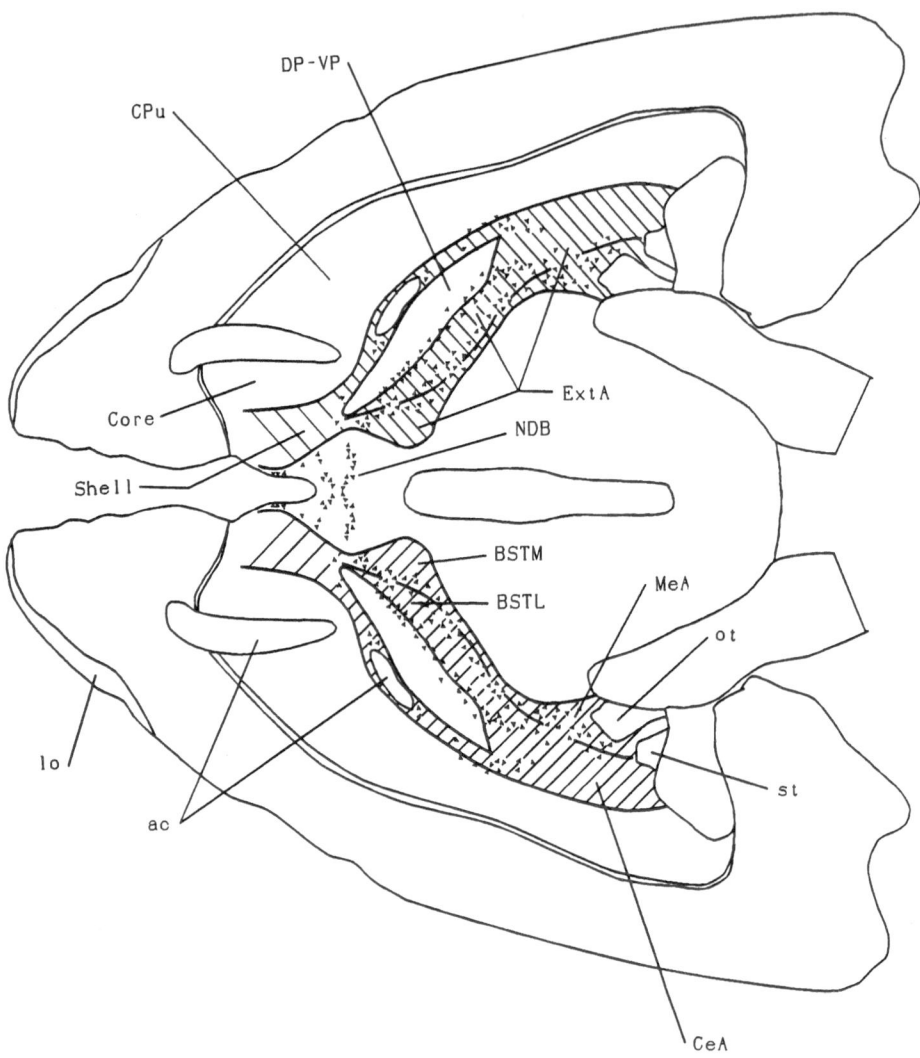

Fig. 7. Schematic illustration of the large forebrain continuum that we
have designated the "extended amygdala" (hatched area). This
drawing was composed with the aid of several immunostained
(Angiotensin II and choline acetyltransferase) horizontal
sections of the rat brain ventral to the crossing of the anterior
commissure. The cholinergic corticopetal cell population,
portions of which are coextensive with, or immediately adjacent
to the extended amygdala, is indicated by small black triangular
neurons. Note how the shell of the accumbens is directly
continuous with the rostral portion of the extended amygdala
(ExtA). Abbrev. ac = anterior commissure, BSTL = lateral division
of bed nucleus of stria terminalis, BSTM = medial division of bed
nucleus of stria terminalis, CeA = central amygdaloid nucleus,
CPu = caudate-putamen, NDB = nucleus of diagonal band, DP-VP =
dorsal pallidum-ventral pallidum, lo = lateral olfactory tract,
MeA = medial amygdaloid nucleus, ot = optic tract, st = stria
terminalis.

20

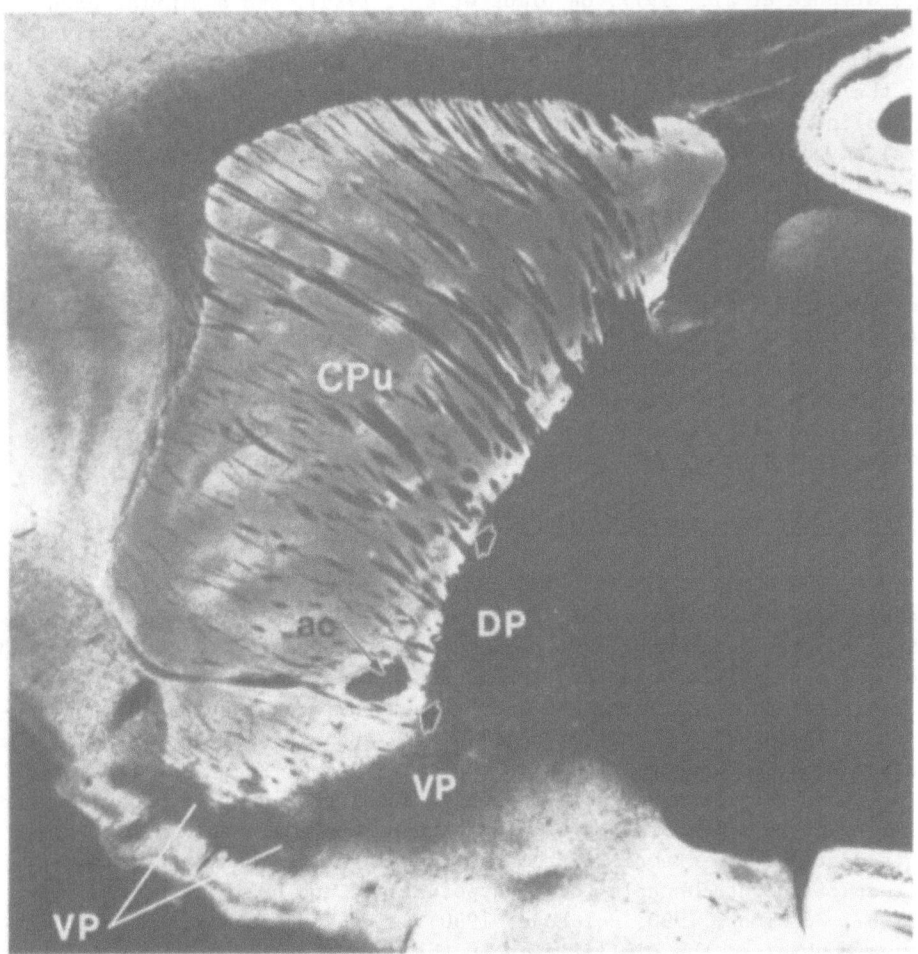

Fig. 8.   Timm's stain for zinc (sodium selenite modification, Danscher, 1982) in sagittal section of the rat brain. Note the bright band of zinc-rich terminals in the marginal zone of the caudate-putamen (CPu) which is continuous with similar dense zinc staining in the caudal accumbens (wide arrowheads). <u>Abbrev</u>. ac = anterior commissure, DP = dorsal pallidum, VP = ventral pallidum.

suggests that one might examine the entire "marginal division" for a similar relationship with the amygdala. It is known, for example in the monkey, that errant axons of the stria terminalis leave the central amygdaloid nucleus and traverse the medullary lamina of the globus pallidus in order to reach the bed nucleus of the stria terminalis among other targets (Price and Amaral, 1981). Similarly, in the rat some axons anterogradely labeled from the sublenticular part of the extended amygdala also appear to enter the medullary laminae and course dorsally (Grove, 1988b, Fig. 1C and 1D), presumably as errant stria terminalis fibers. Path neurons appear to accompany the supracapsular course of the main bundle of the stria terminalis (e.g., Strenge et al., 1977; de Olmos et al., 1985), and a similar case might be made for cells in the marginal division. Since Shu et al. (1988) observed collaterals of axons from cells in the marginal division ramifying among the cholinergic neurons in caudal pallidum, it should be remembered that cholinergic basal forebrain neurons are a probable target for the central amygdala (Price and Amaral, 1981; Grove and Nauta, 1984; Russchen et al., 1985; Grove, 1988a). Needless to say, striatal terminations are also probable for cholinergic cells in pallidal areas (Grove et al., 1986).

Instead of the favorable comparison between the marginal division and the extended amygdala, an alternative hypothesis would be to somehow account for this division as one of the specializations of the striatum as represented by striosomes, or various pallidal subdivisions, an approach favored by Shu and colleagues (1988). This, however, does not account for its unusual pattern of connections including its special complement of neurotransmitters when compared with the rest of striatum.

Significance of the Extended Amygdala

The extended amygdala, together with its efferents to the corticopetal neurons, emerges as a formidable forebrain system that seems ideal for evaluating the significance of external and internal events, and to effect and coordinate appropriate behavioral responses through its widespread and highly organized projections, not only to autonomic and neuroendocrine centers in the hypothalamus, but also to somato- and visceromotor areas in the brainstem with subsequent effects on the spinal cord. Such a system would seem designed to play a key role in motivational and adaptive behavior, and some parts of this forebrain system have been studied intensely for years in this context. One example of this is the amygdaloid body, which has a multitude of output channels to the hypothalamus and brainstem (Price et al., 1987; de Olmos, 1990; Holstege, 1990). A growing number of investigators have emphasized the similarities in the downstream projections from the centromedial amygdala and the bed nucleus of stria terminalis. This implies functional similarities between the centromedial amygdala and the bed nucleus of stria terminalis, and against the background of our present discussion, such functional similarities can be expected for the remainder of the extended amygdala. Specifically, it is reasonable to expect that the bed nucleus of stria terminalis and its related sublenticular cell columns, and maybe to some extent even the shell of the accumbens and related medial parts of the olfactory tubercle, have functional attributes that are similar to those of the centromedial amygdala. If the centromedial part of the amygdaloid body is crucially involved in evaluating the biological relevance of external events, this role should apply to the much larger forebrain continuum, of which it is a part. The anatomy suggests that processing within the extended amygdala may be distributed across the entire continuum as evidenced by a highly developed internal association system for each of its major divisions (de Olmos et al., 1985; Grove, 1988b; de Olmos, 1990).

The re-conceptualization of the basal forebrain that has gradually taken place during the last 15 years, from the rediscovery of kinship between centromedial amygdala and bed nucleus of stria terminalis, and the elucidation of their diversified and far-reaching output channels, to the notions of the ventral striatopallidal system and extended amygdala, has provided a new theoretical framework for studies of basal forebrain functions. In the light of the new structural concepts, the results of some previous studies are likely to take on a different significance. For instance, Mogenson and colleagues (Mogenson, 1984, 1987; Mogenson et al., 1980,1983; Swanson et al., 1984) have argued cogently that the locomotor effects of stimulation within the nucleus accumbens of the rat are effected through the projections of this area to the ventral pallidum and consequent projections of this part of the pallidum to the locomotor areas of the brainstem. Anatomical observations, (e.g. Alheid, unpublished; Jackson and Crossman, 1981) including their own (Swanson et al., 1984) suggest that the projections of the ventral pallidum to the brainstem caudal to the substantia nigra are sparse if they exist at all. However, Mogenson and co-workers regularly include within "ventral pallidum" the more caudal sublenticular region which is an integral portion of the extended amygdala rather than part of the striatopallidal system. Since it is known that the extended amygdala projects massively to the brainstem (see review in Heimer et al., 1991a), it is possible to interpret their physiological results as an aspect of the extended amygdala, rather than of the ventral striatopallidal system. It should be recalled that the accumbens projects massively to the ventral pallidum, but it is only the medial shell area that appears to provide any significant input to the sublenticular part of extended amygdala, as do the bed nucleus of the stria terminalis and the centro-medial amygdala. This implies a more direct action of a structure with obvious motivational importance on the psychomotor activity associated with arousal (e.g., Garcia-Rill et al., 1983). An additional complication, however, is provided by the realization that both the shell and the core of the accumbens may target brainstem projecting neurons within the most medial aspects of the entopeduncular nucleus; this provides an alternative route for modulation of somatomotor activity. Thus, the accumbens is well endowed with efferents capable of impacting on locomotion but neither of these two potential descending corridors for this effect seems to significantly involve the ventral pallidum. On the other hand, ventral pallidum could act out a role in orchestrating locomotor responses via the poorly documented, but likely projections to the entopeduncular nucleus itself.

THE MAGNOCELLULAR CORTICOPETAL COMPLEX

Distribution and Connections of the Magnocellular Corticopetal Cell Groups

Scattered across the forebrain from the medial septum and diagonal band rostrally through the ventral pallidum and sublenticular part of the extended amygdala to the caudal globus pallidus, are the cells of the magnocellular corticopetal forebrain system (e.g., Divac, 1975; Levey et al., 1983; Mesulam et al., 1983ab; Butcher and Woolf, 1986; Mesulam et al., 1986; Saper, 1987; Everitt et al., 1988; Grove, 1988a; see Záborszky, this book). While the majority of the magnocellular corticopetal cells with axonal destinations in neocortex appear to be cholinergic, those with projections to proiso-, periallo- and allocortex as well as basolateral amygdala use acetylcholine, but also galanin, GABA and possibly other peptides as transmitters (Köhler et al., 1984; Rye et al., 1984; Amaral and Kurtz, 1985; Carlsen et al., 1985; Melander et al., 1985; Vincent et al., 1985; Záborszky et al., 1986; Schwaber et al., 1987; Caffé et al., 1989; Fisher et al., 1988; Walker et al., 1989; Ulfig et al., 1990). Since it is to these latter cortical areas that the most dense forebrain projections are targeted, it is inappropriate to consider only the cholinergic cells when analyzing the corticopetal system.

The distribution of cholinergic neurons in the forebrain of the rat is demonstrated in Fig. 9. The corticopetal cells, more or less aggregated, can be seen to occupy a large contiguous territory beneath the striatal complex, including diagonal band areas, pallidal and peripallidal regions, as well as the extended amygdala. A number of more evenly dispersed large cholinergic neurons are also found in dorsal striatum (caudate-putamen) and ventral striatum (accumbens and olfactory tubercle), but unlike the corticopetal neurons which are typical projection neurons, these are part of the striatal interneuronal system. This difference, however, may be more illusory than real, and we have argued that the basic morphological similarity that seems to exist between the striatal interneurons and the corticopetal magnocellular neurons (Schwaber et al., 1986; see also Fig. 9), includes also the basic sequence of steps in their synaptology (Alheid and Heimer, 1988; Alheid et al., 1989b, 1990). Whereas the feedback target for the corticopetal cells is represented by cells in various cortical areas, the feedback target for the striatal interneurons is located within the striatum itself. In regard to the corticopetal neurons, it is apparent that significant numbers of these cells invade the ventral pallidum, but even more are found aggregated in more caudal subpallidal areas (Fig. 9). This region just caudal to the ventral pallidum is occupied most prominently by the sublenticular portion of the extended amygdala and the horizontal limb of the diagonal band, both of which contain various clusters of cholinergic and non-cholinergic corticopetal cells. The intimate relations between the corticopetal cells and the extended amygdala are discussed further below.

The connections of the corticopetal complex are the subject of intense scrutiny. This has been driven in part by the observation of pathological changes in the cholinergic innervation of cortex in patients with Alzheimer's disease (e.g., Davis and Maloney, 1976; Whitehouse et al., 1983). As a result, the topography of the corticopetal projections from this complex are known in some detail in rats and primates (Bigl et al., 1982; Ribak and Kramer, 1982; Mesulam et al., 1983ab; Pearson et al., 1983; Rye et al., 1984; Saper, 1984; Butcher and Woolf, 1986; Mesulam et al., 1986; Everitt et al., 1988; Mesulam and Geula, 1988) along with progress in defining a variety of their subcortical targets. These include significant connections to the thalamus and habenula (Hallanger et al., 1987; Parent et al., 1988) as well as efferents to the cortical-like basolateral amygdala complex (Woolf and Butcher, 1982; Woolf et al., 1984; Carlsen et al., 1985). The afferents to the corticopetal cells are more problematic, but also under active investigation. Since these are discussed in another portion of this book by Dr. Záborszky, we will not reiterate this data in any detail here. One aspect of their innervation, however, is worthy of comment.

The Magnocellular Corticopetal System and the Extended Amygdala

The problem of analyzing the functional relations of the corticopetal cells in basal forebrain often is complicated by considering the entire complex as a unit structure. Separate populations of these cells, however, also appear to form integral parts of other basal forebrain systems. This is perhaps best seen within the context of the hippocampal-septal-diagonal band complex, where the population of cholinergic and GABA-ergic hippocampopetal neurons are most often considered together with the septo-hippocampal system as a unit. This stands in contrast to the remainder of the corticopetal complex, whose functional role is most often considered independently from its apparent relations to the extended amygdala or the striatopallidal system. For instance, the parallel distribution of magnocellular cholinergic nerve cells with extended amygdaloid cell columns or axons is especially prominent in the region of the sublenticular areas of both rat (Fig. 9) and primate. The extended amygdala is characterized by a highly developed system of internal association fibers and the sublenticular part of the extended amygdala receives massive fiber systems, not only from the central and basolateral amygdaloid nuclei, but from a variety of structures in hypothalamus and brainstem areas as well. This places the extended amygdala in a strategic position, and we have recently proposed (Alheid and Heimer, 1988; Alheid et al., 1989b, 1990) that the large corticopetal cells in the sublenticular part of the extended amygdala may be part of an integrated feedback system of special relevance for the well-known influence of motivational and emotional states on cognition (Rolls et al., 1979; Richardson et al., 1988).

CONCLUDING REMARKS

The Changing Boundaries in the Basal Forebrain

It appears as if the new anatomy discussed in this chapter has effectively deprived structures like the olfactory tubercle, nucleus accumbens, and substantia innominata of legitimacy as independent functional-anatomical units. These terms are justifiable only to the extent that they indicate a general topographical arrangement of these areas. Various territories of the olfactory tubercle, nucleus accumbens, and rostral part of substantia innominata belong to ventral extensions of the basal ganglia, termed ventral striatum and ventral pallidum. For instance, both the striatal and the pallidal complex extend ventrally into the olfactory tubercle in a continuous fashion. In other words, the two main functional-anatomical components of the tubercle, which are striatal and pallidal in character, are continuous with more dorsal parts of the basal ganglia through cell columns between the fascicles of the medial forebrain bundle. This realization is the result of the remarkable progress that has taken place in anatomical mapping strategies during the last 20 years. By exclusive reliance on Nissl sections (e.g. Geeraedts et al., 1990), one might easily gain the wrong impression that a meaningful dorsal boundary can be drawn for the olfactory tubercle.

Another part of the substantia innominata, its sublenticular portion, can best be understood as an extension of the centromedial amygdaloid nuclei, i.e. extended amygdala. The histology and connections of this sublenticular area are so distinctive from the adjoining more rostrally located ventral pallidum that it would be a disservice to lump them together by the careless use of the term substantia innominata or subpallidal region. In other words, the outdated boundaries of the traditionally defined substantia innominata are of little relevance, whereas the territorial boundaries that separate the ventral striatum from ventral pallidum, or the striatopallidal system from neighboring functional-anatomical systems, e.g. the extended amygdala, are of great functional relevance. A similar situation exists in regard to the nucleus accumbens. Whereas the border between the caudate-putamen and nucleus accumbens cannot be drawn with certainty (e.g., Nauta et al., 1978; Beckstead et al, 1979; Chronister et al., 1981; Hedreen, 1981), and is probably of little functional relevance, the rather more distinct border between the core and shell of the nucleus accumbens

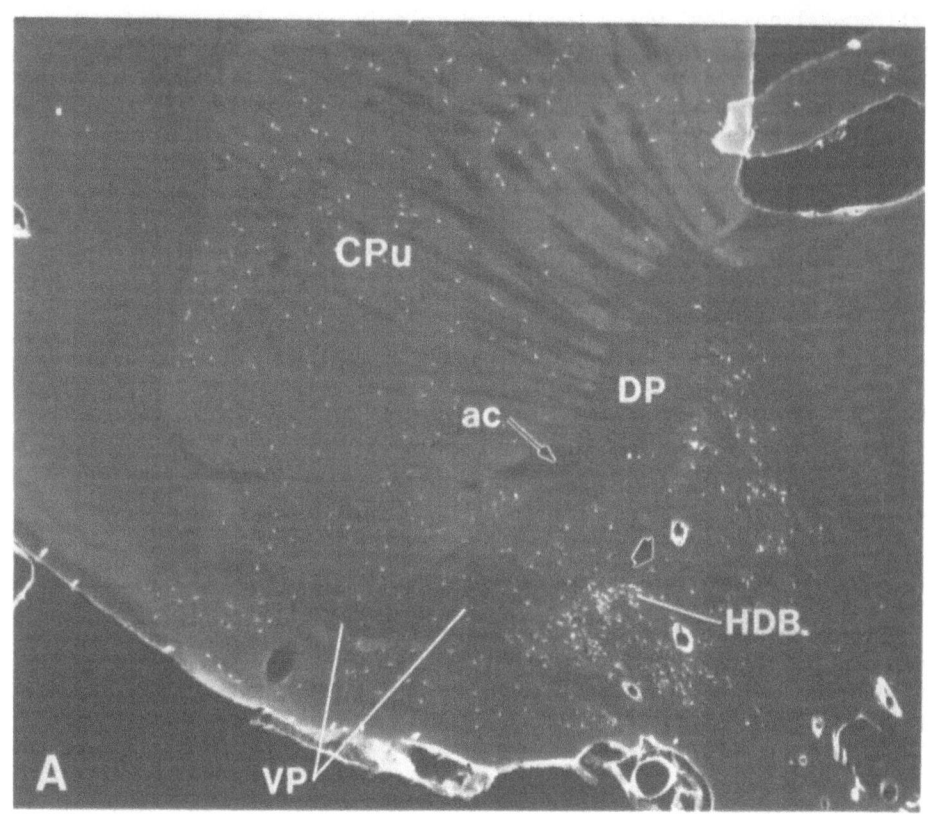

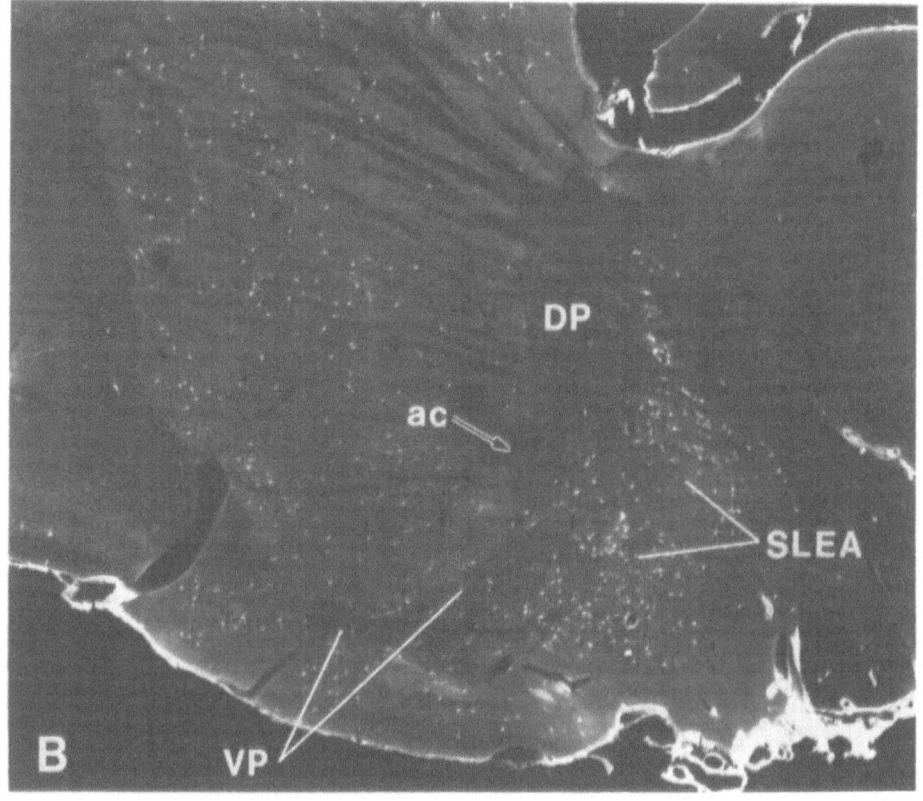

reflects a shift between two different functional-anatomical forebrain systems. In other words, it seems that only the core of the accumbens is anatomically homogeneous with the striatal complex or caudate-putamen (see Fig. 10); for every element found in the core of the accumbens, a corresponding element exists in dorsal parts of the striatum. With regard to the shell, the situation is far more complex; it is reminiscent of both the striatum and the extended amygdala.

Strategic boundaries within the basal forebrain do not necessarily coincide with those derived from classic cytoarchitectonic studies. The ignorance of this fact has precluded a rational design of many functional experiments, and in all likelihood accounts for some of the uncertainties that have persistently surrounded functional studies in the basal forebrain. On the brighter side, the new anatomy of the basal forebrain can provide new and more concrete opportunities for lasting progress in basal forebrain neuroscience.

De-emphasizing the Limbic System Concept

Normally, the amygdaloid body is considered a key structure in the limbic system, and it has been suggested for the rat that the amygdalo-striatal projections define a "limbic striatum" (Kelly et al., 1982). This implies that the predominant part of the striatal complex is "limbic", except its extreme dorsolateral quadrant, which corresponds to the putamen of the primate. Considering the same amygdaloid pathway in the primate (Russchen et al., 1985), much of the head, body and tail of the caudate nucleus could then be considered a "limbic" structure. This highlights one of the perennial problems of the limbic system concept, i.e. lack of precise and generally accepted rules for the inclusion of different structures. Whereas the amygdaloid input to the striatum comes primarily from the basolateral complex, it is the centromedial portion of the amygdala, or for that matter, the entire extended amygdala that is intimately and reciprocally related to the hypothalamus. It is also the centromedial portion of the

---

Fig. 9.   ChAT-immunostained neurons in sagittal rat brain sections. These cells are scattered across most areas of the forebrain including the caudate-putamen (CPu) diagonal band, dorsal and ventral pallidum and sublenticular zones. Note the dense accumulation of cholinergic neurons in the areas just caudal to the ventral pallidum(VP), for example in the horizontal limb of the diagonal band (HDB) in (A), and more laterally in B. The sublenticular portion of the extended amygdala (SLEA) occupies an area just dorsal to the horizontal limb of the diagonal band (wide arrowhead in A) and lateral to it (B) as it sweeps caudoventrally toward the centromedial portions of the temporal lobe. In (B) one can note that the sublenticular portions of the extended amygdala is contiguous not only with the sublenticular cholinergic cell populations, but also with the dorsal column of cholinergic neurons associated with the caudal part of the dorsal pallidum(DP). This proximity is further reinforced by the observation that efferents from the sublenticular area ascend within this caudal column of corticopetal cells (see text and Grove, 1988b). Sections courtesy of W. Cullinan and L. Záborszky.

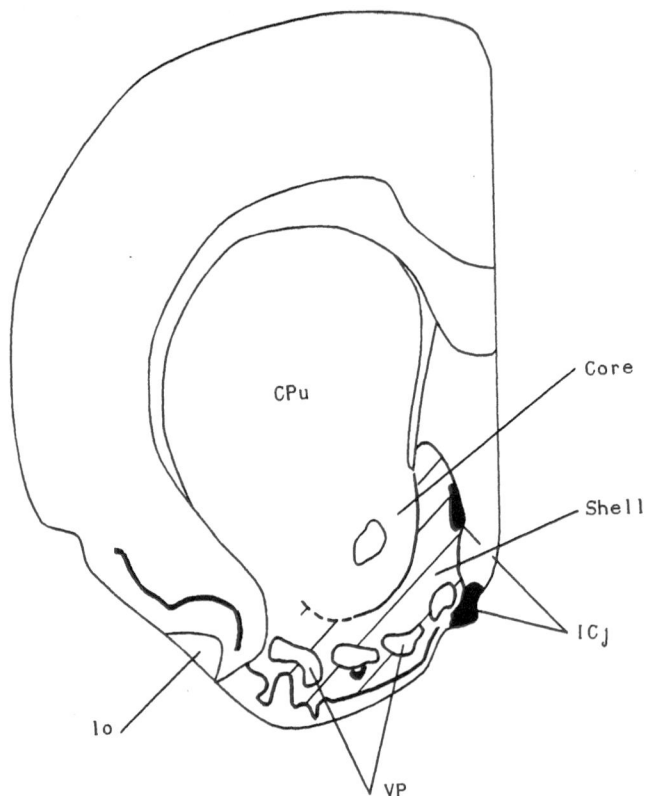

Fig. 10. Schematic diagram to emphasize the distinction between the core
and the shell of the accumbens. Note that the core is an integral part of the caudate-putamen whereas the shell of the accumbens, together with contiguous parts of the olfactory tubercle,
represents a specialised part of the striatal complex which also
appears to contain elements of the "extended amygdala". <u>Abbrev</u>.
CPu = caudate-putamen, ICj = Island of Calleja, Tu = olfactory
tubercle, VP = ventral pallidum.

amygdala that appears to contain most of the neurons that are responsive to
stimuli related to reinforcement (Wilson and Rolls, 1985; Nakano et al.,
1986; Nishijo et al., 1988). Moreover, one of the major telencephalic targets of the extended amygdala seems to be the cells of the magnocellular
corticopetal complex, which themselves are even more responsive to stimuli
related to the motivational contingencies of the environment (Rolls et al.,
1979; Richardson et al., 1988, and this volume; Wilson and Rolls, this
volume). Only few neurons within the basolateral complex show these types
of responses, and as indicated earlier, it may be better to conceive of the
large basolateral complex as a polysensory cortical-like area that is afferent to both the extended amygdala and the striatum.

We are not arguing for a re-definition of the limbic system concept to
harmonize with the notion of the extended amygdala. It is essential, how-

ever, to recognize the fundamental differences that exist between the centromedial and basolateral part of the amygdaloid body, and in a broader sense between the extended amygdala and the ventral and dorsal striatopallidal system. The limbic system concept tends to camouflage these important distinctions. In fact, the popularity of the limbic system concept may depend on the lack of a strict anatomical definition. In the same vein, Brodal and Kaada argued more than twenty years ago, that the concept of the limbic system was not only unnecessary, but tended to obscure the distinct functional-anatomical relations that characterize different units of the basal forebrain (Kaada, 1960, p. 1346; Brodal, 1969, p. 538).

In this chapter we have emphasized the existence of two separate basal forebrain systems, the ventral striatopallidal system and the extended amygdala (Fig. 11) at the expense of another equally important system, the septal-diagonal band system. These three large telencephalic subcortical systems encompass most of the extrahypothalamic anatomy of basal forebrain. Although these forebrain systems show similarities, e.g. at the cellular level and in their relation to prominent non-thalamic corticopetal feedback systems (Alheid and Heimer, 1988; Alheid et al., 1989b, 1990), they can be defined and separated from each other, most clearly on the basis of transmitter content and "output channels". As indicated earlier, there are admittedly regions, e.g. at the boundaries between the extended amygdala and striatum, where these systems may overlap. But such regions, important as they may be, are exceptions to the rule. It is a fair guess that the three distinct forebrain units, e.g. the striatopallidum, septum-diagonal band, and extended amygdala, working in harmony with the cerebral cortex, have differing functional objectives. The systematic anatomy, that finds its expression in these three parallel forebrain systems cannot easily be reconciled with the limbic system concept.

Future Challenges

The discovery and the subsequent elaboration of the ventral striatopallidal system and extended amygdala has provided a new anatomical framework for the planning and interpretation of experiments designed to elucidate some of the more complex basal forebrain functions. These fundamental organizational principles are of immediate clinical relevance, especially in this new era of in vivo brain imaging. But this new anatomy has also presented us with new challenges. One intriguing development relates to the core-shell differentiation of the accumbens. The shell is characterized by a mixture of striatal and extended amygdaloid components, which may reflect an overlap or transition between two functional-anatomical systems. The general importance of this subject is emphasized by the fact that such special features are characteristic, not only for accumbens, but for most of the fringes of the striatal complex where it adjoins the extended amygdala. The exact relation of all these areas with either the striatum or the extended amygdala is far from clear and will demand the most powerful anatomical methods for their dissection.

Other crucially important questions relate to the "output channels" of these various forebrain systems, especially those directed through the lateral hypothalamus and towards various parts of the magnocellular corticopetal neuronal system. For instance, the synaptic arrangements in the lateral hypothalamus, where especially the shell of the accumbens and the rest of the extended amygdala have intimate relations, are largely unknown, and before the specific neuronal targets in the lateral hypothalamus and their

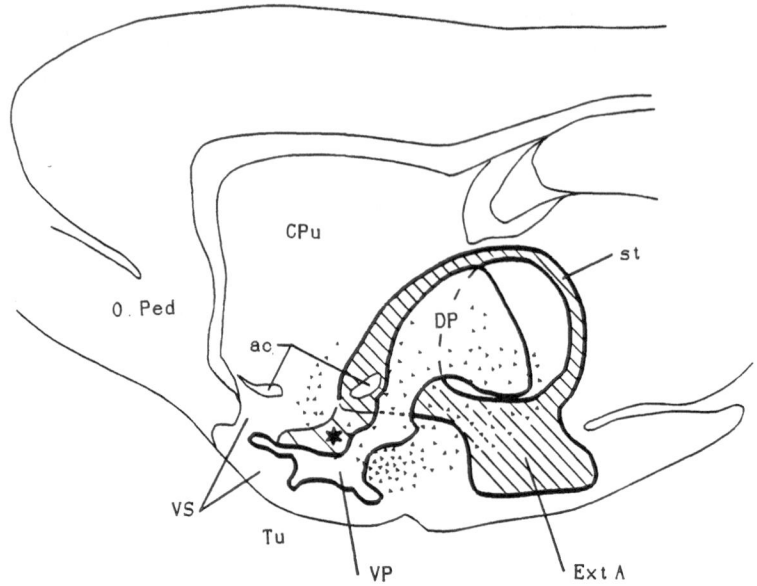

Fig. 11. Schematic diagram showing the topographical relationship of the
dorsal (DP) and ventral (VP) pallidum to the extended amygdala
(ExtA) and the cholinergic corticopetal cell populations (indicated by small black triangular neurons). Note that for simplicity, the mediolateral extent of the pallidal, extended amygdaloid, and corticopetal systems have been compressed into a single idealized sagittal section of the rat brain. Black star indicates the shell area of the accumbens. The interstitial nucleus
of the posterior limb of the anterior commissure is not shown.
Abbrev. ac = anterior commissure, BST = bed nucleus of stria
terminalis, CPu = caudate-putamen, O.Ped = olfactory peduncle, st
= stria terminalis, Tu = olfactory tubercle, VS = ventral
striatum.

ultimate axonal projections have been identified, it is not possible to
assess the full functional significance of the various forebrain systems
referred to in this chapter.

Elucidating the systematic organization of ascending feedback systems
in the basal forebrain is crucial to understanding the function of the
forebrain, and is certain to impact physiology, pharmacology, and clinical
medicine. We have argued that separate portions of the magnocellular cell
groups in the forebrain can best be understood as part of the individual
systems with which they are most closely associated topographically. For
instance, significant aggregates of corticopetal cells are located, not
only within the basal forebrain, but also in the posterolateral hypothalamus (e.g., Köhler and Swanson, 1984; Köhler et al., 1984, 1985; Saper 1985,
1987; Ericson et al., 1989) and brainstem (Wallace et al., 1989), where
they are probable targets of descending pathways from, for instance, the
extended amygdala. Considerable ambiguity exists about the numerous cholinergic neurons in the intrapallidal or peripallidal zones, including the
caudal and intralaminar parts of the globus pallidus. It may well be that
some of these are strung out as path neurons along the trajectory of the

dispersed fibers of the stria terminalis, or alternatively, that these receive striatal input as a special condensation of the cholinergic cells found within the striatum, but yet provide feedback to cortex. It is clear that the answer to problems such as these may be critical to understanding the functional operations of the basal ganglia, but will require further detailed study of the synaptic relations of these cells.

ACKNOWLEDGEMENTS

We are grateful for the generous support that NIH has provided through USPHS Grant Nr. **17743**. Ms. Carol Haselton, Linda S. Vega, Zhou Wang, and Mr. Lee Snavely have provided skillful assistance. We would also like to acknowledge the many informative and thought-provoking discussions we have had through many years on the subject of basal forebrain anatomy with Drs. José de Olmos, Laszlo Záborszky, D. Scott Zahm, Jørn Carlsen, Robert Switzer III, Larry Schmued, H. Robert Brashear, and Robert Chronister. Our collaboration with Dr. Alex Braun is particularly acknowledged with respect to sections of the rat brain immunohistochemically stained for angiotensin II and epidermal growth factor. In the same vein we would like to acknowledge the gift of these antibodies from Drs. C. Deschepper (angiotensin II) and D. Knauer (epidermal growth factor).

REFERENCES

Aggleton, J.P., 1985, A description of intra-amygdaloid connections in old world monkeys, Exp. Brain Res., 515-526.

Alexander, G.E., DeLong, M.R., and Strick, P.L., 1986, Parallel organization of functionally segregated circuits linking basal ganglia and cortex, Ann. Rev. Neurosci., 9:357-381.

Albert, D.J., Petrovic, D.M., Walsh, M.L., and Jonik, R.H., 1989, Medial accumbens lesions attenuate testosterone-dependent aggression in male rats, Physiol. Behav., 46:625-631.

Alheid, G.F., Haselton, C.L., and Heimer, L., 1989a, Accumbens projections to dorsal and ventral pallidum and to the extended amygdala in the monkey using PHA-L, Soc. Neurosci. Abstr., 15:904.

Alheid, G.F. and Heimer, L., 1988, New perspectives in basal forebrain organization of special relevance for neuropsychiatric disorders; the striatopallidal, amygdaloid, and corticopetal components of substantia innominata, Neurosci., 27:1-39.

Alheid, G.F., Heimer, L., and Switzer, R.C., 1990, The basal ganglia, in: "The Human Nervous System," G. Paxinos, ed., Academic Press, San Diego, pp. 483-582.

Alheid, G.F., Van Hoesen, G., and Heimer, L., 1989b, Functional neuroanatomy, in: "Comprehensive Textbook of Psychiatry," H.I. Kaplan and J. Sadock, ed., Williams & Wilkins, Baltimore, pp. 26-45.

Amaral, D.G., Avendano, C., and Benoit, R., 1989, Distribution of somatostatin-like immunoreactivity in the monkey amygdala, J. Comp. Neurol., 284:294-313.

Amaral, D.G. and Kurtz, J., 1985, An analysis of the origins of the cholinergic and non-cholinergic septal projections to the hippocampal formation in the rat, J. Comp. Neurol., 240:37-59.

Bartus, R.T., Dean III, R.L., Beer, B., Lippa A.S., 1982. The cholinergic hypothesis of geriatric memory dysfunction. Science 217: 408-417.

Beccari, M., 1910, Il lobo parolfattoro nei mammiferi, Arch. Ital. Anat. Embryol., 9:173-220.

Beccari, M., 1911, La sostanza perforata anteriore e i suoi rapporti col rinencefalo nel cerbello dell'uomo, Arch. Ital. Anat. Embryol., 10:261-328.

Beckstead, R.M., Domesick, V.B., and Nauta, W.J.H., 1979, Efferent connections of the substantia nigra and ventral tegmental area in the rat, Brain Res., 175:191-217.

Biesold, D., Inanami, O., Sato, A., and Sato, Y., 1989, Stimulation of the nucleus basalis of Meynert increases cerebral cortical blood flow in rats, Neurosci. Lett., 98:39-44.

Bigl, V., Woolf, N.J., and Butcher, L.L., 1982, Cholinergic projections from the basal forebrain to frontal, parietal, temporal, occipital, and cingulate cortices: a combined fluorescent tracer and acetylcholinesterase analysis, Brain Res. Bull., 8:727-749.

Blackstad, T., 1967, Cortical gray matter; a correlation of light and electron microscopic data, in: "The Neuron," H. Hyden, ed., Elsevier, Amsterdam, pp. 49-118.

Brodal, A., 1969. Neurological Anatomy. Oxford University Press, London.

Butcher, L.L. and Woolf, N.J., 1986, Central cholinergic systems; synopsis of anatomy and overview of physiology and pathology, in: "The Biological Substrates of Alzheimer's Disease," A.B. Scheibel and A.F. Wechsler, eds., Academic Press, New York, pp. 73-86.

Caffé, A.R., Van Ryen, P.C., Van der Woude, T.P., and Van Leeuwen, F.W., 1989, Vasopressin and oxytocin systems in the brain and upper spinal cord of macaca fascicularis, J. Comp. Neurol., 287:302-325.

Carlsen, J. and Heimer, L., 1988, The basolateral amygdaloid complex as a cortical-like structure, Brain Res., 441:377-380.

Carlsen, J., Záborszky, L., and Heimer, L., 1985, Cholinergic projections from the basal forebrain to the basolateral amygdaloid complex; a combined retrograde fluorescent and immunohistochemical study, J. Comp. Neurol., 234:155-167.

Chang, H.T., 1989, Noradrenergic innervation of the substantia innominata; a light and electron microscopic analysis of dopamine ß-hydroxylase immunoreactive elements in the rat, Exp. Neurol., 104:101-112.

Chang, H.T. and Kuo, H., 1989, Calcitonin gene-related peptides (CGRP) in the rat substantia innominata and globus pallidus; a light and electron microscopic immunocytochemical study, Brain Res., 495:167-172.

Chronister, R.B., Sikes, R.W., Trow, T.W., and DeFrance, J.F., 1981, The Organization of Nucleus Accumbens, in: "The Neurobiology of the Nucleus Accumbens," R.B. Chronister and J.F. De France, eds., Haer Institute for Electrophysiological Research, Maine, pp. 97-146.

Cools, A.R., Lohman, A.H.M., and Van der Bercken, eds., 1977, Psychobiology of the Striatum, Elsevier, Amsterdam.

Crosby, E.C. and Humphrey, T., 1941, Studies of the vertebrate telencephalon, II, the nuclear pattern of the anterior olfactory nucleus, tuberculum olfactorium, and the amygdaloid complex in adult man, J. Comp. Neurol., 71:121-213.

Danscher, G., 1982, Exogenous selenium in the brain; a histochemical technique for light and electron microscopical localization of catalytic selenium bonds, Histochem., 76:281-293.

Davis, P. and Maloney, A.J., 1976, Selective loss of cholinergic neurons in Alzheimer's disease, Lancet, 2:1403.

Delacour, J., Houcine, O., and Costa, J.C., 1990, Evidence for a cholinergic mechanism of "learned" changes in the responses of barrel field neurons of the awake and undrugged rat, Neurosci., 34:1-8.

de Olmos, J.S., 1972, The amygdaloid projection field in the rat as studied with the cupric-silver method, in: "The Neurobiology of the Amygdala," B.E. Elefteriou, ed., Plenum, New York, pp. 145-204.

de Olmos, J.S., 1990, The amygdala, in: "The Human Nervous System," G. Paxinos, ed., Academic Press, New York, pp. 583-710.

de Olmos, J.S., Alheid, G.F., and Beltramino, C.A., 1985, Amygdala, in: "The Rat Nervous System," G. Paxinos, ed., Academic Press, New York, pp. 223-334.

Dejerine, J.J., 1901, Anatomie des centres nerveux (2 Vols.), Rueff, Paris.

Divac, I., 1975, Magnocellular nuclei of the basal forebrain project to neocortex, brainstem and olfactory bulb; review of some functional correlates, Brain Res., 93:385-398.

Divac, I. and Öberg, R.G.E., eds., 1979a, The Neostriatum, Pergamon Press, Oxford.

Divac, I. and Öberg R.G.E., 1979b, Current conceptions of neostriatal functions history and an evaluation, in: "The Neostriatum (I)," I. Divac and R.G.E. Öberg, eds., Pergamon Press, Oxford, pp. 215-230,.

Ericson, H., Blomqvist, A., and Köhler, C., 1989, Origin of neuronal inputs to the tuberomamillary nucleus of the rat brain, in: "Neurons of the Tuberomamillary Nucleus," Ericson H. Doctoral dissertation, Uppsala University, Sweden. (Acta Universitatis Uppsaliensis 208.)

Evered, D. and O'Conner, M., 1984, Functions of the Basal Ganglia, Pitman, London (Ciba Symposium 107).

Everitt, B.J., Cador, M., and Robbins, T.W., 1989, Interactions between the amygdala and ventral striatum in stimulus-reward associations; studies using a second-order schedule of sexual reinforcement. Neurosci., 30:63-75.

Everitt, B.J., Sirkiä, T.E., Roberts, A.C., Jones, G.H., and Robbins, T.W. 1988, Distribution and some projections of cholinergic neurons in the brain of the common marmoset, Callithrix jacchus, J. Comp. Neurol., 271:533-558.

Fallon, J.H., 1983, The islands of Calleja complex of rat basal forebrain, II, Connections of medium and large sized cells, Brain Res. Bull., 10:775-793.

Farley, I.J. and Hornykiewicz, O., 1977, Noradrenaline distribution in subcortical areas of the human brain, Brain Res., 126:53-62.

Fibiger, H.C. and Phillips, A.G., 1986, Reward, motivation, cognition; psychobiology of mesotelencephalic dopamine systems, in: "Handbook of Physiology", Section 1, The Nervous System, Vol. 4, V.B. Mountcastle, F. Plum, S.R. Geiger, eds., American Physiological Society, Bethesda, pp. 647-674.

Fisher, R.S., Buchwald, N.A., Hull, C.D., and Levine, M.S., 1988, GABAergic basal forebrain neurons project to the neocortex; the localization of glutamic acid decarboxylase and choline acetyltransferase in feline corticopetal neurons, J. Comp. Neurol., 272:489-502.

Freund-Mercier, M.J., Dietl, M.M., Stoeckel, M.E., Palacios, J.M., and Richard, P.H., 1988, Quantitative autoradiographic mapping of neurohypophysial hormone binding sites in the rat forebrain and pituitary gland, II, Comparative study on the Long-Evans and Brattleboro strains, Neurosci., 26:273-281.

Garcia-Rill, E., Skinner, R.D., Gilmore, S.A., and Owings, R., 1983, Connections of the mesencephalic locomotor region (MLR), II, Afferents and efferents, Brain Res. Bull., 10:63-71.

Geeraedts, L.M.G., Nieuwenhuys, R., and Veening, J.G., 1990, Medial forebrain bundle of the rat, III, Cytoarchitecture of the rostral (telencephalic) part of the medial forebrain bundle bed nucleus, J. Comp. Neurol., 294: 507-536.

Gerfen, C.R., 1984, The neostriatal mosaic: compartmentalization of corticostriatal input and striatonigral output system, Nature, 311:461-464.

Gerfen, C.R., 1985, The neostriatal mosaic, I, Compartmental organization of projections from the striatum to the substantia nigra in the rat, J. Comp. Neurol., 236:454-476.

Gloor, P., 1955, Electrophysiological studies on the connections of the amygdaloid nucleus in the cat, I, The neuronal organization of the amygdaloid projection system, EEG Clin. Neurophysiol., 7:223-242.

Gray, T.S., 1987, Autonomic neuropeptide connections of the amygdala. in: "Hans Selye Symposium: Neuropeptides and Stress," Y. Tache, J.E. Morley, and M.R. Brown, eds., Springer-Verlag, New York, pp. 92-105.

Gray, T.S. and Magnuson, D.J., 1987, Neuropeptide neuronal efferents from the bed nucleus of the stria terminalis and central amygdaloid nucleus to the dorsal vagal complex in the rat, J. Comp. Neurol., 262:365-374.

Graybiel, A.M., 1990, Neurotransmitters and neuromodulators in the basal ganglia, Trends Neurosci., 13:244-254.

Graybiel, A.M., Ragsdale, C.W. Jr., Moon Edley, S., 1979, Compartments in the striatum of the cat observed by retrograde cell-labeling, Brain Res., 34:189-195.

Groenewegen, H.J., 1982, Organization of the afferent connections of the mediodorsal thalamic nucleus in the rat, related to the mediodorsal-prefrontal topography, Neurosci., 24:379-431.

Groenewegen, H.J. and Berendse, H.W., 1990, Connections of the subthalamic nucleus with ventral striatopallidal parts of the basal ganglia in the rat, J. Comp. Neurol., 294:607-622.

Groenewegen, H.J., Berendse, H.W., Meredith, G.E., Haber, S.N., Voorn, P., Wolters, J.G., and Lohman, A.H.M., 1991, Functional anatomy of the ventral limbic-innervated striatum, in: "The Mesolimbic Dopamine System: From Motivation to Action," P. Willner and J. Scheel-Kruger, eds., John Wiley and Sons Ltd., England, Chichester, pp. 19-59.

Groenewegen, H.J., Meredith, G.E., Berendse, H.W., Voorn, P., and Wolters, J.G., 1989, The compartmental organization of the ventral striatum in the rat, in: "Neural Mechanisms in Disorders of Movement," A.R. Crossman and M.A. Sambrook, eds., Libbey and Co., London, pp. 45-54.

Groenewegen, H.J., Russchen, F.T., 1984, Organization of the efferent projections of the nucleus accumbens to pallidal, hypothalamic and mesencephalic structures; a tracing and immunohistochemical study in the cat, J. Comp. Neurol., 223:347-367.

Grove, E.A., 1988a, Neural associations of the substantia innominata in the rat; afferent connections, J. Comp. Neurol., 277:315-346.

Grove, E.A., 1988b, Efferent connections of the substantia innominata in the rat, J. Comp. Neurol., 277:347-364.

Grove, E.A., Domesick, V.B., and Nauta, W.J., 1986, Light microscopic evidence of striatal input to intrapallidal neurons of cholinergic cell group Ch4 in the rat; a study employing the anterograde tracer Phaseolus vulgaris leucoagglutinin (PHA-L), Brain Res., 367:379-384.

Grove, E.A. and Nauta, W.J.H., 1984, Light microscopic evidence for striatal and amygdaloid input to cholinergic cell group CH4 in the rat, Soc. Neurosci. Abstr., 10:7.

Haber, S.N., 1987, Anatomical relationship between the basal ganglia and the basal nucleus of Meynert in human and monkey forebrain, Proc. Natl. Acad. Sci., 84:1408-1412.

Haber, S.N. and Elde, R., 1981, Correlation between met-enkephalin and substance P immunoreactivity in the primate globus pallidus, Neurosci., 6:1291-1297.

Haber, S.N., Groenewegen, H.J., Grove, E.A., and Nauta, W.J.H., 1985, Efferent connections of the ventral pallidum; evidence of a dual striatopallidofugal pathway, J. Comp. Neurol., 235:322-335.

Haber, S.N., Lind, E., Klein, C., and Groenewegen, H.J., 1990, Topographic organization of the ventral striatal efferent projections in the Rhesus monkey; an anterograde tracing study, J. Comp. Neurol., 293:282-298.

Haber, S.N. and Nauta, W.J.H., 1983, Ramifications of the globus pallidus in the rat as indicated by patterns of immunohistochemistry, Neurosci., 9:245-260.

Haber, S.N. and Watson, S.J., 1985, The comparative distribution of enkephalin, dynorphin and substance P in the human globus pallidus and basal forebrain, Neurosci., 14:1011-1024.

Haberly, L.B. and Price. J.L., 1978, Associational and commissural fiber systems of the olfactory cortex of the rat, I, Systems arising in the piriform cortex and adjacent areas, J. Comp. Neurol., 178:711-740.

Hall, E., 1972, The amygdala of the cat; a Golgi study, Z. Zellorsch, 134:439-458.

Hallanger, A.E., Levey, A.I., Henry, J.L., Rye, D.B., and Wainer, B.H., 1987, The origins of cholinergic and other subcortical afferents to the thalamus in the rat, J. Comp. Neurol., 262:105-124.

Hallström, A., Sato, A., Sato, Y., and Ungerstedt, U., 1990, Effect of stimulation of the nucleus basalis of Meynert on blood flow and extracellular lactate in the cerebral cortex with special reference to the effect of noxious stimulation of skin and hypoxia, Neurosci. Lett., 116:227-232.

Heimer, L., 1972, The olfactory connections of the diencephalon in the rat, Brain Behav. Evol., 6:484-523.

Heimer, L., 1978, The olfactory cortex and the ventral striatum, in: "Limbic Mechanisms," K.E. Livingston and O. Hornykiewicz, eds., Plenum, New York, pp. 95-187.

Heimer, L., Alheid, G.F., and Záborszky, L., 1983, Microinjections of retrograde fluorescent tracers in the ventral pallidum of rat label neurons at the medial edge of the subthalamic nucleus. Soc. for Neurosci. Abstr., 9:1230.

Heimer, L., Alheid, G.F., and Záborszky, L., 1985, The basal ganglia, in: "The Rat Nervous System," G.Paxinos, ed., Academic Press, Sydney, pp. 37-74.

Heimer, L., de Olmos, J.S., Alheid, G.F., and Záborszky, L., 1991a, "Perestroika" in the basal forebrain; opening the borders between neurology and psychiatry,in: "Role of the Forebrain in Sensation and Behaviour", G. Holstege, ed., pp. 109-165.

Heimer, L., Switzer, R.D., and Van Hoesen, G.W., 1982, Ventral striatum and ventral pallidum; components of the motor system? Trends in Neurosci., 5:83-87.

Heimer, L. and Wilson, R.D., 1975, The subcortical projections of allocortex; similarities in the neural associations of the hippocampus, the piriform cortex and the neocortex, in: "Golgi Centennial Symposium Proceedings," M. Santini, ed., Raven Press, New York, pp. 177-193.

Heimer, L., Záborszky, L., Zahm, D.S., and Alheid, G.F., 1987, The ventral striatopallidothalamic projection. I. The striatopallidal link originating in striatal parts of the olfactory tubercle. J. Comp. Neurol., 255:571-591.

Heimer, L., Zahm, D.S., Churchill, L., Kalivas, P.W., and Wohltmann, C., 1991b, Specificity in the projection patterns of accumbal core and shell in the rat, Neurosci., (in press).

Heimer, L., Zahm, D.S., and Schmued, L.C., 1990, The basal forebrain projection to the region of the nuclei gemini in the rat; a combined light and electron microscopic study employing horseradish peroxidase, fluorescent tracers and phaseolus vulgaris-leucoagglutinin, Neurosci., 34(3):707-731.

Herkenham, M., Moon Edley, S., and Stuart, J., 1984, Cell clusters in the nucleus accumbens of the rat and the mosaic relationship of opiate receptors, acetylcholinesterase and subcortical afferent terminations, Neurosci., 11:561-593.

Herrick, C.J., 1910, The morphology of the forebrain in amphibia and reptilia, J. Comp. Neurol. and Psychol., 20:413-546.

Holstege, G., 1990, Subcortical limbic system projections to caudal brainstem and spinal cord, in: "The Human Nervous System," G. Paxinos, ed., Academic Press, San Diego, pp. 261-286.

Holstege, G., Meiners, L., and Tan, K., 1985, Projections of the bed nucleus of the stria terminalis to the mesencephalon, pons, and medulla oblongata in the cat, Expl. Brain Res., 58:379-391.

Insel, T.R., Miller, L.P., and Gelhard, R.E., 1990, The ontogeny of excitatory amino acid receptors in rat forebrain, I, N-methyl-D-aspartate and quisqualate receptors, Neurosci., 35:31-43.

Jackson, A. and Crossman, A.R., 1981, Basal ganglia and other afferent projections to the peribrachial region in the rat; a study using retrograde and anterograde transport of horseradish peroxidase, Neurosci., 6:1537-1549.

Johnston, J.B., 1923, Further contributions to the study of the evolution of the forebrain, J. Comp. Neurol., 35:337-481.

Kaada, B., 1960, Cingulate, posterior orbital, anterior insular and temporal pole cortex, in: J. Field, H.W. Magoun, V.E. Hall (eds), "Handbook of Physiology", Section 1. Neurophysiology, Vol. II. American Physiological Society, Washington, D.C., pp 1345-1372.

Kelley, A.E., Domesick, V.B., and Nauta, W.J.H., 1982, The amygdalostriatal projection in the rat; an anatomical study by anterograde and retrograde tracing methods, Neurosci., 7:615-630.

Koikegami, H., Hirata, Y., and Oguma, J., 1967, Studies on the paralimbic brain structures, I, definition and delimitation of the paralimbic brain structures and some experiments on the nucleus accumbens, Folia Psychiatica et Neurologica Japonica, 21:151-180.

Krieger, N.R., 1981, Neurochemistry of the olfactory tubercle, in: "Biochemistry of Taste and Olfaction," Cagan R.H. and Kare M.R., eds., Academic Press, New York, pp. 417-441.

Köhler, C., Chan-Palay, V., and Wu, J.Y., 1984, Septal neurons containing glutamic acid decarboxylase immunoreactivity project to the hippocampal region in the rat brain, Anat. Embryol., 169:41-44.

Köhler, C., Haglund, L., and Swanson, L.W., 1984, A diffuse αMSH-immunoreactive projection to the hippocampus and spinal cord from individual neurons in the lateral hypothalamic area and zone incerta, J. Comp. Neurol., 223:501-514.

Köhler, C. and Swanson, L.W., 1984, Acetylcholinesterase-containing cells in the lateral hypothalamic area are immunoreactive for alpha-melanocyte stimulating hormone (alpha-MSH) and have cortical projections in the rat, Neurosci. Lett., 49:39-43.

Köhler, C., Swanson, L.W., Haglund, L., and Wu, J.Y., 1985, The cytoarchitecture, histochemistry and projections of the tuberomammillary nucleus in the rat, Neurosci., 16:85-110.

Krettek, J.E. and Price, J.L., 1978, Amygdaloid projections to subcortical structures within the basal forebrain and brainstem in the rat and cat, J. Comp. Neurol., 178:225-254.

Lauer, E.W., 1945, The nuclear pattern and fiber connections of certain basal telencephalic centers in the macaque, J. Comp. Neurol., 183:785-816.

LeDoux, J.E., Ruggiero, D.A., and Reis, D.J., 1985, Projections to the subcortical forebrain from anatomically defined regions of the medial geniculate body in the rat, J. Comp. Neurol., 242:182-213.

Lesur, A., Gaspar, P., Alvarez, Z., and Berger, B., 1989, Chemoanatomic compartments in the human bed nucleus of the stria terminalis, Neurosci., 32:181-194.

Levey, A.E., Wainer, B.H., Mufson, E.J., and Mesulam, M.M., 1983, Colocalization of acetylcholinesterase and choline acetyltransferase in the rat cerebrum, Neurosci., 9:9-22.

Lind, R.W., Swanson, L.W., and Ganten, D., 1985, Organization of Angiotensin II immunoreactive cells and fibers in the rat central nervous system; an immunohistochemical study, Neuroendocrin., 40:2-24.

Luskin, M.B., and Price, J.L., 1983, The topographic organization of associational fibers of the olfactory system in the rat, including centrifugal fibers to the olfactory bulb, J. Comp. Neurol., 216:264-291.

Lynd, E., Klein, C., Groenewegen, H.J., and Haber, S.N., 1988, Organization of the efferent projections from the primate ventromedial striatum, Soc. Neurosci. Abstr., 14:156.

Ma, W., Höhmann, C.F., Coyle, J.T., and Juliano, S.L., 1989, Lesions of the basal forebrain alter stimulus-evoked metabolic activity in mouse somatosensory cortex, J. Comp. Neurol., 288:414-427.

Macchi, G., 1951, The ontogenetic development of the olfactory telencephalon in man, J. Comp. Neurol., 95:245-305.

Maeda, M., Nakai, M., Krieger, A.J., and Sapru, H.N., 1990, Chemical stimulation of the nucleus tractus solitarii decreases cerebral blood flow in anesthetized rats, Brain Res., 520:255-261.

Mai, J.K., Stephens, P.H., Hope, A., and Cuello, A.C., 1986, Substance P in the human brain, Neurosci., 17:709-739.

Martin, L.J., Koliatsos, V.E., Struble, R.G., Powers, R.E., and Price, D.L., 1988, Chemoarchitectonic patterns of peptides in human basal forebrain; evidence for a system comprising the bed nucleus, substantia innominata, and central amygdala, Soc. Neurosci. Abstr., 14:671.

Matelli, M., Luppino, G., Fitzpatrick, D., and Diamond, I.T., 1988, The pulvinar nucleus of Tupaia; comparative study of its connections with the superior coliculus, the neocortex, and the corpus striatum, in: "Cellular Thalamic Mechanisms", M. Bentivolio and R. Spreafico (eds.), Elsevier Science Publishers, Amsterdam, pp. 207-220.

Matthysse, S., 1973, Antipsychotic drug actions; a clue to the neuropathology of schizophrenia? Federation Proc., 32:200-205.

McDonald, A.J., 1984, Neuronal organization of the lateral and basolateral amygdaloid nuclei in the rat, J. Comp. Neurol., 222:589-606.

McDonald, A.J., 1991, Cell types and intrinsic connections of the amygdala, in: "The Amygdala," J.P. Aggleton, ed., Wiley, New York, (in press).

McKenzie, J.S., Kemm, R.E., and Wilcock, L.N., eds., 1984, The Basal Ganglia, Plenum Press, New York.

Melander, T., Staines, W.A., Hökfelt, T., Rokaeus, A., Eckenstein, F., Salvaterra, P.M., and Wainer, B.H., 1985, Galanin-like immunoreactivity in cholinergic neurons of the septum-basal forebrain complex projecting to the hippocampus of the rat, Brain Res., 360:130-138.

Mesulam, M.M., and Geula, C., 1988, Nucleus basalis (Ch4) and cortical cholinergic innervation in the human brain: Observations based on the distribution of acetylcholinesterase and choline acetyltransferase. J. Comp. Neurol., 275:216-240.

Mesulam, M.M., Mufson, E.J., Levey, A.I., and Wainer, B.H., 1983a, Cholinergic innervation of cortex by the basal forebrain: cytochemistry and cortical connections of the septal area, diagonal band nuclei, nucleus basalis (substantia innominata), and hypothalamus in the Rhesus monkey, J. Comp. Neurol., 214:170-197.

Mesulam, M.M., Mufson, E.J., and Wainer, B.H., 1986, Three-dimensional representation and cortical projection topography of the nucleus basalis (Ch4) in the macaque; concurrent demonstration of choline acetyltransferase and retrograde transport with a stabilized tetramethylbenzidine method for HRP, Brain Res., 367:301-308.

Mesulam, M.M., Mufson, E.J., Wainer, B.H., and Levey, A.J., 1983b, Central cholinergic pathways in the rat; an overview based on an alternative nomenclature, Neurosci., 10:1185-1201.

Millhouse, O.E. and De Olmos, J., 1983, Neuronal configurations in lateral and basolateral amygdala, Neurosci., 10:1269-1300.

Millhouse, O.E. and Heimer, L., 1984, Cell configurations in the olfactory tubercle of the rat, J. Comp. Neurol., 265:1-24.

Miodonski, R., 1967, Myeloarchitectonics and connections of substantia innominata in the dog brain, Acta. Biologiae Expert. (Warszawa), 27:61-84.

Mishkin, M., Malamut, B., and Bachevalier, J., 1984, Memories and habits; two neural systems, in: "Neurobiology of Learning and Memory," G. Lynch, J.L. McGaugh, and N.M. Weinberger, eds., Guildford Press, New York, pp. 65-77.

Mizuno, N., Takahashi, O., Satoda, T., and Matsushima, R., 1985, Amygdalospinal projections in the macaque monkey, Neurosci. Lett., 53:327-330.

Mogenson, G.J., 1984, Limbic-motor integration with emphasis on initiation of exploratory and goal-directed locomotion, in: "Modulation of Sensorimotor Activity During Alternations in Behavioral States," Alan Liss, New York, pp. 121-137.

Mogenson, G.J., 1987, Limbic motor integration, Progr. Psychobiol., 12:117-170.

Mogenson, G.J., Jones, D.L., and Yim, C.Y., 1980, From motivation to action; functional interface between the limbic system and the motor system, Progr. Neurobiol., 14:69-97.

Mogenson, G.J., Swanson, L.W., and Wu, M., 1983, Neural projections from nucleus accumbens to globus pallidus, substantia innominata, and lateral preoptic-lateral hypothalamic area; an anatomical and electrophysiological investigation in the rat, J. Neurosci., 3:189-202.

Mori, S., Ueda, S., Yamad, H., Takino, T., and Sano, Y., 1985, Immunohistochemical demonstration of serotonin nerve fibers in the corpus striatum of the rat, cat, and monkey, Anat. Embryol. (Berl.), 173:1-5.

Mugnaini, E. and Oertel, W.H., 1985, Atlas of the distribution of GABAergic neurons and terminals in the rat CNS as revealed by GAD immunohisto- chemistry, in: "Handbook of Chemical Neuroanatomy: GABA and Neuropeptides in the CNS," A. Björklund and T. Hökfelt, eds., Elsevier, Amsterdam, pp. 436-595.

Nakano, Y., Oomura, Y., Leonard, L., Nishino, H., Aou, S., Yamamoto, T., and Aoyagi K., 1986, Feeding-related activity of glucose and morphine-sensitive neurons in the monkey amygdala, Brain Res., 399:167-172.

Nauta, W.J.H. and Domesick, V.B., 1978, Crossroads of limbic and striatal circuitry; hypothalmo-nigral connections, in: "Limbic Mechanisms: The Continuing Evolution of the Limbic System Concept," Livingston K.E. and O. Hornykiewicz, eds., pp. 75-93.

Nauta, W.J.H. and Domesick, V.B., 1984, Afferent and efferent relationships of the basal ganglia, in: "Functions of the Basal Ganglia," D. Evered and M. O'Conner, eds., (Ciba Foundation Symposium 107), Pitman, London, pp. 3-23.

Nauta, W.J.H., Smith G.P., Faull, R.L.M., and Domesick, V.B., 1978, Efferent connections and nigral afferents of the nucleus accumbens septi in the rat, Neurosci., 3:385-401.

Newman, R. and Winans, S.S., 1980, An experimental study of the ventral striatum of the golden hamster, II, Neuronal connections of the olfactory tubercle, J. Comp. Neurol., 191:193-212,.

Nielsen, E.B. and Scheel-Krüger, J., 1986, Cueing effects of amphetamine and LSD; elicitation by direct microinjection of the drugs into the nucleus accumbens, Eur. J. Pharmacol., 125:85-92.

Nishijo, H. and Ono, T., Nishino, H., 1988, Single neuron responses in amygdala of alert monkey during complex sensory stimulation with affective significance, J. Neurosci., 8:3570-3583.

Parent, A., 1986, Comparative Neurobiology of the Basal Ganglia, John Wiley and Sons, New York.

Parent, A., Paré, D., Smith, Y., and Steriade, M., 1988, Basal forebrain cholinergic and noncholinergic projections to the thalamus and brainstem in cats and monkeys, J. Comp. Neurol., 277:281-391.

Pearson, R.C.A., Gatter, K.C., Brodal, P., and Powell, T.P.S., 1983, The projection of the basal nucleus of Meynert upon the neocortex in the monkey, Brain Res., 259:132-136.

Penney, J.B. and Young, A.B., 1983, Speculations on the functional anatomy of basal ganglia disorders, Ann. Rev. Neurosci., 6:73-94.

Phelps, P.E. and Vaughn, J.E., 1986, Immunocytochemical localization of choline acetyltransferase in rat ventral striatum; a light and electron microscopic study, J. Neurocytol., 15:595-617.

Phillips, A.G. and Carr, G.D., 1987, Cognition and the basal ganglia; a possible substrate for procedural knowledge, Can. J. Neurol. Sci., 14:381-385.

Phillips, P.A., Abrahams, J.M., Kelly, J., Paxinos, G., Grzonka, Z., Mendelsohn, F.A.O., and Johnston,C.I., 1988, Localization of vasopressin binding sites in rat brain by in vitro autoradiography using a radioiodinated V1 receptor antagonist, Neurosci., 27:749-761.

Pioro, E.P., Mai, J.K., and Cuello, A.C., 1990, Distribution of substance P-and enkephalin-immunoreactive neurons and fibers, in: "The Human Nervous System," G. Paxinos, ed., Academic Press, San Diego, pp. 1051-1094.

Price, J.L. and Amaral, D.G., 1981, An autoradiographic study of the projections of the central nucleus of the monkey amygdala, J. Neurosci., 11:1242-1259.

Price, J.L., Russchen, F.T., and Amaral, D.G., 1987, The limbic region, II, The amygdaloid complex, in: "Handbook of Chemical Neuroanatomy," A. Björklund, T. Hökfelt, and L.W. Swanson, eds., Elsevier, Amsterdam, pp. 279-388.

Price, J.L., Slotnick, B.M., and Revial, M.-F., 1991, Olfactory projections to the hypothalamus, J. Comp. Neurol., in press.

Ramón y Cajal, S., 1911, Histologie du système nerveux de l'homme et des vertèbres (Part II), Maloine, Paris.

Ribak, C.E. and Kramer III, W.G., 1982, Cholinergic neurons in the basal forebrain of the cat have direct projections to the sensorimotor cortex, Exper. Neurol., 75:453-465.

Richardson, R.T. and DeLong, M., 1990, Context-dependent responses of primate nucleus basalis neurons in a Go/No-Go task, J. Neurosci., 10:2528-2540.

Richardson, R.T., Mitchell, S.J., Baker, F.H., and DeLong, M.R., 1988, Responses of nucleus basalis of Meynert neurons in behaving monkeys, in: "Cellular Mechanisms of Conditioning and Behavioral Plasticity," C.D. Woody, D.L. Alkon, and J.L. McGaugh, eds., Plenum, New York.

Rolls, E.T., Sanghera, M.K., and Roper-Hall, A., 1979, The latency of activation of neurones in the lateral hypothalamus and substantia innominata during feeding in the monkey, Brain Res., 16:121-135,

Roberts, G.W., Woodhams, P.L., Polak, J.M., and Crow, T.J., 1980, Distribution of neuropeptides in the limbic system of the rat; the amygdaloid complex, Neurosci., 7:99-131.

Russchen, F.T., Bakst, I., Amaral, D.G., and Price, J.L., 1985, The amygdalostriatal projections in the monkey; an anterograde tracing study, Brain Res., 329:241-257.

Russell, V.A., Allin, R., Lamm, M.C.L., and Taljaard, J.J.F., 1989, Increased dopamine D2 receptor-mediated inhibition of [^{14}C]Acetylcholine release in the dorsomedial part of the nucleus accumbens, Neurochem. Res., 14:877-881.

Rye, D.B., Wainer, B.H., Mesulam, M.M., Mufson, E.J., and Saper, C.B., 1984, Cortical projections arising from the basal forebrain; a study of cholinergic and non-cholinergic components employing combined retrograde tracing and immunohistochemical localizations of choline acetyltransferase, Neurosci., 13:627-643.

Sandler, M., Feuerstein, C., and Scatton, B., eds., 1987, Neurotransmitter Interactions in the Basal Ganglia, Raven, New York.

Sandrew, B.B., Edwards, D.L., Poletti, C.E., and Foote, W.E., 1986, Amygdalo-spinal projections in the cat, Brain Res., 373:235-239.

Saper, C.B., 1984, Organization of cerebral cortical afferent systems in the rat, II, Magnocellular basal nucleus, J. Comp. Neurol., 222:313-342.

Saper, C.B., 1985, Organization of cerebral cortical afferent systems in the rat, II, Hypothalamocortical projections, J. Comp. Neurol., 237:21-46.

Saper, C.B., 1987, Diffuse cortical projection systems; anatomical organization and role in cortical function, in: "Handbook of Physiology: The Nervous System," V.B. Mountcastle, R. Plum, and S.R. Geiger, eds., American Physiological Society, Maryland, pp. 169-210.

Schneider, J.S. and Lidsky, T.J., eds., 1986, Basal Ganglia and Behavior: Sensory Aspects of Motor Functioning, Hans Huber Publishers, New York.

Schwaber, J.S., Kapp, B.S., Higgins, G.A., and Rapp, P.R., 1982, Amygdaloid and basal forebrain direct connections with the nucleus of the solitary tract and the dorsal motor nucleus of the vagus, J. Neurosci., 2: 424-1438.

Schwaber, J.S., Rogers, W.T., Satoh, K., and Fibiger, H.C., 1987, Distribution and organization of cholinergic neurons in the rat forebrain demonstrated by computer-aided data acquisition and three-dimensional reconstruction, J. Comp. Neurol., 263:309-325.

Scott, J.W. and Chafin, B.R., 1975, Origin of olfactory projections to lateral hypothalamus and nuclei gemini of the rat, Brain Res., 88:64-6.

Scott, J.W. and Leonard, C.M., 1971, The olfactory connections of the lateral hypothalamus in the rat, mouse, and hamster, J. Comp. Neurol., 141:331-344.

Scott, J.W. and Pfaffmann, C., 1967, Olfactory input to the hypothalamus; electrophysiological evidence, Science., 158:1592-1594.

Scott, J.W. and Pfaffmann, C., 1972, Characteristics of responses of lateral hypothalamic neurons to stimulation of the olfactory system, Brain Res., 48:251-264.

Shu, S.Y., Penny, G.R., and Peterson, G.M., 1988, The "marginal division"; a new subdivision in the neostriatum of the rat, J. Chem. Neuroanat., 1:147-163.

Shu, S.Y., McGinty, J.F., and Peterson, G.M., 1990, High density of zinc-containing and dynorphin B- and substance P-immunoreactive terminals in the marginal division of the rat striatum, Brain Res. Bull., 24:201-205.

Slotnick, B.M., 1990, Olfactory perception, in: "Comparative Perception, Vol. I, Basic Mechanisms", Mark A. Berkley and Williams C. Stebbins, eds., John Wiley and Sons, Inc., pp. 155-214.

Small, R.K. and Leonard, C.M., 1983, Early recovery of function after olfactory tract section correlated with reinnervation of olfactory tubercle, Brain Res., 283:25-40.

Stevens, J.R., 1973, An anatomy of schizophrenia? Arch. Gen. Psychiat., 29:177-189.

Strenge, H., Braak, E., and Braak, H., 1977, Über den Nucleus Striae Terminalis im Gehirn des Erwachsenen Menschen, Z. Mikrosk. Anat. Forsch. (Leipzig), 91:105-118.

Swanson, L.W. and Cowan, W.M., 1975, A note on the connections and development of the nucleus accumbens, Brain Res., 92:324-330.

Swanson, L.W., Mogenson, G.J., Gerfen, C.R., and Robinson, P., 1984, Evidence for a projection from the lateral preoptic area and substantia innominata to the "mesencephalic locomotor region" in the rat. Brain Res., 295:161-178.

Swerdlow, N.R. and Koob, G.F., 1987, Dopamine, schizophrenia, mania and depression; toward a unified hypothesis of cortico-striato-pallido-thalamic function, Behav. Brain Sci., 10:197-245.

Switzer, R.C., Hill, J., and Heimer, L., 1982, The globus pallidus and its rostroventral extension into the olfactory tubercle of the rat; a cyto- and chemoarchitectural study, Neurosci., 7:1891-1904.

Syzmusiak, R. and McGinty, D., 1989, Effects of basal forebrain stimulation on the walking discharge of neurons in the midbrain reticular formation of cats, Brain Res., 498:355-359.

Ulfig, N., Braak, E., Ohm, T.G., and Pool, C.W., 1990, Vasopressinergic neurons in the magnocellular nuclei of the human basal forebrain, (in press).

Vaccarino F.J. and Rankin J. (1989) Nucleus accumbens cholecystokinin (CCK) can either attenuate or potentiate amphetamine-induced locomotor activity; evidence for rostral-caudal differences in accumbens CCK function. Behav. Neurosci., 103:831-836.

Vanderwolf, C.H., 1983, The role of the cerebral cortex and ascending activating systems in the control of behavior, in: "Handbook of Behavioral Neurobiology, Vol. 6", E. Satinoff and P. Teitelbaum, eds., Plenum Publishing Corp., pp. 67-104.

Vincent, S.R., McIntosh, C.H.S., Buchan, A.M.J., and Brown, J.C., 1985,

Central somatostatin systems revealed with monoclonal antibodies, J. Comp. Neurol., 238:169-186.

Von Economo, C. and Koskinas, G.N., 1925, Die Cytoarchitektonik der Hirnrinde des Erwachsenen Menschen, Springer Verlag, Berlin.

Voorn, P., Gerfen, C.R., and Groenewegen, H.J., 1989, Compartmental organization of the ventral striatum of the rat; immunohistochemical distribution of enkephalin, substance P, dopamine and calcium binding protein, J. Comp. Neurol., 289:189-201.

Walker, L.C., Koliatsos, V.E., Kitt, C.A., Richardson, R.T., Rökaeus, and Price, D.L., 1989, Peptidergic neurons in the basal forebrain magnocellular complex of the rhesus monkey, J. Comp. Neurol., 280:272-282.

Wallace, D.M., Magnuson, D.J., and Gray, T., 1989, The amygdalo-brainstem pathway; selective innervation of dopaminergic, noradrenergic, and adrenergic cells in the rat, Neurosci. Lett., 97:252-258.

Whitehouse, P.J., Price, D.L., Clark, A.W., Coyle, J.T., and DeLong, M.R., 1981, Alzheimer disease; evidence for selective loss of cholinergic neurons in the nucleus basalis, Ann. Neurol., 10:122-126.

Williams, D.J., Crossman, A.R., and Slater, P., 1977, The efferent projections of the nucleus accumbens in the rat, Brain Res., 130:217-227.

Wilson, F.A.W. and Rolls, E.T., 1985, Reinforcement-related neuronal activity in the basal forebrain and amygdala, Soc. Neurosci. Abstr., 15:52.

Wood, D.M. and Emmett-Oglesby, M.W., 1989, Mediation in the nucleus accumbens of the discriminative stimulus produced by cocaine, Pharmacol. Biochem. Behav., 33:453-457.

Woolf, N.J. and Butcher, L.L., 1982, Cholinergic projections to the basolateral amygdala; a combined Evans Blue and acetylcholinesterase analysis, Brain Res. Bull., 8:751-763.

Woolf, N.J., Eckenstein, F., and Butcher, L.L., 1984, Cholinergic systems in the rat brain, I, Projections to the limbic telencephalon, Brain Res. Bull., 13:751-784.

Yoshikawa, T., Fukamauchi, F., Shibuya, H., and Takahashi, R., 1989, Regional heterogeneity with the nucleus accumbens concerning the effects of dopaminergic agents on the content of cholecystokinin, Neurochem. Inst., 14:467-469.

Young, W.S. III, Alheid, G.F., and Heimer, L., 1984, The ventral pallidal projection to the mediodorsal thalamus; a study with fluorescent retrograde tracers and immunohistofluorescence, J. Neurosci., 4:1626-1638.

Záborszky, L., Alheid, G.F., Beinfeld, M.L., Eiden, L.E., Heimer, L., and Palkovits, M., 1985, Cholecystokinin innervation of the ventral striatum; amorphological and radioimmunological study, Neurosci., 14:427-453.

Záborszky, L., Carlsen, J., Brashear, H.R., and Heimer, L., 1986, Cholinergic GABAergic afferents to the olfactory bulb in the rat with special emphasis on the projection neurons in the nucleus of the horizontal limb of the diagonal band, J. Comp. Neurol., 243:488-509.

Zahm, D.S., 1989, The ventral striatopallidal parts of the basal ganglia in the rat, II, Compartmentation of ventral pallidal efferents, Neurosci., 30:33-50.

Zahm, D.S. and Heimer, L., 1987, The ventral striatopallido thalamic projection, III, Striatal cells of the olfactory tubercle establish direct synaptic contact with ventral pallidal cells projecting to mediodorsal thalamus, Brain Res., 404:327-331.

Zahm, D.S. and Heimer, L., 1988, Ventral striatopallidal parts of the basal ganglia in the rat, I, Neurochemical compartmentation as reflected by the distributions of neurotensin and substance P immunoreactivity, J. Comp. Neurol., 272:516-535.

Zahm, D.S. and Heimer, L., 1990, Two transpallidal pathways originating in nucleus accumbens, J. Comp. Neurol., 302:437-446.

Zahm, D.S., Záborszky, L., Alheid, G.F., and Heimer, L., 1987, The ventral striatopallidothalamic projection, II, The ventral pallidothalamic link, J. Comp. Neurol., 255:592-605.

Ziehen, T., 1909 (1897), Das Centralnervensystem der Monotremen und Marsupialier, Ein Beitrag zur vergleichenden makroskopischen und mikros-
kopischen Anatomie und zur Vergleichenden Entwicklungsgeschichte des
Wirbelthiergehirns, II. Theil. Mikroskopische Anatomie. Erster Absch-
nitt. Der Faserverlauf im Hirnstamm von Pseudochirus peregrinus. _In_: R.
Semon (Ed.), "Zoologische Forschungsreisen in Australien und dem
Malayischen Archipel, III, Monotremen und Marsupialier, II, I,
Lieferun", Jena, Gustav Fischer, pp. 677-728 (note: actual date is
1901; also listed as "Denkschriften der Medicinisch-Natur-
wissenschaftlichen Gesellschaft Zu Jena, Sechster Band).

AFFERENTS TO BASAL FOREBRAIN CHOLINERGIC PROJECTION NEURONS: AN UPDATE

László Záborszky[1], William E. Cullinan[1] and Alex Braun[2]

[1]Departments of Otolaryngology, Neurosurgery and Neurology
University of Virginia Health Science Center
Charlottesville, VA 22908
[2]Department of Pathology, State University of New York at
Stony Brook, New York 11780

INTRODUCTION

The basal forebrain cholinergic projection (BFC) system has been the focus of considerable attention as a result of evidence implicating it in a number of behavioral functions, including arousal, sensory processing, motivation, emotion, learning, and memory (Deutsch, 1983; Buzsaki et al., 1988; Richardson and DeLong, 1988; Durkin, 1989; Rolls 1989, Steriade and McCarley, 1990). Moreover, neuropathological changes in the BFC have been reported in a surprisingly large number of neurological diseases, including Alzheimer's and Parkinson's diseases (for ref. see Coyle et al., 1983; Mesulam and Geula, 1988; Arendt et al., 1989). BFC neurons in the rat are dispersed across a number of classically defined territories of the basal forebrain, as illustrated from a series of coronal sections in Fig. 1. However, using a computer graphic three-dimensional reconstruction technique (Schwaber et al., 1987) or manual reconstruction from camera lucida drawings (Fig. 2) it is evident that BFC neurons form a continuum, rather than being arranged as distinct nuclear groups.

Although there is considerable species variation in the precise locations of cholinergic projection neurons in the basal forebrain, the efferent projections of these cells follow basic organizational principles in all vertebrate species studied. Thus, neurons within the medial septum and nucleus of the vertical limb of the diagonal band (MS/VDB; also termed Ch1/Ch2 according to the classification of Mesulam et al., 1983b) provide the major cholinergic innervation of the hippocampus; cholinergic cells within the horizontal limb of the diagonal band and magnocellular preoptic nucleus (HDB/MCP; Ch3) project to the olfactory bulb, piriform, and entorhinal cortices; cholinergic neurons located in the ventral pallidum, sublenticular substantia innominata (SI), globus pallidus, internal capsule, and nucleus ansa lenticularis, collectively termed the nucleus basalis (Ch4), project to the basolateral amygdala, and innervate the entire neo-cortex according to a rough medio-lateral and antero-posterior topography (Sofroniew et al., 1982; Armstrong et al., 1983; Mesulam et al., 1983a,b; Lamour et al., 1984b; Rye et al., 1984; Woolf et al., 1984; Amaral and Kurz, 1985; Carlsen et al., 1985; Wainer et al., 1985; Woolf et al., 1986; Záborszky et al., 1986a; Luiten et al., 1987; Nyakas et al., 1987; Sofroniew et al., 1987; Fisher et al., 1988; Gaykema et al., 1990). In the primate the corticopetal cholinergic cells (Mesulam et al., 1983b; Mesulam et al., 1986a; see also Butcher and Semba, 1989; Heimer et al.,

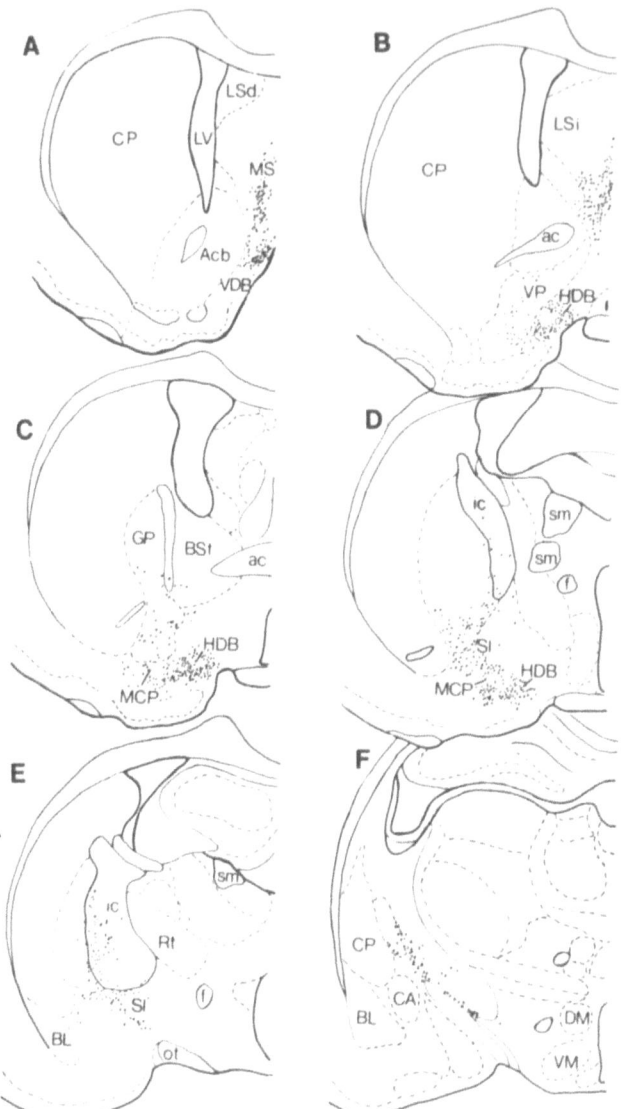

Fig. 1.   A-F: Series of drawings made from coronal sections through a rat brain (rostral to caudal) that have been immunostained for ChAT (choline acetyltransferase), illustrating the distribution of cholinergic neurons (dots). Striatal cholinergic neurons (including those in the ventral striatum) have been omitted for simplicity. Abbrev. ac = anterior commissure; Acb = accumbens nucleus; BL = basolateral amygdaloid nucleus; BSt = bed nucleus of the stria terminalis; CA = central amygdaloid nucleus; CP = caudate putamen; DM = dorsomedial hypothalamic nucleus; f = fornix; GP = globus pallidus; HDB = horizontal limb of the diagonal band; ic = internal capsule; LSd = lateral septal nucleus, dorsal; LSi = lateral septal nucleus, intermediate; LV = lateral ventricle; MCP = magnocellular preoptic nucleus; MS = medial septal nucleus; ot = optic tract; Rt = reticular thalamic nucleus; SI = sublenticular substantia innominata; sm = stria medullaris; VDB = vertical limb diagonal band nucleus; VM = ventromedial hypothalamic nucleus; VP = ventral pallidum.

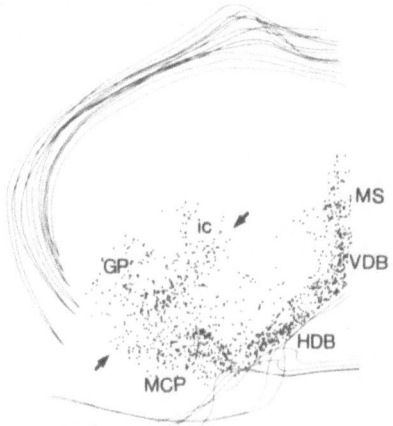

Fig. 2.   Composite map illustrating the distribution of cholinergic projection neurons (dots). This drawing was composed from 6 camera
lucida drawings which were aligned and superimposed to generate
the final figure. Only the ventral outlines of the brain and the
corpus callosum are indicated. The same technique is used to show
the distribution of putative contact sites of different afferent
systems in relation to the cholinergic neurons in Fig. 15. Cells
between the arrows correspond to the area of the SI. Abbrev. See
Fig. 1.

1991a) are subdivided according to the topography of their projections.
Thus, the anteromedial sector (Ch4am) projects to medial cortical areas
including the cingulate gyrus; the anterolateral compartment (Ch4al) to
ventral orbital, frontal, and parietal opercular regions, as well as the
amygdala; the intermediate sector (Ch4i) to lateral frontal, parietal,
peristriate and temporal regions including the insula; and the posterior
sector (Ch4p) to superior temporal and temporopolar areas (Mesulam et al.,
1983a, 1986a; Mesulam and Geula, 1988).

    Despite the wealth of data on the topography of the efferent projections
of BFC neurons, information concerning the afferent connections of these
cells has been more elusive. The reasons for this are easily appreciated in
view of the anatomical complexity of the basal forebrain, in which cholinnergic neurons are intermingled among numerous non-cholinergic cells
(Záborszky et al., 1986a; Walker et al., 1989), and are distributed in close
proximity to several major ascending and descending fiber systems (for
review see Záborszky, 1989a). Therefore the verification of actual synaptic
contact between the afferent fiber systems and the cholinergic projection
neurons requires appropriate combinations of double immunocytochemical
methods at the ultrastructural level, in which the afferent system and the
cholinergic nature of postsynaptic target can be unequivocally determined
(Záborszky and Heimer, 1989). The study of these inputs is further complicated by the morphological characteristics of these neurons. For example,
several studies suggested that the dendrites of BFC neurons extend for very
long distances (Semba et al., 1987; Brauer et al., 1988; Dinopoulos et al.,
1988), and our own experiments involving combinations of Golgi labeling and
choline acetyltransferase (ChAT) immunocytochemistry (Fig. 3) have confirmed
this point. Also, synaptic input to the cell bodies and proximal dendrites
of these cells is notably sparse, although it increases on more distal
dendritic segments (Ingham et al., 1985; Dinopoulos et al., 1986). Thus, a
complete characterization of the afferents of the BFC cells will need to

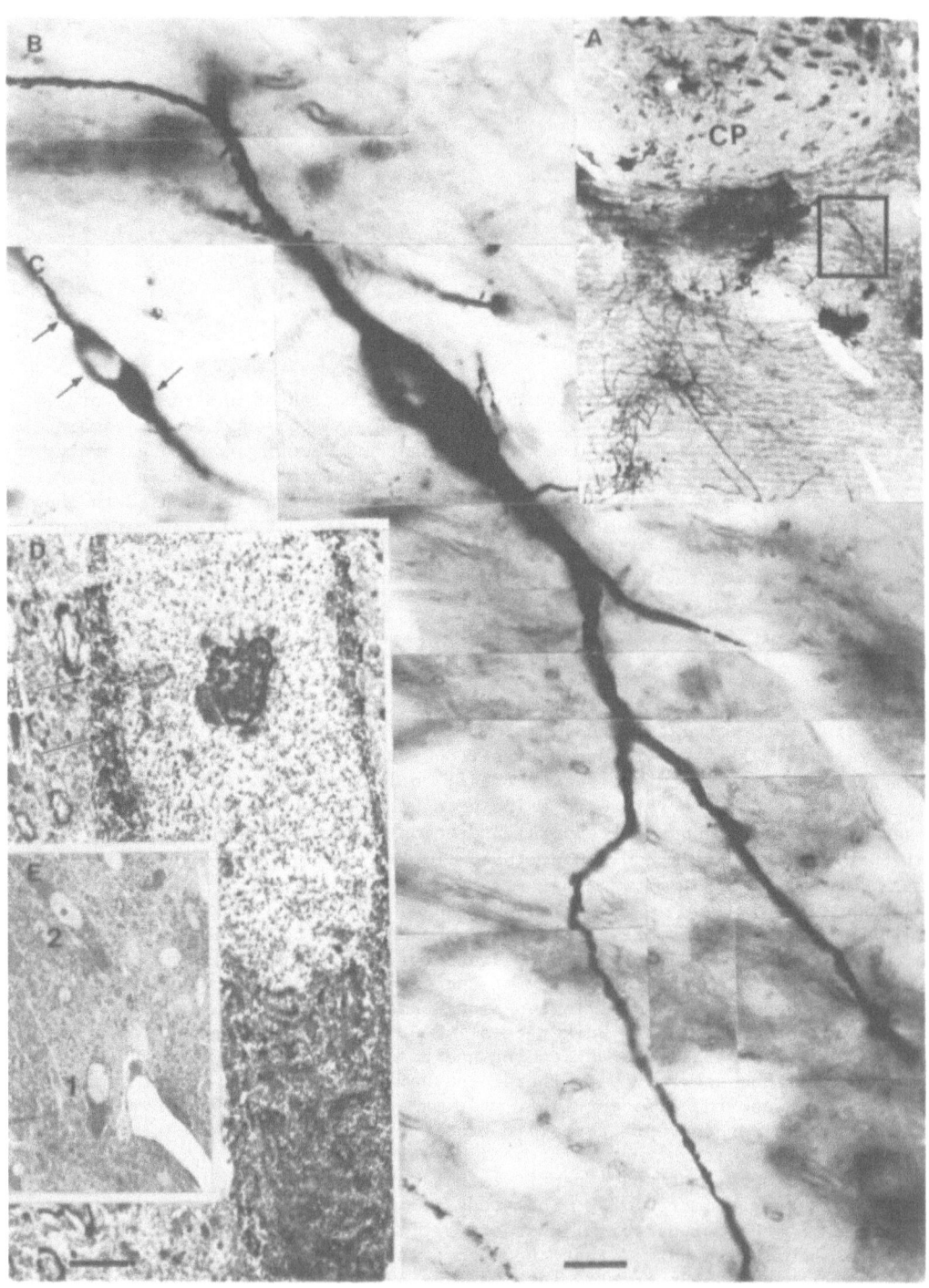

Fig. 3.   A: Golgi impregnated cholinergic neuron (method of Gabbott and
Somogyi, 1984) located medial to the substriatal gray. Neuron in
boxed area is shown under higher magnification in B and C. D:
Electron micrograph showing gold deposits (arrows) in the peri-
karyon, in addition to heavy immunostaining for ChAT. E: Low magnification view of the identified neuron (# 1) in the electron
microscope.   Abbrev. CP = caudate putamen. Scale: B=10μm; E=2μm.

take these factors into account. This information is prerequisite to the design of pharmacological and behavioral investigations of BFC function.

This review will survey data regarding afferents to the basal forebrain with special reference to potential inputs to BFC neurons, including their transmitters, as well as summarize the current state of knowledge from ultrastructural studies of identified afferents to these cells. Finally, we will discuss some possible functional roles of BFC neurons in behaviors as they may relate to specific neural circuits.

CORTICAL AFFERENTS

Studies in monkeys and cats (Mesulam and Mufson, 1984; Irle and Markowitsch, 1986) are consistent with the notion that basal forebrain areas rich in cholinergic neurons receive only a restricted cortical projection, originating from so called paralimbic cortical areas (Mesulam et al., 1986b), including the orbitofrontal cortex, the anterior insular cortex, temporal polar, medial inferotemporal region, entorhinal, piriform, and perirhinal cortices. BFC neurons do not appear to receive direct input from primary sensory and motor cortex or from most higher order association areas. In contrast, in a study in the rat involving wheat germ agglutinin-horseradish peroxidase (WGA-HRP) injections into different cortical areas (Saper, 1984), it was suggested that cortical projections to BFC neurons originate in widespread cortical areas, and that these projections are reciprocal in nature to the corticopetal projection of the BFC system. However, data from recent studies in the rat, in which retrograde tracers were delivered to different regions of the BFC system, do not support this concept (Lamour et al., 1984b; Haring and Wang, 1986; Grove, 1988; Jones and Cuello, 1989; Semba and Fibiger, 1989; Carnes et al., 1990), but rather suggest a restricted origin of cortical input to basal forebrain areas containing cholinergic neurons in rodents, similar to primates. Indeed, our preliminary light microscopic experiments in the rat, involving anterograde tracing with *Phaseolus vulgaris* leucoagglutinin (PHA-L) from different cortical areas (Cullinan and Záborszky, in preparation), are consistent with the notion that cortical inputs to the BFC are restricted to a number of allocortical regions, including orbital areas, insular cortex, and perhaps the piriform cortex. One reason for the discrepancy with the results of Saper (1984) appears to be that many corticofugal axons of neocortical origin descending in the internal capsule, although located in close proximity to corticopetal cholinergic neurons, are in most instances smooth and devoid of varicosities, while axons from allocortical regions showed profuse terminal arborizations in some portions of the basal forebrain, where they could be seen to approximate cholinergic neuronal elements in patterns suggestive of synaptic contact.

In the rat, the pattern of input to the basal forebrain from the various allocortical regions investigated appears to be restricted. For example, terminal arborizations of orbitofrontal axons were found mainly in the ventral portions of the MCP and HDB (Fig. 4). Interestingly, the forebrain regions which receive inputs from allocortical areas correspond well with the location of BFC neurons projecting to the same cortical areas, suggesting a reciprocity between BFC neurons and allocortical areas.

Although the possibility cannot be excluded that a proportion of the cholinergic neurons receiving allocortical input project to cortical regions, other than allocortex, it is clear that large portions of the cortex are not influenced from the allocortex via BFC cells. This is particularly true for many neocortical areas (e.g. sensorimotor, auditory), which receive cholinergic projections from BFC neurons located primarily in

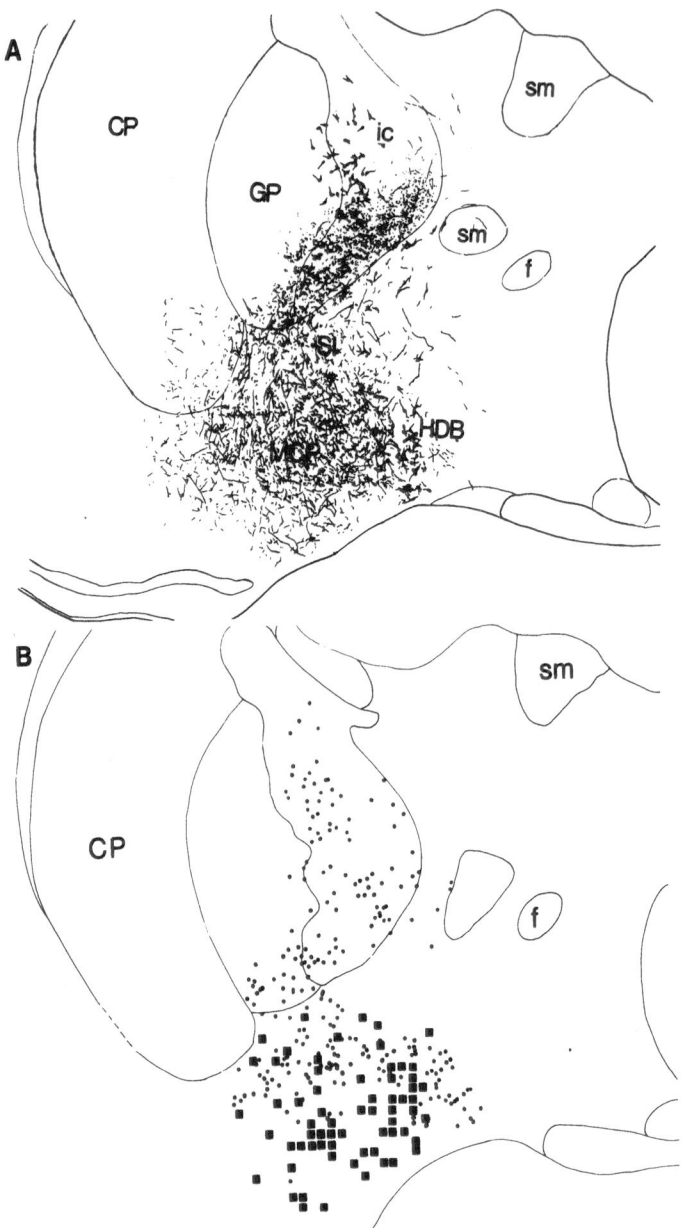

Fig. 4.   A: Camera lucida drawing (10x) from a double-labeled section
           illustrating the distribution of PHA-L labeled fibers in re-
           lation to cholinergic neurons following delivery of the tracer
           to the orbitofrontal cortex. B: Schematic drawing (from section
           adjacent to the one shown in A) illustrating the distribution
           pattern of PHA-L labeled terminals in apposition to cholinergic
           neurons from a section that was analyzed using high magnifi-
           cation light microscopy. Cholinergic neurons are represented by
           dots. Zones of putative contacts between PHA-L-positive termi-
           nals and cholinergic profiles are depicted as squares. Sections
           were screened using an ocular reticle (80x80 μm) at 63x, and
           contact sites were marked on a camera lucida drawing of the
           corresponding section using a proportional grid. Abbrev. see
           Fig. 1.

the internal capsule, globus pallidus, and peripallidal regions, areas which are apparently not reached by projections from allocortical regions. While it has been suggested that in the primate paralimbic cortical areas may influence not only their own cholinergic innervation, but also that of widespread cortical regions by virtue of their projections to the BFC (Mesulam and Mufson, 1984), these conclusions were based on autoradiographic data, in which fibers of passage are not clearly distinguished from terminals. Therefore, the extent to which paralimbic cortical afferents might control the BFC system in the primate remains to be disclosed. On the other hand, cortical areas may indirectly influence their own cholinergic input through the striatopallidal system (see below), or through cortico-cortical connections.

The behavioral specializations of the orbitofrontal cortex are remarkably similar in rodents and primates, including humans, and are thought to include the regulation of autonomic functions, feeding, species-specific social-affective behaviors, motivation, and some aspects of learning and memory (Kolb, 1984; Goldman-Rakic, 1987; Fuster, 1989; Sesack et al., 1989). Monkeys with orbitofrontal lesions were particularly impaired on behavioral tasks requiring the frequent making and breaking of stimulus-reward associations (Jones and Mishkin, 1972). In the absence of the orbitofrontal cortex, these animals continue to respond to inappropriate (unrewarded) stimuli, similar to humans with frontal lobectomy.

Electrophysiological experiments in primates have revealed that basal forebrain, possibly cholinergic neurons, similar to many orbitofrontal units, respond to a range of visual and auditory stimuli that have been reinforced (Richardson and De Long, 1988; Rolls, 1989; Wilson and Rolls, 1990a,b). It has been suggested that this information is conveyed to the basal forebrain through the orbitofrontal cortex (Rolls, 1989).

Behavioral studies in primates have shown that the dorsolateral prefrontal cortex, or in rats the medial prefrontal cortex, is critically concerned with an animal's ability to connect events across time, as ablations of this region result in severe deficits in performance on delay tasks (Kolb, 1984; Goldman-Rakic, 1987; Fuster, 1989). Electrophysiological studies in behaving monkeys revealed that prefrontal units which participate in different phases of behavior organization show a topographical distribution. Units encoding the sensory cue (both its specific sensory nature and its behavioral significance), are located rostrally in the prefrontal cortex; other units, which discharge in relation to the anticipatory motor action, or engaging in the rapid relay between the first two categories of units, are located more caudally towards the premotor and motor areas. Finally, units which are related to reward or punishment concentrate in orbital areas and are intermingled with units responding to gustatory and olfactory inputs (Rosenkilde et al., 1981; Fuster et al., 1982; Thorpe et al., 1983; Innoue et al., 1985). Although the exact mechanism by which these different units participate in the temporal organization of behavior is unclear, the cholinergic system, with afferents from the orbitofrontal cortex and probable projections back to the prefrontal cortex, is likely to influence these processes. Since corticofugal projections are thought to be primarily excitatory (Fonnum et al., 1981; Giuffrida and Rustioni, 1988), and a facilitatory role has been typically ascribed to acetylcholine (ACh) from physiological studies, feedback through the BFC system could conceivably hold the information about previously experienced stimuli current or 'on line' in prefrontal-cortical circuits over the delay period (while the stimulus is absent) to guide appropriate responses. This notion is consistent with the results of pharmacological and lesion studies, as well as neuropsychiatric data, indicating that disruption of the BFC system is associated with deficits in memory functions (Drachman and Leavitt, 1974; Bartus et al., 1982).

Ventral striatal afferents

The nucleus accumbens, together with the ventral part of the caudate putamen, the olfactory tubercle, and the striatal cell bridges that connect these structures, form the ventral striatum (Heimer and Wilson, 1975). A major output of the ventral striatum is directed to the ventral extension of the globus pallidus known as the ventral pallidum, in a topographically organized manner (Heimer and Wilson, 1975; Nauta et al., 1978; Mogenson et al., 1983; Haber et al., 1990; Parent, 1990; Heimer et al., 1991a,b).

In rat, many cholinergic projection neurons invade the ventral pallidum and dorsal pallidum (or globus pallidus), or the narrow spaces between fiber bundles of the internal capsule, and it was suggested based upon light microscopic data (Grove et al., 1986) that cholinergic neurons share to some extent the afferents of neighboring noncholinergic pallidal neurons. In experiments using PHA-L tracing and ChAT immunocytochemistry at the ultrastructural level (Cullinan and Záborszky, in preparation) we confirmed our preliminary electron microscopic degeneration data (Záborszky et al., 1984b; Fig. 5) that nucleus accumbens axons establish synaptic contact with ventral pallidal cholinergic neurons. Since the majority of the PHA-L labeled neurons at the injection site in the nucleus accumbens in our study were also immunopositive for calcium binding protein (calbindin D_{28}), a marker for a subpopulation of GABAergic neurons in this region, it is likely that the neurotransmitter involved in this projection is, in large part, GABA. Indeed, terminals containing the enzyme glutamic acid decarboxylase (GAD), the GABA synthesizing enzyme, have been confirmed to terminate on cholinergic neurons in the ventral pallidum (Záborszky et al., 1986b). Substance P and enkephalin are also transmitter candidates for these inputs, as lesions of the nucleus accumbens result in decreases in immunostaining for these substances in the ventral pallidum (Walaas and Fonnum, 1979; Záborszky et al., 1982; Haber and Nauta, 1983), and substance P and enkephalin containing terminals have been found to contact cholinergic neurons in pallidal regions (Bolam et al., 1986; Martinez-Murillo et al., 1988a). These substances may be co-localized with GABA, since striatal neurons have been shown to co-localize GABA and either substance P or enkephalin (Penny et al., 1986).

In autoradiographic (Mesulam and Mufson, 1984; Haber et al., 1990) and limited PHA-L tracing experiments in primates (Haber et al., 1990) it has been suggested that axons originating from the nucleus accumbens and ventromedial portions of the caudate and putamen innervate the nucleus basalis. The extent to which these fibers terminate in this region or merely pass through, however, remains to be more fully disclosed.

The possibility that cholinergic neurons receive ventral striatal input in primates, together with the observations of a topographical organization within corticostriatal projections (Phillipson and Griffiths, 1985; Goldman-Rakic and Selemon, 1986; McGeorge and Faull, 1989), ventral striatal projections (Haber et al., 1990; Heimer et al., 1991b), and in the BFC system, may be of interest in the light of recent ideas about the functional organization of forebrain circuits that involve the cortex and basal ganglia. Alexander et al. (1986) have suggested that in primates distinct regions of the frontal cortex, basal ganglia, and the thalamus are connected through parallel, functionally segregated circuits. Cholinergic neurons receiving ventral striatal input may be part of such channels, however these would provide an extrathalamic outflow to the cortex. Several of such distinct circuits can be envisaged (Fig. 6, upper row), although more detailed data are needed, particularly with respect to the input-output relations of identified BFC neurons. Since the corticostriatal as well as the hippocampo- or amygdalofugal projections are most probably excitatory (Fonnum et al.,

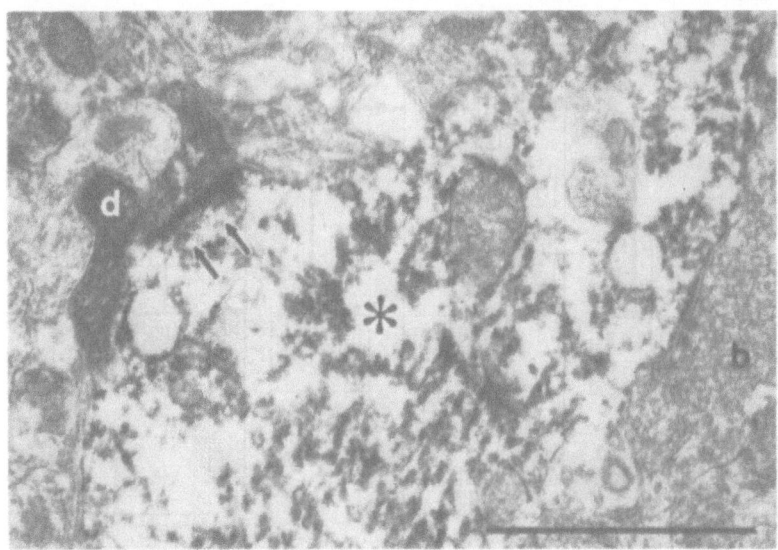

Fig. 5.   Degenerated terminal (d) contacting with a cholinergic dendrite
(asterisk) in the ventral pallidum two days after lesion of the
nucleus accumbens; b: for comparison a normal bouton. Arrows
point to postsynaptic thickening. Bar scale: 1 μm.

1981; Christie et al., 1987; Fuller et al., 1987), and the input from the
ventral striatum to the cholinergic neurons is likely to be inhibitory, the
activation of this corticostriatal link would appear to result in reduced
cholinergic activity in the cortical target areas.

In rat, medial and lateral parts of the prefrontal cortex, (anterior
cingulate, prelimbic, agranular insular) the perirhinal, entorhinal cortices, the amygdala and the hippocampus project to the ventral striatum in a
topographical fashion (Krayniak et al., 1981; Kelley and Domesick, 1982;
Kelley et al., 1982; Phillipson and Griffith, 1985; Swanson and Köhler,
1986; Fuller et al., 1987; Groenewegen et al., 1987; Sesack et al., 1989;
Witter et al., 1989; Groenewegen et al., 1990), and the projections from the
ventral striatum to the pallidal complex also maintain a high degree of
topographical organization (Gerfen, 1985; Heimer et al., 1991a,b). Cholinnergic neurons of the ventral pallidum project to the basolateral amygdala
(Carlsen et al., 1985), and to the medial prefrontal cortex and an adjoining medial strip of the motor cortex as well as allo- (piriform and entorhinal cortex) and periallocortical (insula) regions (Saper, 1984; Haber et
al., 1985; Luiten et al., 1987). Thus it appears that cholinergic neurons
within the ventral pallidum may also participate in cortico-striatopallido-
cortical loops (Fig. 6, lower row). In view of the fact that afferents from
different cortical areas and the amygdala project to the striatum in a
complicated, often overlapping or interdigitating fashion (Groenewegen et
al., 1990) in rats, similar to monkeys (Selemon and Goldman-Rakic, 1985),
and since the topography of the cholinergic projection neurons in the VP is
not well understood, two-chain tracing/immunocytochemical experiments are
required to determine whether such pathways involving cholinergic projections exist as proposed in Fig. 6.

Dorsal striatal input

The significance of striatal input to pallidal cholinergic neurons would
appear to vary according to the species. In primates, only a small number of

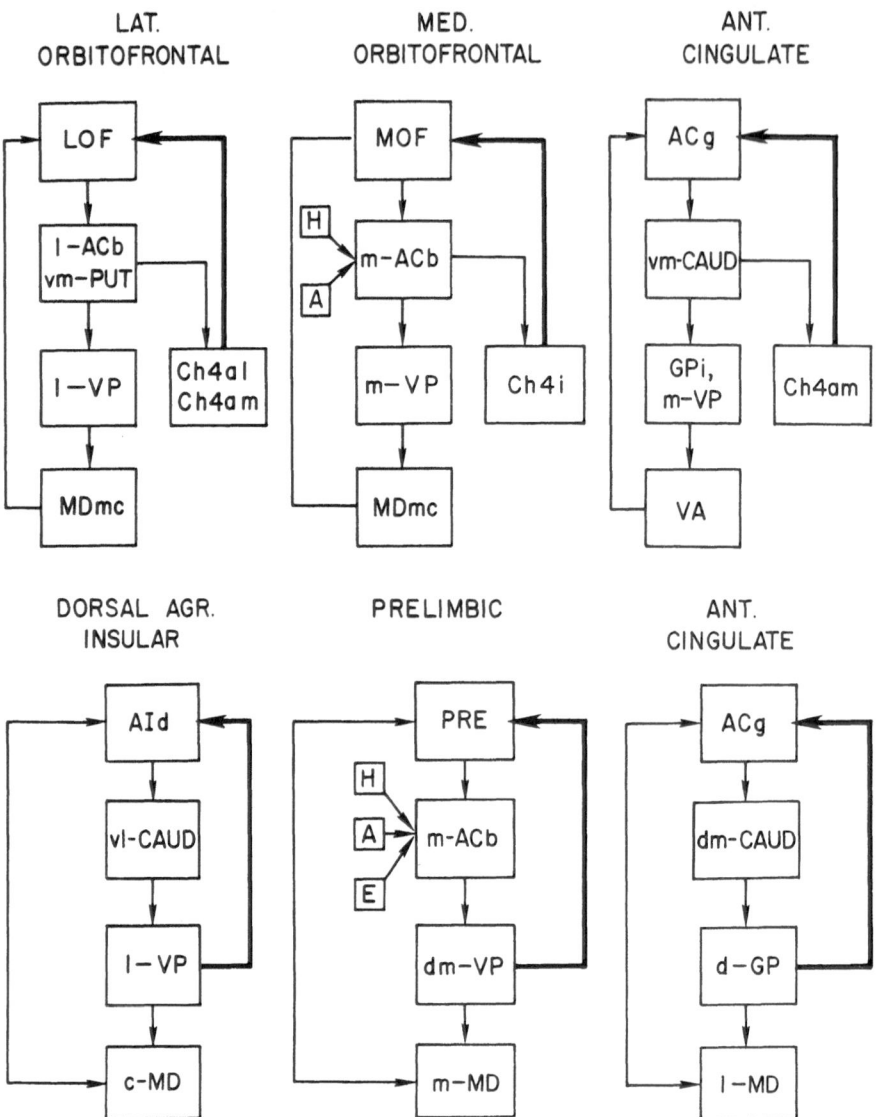

Fig. 6.   Proposed forebrain circuits involving the cortex, basal ganglia
and BFC neurons. Upper row: primate, lower row: rat. Data from
Alexander et al. (1986), Groenewegen et al., (1990), Haber et al.
(1990). Thick arrow marks the cholinergic link. <u>Abbrev.</u> A = amygdala; ACg = anterior cingulate cortex; AId = dorsal agranular
insular cortex; c-MD = mediodorsal thalamic nucleus, central part;
d-GP = globus pallidus, dorsal part; dm-CAUD = caudate nucleus,
dorsomedial part; dm-VP = ventral pallidum, dorsomedial part; E =
entorhinal cortex; GPi = globus pallidus, internal segment; H =
hippocampus; l-ACb = nucleus accumbens, lateral part; l-MD =
mediodorsal thalamic nucleus, lateral part; l-VP = ventral pallidum, lateral part; LOF = lateral orbitofrontal cortex; m-ACb =
nucleus accumbens, medial part; m-MD = mediodorsal thalamic nucleus, medial part; m-VP = ventral pallidum, medial part; MDmc =
mediodorsal thalamic nucleus, magnocellular division; MOF = medial
orbitofrontal cortex; PRE = prelimbic cortex; vl-CAUD = caudate
nucleus, ventrolateral part; vm-CAUD = caudate nucleus, ventromedial part; vm-PUT; putamen, ventromedial part; VA = ventral
anterior thalamus.

cholinergic cells are located in the internal and external medullary laminae of the globus pallidus, or are embedded within the fibers of the internal capsule, in addition to the few neurons found within the globus pallidus proper (Mesulam et al., 1983a; Saper and Chelimsky, 1984; Everitt et al., 1988; Mufson et al., 1989). It is not clear, however, whether these neurons are reached by striatofugal fibers from the dorsal striatum, or only by aberrant fibers from the stria terminalis (Price and Amaral, 1981; Price et al., 1987). In contrast, in rat, a considerable population of corticopetal cholinergic neurons are located in the globus pallidus, internal capsule and peripallidal areas, and it has been suggested that cholinergic neurons in these areas receive striatal input from the medial striatum (Grove et al., 1986; Shu et al., 1988). In rat, corticostriatal projections from pre-frontal, auditory, visual cortices, hippocampus and the amygdala project to the medial part of the dorsal striatum in a topographical manner (e.g. Beck-stead, 1979; McGeorge and Faull, 1989). Due to the complex compartmental terminations of the different afferents in the dorsal striatum (e.g. Gerfen, 1984; Donoghue and Herkenham, 1986; Faull et al., 1986; Gerfen, 1989), it is unclear which afferents might reach those striatal cells, which in turn, project to cholinergic neurons in the globus pallidus.

In summary, striatal afferents to BFC neurons may constitute a way by which different cortical regions indirectly control the cholinergic input they receive. Future studies should identify the specific neuronal elements involved in these circuits, and determine how this indirect cortical control may be related to the more direct cortical control of cholinergic function.

HIPPOCAMPAL AFFERENTS

The majority of hippocampal projections to the septum in rat (Swanson and Cowan, 1979; Groenewegen et al., 1987) course through the fimbria-fornix and terminate on spiny multipolar neurons (Alonso and Frotscher, 1989) and GABAergic neurons (Leranth and Frotscher, 1989) in the lateral septal nucleus. It has been suggested that lateral septal neurons establish synaptic contacts with cholinergic and GABAergic neurons in the medial septum which in turn project back to the hippocampus, but that hippocampopetal neurons in the MS/VDB do not receive afferents from the hippocampus directly (Köhler et al., 1984; Freund and Antal, 1988; Leranth and Frotscher, 1989). Light microscopic tracing studies in primates (Aggleton et al., 1987) and rodents (Carnes et al., 1990) suggested that hippocampal efferents en route to the lateral hypothalamus and amygdala may terminate on neurons in the MS/VDB complex and medial part of the SI, however, the notion that cholinnergic hippocampopetal neurons receive direct input from the hippocampus remains questionable and requires confirmation.

AFFERENTS FROM THE AMYGDALA

Autoradiographic studies in primates and rodents have shown that fibers originating from different amygdaloid nuclei course through the HDB and SI as part of the ventral amygdalofugal pathway en route to other brain regions, including the hypothalamus, thalamus, striatum, and prefrontal cortical areas (Krettek and Price, 1978; Price and Amaral, 1981; Russchen et al., 1985a,b; de Olmos, 1990). Although the extent to which these projections terminate in the SI was unclear due to the limitations of the autoradiographic method, experiments using the PHA-L technique have shown that amygdaloid fibers have varicosities along their trajectory through the SI (Russchen and Price, 1984; Russchen et al., 1985a). In double-labeling experiments at the electron microscopic (EM) level in the rat, we have shown that fibers from the basolateral amygdaloid nucleus form synapses on dendrites of both cholinergic and noncholinergic cells in the ventral pallidum (Záborszky et al., 1984a). The extent to which BFC neurons outside of the ventral pallidum receive amygdaloid afferents remains to be determined, as do the projection targets of those cells which may receive this input. The available evidence, however, suggests that the amygdaloid complex as a whole is reciprocally connected with the BFC system in both rat and primate, but that this reciprocity is not exact.

Neurons of the amygdala receive specific multimodal sensory information (Aggleton et al., 1980; Turner et al., 1980; van Hoesen, 1981; Ruschen et al., 1985a,b; Amaral, 1987; LeDoux, 1987) and participate in stimulus-reward-associative learning (Gaffan and Harrison, 1987; Nishijo et al., 1988; LeDoux et al., 1990). It is likely that a subpopulation of BFC neurons participate in similar processes through amygdaloid input.

It is interesting to note that polysensory paralimbic cortical areas, including the temporal pole, inferotemporal, orbitofrontal, perirhinal and entorhinal cortices, which project to the amygdaloid nuclei, also project to the SI in primates (Aggleton et al., 1980; Turner et al., 1980; van Hoesen, 1981; Russchen et al., 1985a,b). Similar interconnections are also known in rat (Witter et al., 1989; Groenewegen et al., 1990). Therefore, from a functional point of view, it will be important to examine the possible convergence of cortical and amygdaloid inputs to the same cholinergic neuron.

Based upon hodological and chemoarchitectural characteristics, certain parts of the amygdala, the SI, and bed nucleus of the stria terminalis (BSt) constitute a morphological and perhaps functional entity, as originally suggested by de Olmos et al. (1985), and discussed recently in several papers (Alheid and Heimer, 1988; Moga et al., 1990; Heimer et al., 1991a; Heimer and Alheid, this volume). In particular, the lateral BSt-SI-central nucleus of the amygdala share many connections. Similarly, an affiliation between the medial segment of the BSt and medial nucleus of the amygdala and an intervening portion of the SI has been suggested. The results of a PHA-L tracing study (Grove, 1988) have suggested that the lateral BSt and the central amygdala project to the more dorsal SI/ventral part of globus pallidus, while the medial BSt, medial and basomedial amygdaloid nuclei project to more ventral parts of the SI. Cholinergic projection neurons located within these subdivisions may thus subserve different functional roles accordingly.

HYPOTHALAMIC AFFERENTS

The possibility of hypothalamic input to the general forebrain regions containing cholinergic projection neurons was suggested from a number of autoradiographic studies (Conrad and Pfaff, 1976a,b; Saper et al., 1976; Swanson, 1976; Saper et al., 1978; Krieger et al., 1979; Saper et al., 1979; Berk and Finkelstein, 1982; Mesulam and Mufson, 1984; Saper, 1985), although this issue remained somewhat unclear due to the difficulty in distinguishing fibers from terminals with the autoradiographic technique. More recent experiments using the anterograde tracer PHA-L have confirmed the presence of projections to these regions from several hypothalamic nuclei (ter Horst and Luiten, 1986; Simerly and Swanson, 1988) and a light microscopic double-labeling study investigated the distribution of hypothalamic afferents to cholinergic cells in the SI (Grove, 1988). We recently have mapped the distribution of hypothalamic axons originating from various portions of the caudal lateral hypothalamus and from different medial hypothalamic cell groups in relation to the BFC system in its entirety (Cullinan and Záborszky, 1991). Inputs to the cholinergic projection system were distributed in a manner reflecting the gross topography of the ascending hypothalamic projections. Axons originating from neurons in the far-lateral hypothalamus

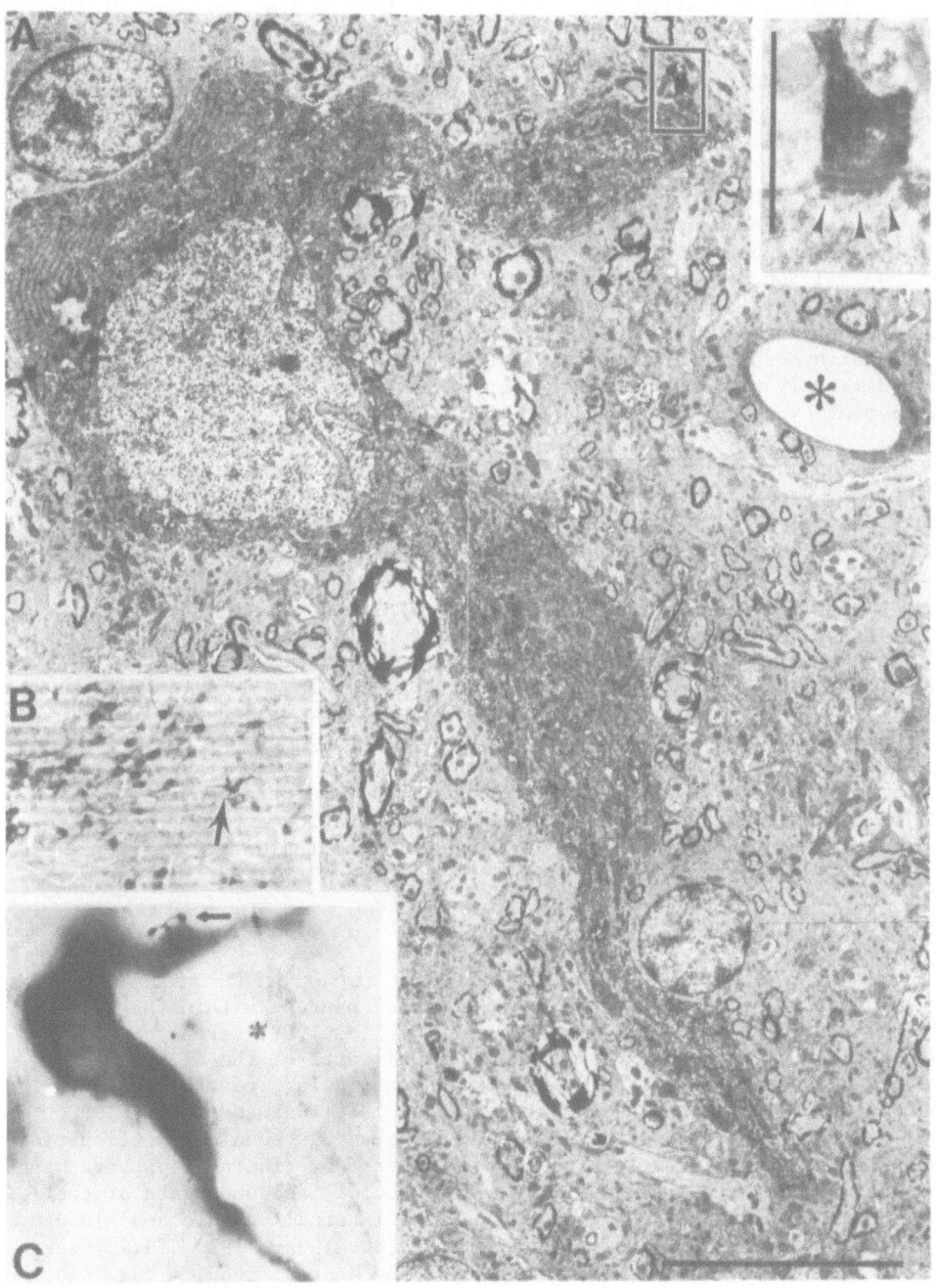

Fig. 7.   A: Low power electron micrograph of a reconstructed cholinergic
neuron from the substantia innominata (arrow in B). Boxed area in
A is enlarged at inset in upper right corner, and show the synaptic contact of the PHA-L varicosity, indicated by an arrow in C.
Arrowheads in the inset denote sub-synaptic dense bodies. PHA-L
was injected into the lateral hypothalamus. Asterisks in A and C
mark the same vessel for orientation. Bar scale: A = 10 μm, inset
= 1 μm. Reproduced from Záborszky and Cullinan (1989) by permission of Elsevier Science Publishers.

reach cholinergic neurons in a zone which extends from the dorsal part of the SI caudolaterally, to the lateral portion of the bed nucleus of the stria terminalis rostromedially, encompassing a narrow band along the ventral part of the globus pallidus and medial portion of the internal capsule. Axons originating from cells in the more medial portions of the lateral hypothalamus reach cholinergic cells primarily in more medial and ventral parts of SI, and in the magnocellular preoptic nucleus and HDB. Finally, axons from medial hypothalamic cells, particularly the anterior hypothalamic and medial preoptic areas, appear to contact cholinergic neurons primarily in the medial part of HDB, and in the MS/VDB complex (Fig. 15D-F). Electron microscopic double-labeling experiments confirmed contacts between labeled terminals and cholinergic neurons in the HDB and SI, both from the lateral and medial hypothalamus (Záborszky and Cullinan, 1989, Cullinan and Záborszky, 1991; Fig. 7). A comparison of the topography of corticopetal and hippocampopetal neurons (Mesulam et al., 1983a; Rye et al., 1984; Woolf et al., 1984; Amaral and Kurz, 1985; Woolf et al., 1986) with our findings suggests that corticopetal cholinergic neurons are innervated by lateral hypothalamic neurons, while hippocampopetal cholinergic neurons located largely within the MS/VDB complex are likely to receive input mainly from medial hypothalamic groups.

Considering the fact that lateral hypothalamic neurons receive both general and special viscerosensory information (Fulwiller and Saper, 1984) either directly from the nucleus of the solitary tract or through the parabrachial nucleus, it is possible that this integrated viscerosensory input reaches BFC neurons through lateral hypothalamic projections. The significance of afferents from the medial hypothalamus is less clear, as this region is thought to participate in the control of a number of neuroendocrine, autonomic, and behavioral mechanisms (Záborszky, 1982; Swanson, 1987). A recent review (McGinty and Szymusiak, 1990) suggested that inhibitory input from preoptic neurons to basal forebrain cholinergic neurons could be involved in thermoregulatory and hypnogenic mechanisms. Interestingly, in our combined electron microscopic study (Cullinan and Záborszky, 1991), we have identified symmetric synaptic contacts on cholinergic neurons in the basal forebrain originating from the medial preoptic-anterior hypothalamic area, suggesting an inhibitory influence.

THALAMIC AND SUBTHALAMIC PROJECTIONS

Studies in the rat involving the delivery of retrograde tracers into the SI/globus pallidus area resulted in labeling of neurons within the parafascicular nucleus (Grove, 1988; Jones and Cuello, 1989; Carnes et al., 1990), while injections into more caudoventral parts of the SI produced retrograde labeling in the paratenial, paraventricular, central medial, rhomboid and reuniens thalamic nuclei (Grove, 1988; Carnes et al., 1990). The interpretation of these results is confounded by technical difficulties, however, since these tracers often involve uptake by fibers of passage, and axons from intralaminar and medial thalamic nuclei, as components of the inferior thalamic radiation, pass through the internal capsule and globus pallidus en route to the striatum and cortex (Herkenham, 1978, 1979; Beckstead, 1984; Kelley and Stinus, 1984; Royce and Mourey, 1985).

Using the autoradiographic tracing method, Ricardo (1981) described a significant projection from the zona incerta to several rostral forebrain structures including the SI, ventral pallidum, and globus pallidus. Since the labeling in these structures appeared to represent an end-point, it is likely that many fibers actually establish synaptic contact in these regions. Both the zona incerta and the intralaminar thalamic nuclei are known to receive significant projections originating from several levels of the brainstem reticular formation, to project upon widespread regions of the

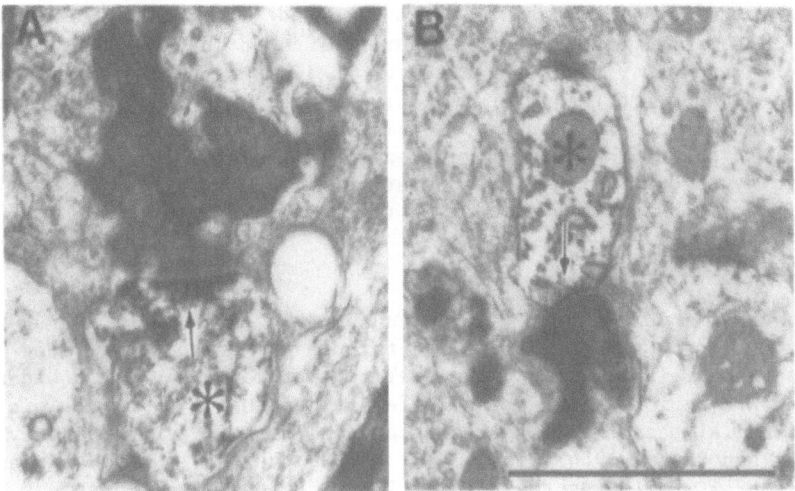

Fig. 8.   A,B: Degenerated terminals after unilateral meso-diencephal-
ic knife cut on cholinergic dendrites (asterisks) in basal
forebrain. Arrows point to postsynaptic thickenings. Bar
scale: 1 μm.

cortex (Macchi and Bentivoglio, 1986), and have been implicated in the
modulation of cortical activity (Steriade and McCarley, 1990). In view of
evidence implicating the BFC system in arousal (see below), it will be
important to determine whether brainstem reticular influences upon the
cortex are also mediated through a subthalamic relay to the BFC system.

BRAINSTEM AFFERENTS

    Following knife cuts at the meso-diencephalic border, degenerating
terminals were found in synaptic contact with BFC neurons (Fig. 8A,B), and
this intervention also resulted in a decrease of ChAT (choline acetyltransferase) activity in forebrain areas rich in cholinergic cell bodies
(Záborszky et al., 1986c). Although this preliminary study did not reveal
the origin or chemical specificity of afferents, it was the first direct
evidence of brainstem input to BFC neurons, and further suggested that
removal of brainstem afferents may induce transynaptic changes in these
cells. Since this report we have confirmed the presence of locus coeruleus
terminals on BFC neurons. Several other brainstem afferent systems are also
potentially important, since they might mediate the effects of reticular
activation upon the cortex.

Peripeduncular Area

    The peripeduncular area (PPa) is located ventrally and medially to the
anterior part of the medial geniculate body (MGB), abutting the dorsal end
of the cerebral peduncle and the substantia nigra pars lateralis. It contains several cell groups such as the peripeduncular nucleus (PP) and the
posterior intralaminar nucleus, which as parts of the "adjunct" auditory
system receive acoustic signals from different auditory relay stations,
although with less topographic and modality specificity (LeDoux et al.,
1985; Arnault and Roger, 1987; LeDoux et al., 1990). Using the autoradiographic tracing method in primates, Jones et al. (1976) suggested that
the PP projects to the large aggregated neurons of the SI and medullary

laminae of the globus pallidus. In the rat, retrograde tracing experiments (Jones and Cuello, 1989), as well as anterograde tracing studies using PHA-L (Grove, 1988), have confirmed that neurons located in the PPa project rather specifically to the vicinity of BFC neurons in the caudal part of the globus pallidus, adjacent to the internal capsule. Since the same group of cholinergic neurons which are likely to receive input from the PPa have been shown to project to the auditory cortex (Rye et al., 1984), the PPa may provide a relay through which a subpopulation of cholinergic neurons participates in auditory sensory processing (Fig. 9A).

Parabrachial Nucleus

The brachium conjunctivum is surrounded by several cytoarchitectonically distinct cell groups along its course through the dorsolateral pons (Fulwiler and Saper, 1984). Both the medial and lateral nuclei have widespread ascending projections innervating different hypothalamic, amygdaloid and thalamic nuclei, basal forebrain and cortical areas (Saper and Loewy, 1980; Fulwiller and Saper, 1984; Semba et al., 1988; Vertes, 1988; Jones and Cuello, 1989). Following PHA-L injections into the lateral segment of the parabrachial nucleus (PB), varicose fibers were evident throughout the dorsal part of SI, lateral BSt, and lateral part of the central nucleus of the amygdala (Grove, 1988).

The parabrachial nucleus relays both specific (gustatory) and general (respiratory, cardiovascular, gastrointestinal) viscerosensory input from the nucleus of the solitary tract (NTS) to the viscerosensory nuclei of the thalamus, and the insular cortex (Saper, 1982; Block and Schwartzbaum, 1983; Cechetto and Saper, 1987; Ruggiero et al., 1987). Cholinergic cells in the SI may receive specific viscerosensory input through the medial PB, since it is this region of the PB that receives afferents from the rostral, gustatory part of the NTS (Norgren, 1978), and is where many of the labeled cells were found following injections of retrograde tracers in the SI (Fullwiller and Saper, 1984). However, BFC neurons in the SI may also receive general viscerosensory input from the lateral PB, since the lateral PB receives input from the caudomedial, general viscerosensory portion of the NTS (Ricardo and Koh, 1978; Milner et al., 1984), and this portion of the parabrachial nucleus contained retrogradely labeled cells backfilled from the SI (Fullwiller and Saper, 1984; Moga et al., 1990). Interestingly, neurons in the SI also appear to receive direct input from the caudal NTS area, therefore cholinergic neurons may receive general viscerosensory input both directly, and to some extent indirectly through the PB, but are likely to receive gustatory input mainly indirectly through the PB relay.

Information relayed through the BFC system from the PB is likely to reach, among other cortical regions, the insular cortex. The insular cortex receives highly processed viscerosensory information from the thalamus (Cechetto and Saper, 1987). Viscerosensory information relayed through the BFC system to the insular cortex might thus provide a primitive representation of the stimulus. These relationships are illustrated in the simplified circuit diagram of Fig. 9. It should be added that the PB and caudal NTS also project to the central nucleus of the amygdala (Saper, 1982; Shipley and Geinisman, 1984; de Olmos, 1985), which is itself a likely source of input to BFC neurons in the SI. Thus, a certain population of cholinergic neurons in the SI might participate in a complex neural circuit implicated in viscerosensory information processing, conditioned taste aversion, (Lasiter et al., 1985) and affective evaluation of the stimulus (LeDoux, 1987). According to recent studies, information from the NTS is relayed through the PB in a much more complicated fashion (Herbert et al., 1990; Moga et al., 1990). It will be a challange for future studies to identify whether cholinergic cells indeed participate in such circuits and if so what type of specific information they receive.

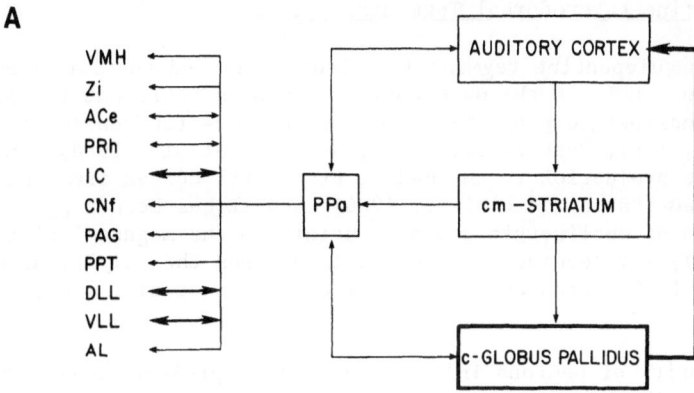

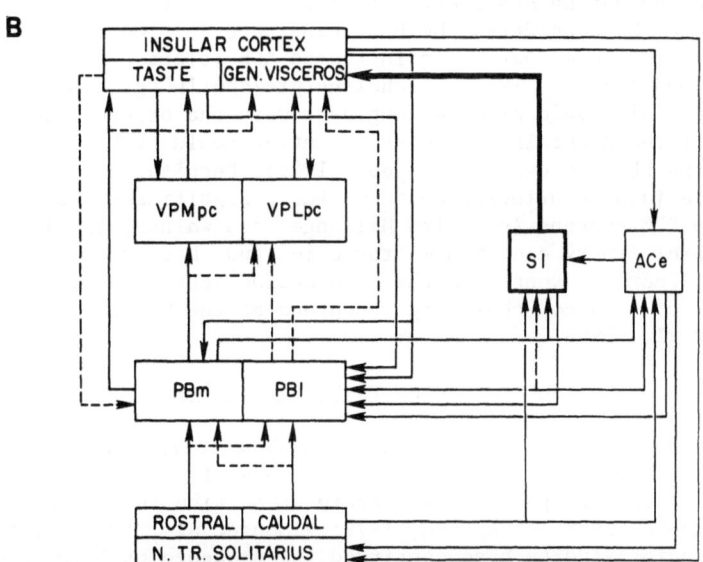

Fig. 9.   A: Hypothetical circuit involving the auditory cortex, peripeduncular area (PPa), caudomedial (cm) striatum and cholinergic
and noncholinergic neurons in the caudal (c) globus pallidus in
rat. Heavy line on the right marks the putative cholinergic link.
Other connections of the PPa are indicated on the left, including
those with auditory subcortical relay stations (heavy lines). Data
from Arnault and Roger (1987) and LeDoux et al. (1990). Abbrev.
ACe = central amygdaloid nucleus; AL = lateral amygdaloid nucleus;
CNf = cuneiform nucleus; DLL = dorsal nucleus of the lateral
lemniscus; IC = inferior colliculus; PAG = periaqueductal gray;
PPT = pedunculopontine tegmental nucleus; PRh = perirhinal cortex;
VLL = ventral nucleus of the lateral lemniscus; VMH = ventromedial
hypothalamic nucleus; Zi = zona incerta. B: Hypothetical circuit
involving the insular cortex, thalamic viscerosensory nuclei,
parabrachial nucleus, nucleus of the solitary tract, the central
amygdaloid nucleus and cholinergic and noncholinergic neurons in
the SI in the rat. Hyphenated lines represent weaker connections,
heavy line the putative cholinergic link. Data from van der Kooy
et al. (1984), Shipley and Geinisman (1984), Cechetto and Saper
(1987), and Moga et al. (1990). Abbrev. PBl = lateral parabrachial
nucleus; PBm = medial parabrachial nucleus; VPLpc = parvicellular
ventroposterior lateral nucleus of the thalamus; VPMpc = parvicellular ventroposterior medial nucleus of the thalamus.

Pedunculopontine-Laterodorsal Tegmental Nuclei

The pedunculopontine tegmental nucleus (PPT) and the laterodorsal tegmental nucleus (LDT) in the mesopontine tegmentum represent the primary sources of cholinergic projections to the thalamus (Sofroniew et al., 1985; Levey et al., 1987; Pare et al., 1988; Steriade et al., 1988). In addition, a cholinergic projection to the medial prefrontal cortex from the LDT has been described (Satoh and Fibiger, 1986), and it has been suggested that a subpopulation of cholinergic neurons, mainly in the magnocellular preoptic nucleus (MCP), may receive cholinergic input from the LDT/PPT area (Woolf and Butcher, 1986; Satoh and Fibiger, 1986; Semba et al., 1988; Jones and Cuello, 1989).

The majority of neurons in these areas that project to the thalamus display tonic discharge patterns and increment firing rates that precede the earliest change from EEG synchronization during quiet sleep to EEG desynchronization during REM sleep (Steriade and McCarley, 1990). The PPT/LDT may thus be considered the best candidate for inducing EEG desynchronization, which parallels increased ACh release (Kanai and Szerb, 1965; Jasper and Tessier, 1971). A cholinergic/cholinergic interaction in the basal forebrain would also be congruent with the original concept that stimulation of the midbrain reticular formation could evoke desynchronization of the EEG (Moruzzi and Magoun, 1949). Despite intensive efforts, however, there is no convincing evidence for a significant cholinergic projection to BFC neurons (see also Hallanger and Wainer, 1988). In contrast, a preliminary electron microscopic study has suggested that the ascending cholinergic axons from the mesopontine tegmentum establish synaptic contact with non-cholinergic neurons of the basal forebrain (Hallanger et al., 1988).

Ventral Tegmental Area-Substantia Nigra-Retrorubral Field

Using autoradiographic tracing from the dopaminergic ventral tegmental area (VTA), substantia nigra (SN) and retrorubral field (RRF), labeled fibers were traced through basal forebrain areas rich in cholinergic cells (Fallon and Moore, 1978; Beckstead, 1979; Simon et al., 1979; Vertes, 1988). Retrograde tracing studies have confirmed the presence of labeled cells primarily in the VTA after tracer injections into the VDB/HDB area, while tracer injections into the cholinergic rich region of the globus pallidus (Haring and Wang, 1986; Hallanger and Wainer, 1988; Martinez-Murillo et al., 1988b; Semba et al., 1988; Jones and Cuello, 1989) resulted in a large number of retrogradely labeled cells in the SN and RRF. In the substantia nigra the majority of the cells were in the zona compacta (SNc), with few seen in the pars reticulata. The simultaneous detection of retrogradely labeled cells and dopaminergic neurons using an antibody against tyrosine hydroxylase (TH) revealed that almost all retrograde neurons in the SNc were double labeled, while in the VTA and RRF 75-82% of the retrogradely labeled cells contained TH (Jones and Cuello, 1989). Since many dopaminergic axons in the globus pallidus and other forebrain areas represent fibers of passage toward the striatum, septum, and cortex, further studies are needed to determine the origin and distribution of putative dopaminergic input to BFC neurons.

Locus Coeruleus

Evidence from autoradiographic and retrograde tracing studies has suggested projections to forebrain areas rich in cholinergic projection neurons from the locus coeruleus (Semba et al., 1988; Jones and Cuello, 1989). However, since axons from the locus coeruleus are known to pass through the basal forebrain areas en route to the cortex (Jones and Moore, 1977; Jones and Yang, 1985), these data are of limited value in determining the extent to which these projections terminate in the basal forebrain and contact BFC neurons. In a PHA-L study (Záborszky et al., in preparation) in which the majority of the labeled cells at the injection site were localized within the locus coeruleus, labeled varicosities were detected in direct apposition to cholinergic projection neurons in extensive basal forebrain areas, including the MS/VDB complex, HDB and SI. In contrast, cholinergic neurons in peripallidal regions appear to receive very few contacts. Results of experiments at the electron microscopic level indicate that cholinergic neurons are indeed contacted by locus coeruleus axons, and that an individual locus axon can establish synaptic contacts with both cholinergic and non-cholinergic neuronal elements (Záborszky et al., in preparation).

Locus coeruleus neurons are activated by a wide variety of afferent stimuli, including auditory, visual, somato- and viscerosensory (Aston-Jones and Bloom, 1981; Foote et al., 1983), as well as stressful and/or aversive stimuli (Abercrombie and Jacobs, 1987, Jacobs, 1987), although there is controversy concerning whether this nucleus receives afferent information from widespread (Cederbaum and Aghajanian, 1978) or restricted (Aston-Jones et al., 1986) areas. Locus coeruleus cells, similar to dorsal raphe serotonergic neurons, show pacemaker-like activity and state-related modulation of discharge, which is highest in the waking state and decreases in slow wave sleep. Arousing stimuli are accompanied by bursting activity of these neurons (Steriade and McCarley, 1990). Electrophysiological studies have shown that noradrenergic activation often leads to enhanced efficacy of synaptic transmission for other afferents that converge upon the same postsynaptic neuron (Foote et al., 1983; Björklund and Lindvall, 1986). Locus coeruleus axons may therefore participate in selective attention by filtering out irrelevant stimuli and thus increase the signal-to-noise ratio of behaviorally significant stimuli. In view of the fact that similar physiological mechanisms have been described generally for the action of acetylcholine on its target neurons, as well as morphological evidence that locus coeruleus axons terminate on BFC neurons, these effects may be mediated both indirectly through the connections of the locus coeruleus with the BFC system, as well as directly through its widespread corticopetal projections. An important difference between BFC and locus coeruleus mediated arousal is that locus coeruleus neurons virtually cease discharge with the advent of REM sleep (Steriade and McCarley 1990).

Phasic activation of cholinergic neurons by noradrenergic afferents has been implicated in certain memory processes (Durkin, 1989), which is compatible with the observation that LC neuronal activity is increased during conditioning (Jacobs, 1987).

Raphe Nuclei

The results of retrograde tracing experiments have suggested projections from the raphe nuclei to areas containing BFC neurons (Semba et al., 1988; Vertes, 1988; Jones and Cuello, 1989; Carnes et al., 1990). Consistent with data from autoradiographic studies (Azmitia and Segal, 1978; Vertes and Martin, 1988), it appears that the MS/VDB complex receives projections mainly from the median raphe nucleus (MR; B8 serotonergic cell group of Dahlström and Fuxe, 1964), while inputs to the MCP/HDB and SI/globus pallidus regions originate from the dorsal raphe nucleus (DR; B7 cell group). Additional projections have been described from the caudal linear, interfascicular, magnus, pontine raphe nuclei, and from neurons within the B9 serotonergic cell group in the mesopontine ventral tegmentum (Molliver, 1987; Semba et al., 1988; Vertes, 1988; Jones and Cuello, 1989). Using concurrent immunostaining for serotonin (5-HT), Jones and Cuello (1989) revealed that after large WGA-HRP injections into the globus pallidus/SI area, about 85% and 61% of retrogradely labeled neurons were also 5-HT-positive in the DR and MR, respectively. On the other hand, in the B9 cell group only 15% of the retrogradely labeled neurons contained 5-HT. Interestingly, in the same study, the remaining neurons (15%) in the DR were positive for tyrosine hydroxylase (TH), indicating a dopaminergic input to the area of globus pallidus/SI from this region. In double-labeling experiments involving ChAT and PHA-L, we have found light microscopic evidence suggesting input to BFC neurons from the DR (Fig. 10). Confirmation of synaptic input remains to established at the electron microscopic level.

There have been several reports that MR stimulation produces hippocampal desynchronization and inhibition of the bursting discharge of the septal pacemaking cells (i.e. those neurons directly involved in controlling the rhythmical slow-wave [*theta*] activity of the hippocampus; Assaf and Miller, 1978; Vertes, 1981). MR-elicited desynchronization appears to be a serotonin mediated effect (McNaughton et al., 1980). Although the MR can influence the hippocampus directly, it is possible that it also may indirectly influence it through GABAergic or cholinergic septohippocampal neurons, both of which may be involved in pacing hippocampal *theta* (Stewart and Fox, 1990). However, there is no direct morphological evidence to date for such connections.

DR presumably serotonergic neurons show pacemaker-like activity and state-related modulation of discharge (for ref. see Jacobs, 1987; Jacobs et al., 1990; Steriade and McCarley, 1990). Phasic sensory (visual and auditory) stimulation produces an excitation followed by inhibition in these units (Jacobs, 1987). Injections of 5,7-DHT into brainstem serotonergic cell groups produced a severe reduction of the low-voltage fast activity in the neocortex that paralleled the extent of serotonin depletion in the forebrain (Vanderwolf et al., 1990). Moreover, Vanderwolf and colleagues (Vanderwolf, 1988; Vanderwolf et al.,1990) and Riekkinen et al. (1990) have reported that combined cholinergic and serotoninergic blockade in rat have potentiating effects on suppressing cortical arousal. Such animals also show severe memory impairments (Nilsson et al., 1988; Richter-Levin and Segal, 1989). An interaction between the serotonergic and cholinergic systems could take place at the level of the target in the cortex and/or at the cholinergic cell body level in the basal forebrain. Our preliminary morphological (Fig.10C) and biochemical data (Záborszky and Luine, 1987) are suggestive of a basal forebrain interaction.

Reticular Formation

The demonstration by Moruzzi and Magoun (1949) that the reticular formation (RF) is involved in controlling the cortical EEG focused attention on the course and termination of ascending reticular projections. After the initial Golgi and degeneration studies (Nauta and Kuypers, 1958; Scheibel and Scheibel, 1958), several subsequent papers used autoradiography (Edwards and de Olmos, 1976; Zemlan et al., 1984; Eberhart et al., 1985; Jones and Yang, 1985; Vertes et al., 1986; Vertes and Martin, 1988) or retrograde techniques (Semba et al., 1988; Vertes, 1988; Jones and Cuello, 1989) to trace the ascending projections from the brainstem RF. The reader should consult with the original papers or the review by Vertes (1990a) for further details. Here we will summarize only data relevant to basal forebrain projections.

In general, long ascending reticular fibers travel through the central tegmentum and are collected at the caudal diencephalon in Forel's fields. From this broad band of fibers, three main fiber systems emanate: a dorsal one coursing into the thalamus to innervate the intralaminar (parafascicular, paracentral, central, lateral), midline (rhomboid, reuniens, paraventricular, intermediodorsal), posterior, ventromedial, and dorsomedial nuclei; an intermediate group of fibers, as the rostral continuation

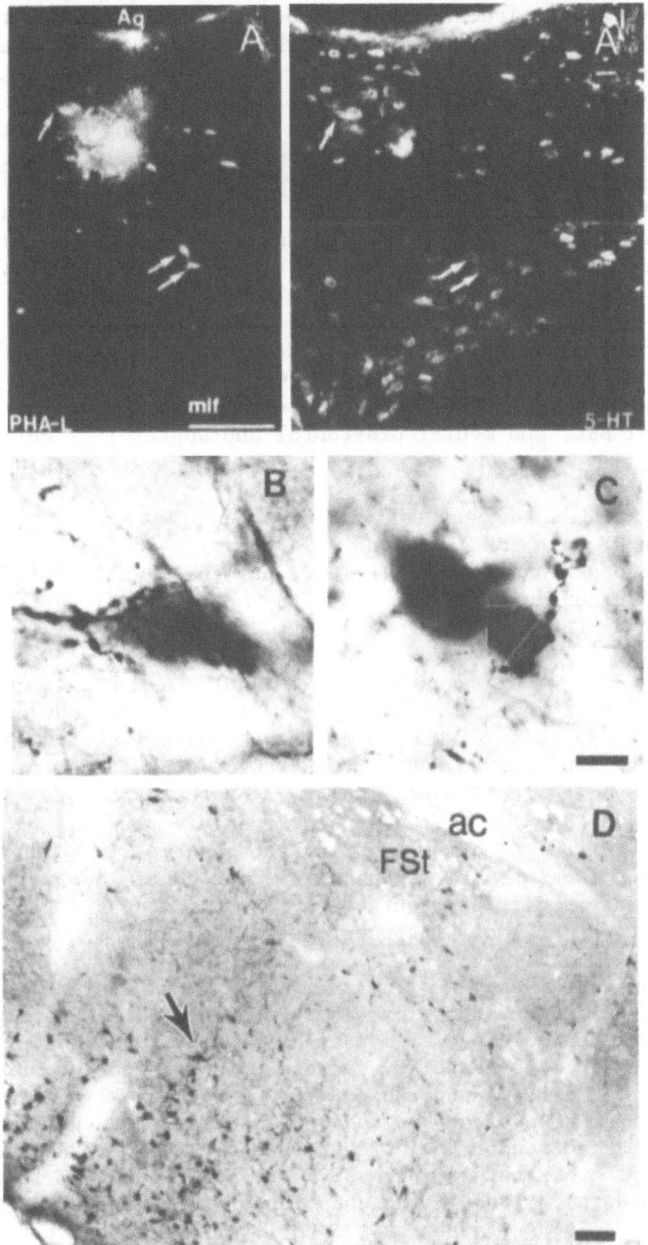

Fig. 10.   A: PHA-L injection site in the dorsal raphe nucleus. The same
section stained for serotonin (5-HT) (A') reveals that many of the
labeled neurons contain serotonin. B: A cholinergic neuron from
the dorsal aspect of the HDB (arrow in D) is approached by PHA-L
fibers with varicosities. C: Serotonin axon with varicosities in
juxtaposition to cholinergic neurons in the same area of intact
animal. Note the similar pattern of axonal arborization with
respect to cholinergic neurons in B and C. Scale; A, A', and D =
100μm; B and C = 10μm. Abbrev. ac = anterior commissure; Aq =
aqueductus cerebri; FStr = fundus striati; mlf = medial longitudinal fascicle.

of Forel's fascicle, passes through the subthalamus, giving off fibers to the zona incerta and reticular thalamic nucleus, with some fibers reaching the basal ganglia; and a ventrolateral pathway, which travels mainly within the medial forebrain bundle (MFB) to innervate the basal forebrain. Although the different subnuclei of the RF show overlapping projections, the more rostral the fibers originate, the stronger the ascending projection. For example, the medullary parvicellular reticular nucleus innervates mostly the lateral pontomedullary tegmentum, with only a few fibers reaching the mesencephalic RF, and none reaching further rostrally. In contrast, the nuclei gigantocellularis (Gc), pontis caudalis (RPc), pontis oralis (RPo), and the midbrain reticular formation (MRF), give rise to progressively heavier projections. Thus, only a few labeled fibers were detected in the medial portion of the HDB and VDB from the Gc. In contrast, the whole MS/VDB and medial two thirds of the ventral pallidum/HDB area are innervated by the RPo. Finally, the MRF innervates the entire ventral pallidum, globus pallidus, MS/VDB complex, parts of the caudate-putamen, the central amygdaloid nucleus, lateral BSt, and medial prefrontal and suprarhinal cortices (Jones and Yang, 1985).

Retrograde studies suggest that ascending projections from more rostral portions of the RF (e.g. MRF) project more dorsally in the basal forebrain, and may reach BFC neurons in the SI/globus pallidus, while more caudally located RF cell groups project mainly to the more ventral parts of the basal forebrain, such as the HDB.

The pontine RF appears to serve a prominent role in the generation of the hippocampal *theta* rhythm. Petsche et al. (1962) first suggested that discharges of a population of MS/VDB cells are phase-locked with the hippocampal *theta*, and showed further that these septal "pacemaking" cells were activated by high frequency stimulation of the brainstem RF. Vertes (1977, 1979, 1990b) has identified, in the freely moving rat, a subset of RPo neurons that fire selectively during those states in which the *theta* rhythm is present (waking, REM sleep). It is not clear, however, whether these reticular axons directly contact cholinergic or GABAergic septo-hippocampal neurons, or local interneurons in the septum.

On the other hand, Steriade and colleagues (Steriade, 1970; Steriade and Llinas, 1988; Steriade and McCarley, 1990) proposed that MRF projections to the intralaminar thalamic nuclei may be critically involved in the tonic activation processes related to EEG desynchronization during waking and REM sleep. Although there is considerable evidence for thalamic involvement, it is likely that the MRF projections to areas rich in BFC neurons may be a route by which MRF acts upon the cortex (Buzsaki et al., 1988). PHA-L injections into the MRF result in massive projections in the lateral part of SI (unpublished data), although the question of whether these fibers contact BFC neurons awaits investigation at the EM level. Brainstem activation of BFC might also come from the RPc, which has been implicated in REM sleep (Greene et al., 1989), although this possibility has yet to be investigated.

Lateral Tegmental and Dorsal Medullary Catecholaminergic Cell Groups

Intermingled with other transmitter-containing neurons in the lateral and dorsal medullary-pontine RF, several catecholaminergic cells were identified by Dahlström and Fuxe (1964). The basal forebrain projections from these areas are summarized below.

A7 area. The A7 noradrenergic cell group, which is located between the ventrolateral border of the superior cerebellar peduncle and the lateral lemniscus, constitutes a continuation of the A5 cell group. Although in combined lesion-biochemical experiments it was reported that ascending noradrenergic fibers from the area of the A7 cell group contribute to the innervation of the hypothalamus (Palkovits et al., 1980), due to the fact that fibers originating from more caudal noradrenergic cell groups project through the A7 area, these results must be interpreted with caution. It is thus not clear at present whether the A7 cell group projects to the basal forebrain.

A5 area.   The A5 noradrenergic cell group area (Dahlström and Fuxe, 1964) is located in the caudal pons, dorsolateral to the superior olive. The projections of the A5 noradrenergic cell groups have been described recently by Byrum and Guyenet (1987). Noradrenergic neurons from this region project to several hypothalamic, thalamic, and limbic nuclei, and it is possible that a subpopulation of BFC neurons in the caudal SI receive such input. In view of the widespread interconnections of the A5 group with basal forebrain regions involved in cardiovascular regulation (Guyenet and Byrum, 1985), it is possible that this information also reaches the BFC system.

A2 area.   The A2 noradrenergic cell group is located in the caudal portion of the nucleus of the solitary tract (NTS). It has been estimated that at least 90% of all nucleus commissuralis (caudal, noradrenergic part of the NTS) neurons projecting through the MFB are catecholaminergic (Moore and Guyenet, 1983). Projections from this caudal part of the NTS in rat were followed to the lateral PB area, SI, central amygdala and lateral BSt (Norgren, 1978; Ricardo and Koh, 1978). It is thus possible that BFC cells receive general viscerosensory input from the vagal nerve mediated through noradrenergic afferents. On the other hand, different peptides such as enkephalin, somatostatin, or substance P have been localized in ascending projections from the caudal NTS (Riche et al., 1990; Sawchenko et al., 1990), and are known to be present in fibers within forebrain regions containing BFC neurons. Therefore, it is possible that BFC neurons receive some peptidergic projections from the NTS.

Several studies have shown that serotonin applied directly to the area of the NTS and immediate surrounding region of the lower medulla elicits cortical synchronization and/or sleep. It has further been proposed that serotonin afferents to the NTS responsible for triggering slow wawe sleep may primarily arise from the area postrema (for ref. see Vertes, 1990b). It is unclear whether information relayed to the cortex from this "medullary sleep center" is mediated via BFC or other basal forebrain neurons.

A1 area.   The A1 noradrenergic cell group area (Dahlström and Fuxe, 1964) is located caudally within the ventrolateral medulla. The projections from the A1 area differ significantly from those of each of the reticular nuclei discussed above (McKellar and Loewy, 1982; Vertes et al., 1986). Although some fibers project to the midline thalamus, zona incerta and rostral intralaminar thalamic nuclei, a considerable proportion of fibers ascend in the medial forebrain bundle (MFB) and terminate in various hypothalamic nuclei, as well as in areas containing BFC neurons such as the SI, and MS/VDB complex. Using PHA-L tracing from a region which included the A1 catecholaminergic cell group (Fig. 11), labeled fibers could be traced to the vicinity of BFC neurons in the medial HDB and ventral part of the globus pallidus, although confirmation of synaptic contacts awaits ultrastructural examination.

TRANSMITTER-SPECIFIC AFFERENTS

GABA

GABAergic synapses on cholinergic projection neurons were first described in the ventral pallidum (Záborszky et al., 1986b), and have since been identified on corticopetal, presumably cholinergic neurons in the

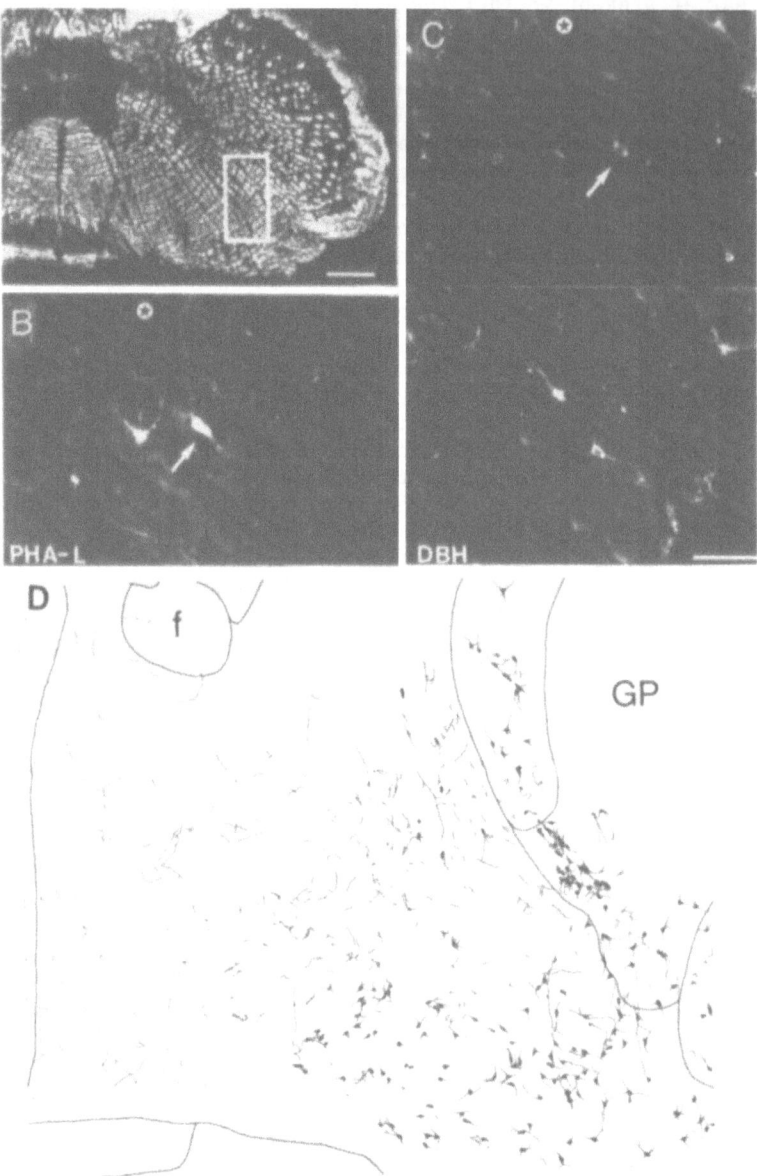

Fig. 11.   A: PHA-L injection site in the ventral medullary reticular
formation. B: Enlarged view of the upper part of the box of A
showing several PHA-L labeled neurons. Arrow points to a neuron
which is double-labeled for dopamine-ß-hydroxylase (DBH). C: The
same section as in B stained for DBH. A number of noradrenergic
cells of the A1 group are visible in the lower half of the picture. D: Combined PHA-L/ChAT staining from the same case showing
the distribution of PHA-L labeled fibers in the forebrain in
relation to cholinergic neurons. Bar Scale; A = 500µm; C = 100µm.
Star indicates the same vessel in B and C. Abbrev. f = fornix; GP
= globus pallidus.

globus pallidus (Ingham et al., 1988), on septal cholinergic neurons (Leranth and Frotscher, 1989), and on cholinergic cells in the anterior amygdaloid area (Nitecka and Frotscher, 1989). GABAergic boutons form symmetric membrane specializations (Fig. 14D), which in some areas comprise over half of the input to the perikarya and proximal dendrites (Ingham et al., 1988). In the ventral pallidum, a significant proportion of GABAergic terminals comes from the nucleus accumbens, since lesions of this structure result in a significant decrease of GAD in the ventral pallidum (Walaas and Fonnum, 1979; Záborszky et al., 1982). Another source of GABAergic terminals may be local GABAergic neurons, such as postulated for the septum (Leranth and Frotscher, 1989).

There is ample evidence that the turnover rate of hippocampal acetyl-choline (ACh), as well as *theta* activity, can be manipulated by intraseptal administration of GABAergic agents (Wood and Cheney, 1979; Costa et al., 1983; Allen and Crawford, 1984; Blaker et al., 1986). Changes in cortical acetylcholine turnover rate (ACh_{TR}), sodium dependent high affinity choline uptake (SDHACU), or cortical ACh output, have also been reported following muscimol injections into the SI (Wood and Richard, 1982; Wenk, 1984; Blaker, 1985; Casamenti et al., 1986; Wood and McQuade, 1986). The pharmacological studies have indicated further that muscimol acts through GABA-A receptors located on the cholinergic neurons (Blaker et al., 1986). It has been also suggested that other septal neurotransmitters such as dopamine, glutamate, and ß-endorphin modulate ACh_{TR} indirectly through their influences on GABAergic interneurons (Costa et al. 1983). Whether the GABAergic link indeed acts as a "final common pathway" (Chrobak et al., 1989) for these afferents in the regulation of septohippocampal neurons is unresolved, since it is unknown whether GABAergic interneurons which receive such afferents project, in turn, to septohippocampal or corticopetal cholinergic neurons.

GABAergic input to BFC neurons may have functional significance with respect to learning and memory processes, as pharmacological manipulation of GABA-A receptors within the medial septum or nucleus basalis has been shown to affect performance on working memory tasks, as well as in inhibitory avoidance behavior (Nagel and Houston, 1988; Chrobak et al., 1989). Also, enhancement of cholinergic function through pharmacological disinhibition of the GABAergic-cholinergic interaction has been suggested as a therapy for dementia of the Alzheimer's type (Sarter et al., 1988, 1990).

Acetylcholine

It has been generally noted that cholinergic neurons do not receive significant cholinergic input (Ingham et al., 1985; Armstrong, 1986; Dinopoulos et al., 1986; Záborszky et al., 1986b; Bialowas and Frotscher, 1987; Martinez-Murillo et al., 1990). We have occasionally found ChAT-immunoreactive boutons in contact with cholinergic cell bodies or proximal dendrites in the SI, and these synapses were usually of the symmetric type (Fig. 14). Since symmetric type synapses have generally been associated with inhibitory synaptic effects (Peters et al., 1976), this finding is not readily compatible with electrophysiological studies which have indicated that the majority of septohippocampal or corticopetal neurons are excited by the iontophoretic application of ACh (Lamour et al., 1984a; 1986). These effects might be explained by an action through inhibitory interneurons, however. Also, physiological studies have indicated that the action of ACh within the CNS is quite diverse, and may largely depend upon the nature of the postsynaptic receptor (Nicoll, 1988).

The origin of cholinergic axons in the basal forebrain remains unclear. One possible source is the ascending cholinergic projection from the PPT/LDT (Satoh and Fibiger, 1986), although a PHA-L study did not find evidence for terminations on BFC neurons, but rather, on non-cholinergic cells (Hallanger et al., 1988). In another study, physiologically identified corticopetal neurons in the SI were observed to have local axon collaterals (Semba et al., 1987), however, since the chemical nature of these neurons was not confirmed it is unclear whether they were indeed cholinergic.

Glutamate

It has been suggested that amygdalofugal axons use glutamate as transmitter, as evidenced by decreases in glutamate uptake in the SI following amygdaloid lesions (Francis et al., 1987), as well as by the presence of ^{3}H-D-Asp containing cell bodies in the amygdala after injection of this tracer into the ventral pallidum (Fuller et al., 1987). We have found that asymmetric contacts are formed on BFC neurons by amygdalofugal axons (Záborszky et al., 1984a), although the transmitter of these terminals was not identified. In addition, hippocampal efferents to the lateral septum apparently use also glutamate as a transmitter, as evidenced by decreases in glutamate uptake after surgical lesions of the fimbria (Fonnum and Walaas, 1978). Injections of glutamate in the SI significantly increased cortical SDHACU (Wenk, 1984), suggesting that corticofugal axons are also likely to be glutamatergic. A recent retrograde tracing study injecting ^{3}H-D-Asp into different basal forebrain areas containing BFC neurons suggests a more widespread origin of glutamatergic projection to basal forebrain areas, including the intralaminar thalamic nuclei, lateral septum, habenula, and several hypothalamic and brainstem sites (Carnes et al., 1990). Glutamic acid has been shown to cause rapid depolarization of cultured cells from the nucleus basalis (Nakajima et al., 1985), although the identification of glutamate in synapses on BFC neurons awaits ultrastructural double-labeling studies.

Substance P

Substance P-containing terminals have been found to contact BFC neurons in the ventral pallidum and ventromedial globus pallidus (Bolam et al., 1986). Such synapses were detected on the proximal dendrites and cell bodies of cholinergic neurons. It has been estimated that as many as one-third of the observed contacts contained substance P. The majority of immunoreactive terminals formed symmetrical synaptic specializations. However, a few boutons formed asymmetric contacts with the proximal dendrites of cholinergic neurons. Individual cholinergic neurons apparently receive multiple substance P-containing terminals. A light microscopic study performed on human tissue has suggested that substance P innervation of acetylcholine esterase (AChE)-positive nucleus basalis neurons is heterogeneous, since in some areas all AChE-positive cell bodies were in possible contact with approximately 5-20 substance P-immunoreactive terminal-like structures, while in other areas AChE positive neurons were entirely devoid of such putative contacts (Beach et al., 1987). One potential site of origin of substance P-containing terminals to BFC neurons in the ventral pallidum or ventromedial globus pallidus is the nucleus accumbens, since lesions of this region cause a reduction of substance P immunostaining in the ventral pallidum (Záborszky et al., 1982; Haber and Nauta, 1983). However, the different morphological types of substance P synapses, as well as the fact that several other structures which project to the areas containing BFC neurons also contain substance P-positive cells, suggests other potential input sources, including the amygdala, BSt, hypothalamus, and the pontine tegmentum (see Beach et al., 1987).

Iontophoretic application of substance P has been shown to produce increased excitability of cholinergic neurons from cultures of the medial septum and diagonal band, an effect caused by inhibition of inward rectifying K-channels (Nakajima et al., 1985, 1988). Moreover, intraseptal application of substance P facilitates performance in passive avoidance

behavior (Staubli and Huston, 1980). These findings are in apparent con-
tradiction to the effects of intraseptal administration of substance P on
hippocampal ACh_{TR}, which has been found to be inhibitory (Malthe-Sorensen et
al., 1978a). Studies on the effects of muscimol and substance P on the ACh_{TR}
in different regions of the hippocampus suggest that these two substances
control two different projections from the septum to the hippocampus (Blaker
et al., 1984). Namely, septal administration of substance P in the MS/VDB
resulted in reduction of ACh_{TR} only in the dorsal hippocampus, while muscimol
affected only ACh_{TR} in the ventral hippocampus. Considering the topographical
arrangement of cholinergic neurons projecting to the dorsal versus the ventral hippocampus (Amaral and Kurz, 1985), it is possible that distinct populations of cholinergic neurons in the septum receive GABA or substance P
input. At present, however, direct morphological data are lacking for such a
possibility. Similar to septal injections, administration of substance P
into the region of the nucleus basalis in rat facilitates performance of an
inhibitory avoidance task (Kafetzopoulos et al., 1986; Nagel and Huston,
1988).

Enkephalin

   An immunocytochemical double-labeling experiment has shown that
enkephalin-positive terminals contact with cholinergic dendrites in the
globus pallidus (Chang et al., 1987), although these were reported to be
few. Another study (Martinez-Murillo et al., 1988a) found somewhat more
enkephalin-positive terminals in contact with AChE-positive neurons in the
globus pallidus and internal capsule. Although technical factors can limit
the detection of enkephalinergic/cholinergic interactions, these studies
appear to suggest that there is little monosynaptic interaction between
enkephalinergic terminals and cholinergic neurons.

   Injections of enkephalin or its derivatives to the SI have been shown to
reduce neocortical SDHACU by approximately 50% (Wenk, 1984), and produce
naloxone-reversible locomotor hyperactivity (Baud et al., 1988). In another
study, however, local delivery of enkephalin derivatives into the SI failed
to induce changes in cortical ACh_{TR}, although a significant decrease resulted
after parenteral administration (Wood and McQuade, 1986). On the other hand,
septal injections of enkephalin have been found to cause a significant
decrease in hippocampal ACh_{TR} (Costa et al., 1983). Clearly, the role of
enkephalin in modulating cortical or hippocampal cholinergic activity
requires clarification in further anatomical and pharmacological studies.

Somatostatin

   Somatostatin has been found in axon terminals in contact with BFC
neurons in the SI (Záborszky, 1989a). The synapses observed were of the
symmetric type, and were found primarily on proximal dendrites (Fig. 14B).
The results of a high magnification light microscopic analysis of the
distribution of putative contact sites suggest that BFC neurons in the SI
and MCP/HDB may receive a rather diffuse innervation (Záborszky et al., in
preparation). At least a proportion of such terminals are likely to
originate from local interneurons.

   Cysteamine-induced release of somatostatin has been shown to lead to
significant impairment of cognitive processes such as on retention in
passive avoidance tasks (Vecsei et al., 1984; Haroutunian et al., 1987).
Since combinations of nucleus basalis lesions and cysteamine depletion in
rats did not lead to any greater impairment of mnemonic function than that
produced by ablation of the nucleus basalis alone, it was suggested that the
amnesia produced by the cysteamine-induced release of somatostatin is
mediated through the BFC system (Haroutunian et al., 1989). Our morphological demonstrations of symmetric synapses on cholinergic corticopetal

neurons suggests that somatostatin released at these synapses may inhibit cholinergic function in target areas. Somatostatin has been reported to inhibit ACh release from cholinergic neurons of the myenteric plexus (Yau et al., 1983), although it has been shown that intraventricular injection of somatostatin increases hippocampal Ach_{TR} (Wood et al., 1979). The functional impact of such somatostatinergic/cholinergic interactions in the basal forebrain remains to be elucidated.

Neuropeptide Y

Neuropeptide Y (NPY) has been found in axon terminals in synaptic contact with BFC neurons in the SI (Záborszky and Braun, 1988). In many cases, multiple NPY-containing boutons were found to encompass cholinergic cell bodies. The synapses were always of the symmetric type. Similar to somatostatin, at least a proportion of NPY terminals are likely to be derived from local interneurons. Light microscopic observations have suggested that a single peptidergic neuron (NPY or somatostatin), with its locally arborizing axon collaterals, may innervate a number of BFC neurons. Conversely, a single BFC neuron may receive axon terminals from several NPY or somatostatin cells (Záborszky, 1989b).

It has recently been demonstrated that intraventricular NPY administration modulates retention of foot-shock avoidance in a dose and time-dependent manner (Flood et al., 1987). Post-training administration of NPY to the septum or hippocampus also has been demonstrated to affect performance on a learning task (Flood et al., 1989). Whether these effects are mediated in part through the BFC system is presently unknown. Clearly, detailed anatomical and pharmacological studies are necessary to reveal the possible significance of NPY-cholinergic interactions in the basal forebrain.

Other Peptides

Various peptide-containing fiber systems are differentially distributed in the basal forebrain (for ref. see Palkovits, 1984). As a result, different populations of BFC neurons are likely to be contacted by different afferent peptidergic fibers (Záborszky, 1989b). For example, the majority of BFC neurons in the rostral globus pallidus and ventral pallidum may receive a rich neurotensin innervation (see Fig. 12), but appear to be contacted only occasionally by other peptidergic afferents. BFC neurons in the SI appear to receive a substantial input from a number of different peptidergic systems, including NPY, somatostatin and neurotensin.

The possibility of a rich innervation of a subpopulation of BFC neurons by neurotensin fibers is supported by a study reporting localization of radiolabeled neurotensin binding sites on AChE-positive neurons in the area of the nucleus basalis (Szigethy et al. 1990). Since intraventricular administration of neurotensin in the rat has been shown to modulate hippocampal ACh_{TR} (Malthe-Sorenssen et al., 1978b), as well as to affect performance in a conditioned avoidance paradigm (van Wimersma Greidanus et al., 1982), it is possible that neurotensin modulation of BFC output may have a role in memory processes.

Vasopressin axons emanating from the hypothalamic paraventricular nucleus en route to the posterior pituitary cross cholinergic dendrites of the SI or HDB, and boutons containing large neurosecretory vesicles are occasionally found in synaptic contact with distal cholinergic dendrites (unpublished data, Fig. 14C). The functional significance of such connections is presently unclear.

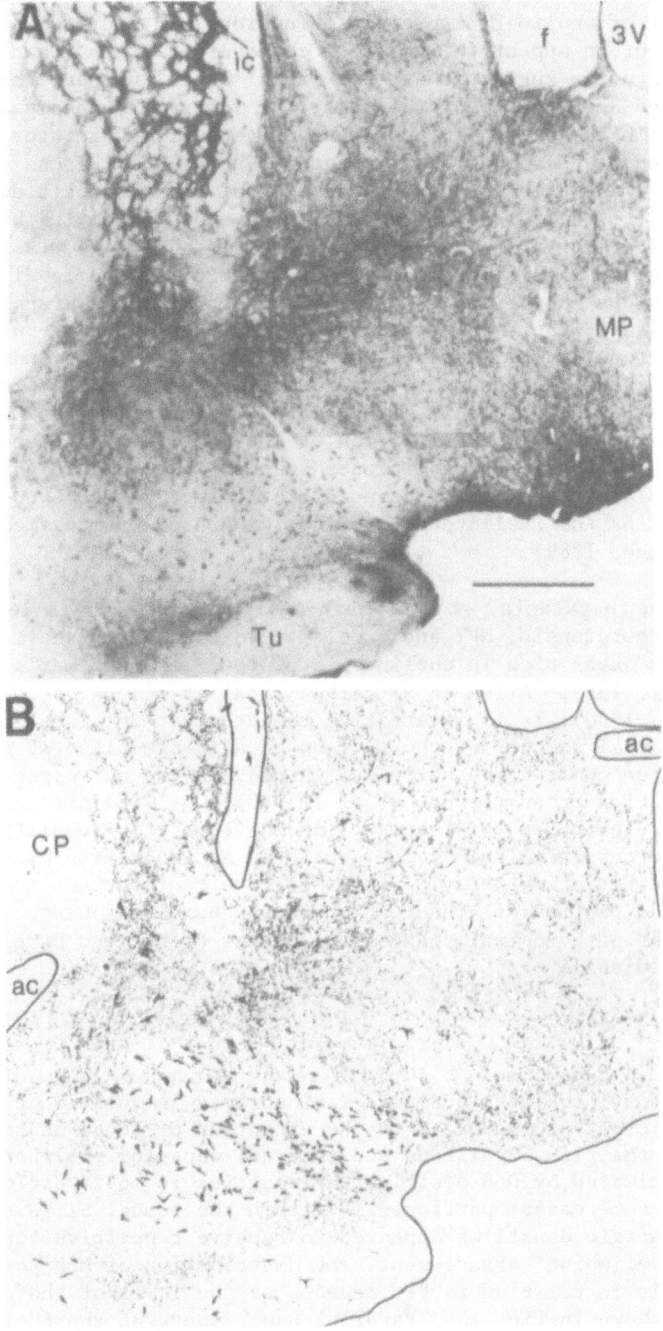

Fig. 12.   A: Distribution of neurotensin-containing fibers/terminals and
cholinergic cells at the level of the posterior part of the
crossing of the anterior commissure. B: Schematic drawing from
the same section showing that the ventral part of the globus
pallidus contains a number of cholinergic projection neurons which
are embedded in a heavy neurotensin-containing network. Bar Scale:
500μm. Abbrev. ac = anterior commissure; CP = caudate putamen; f =
fornix; ic = internal capsule; MP = medial preoptic nucleus; Tu =
olfactory tubercle; 3V = third ventricle.

Galanin has been shown to be present in neurons in the area of the basal nucleus of Meynert in the normal human brain, both in local circuit neurons, and in a number of cholinergic projection neurons. In addition, local galanin interneurons appear to innervate cholinergic cells (Chan-Palay, 1988). Lesion studies suggest that most medial septal neurons that contain both galanin and acetylcholine project to the ventral hippocampus (Melander et al., 1986). In the ventral hippocampus exogenously administered galanin appears to attenuate scopolamine-stimulated ACh release (Fisone et al., 1987), and inhibit the muscarinic stimulation of phosphoinositide turnover (Palazzi et al., 1988). Galanin also has been shown as an inhibitory peptide affecting myenteric cholinergic neurons (Yau et al., 1986), although no data are yet available for their central action at the level of the cholinergic neurons in the basal forebrain. In Parkinson's disease and Alzheimer's disease, there is an apparent hyperinnervation of surviving cholinergic neurons, and it has been proposed that this hyperinnervation reflects plasticity in response to altered (diminished) cholinergic function. However, since galanin might act as an inhibitory transmitter, the effect might be a further deterioration of the cholinergic metabolism (Chan-Palay, 1988). Behavioral studies in rats suggest that galanin acts as an inhibitory modulator of ACh, having no direct effect of its own, but partially blocking the facilitatory actions of ACh on working memory (see Crawley and Wenk, 1989).

In addition to galanin, studies have indicated changes in levels of somatostatin, neurotensin, NPY and α-melanocyte-stimulating-hormone in the basal forebrain areas rich in cholinergic neurons in Alzheimer's disease (Ferrier et al., 1983; Allen et al., 1984; Arai et al., 1986; Constantinidis et al., 1988). It is interesting to note that in Alzheimer's disease, individual sectors of the nucleus basalis are affected differentially, with concomitant pathological changes in the corresponding cortical projection areas (Arendt et al., 1985). Our findings (Záborszky, 1989b) may be relevant in this context, in that specific populations of cholinergic neurons are likely to be contacted and modulated by specific sets of peptidergic afferents. It will therefore be important to determine whether altered peptidergic innervation in the basal forebrain is correlated with the pathological changes in subsets of nucleus basalis neurons in Alzheimer's disease.

Catecholamines

Noradrenaline/Adrenaline. We have recently mapped the distribution of dopamine-β-hydroxylase (DBH) positive varicosities in relation to BFC system at the light microscopic level. With the exception of those neurons in the dorsal part of the globus pallidus and internal capsule, cholinergic neurons are approximated by DBH-positive varicosities in most portions of the BFC system. In some cases, particularly within the caudal SI, distal segments of cholinergic dendrites appeared to receive repetitive contacts in the form of a "climbing" arrangement. The distribution of DBH-positive fibers/terminals in relation to BFC neurons at the level of the anterior commissure is shown in Fig. 13A. Parallel experiments at the EM level confirmed synaptic contact between DBH-positive terminals and cholinergic neurons (Záborszky et al., 1991). Fig. 13B illustrates the distribution of PHA-L labeled fibers in relation to BFC neurons following PHA-L injection in the locus coeruleus, suggesting that at least part of the noradrenaline innervation of the basal forebrain originates in the locus coeruleus. The results of these studies have suggested that the overall noradrenergic innervation of the BFC system is rather diffuse (see Fig. 15C), particularly from the locus coeruleus (Fig. 15B), although the input is apparently not entirely uniform. Regional variations in the type and distribution of labeled fibers further suggested contributions to the innervation of the BFC from noradrenergic cell groups such as the A1 and A2 groups.

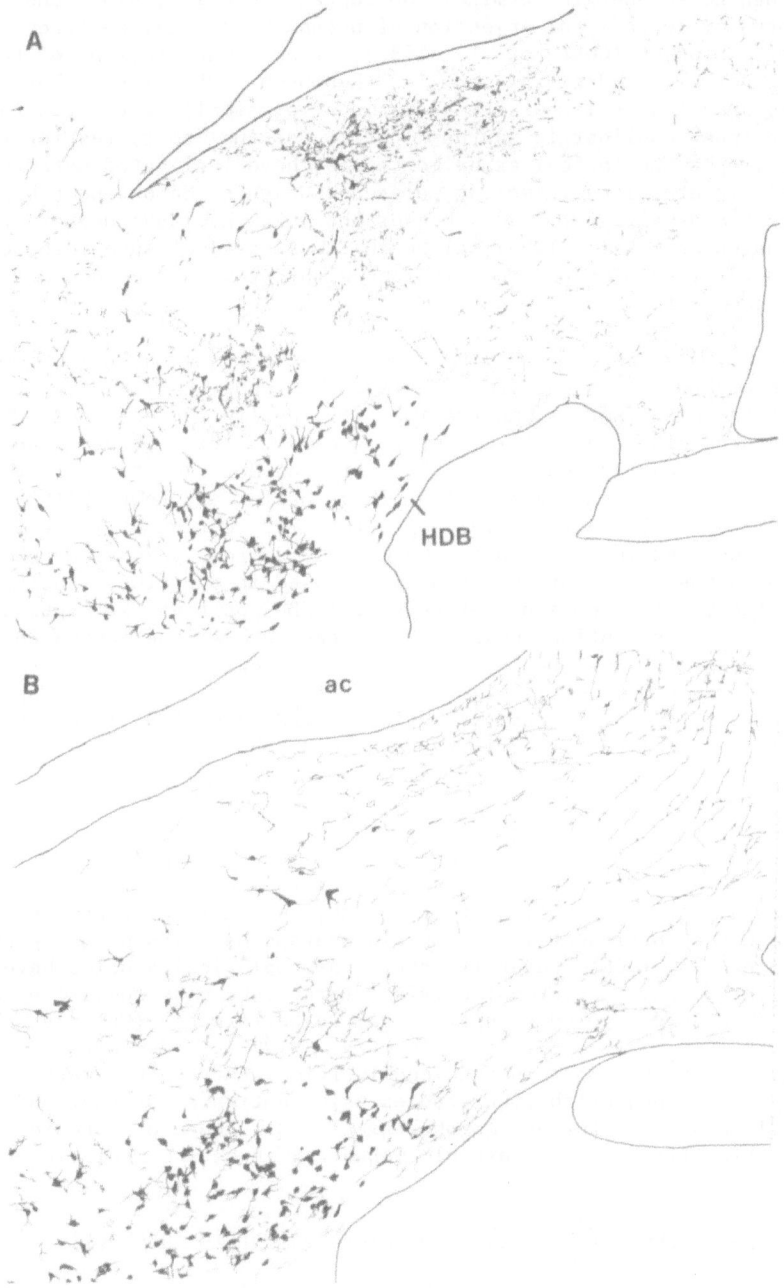

Fig. 13.   A: Camera lucida drawing from a frontal section at the level of
the anterior commissure stained for dopamine-ß-hydroxylase (DBH)
fibers/terminals and cholinergic neurons using the nickel enhanced
DAB/DAB technique. Only the most proximal portions of dendrites of
cholinergic cells are drawn. B: Distribution of PHA-L labeled
fibers in relation to cholinergic neurons at the same level as A
after PHA-L injection in the locus coeruleus. Note that PHA-L
labeled fibers/terminals from locus coeruleus may represent a
portion of those found in the DBH/ChAT material. Abbrev. ac =
anterior commissure; HDB = horizontal limb of the diagonal band.

Parenteral administration of amphetamine increases ACh_{TR} in the hippocampus (Costa et al., 1983). This response is likely to be mediated in part through noradrenergic terminals on septal BFC cells, since the response is prevented by intraseptal injection of phenoxybenzamine, an irreversible α-adrenergic blocker (Costa et al., 1983), or by 6-OHDA lesions of the ventral or dorsal noradrenergic bundles (Robinson, 1986, 1989). Furthermore, it is suggested that this noradrenergic input phasically activates the septohippocampal cholinergic neurons during working memory testing (Durkin, 1989). Other pharmacological experiments also have implicated noradrenergic-cholinergic interactions in memory related processes (Mason and Fibiger, 1979; Decker and Gallagher, 1987; Decker and McGaugh, 1989; Decker et al., 1990; Haroutunian et al., 1990), although the site(s) of such interactions were not determined. The present findings might provide a morphological basis for such effects.

Dopamine. Biochemical, pharmacological, electrophysiological, histofluorescence, and combined retrograde immunohistochemical/tracing studies have suggested dopaminergic innervation of forebrain regions containing BFC neurons (Brownstein et al., 1974; Versteeg et al., 1976; Lindvall and Björklund, 1979; Haring and Wang, 1986; Martinez-Murillo et al., 1988b; Semba et al., 1988; Jones and Cuello, 1989; Napier and Potter, 1989; Napier et al., this volume). We recently mapped the distribution of TH-positive axonal varicosities in relation to cholinergic neurons (Záborszky et al., 1991). Due to the fact that the TH antibody used in our study labeled many fibers in the basal forebrain that were also DBH-positive, interpretation of these data requires caution. However, in areas such as the ventromedial globus pallidus and the internal capsule, where few or no DBH-positive terminals were present, it is likely that TH varicosities represent dopaminergic terminals, and preliminary EM evidence indicates that TH axons in these regions establish symmetric synapses with BFC cells.

Pharmacological studies suggested that septohippocampal BFC neurons are under a tonic inhibitory dopminergic influence from the ventral tegmental area, and that this inhibitory effect of dopamine on hippocampal ACh_{TR} is mediated through the septum via GABAergic interneurons (Robinson et al., 1979; Costa et al., 1983; Gilad et al., 1986). An indirect dopaminergic influence on the corticopetal cholinergic neurons has also been suggested (Casamenti et al., 1986). Several recent pharmacological studies have suggested that dopaminergic/cholinergic interactions play some role in cognitive functions, including spatial memory performance (Galey et al., 1985; McGurk et al., 1988; 1989; Levin et al., 1990). A definitive answer to the question of direct dopaminergic/cholinergic interactions in the basal forebrain awaits double-labeling studies using antisera for ChAT and dopamine. In addition, further morphological and biochemical studies may reveal the significance of a possible GABAergic mechanism mediating this interaction.

Clinical Significance of Catecholaminergic/Cholinergic Interaction

Studies on aging and age related disorders such as Alzheimer's disease and Parkinson's disease suggest that the disruption of catecholaminergic/cholinergic interaction in these conditions may contribute to the cognitive decline observed. In aged rodents, nonhuman primates, and humans, reductions in forebrain catecholaminergic markers have been reported (Carlsson, 1987; Morgan et al., 1987), as has cell loss within the substantia nigra and locus coeruleus (Brody, 1976; McGeer et al., 1977; Chan-Palay and Asan, 1989a). Forebrain catecholaminergic markers are also reduced in Alzheimer's disease (for ref. see Mann, 1988) and neuronal loss has been reported within the locus coeruleus and ventral tegmental area (Forno, 1978; Tomlinson et al., 1981; Ichimaya et al., 1986; Price et al., 1986; Palmer et al., 1987; Chan-Palay and Asan, 1989b). In Parkinson's disease, in addition to the well known degeneration of the substantia nigra, neuronal loss within the locus coeruleus and reductions of forebrain noradrenaline levels have been demonstrated (Mann and Yates, 1983; Chan-Palay and Asan, 1989b). Reductions in forebrain presynaptic cholinergic markers and cell loss in the nucleus basalis have been reported not only in Alzheimer's diseases but also in aging, as well as in Parkinson's disease (McGeer et al., 1984; Decker, 1987; Fisher et al., 1989; Jellinger, 1990). The extent to which reductions in catecholaminergic and cholinergic markers are independent or related processes is unclear. However, the fact that 6-OHDA lesions of the ascending catecholaminergic bundle result in a decrease in ChAT activity in forebrain areas rich in cholinergic cell bodies (Záborszky and Luine, 1987; Záborszky et al., 1991) suggests a catecholaminergic influence on BFC neurons. The lack of this influence could be a factor in the metabolic deterioration of the BFC in Alzheimer's disease, Parkinson's disease, or aging, a notion consistent with a recently proposed theory of transynaptic systems degeneration in neurological diseases (Saper et al., 1987).

INFORMATION PROCESSING IN THE BASAL FOREBRAIN

Recent attempts to assess inputs to BFC neurons have employed double-labeling approaches involving ChAT immunocytochemistry combined with either tracing or transmitter identification at the electron microscope level (Záborszky et al., 1984a; Ingham et al., 1985; Bolam et al., 1986; Záborszky et al., 1986b; Chang et al., 1987; Ingham et al., 1988; Martinez-Murillo, et al., 1988a; Záborszky and Cullinan, 1989; Záborszky et al., 1991). Fig. 14 shows several examples of synapses on identified cholinergic neurons. As suggested in the schematic drawing of Fig. 14A, various inhibitory synapses appear to show a preferential distribution on the cell body or proximal part of the dendrite. These terminals are likely to originate mainly from local interneurons. On the other hand, excitatory inputs from more distant areas apparently concentrate on more distal dendritic segments. This afferent "synaptic topography" (Smith and Bolam, 1990) based on the available data may represent a first level of integration of afferent information. However, since the number of synapses and/or their types may vary according to the location and/or the target of the cholinergic neuron, the minimum operational network for corticopetal cholinergic neurons remains to be established. Another level of interaction between inputs may be represented through local GABAergic and peptidergic interneurons (Záborszky, 1989a,b), which project to cholinergic neurons. Although the exact divergence-convergence relationships of these interneurons, as well as their afferents are unknown, it is likely that such an interactive system signifies complex processing of information through the cholinergic system prior to its reaching the cortex.

ORGANIZATION OF AFFERENTS

For practical purposes only a limited number of neurons with their afferent inputs can be reconstructed at the ultrastructual level. To assess the organization of different afferent systems in relation to the cholinergic projection system in its entirety, we mapped the potential sites of contact under high resolution light microscopy (e.g. Fig. 15). Although this technique is subject to several limitations (see Cullinan and Záborszky, 1991), it does appear to provide insight into the pattern of innervation from a given brain region (see Fig. 15).

On the basis of data using this "double strategy" of identifying terminals on single cells, as well as mapping their quasi 3-D distribution, (Záborszky and Cullinan, 1989; Cullinan and Záborszky, 1991; Záborszky et al., 1991), a number of organizational principles have emerged that are

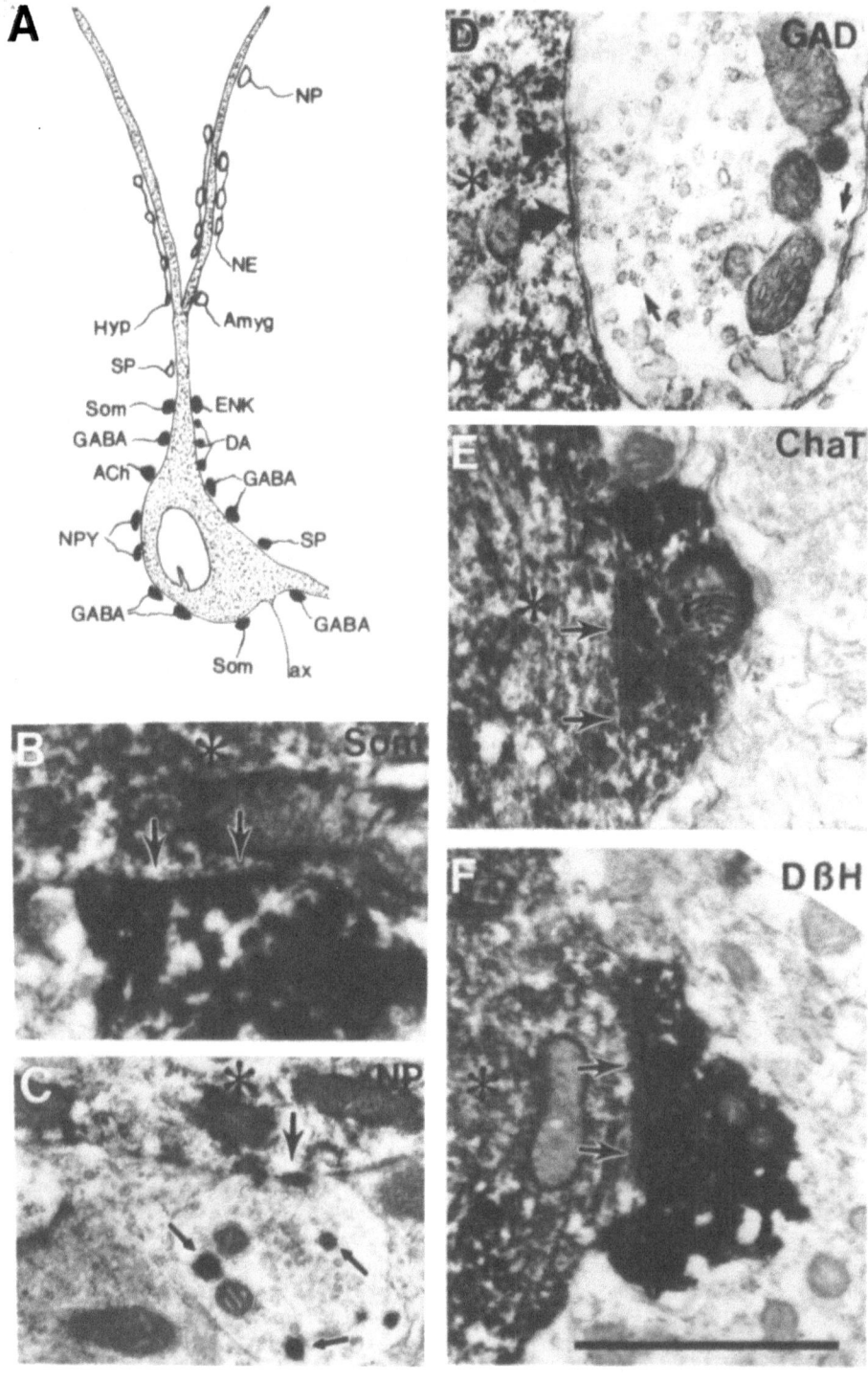

likely to be relevant not only to the cases studied, but with respect to afferents to the BFC system in general: 1) Inputs to the BFC system are apparently non-specific. In all cases examined, labeled terminal varicosities detected in the basal forebrain were related to both cholinergic and non-cholinergic elements. Indeed, the vast majority appeared to be associated with elements that were non-cholinergic. These observations support the notion that cholinergic neurons do not maintain afferent connections distinct from neighboring non-cholinergic cells, but rather, participate to some extent in the circuitry of the forebrain regions in which they are located, as has been suggested by Grove et al., (1986). 2) The distribution patterns of various terminals on the BFC sytem correspond to the general topographical arrangement of basal forebrain fibers. For example, fiber contingents coursing through certain compartments of the MFB tend to have terminations in the rostral forebrain corresponding to the same compartment. Moreover, projections in which fibers ascend through multiple MFB compartments tend to innervate the forebrain in a more diffuse fashion. A precise localization of different ascending brainstem fiber systems in relation to the compartments of the MFB (see Satoh and Fibiger, 1986) may therefore be of value in defining inputs to chemically specific cells in the basal forebrain, including BFC neurons. 3) Afferents to the BFC system may be restricted or relatively diffuse. The majority of afferents examined in our experiments showed a preferential distribution towards subsets of BFC neurons. For example, inputs from paralimbic cortical areas were confined to subterritories of the basal forebrain, and similarly, those from the hypothalamus or the ventral striatum reached subpopulations of these neurons. In addition, the distribution of several peptides and other afferents (see Semba and Fibiger, 1989; Záborszky, 1989b) to the basal forebrain suggest that they might contact subpopulations of BFC cells. Thus, the emerging view is that different subsets of these cells receive different combinations of afferents according to their location in the basal forebrain. On the other hand, noradrenergic afferents, particularly those from the locus coeruleus, apparently contact extended portions of the BFC system. Other afferents, particularly those which comprise "diffuse corticopetal systems" (Saper, 1987), such as the raphe nuclei, might similarly be expected to maintain a diffuse or generalized relationship to the BFC system.

It has been suggested by several authors that BFC neurons receive afferents from those fiber systems with which they are associated in the

---

Fig. 14.   A: Topography of synaptic inputs to a typical cholinergic neuron in the SI. Compiled from our data as well as that of Bolam et al. (1986); Chang et al. (1987) and Martinez-Murillo et al., (1988a, 1990). Putative inhibitory synapses are labeled by solid, excitatory contacts by open symbols. B: Somatostatin-containing bouton in synaptic connection with the dendrite. Double labeling with nickel enhanced DAB/DAB. C: Bouton containing large neurosecretory granula (small arrows) establishing synaptic contact with the distal dendrite. Single immunostaining for ChAT. D: GABAergio bouton labeled with ferritin (small arrow) establishes symmetrical contact with the cholinergic cell body. From the material of Záborszky et al. (1986) by permission of Willey and Liss. E: ChAT-positive bouton contacts proximal ChAT-positive dendrite. Single immunostaining enhanced with cobalt. F: DBH-positive bouton establishes asymmetric synapse with a cholinergic dendrite. Double immunolabeling with NiDAB/DAB. Arrowheads show subsynaptic dense bodies. In all micrographs large arrows point to the postsynaptic side of the synapse, asterisks label the immuno- stained (ChAT) profile. Bar scale for all micrographs: 1 μm.

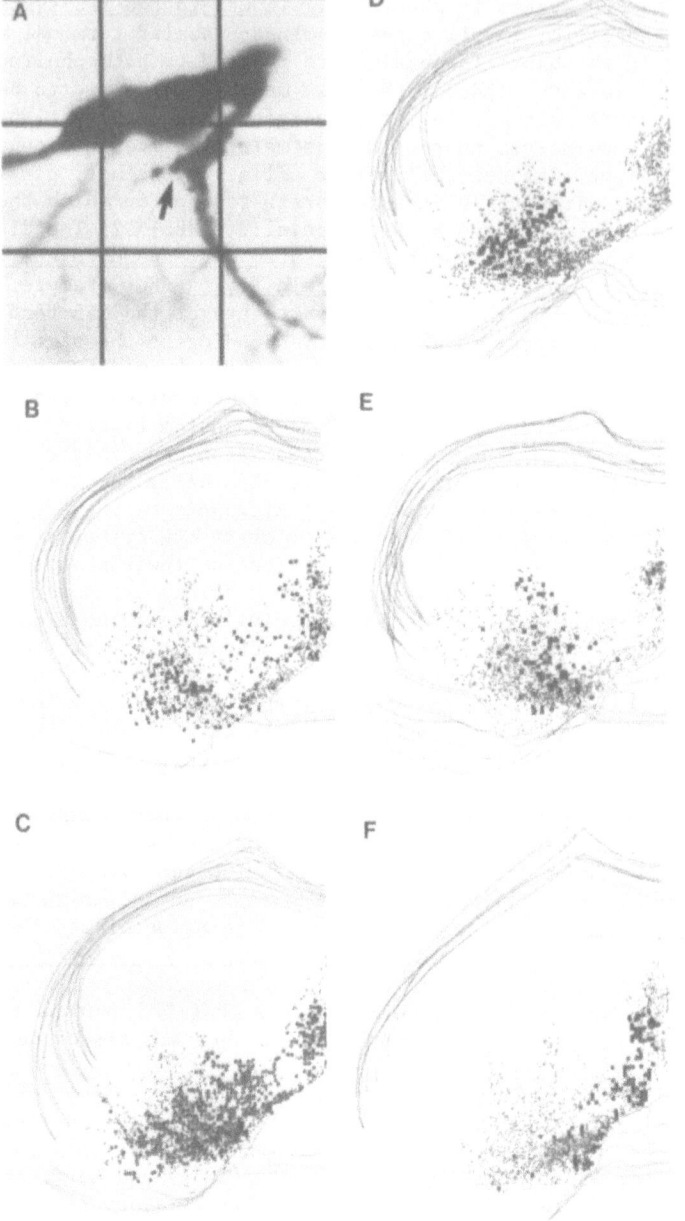

Fig. 15. A: PHA-L labeled terminal varicosities in close apposition to a
proximal dendrite of a cholinergic neuron. The grid simulates the
proportions of the ocular reticle used to screen sections from
high magnification (63x) light microscopic analysis. One division
of grid = 16 μm. B-F: Composite maps illustrating putative zones
of contact between afferent fibers and cholinergic neuronal
elements following PHA-L injections into the (B) locus coeruleus,
(D) far-lateral hypothalamus, (E) mid-lateral hypothalamus, (F)
medial hypothalamus. C: shows the distribution of putative contact
sites from a material stained for DBH/ChAT. Cholinergic neurons
are represented by dots. Zones of putative contacts between
cholinergic elements and terminal varicosities are depicted as
solid squares (corresponding to 80x80 μm areas in the section).

basal forebrain. Recent morphological data obtained with the PHA-L technique, however, have underscored the distinction between terminal fields and fibers of passage in these areas (Grove, 1988; Cullinan and Záborszky, 1991). For example, in some cases, fibers completely devoid of terminal varicosities were noted to course through areas rich in BFC cells. On the other hand, the detectability of afferents to cholinergic neurons was usually proportional to the density of terminals present in a given area. Clearly, a more complete understanding of the afferent connections of the BFC system will require the disclosure of those sources which elaborate significant terminal networks in the forebrain areas containing these neurons.

BEHAVIORAL AFFILIATIONS OF BFC NEURONS AS RELATED TO SPECIFIC CIRCUITS

Various authors have emphasized the role of the BFC system in arousal, sensory processing, emotion, motivation, learning, memory and motor functions. These processes may occur serially, even if they are often interactive. Much connectional information is needed to establish the possible role of the BFC in these functions, although as discussed below, these processes may be mediated by distinct neural circuits involving BFC neurons.

Arousal

Moruzzi and Magoun (1949) demonstrated that electrical stimulation of the midbrain reticular formation (MRF) evokes a general cortical EEG desynchronization similar to natural arousal. Since no direct connections exist between the MRF and the cortex, the facilitatory effect of MRF stimulation on the cortex was thought to be mediated partially through the thalamus and through an ill-defined extrathalamic route (Lindsley et al., 1949; Starzl et al., 1951). Although there is considerable evidence for the involvement of the intralaminar thalamic nuclei on cortical arousal (see reviews by Steriade, 1970; Steriade and Llinas, 1988; Steriade and McCarley, 1990), a growing number of studies have suggested that at least part of the effect of MRF stimulation on cortical or hippocampal activation is mediated through the BFC system (Dudar, 1977; Buzsaki et al., 1988; Semba, this volume).

Several data have been taken in support of a role of the BFC in cortical activation. 1) Cortical release of ACh is closely related to level of EEG arousal (Kanai and Szerb, 1965; Celesia and Jasper, 1966; Collier and Mitchell, 1967; Szerb, 1967; Jasper and Tessier, 1971). 2) Data correlating behavioral states with EEG-related discharge profiles of nucleus basalis neurons (Pirch et al., 1986; Detari and Vanderwolf, 1987; Buzsaki et al., 1988; Szymusiak and McGinty, 1989), together with electrophysiological evidence that ACh acts as a slow excitatory neurotransmitter in the neocortex (Sillito and Kemp, 1983), have been taken as support for the hypothesis that the nucleus basalis provides a steady background of neocortical activity that may enhance the effects of other afferents to the neocortex. 3) Lesions of the basal forebrain lead to decreases in ACh activity in the cortex and reductions of electrical activity over the lesioned hemisphere (LoConte et al., 1982). 4) EEG alterations with an increased tendency toward slow waves are seen in patients suffering from Alzheimer's disease (Coben et al., 1983), in which loss of neurons in the nucleus basalis is reliably observed.

An important issue is thus what drives the BFC system, particularly since these neurons do not show pacemaker-like activity (Griffith, 1986, 1988, this volume). The question of how messages from the outside world (during the waking state) or from the *millieu intérieur* (as during REM sleep) drive the BFC system remains largely unanswered. Since MRF stimulation evidently involves not only activation of cells, but of several passing fiber systems, various potential sources of afferents might be considered based upon the correlation between their discharge patterns and the EEG. For example, locus coeruleus axons appear to contact extended portions of the BFC system, and physiological studies have implicated the locus coeruleus in cortical arousal. Although direct projections from various other brainstem sites to BFC neurons are possible, the effects of MRF stimulation upon the cortex could be mediated in part through the lateral hypothalamus, which is known to receive brainstem afferents, as well as contact BFC neurons.

Sensory Processing, Motivation, Emotion, and Learning

Sensory information passes through multi-synaptic pathways from primary sensory cortical areas, through multi-modal association areas, to the amygdala and hippocampus (Jones and Powell, 1970; van Hoesen and Pandya, 1975; Turner et al., 1980; Pandya and Yeterian, 1985; Amaral, 1987; Heilman et al., 1987). It has been suggested that cortico-amygdaloid connections are particularly important in the affective processing of sensory signals (see LeDoux, 1987) and in stimulus-reinforcement associations (Jones and Mishkin, 1972). The results of biochemical measurements of ChAT activity in various cortical regions suggested that sensory information is likely to come under progressively greater cholinergic influence as it is transferred from sensory association areas, to the multimodal association areas; and finally to the amygdala (Mesulam et al., 1986b). In other words, BFC neurons may gate cortico-limbic interactions. More specifically, Rolls and colleagues (Rolls, 1989; Wilson and Rolls, 1990a,b) proposed that cholinergic neurons from the orbitofrontal cortex and the amygdala receive information on the expected availability of reinforcement brought about through learning, and relay this information to widespread cortical areas to facilitate sensory, motor, or associative functions. Indeed, neuronal responsiveness in sensory cortex appears to be related to similar motivational changes in basal forebrain neurons (Steriade and McCarley, 1990). It is unclear, however, how these motivational related events are transmitted to widespread cortical areas, since it appears that those BFC neurons that receive orbitofrontal input project to limbic cortical areas, rather than sensory or motor cortical regions.

Studies involving pairing of sensory stimulation with cortical appli-cation of ACh suggest that ACh can produce stimulus-specific modification of information processing in sensory cortical areas (Ashe et al., 1989; McKenna et al., 1989; Metherate and Weinberger, 1990). Furthermore, studies using sensory stimulation in combination with nucleus basalis lesions (Satoh et al., 1987; Ma et al.,1989; Juliano et al.,1990) or basal forebrain stimulation (Pirch et al., 1986; Rigdon and Pirch, 1986; Rasmusson and Dykes, 1988; Pirch et al., this volume), suggest that the enhancement of sensory-related cortical responses is correlated with activation of the BFC system. Although more elaborate functional studies are necessary, the connectional data reviewed here are consistent with the notion that sensory stimuli may reach specific cortical areas through relatively localized portions of the BFC system. For example, as discussed in the section on brainstem connections, auditory signals with weak tuning properties may reach a subpopulation of cholinergic neurons through the peripeduncular area. Such cholinergic neurons then would project to the auditory cortex (see Fig. 9A). A similar situation may exist for viscerosensory inputs as discussed and summarized in Fig. 9B. In this way, specific sensory cortical areas might receive a relatively primitive representation of the peripheral stimulus through the cholinergic system, and more precise information through the thalamic relay. This mechanism might allow for the release of ACh that is spatially and temporally coupled with the arrival of specific sensory signal through the thalamic relay nuclei.

A similar coupling of ACh release with increased cell firing in the motor cortex may form the basis for learned motor performance (Woody, 1982; Richardson et al., 1988; Richardson and DeLong, 1991), although the connectional basis for this is more uncertain.

CONCLUDING REMARKS

Recent anatomical studies has suggested that afferents to BFC neurons may contact widespread portions of this system, or be relatively restricted. Generalized behavioral functions such as arousal may be mediated in part through relatively diffuse inputs, such as the noradrenergic afferents from the locus coeruleus. Restricted afferents may be related to more specific functions. However, generalized versus specific functions of this system may not be mutually exclusive alternatives, in that either mechanism might predominate depending upon the current or prevailing state of afferent control.

It is also important to note that afferents to BFC neurons are non-specific, since neighboring non-cholinergic neurons are also apparently the recipients of a given afferent projection. These non-cholinergic cells may have efferent projections that parallel those of BFC neurons, or may constitute interneuronal populations, which in turn, innervate BFC cells. Thus, it is conceivable that such arrangements represent local integrative processing units related to specific functions, of which cholinergic neurons comprise a part.

Thus, future anatomical studies should be directed at distinguishing subpopulations of BFC neurons with their input-output relationships and to define to what extent such cholinergic neurons together with other nearby non-cholinergic neurons participate in specific forebrain circuits. Such detailed anatomical data, together with the information outlined above and from other recent functional studies (e.g. Olton et al., 1991; Richardson and DeLong, 1991), should aid in the design of pharmacological, electro-physiological, and behavioral investigations of the functional roles of specific basal forebrain circuits.

ACKNOWLEDGEMENTS

Particular thanks are due to Dr. Lennart Heimer, who gave continuous suppport and encouragement over many years. Mr. F. Lee Snavely and Ms. Vinessa Alones have provided skillful assistance. The original research summarized in this review is supported by USPHS Grants Nos. 23945 and 17743.

REFERENCES

Abercrombie, E.D. and Jacobs, B.L., 1987, Single-unit response of noradrenergic neurons in the locus coeruleus of freely moving cats, I. Acutely presented stressful and nonstressful stimuli, J. Neurosci., 7:2837-2842.

Aggleton, J.P., Burton, M.J., and Passingham, R.E., 1980, Cortical and subcortical afferents to the amygdala of the rhesus monkey (Macaca mulatta), Brain Res., 190:347-368.

Aggleton, J.P., Friedman, D.P., and Mishkin, M., 1987, A comparison between the connections of the amygdala and hippocampus with the basal forebrain in the macaque, Exp. Brain Res., 67:556-568.

Alexander, G.E., DeLong, M.R., and Strick, P.L., 1986, Parallel organization of functionally segregated circuits linking basal ganglia and cortex, Ann. Rev. Neurosci., 9:357-381.

Alheid, G.F. and Heimer, L., 1988, New perspectives in basal forebrain organization of special relevance for neuropsychiatric disorders: the striatopallidal, amygdaloid, and corticopetal components of substantia innominata, Neurosci., 27:1-39.

Allen, C.N. and Crawford, I.L., 1984, GABAergic agents in the medial septal nucleus affect hippocampal *theta* rhythm and acetylcholine utilization, Brain Res., 322:261-267.

Allen, J.M., Ferrier, I.N., Roberts, G.W., Cross, A.J., Adrian, T.E., Crow, T.J., and Bloom, S.R., 1984, Elevation of neuropeptide Y (NPY) in substantia innominata in Alzheimer's type dementia, J. Neurol. Sci., 64:325-331.

Alonso, J.R. and Frotscher, M., 1989, Hippocampo-septal fibers terminate on identified spiny neurons in the lateral septum: a combined Golgi/electron-microscopic and degeneration study in the rat, Cell Tiss. Res., 258:243-246.

Amaral, D.G., 1987, Memory: anatomical organization of candidate brain regions, in: "Handbook of Physiology, Section I, The Nervous System, Vol. V.", V.B. Mountcastle, F. Plum, and S.R. Geiger, eds., American Physiological Society, Maryland, pp. 211-294.

Amaral, D.G. and Kurz, J., 1985, An analysis of the origins of the cholinergic and non-cholinergic septal projections to the hippocampal formation in the rat, J. Comp. Neurol., 240:37-59.

Arai, H., Moroji, T., Kosaka, K., and Iizuka, R., 1986, Extrahypophyseal distribution of α-melanocyte stimulating hormone (α-MSH)-like immunoreactivity in postmortem brains from normal subjects and Alzheimer-type dementia patients, Brain Res., 377:305-310.

Arendt, T., Bigl, V., Tennstedt, A., and Arendt, A., 1985, Neuronal loss in different parts of the nucleus basalis is related to neuritic plaque formation in cortical target areas in Alzheimer's disease, Neurosci., 14:1-14.

Arendt, T., Allen, Y., Marchbanks, R.M., Schugens, M.M., Sinden, J., Lantos, P.L., and Gray, J.A., 1989, Cholinergic system and memory in the rat: effects of chronic ethanol, embryonic basal forebrain brain transplants and excitotoxic lesions of cholinergic basal forebrain projection system, Neurosci., 33:435-462.

Armstrong, D.M., 1986, Ultrastructural characterization of choline acteylytransferase-containing neurons in the basal forebrain: evidence for a cholinergic innervation of intracerebral blood vessels, J. Comp. Neurol., 250:81-92.

Armstrong, D.M., Saper, C.B., Levey, A.I., Wainer, B.H., and Terry, R.D., 1983, Distribution of cholinergic neurons in the rat brain demonstrated by immunohistochemical localization of choline acetyltransferase, J. Comp. Neurol., 216:53-68.

Arnault, P. and Roger, M., 1987, The connections of the peripeduncular area studied by retrograde and anterograde transport in the rat, J. Comp. Neurol., 258:463-476.

Ashe, J.H., McKenna, T.M., and Weinberger, N.M., 1989, Cholinergic modulation of frequency receptive fields in auditory cortex, II, Frequency-specific effects of acetylcholinesterase provide evidence for a modulatory action of endogenous ACh, Synapse, 4:44-54.

Assaf, S.Y. and Miller, J.J., 1978, The role of a raphe serotonin system in the control of septal unit activity and hippocampal desynchronization, Neurosci., 3:539-550.

Aston-Jones, G. and Bloom, F.E., 1981, Norepinephrine-containing locus coeruleus neurons in behaving rats exhibit pronounced responses to non-noxious environmental stimuli, J. Neurosci., 1:887-900.

Aston-Jones, G., Ennis, M., Pieribone, V.A., Nickell, W.T., and Shipley, M.T., 1986, The brain nucleus locus coeruleus: restricted afferent control of a broad efferent network, Science, 234:734-737.

Azmitia, E.C. and Segal, M., 1978, An autoradiographic analysis of the differential ascending projections of the dorsal and median raphe nuclei in the rat, J. Comp. Neurol., 179:641-688.

Bartus, R.T., Dean, R.L., Beer, B. and Lippa, A.S., 1982, The cholinergic hypothesis of geriatric memory dysfunction, Science, 217:408-417.

Baud, P., Mayo, W., LeMoal, M., and Simon, H., 1988, Locomotor hyperactivity in the rat after infusion of muscimol and [D-Ala^2]Met-enkephalin into the nucleus basalis magnocellularis: possible interaction with cortical cholinergic projections, Brain Res., 452:203-211.

Beach, T.G., Tago, H., and McGeer, E.G., 1987, Light microscopic evidence for a substance P-containing innervation of the human nucleus basalis of Meynert, Brain Res., 408:251-257.

Beckstead, R.M., 1979, An autoradiographic examination of corticocortical and subcortical projections of the mediodorsal-projection (prefrontal) cortex in the rat, J. Comp. Neurol., 184:43-62.

Beckstead, R.M., 1984, The thalamostriatal projection in the cat, J. Comp. Neurol., 223:313-346.

Berk, M.L. and Finkelstein, J.A., 1982, Efferent connections of the lateral hypothalamic area of the rat: an autoradiographic investigation, Brain Res. Bull., 8:511-526.

Bialowas, J. and Frotscher, M., 1987, Choline acetyltransferase-immunoreactive neurons and terminals in the rat septal complex: a combined light and electron microscopic study, J. Comp. Neurol., 259:298-307.

Björklund, A. and Lindvall, O., 1986, Catecholaminergic brainstem regulatory systems, in: "Handbook of Physiology: The Nervous System", V.B. Mountcastle, F.E. Bloom, and S.R. Geiger, eds., American Physiological Society, Maryland, pp. 155-235.

Blaker, W.D., 1985, GABAergic control of the cholinergic projections to the frontal cortex is not tonic, Brain Res., 325:389-390.

Blaker, W.D., Cheney, D.L., and Costa, E., 1986, GABA$_A$ vs. GABA$_B$ modulation of septal-hippcampal interconnections, Adv. Behav. Biol., 30:953-961.

Blaker, W.D., Peruzzi, G., and Costa, E., 1984, Behavioral and neurochemical differentiation of specific projections in the septal-hippocampal cholinergic pathway of the rat, Proc. Natl. Acad. Sci. USA, 81:1880-1882.

Block, C.H. and Schwartzbaum, J.S., 1983, Ascending efferent projections of the gustatory parabrachial nuclei in the rabbit, 1983, Brain Res., 259:1-9.

Bolam, J.P., Ingham, C.A., Izzo, P.N., Levey, A.I., Rye, D.B., Smith, A.D., and Wainer, B.H., 1986, Substance P-containing terminals in synaptic contact with cholinergic neurons in the neostriatum and basal forebrain; a double immunocytochemical study in the rat, Brain Res., 397:279-289.

Brauer, K., Schober, W., Werner, L., Winkelman, E., Lungwitz, W., and Hajdu, F., 1988, Neurons in the basal forebrain complex of the rat: a Golgi study, J. Hirnforsch., 29:43-71.

Brody, H., 1976, An examination of the cerebral cortex and brain stem in aging, in: "Neurobiology of Aging", R.D. Terry and S. Gershon, eds., Raven Press, New York, pp. 177-182.

Brownstein, M., Saavedra, J.M., and Palkovits, M., 1974, Norepinephrine and dopamine in the limbic system of the rat, Brain Res., 79:431-436.

Butcher, L.L. and Semba, K., 1989, Reassessing the cholinergic basal forebrain; nomenclature, schemata, and concepts, TINS, 12:483-485.

Buzsaki, G., Bickford, R.G., Ponomareff, G., Thal, L.J., Mandel, R., and Gage, F.H., 1988, Nucleus basalis and thalamic control of neocortical activity in the freely moving rat, J. Neurosci., 8:4007-4026.

Byrum, C.E. and Guyenet, P.G., 1987, Afferent and efferent connections of A5 noradrenergic cell group in the rat, J. Comp. Neurol., 261:529-542.

Carlsen, J., Záborszky, L., and Heimer, L., 1985, Cholinergic projections from the basal forebrain to the basolateral amygdaloid complex: a combined retrograde fluorescent and immunohistochemical study, J. Comp. Neurol., 234:155-167.

Carlsson, A., 1987, Brain neurotransmitters in aging and dementia: similar changes across diagnostic dementia groups, Gerontology, 33:159-167.

Carnes, K.M., Fuller, T.A., and Price, J.L., 1990, Sources of presumptive glutamaergic/aspartatergic afferents to the magnocellular basal forebrain in the rat, J. Comp. Neurol., 302:824-852.

Casamenti, F., Deffenu, G., Abbamondi, A.L., and Pepeu, G., 1986, Changes in cortical acetylcholine output induced by modulation of the nucleus basalis, Brain Res. Bull., 16:689-695.

Cechetto, D.F. and Saper, C.B., 1987, Evidence for a viscerotopic sensory representation in the cortex and thalamus in the rat, J. Comp. Neurol., 262:27-45.

Cederbaum, J.M. and Aghajanian, G.K., 1978, Afferent projections to the rat locus coeruleus as determined by a retrograde tracing technique, J. Comp., Neurol., 178:1-16.

Celesia, G.G., and Jasper, H.H., 1966, Acetylcholine released from cerebral cortex in relation to state of activation, Neurology, 16:1053-1064,

Chang, H.T., Penny, G.R., and Kitai, S.T., 1987, Enkephalinergic-cholinergic interaction in the rat globus pallidus: a pre-embedding double-labeling immunocytochemistry study, Brain Res., 426:197-203.

Chan-Palay, V., 1988, Neurons with galanin innervate cholinergic cells in the human basal forebrain and galanin and acetylcholine coexist, Brain Res. Bull., 21:465-472.

Chan-Palay, V. and Asan, E., 1989a, Quantitation of catecholamine neurons in the locus coeruleus in human brains of normal young and older adults and in depression, J. Comp. Neurol., 287:357-372.

Chan-Palay, V. and Asan, E., 1989b, Alterations in catecholamine neurons of the locus coeruleus in senile dementia of the Alzheimer type and in Parkinson's disease with and without dementia and depression, J. Comp. Neurol., 287:373-392.

Christie, M.J., Summers, R.J., Stephenson, J.A., Cook, C.J., and Beart, P.M., 1987, Excitatory amino acid projections to the nucleus accumbens septi in the rat: a retrograde transport study utilizing d[³H]GABA, Neurosci., 22:425-439.

Chrobak, J.J., Stackman, R.W., and Walsh, T.J., 1989, Intraseptal administration of muscimol produces dose-dependent memory impairments in the rat, Behavl. Neural Biol., 52:357-369.

Coben, L.A., Danziger, W.L., and Berg, L., 1983, Frequency analysis of the resting awake EEG in mild senile dementia of Alzheimer type, Electroenceph. Clin. Neurophys., 55:372-380.

Collier, B., and Mitchell, J.F., 1967, The central release of acetylcholine during consciousness and after brain lesions, J. Physiol., 188:83-98.

Conrad, L.C.A. and Pfaff, D.W., 1976a, Efferents from medial basal forebrain and hypothalamus in the rat, I. An autoradiographic study of the medial preoptic area, J. Comp. Neurol., 169:185-220.

Conrad, L.C.A. and Pfaff, D.W., 1976b, Efferents from medial basal forebrain and hypothalamus, II. An autoradiographic study of the anterior hypothalamic area, J. Comp. Neurol., 169:221-262.

Constantinidis, J., Bouras, C., and Vallet, P.G., 1988, Neuropeptides in Alzheimer's and in Parkinson's disease, Mt. Sinai J. Med., 55:102-115.

Costa, E., Panula, P., Thompson, H.K., and Cheney, D.L., 1983, The transsynaptic regulation of the septal-hippocampal cholinergic neurons, Life Sci., 32:165-179.

Coyle, J.T., Price, D.L., and DeLong, M.R., 1983, Alzheimer's disease: a disorder of cortical cholinergic innervation, Science, 219:1184-1190.

Crawley, J.N. and Wenk, G.L., 1989, Co-existence of galanin and acetylcholine: is galanin involved in memory processes and dementia? TINS, 12:278-282.

Cullinan, W.E. and Záborszky, L., 1991, Organization of ascending hypothalamic projections to the rostral forebrain with special reference to the innervation of cholinergic projection neurons, J. Comp. Neurol., (in press).

Dahlström, A. and Fuxe, K., 1964, Evidence for the existence of monoamine-containing neurons in the central nervous system, I, Demonstration of monoamines in the cell bodies of brain stem neurons, Acta Physiol. Scand., 62(Suppl. 232):1-55.

Decker, M.W., 1987, The effects of aging on hippocampal and cortical projections of the forebrain cholinergic system, Brain Res. Rev., 12:423-438.

Decker, M.W. and Gallagher, M., 1987, Scopolamine-disruption of radial arm maze performance: modification by noradrenergic depletion, Brain Res., 417:59-69.

Decker, M.W., Gill, T.M., and McGaugh, J.L., 1990, Concurrent muscarinic and ß-adrenergic blockage in rats impairs place-learning in a water maze and retention of inhibitory avoidance, Brain Res., 513:81-85.

Decker, M.W. and McGaugh, J.L., 1989, Effects of concurrent manipulations of cholinergic and noradrenergic function on learning and retention in mice, Brain Res., 477:29-37.

de Olmos, J.S., 1990, The amygdala, in: "The Human Nervous System," G. Paxinos, ed., Academic Press, New York, pp 583-710.

de Olmos, J.S., Alheid, G.F., and Beltramino, C.A., 1985, Amygdala, in: "The Rat Nervous System", G. Paxinos, ed., Academic Press, New York, pp. 223-334.

Détári, L. and Vanderwolf, C.H., 1987, Activity of identified cortically projecting and other basal forebrain neurones during large slow waves and cortical activation in anaesthetized rats, Brain Res., 437:1-8.

Deutsch, J.A., ed., 1983, The cholinergic synapse and the site of memory, in: "The Physiological Basis of Memory", J.A. Deutsch, ed., Academic Press, New York, pp. 367-385.

Dinopoulous, A., Parnavelas, J.G., and Eckenstein, F., 1986, Morphological characterization of cholinergic neurons in the horizontal limb of the diagonal band of Broca in the basal forebrain of the rat, J. Neurocyto., 15:619-628.

Dinopoulous, A., Parnavelas, J.G., Uylings, H.B.M., and Van Eden, C.G., 1988, Morphology of neurons in the basal forebrain nuclei of the rat; a Golgi study, J. Comp. Neurol., 272:461-474.

Donoghue, J.P. and Herkenham, M., 1986, Neostriatal projections from individual cortical fields conform to histochemically distinct striatal compartments in the rat, Brain Res., 365:397-403.

Drachman, D.G. and Leavitt, J., 1974, Human memory and the cholinergic system, Arch Neurol., 30:113-121.

Dudar, J.D., 1977, The role of the septal nuclei in the release of acetylcholine from the rabbit cerebral cortex and dorsal hippocampus and the effect of atropine, Brain Res., 129:237-246.

Durkin, T., 1989, Central cholinergic pathways and learning and memory processes: presynaptic aspects, Comp. Biochem. Physiol., 93A:273-280.

Eberhart, J.A., Morrell, J.I., Krieger, M.S., and Pfaff, D.W., 1985, An autoradiographic study of projections ascending from the midbrain central gray, and from the region lateral to it, in the rat, J. Comp. Neurol., 241:285-310.

Edwards, S.B. and de Olmos, J., 1976, Autoradiographic studies of the projections of the midbrain reticular formation: ascending projections of nucleus cuneiformis, J. Comp. Neurol., 165:417-432.

Everitt, B.J., Sirkiä, T.E., Roberts, A.C., Jones, G.H., and Robbins, T.W., 1988, Distribution and some projections of cholinergic neurons in the brain of the common marmoset, Callithrix jacchus, J. Comp. Neurol., 271:533-558.

Fallon, J.H. and Moore, R.Y., 1978, Catecholamine innervation of the basal forebrain. IV. Topography of the dopamine projection to the basal forebrain and neostriatum, J. Comp. Neurol., 180:545-580.

Faull, R.L.M., Nauta, W.J.H., and Domesick, V.B., 1986, The visual cortico-striato-nigral pathway in the rat, Neurosci., 19:1119-1132.

Ferrier, L.N., Cross, A.J., Johnson, J.A., Roberts, G.W., Crow, T.J., Corsellis, J.A.N., Lee, Y.C., O'Shaughnessy, Adrian, T.E., McGregor, G.P., Baracesb-Hamilton, A.J., and Bloom, S.R., 1983, Neuropeptides in Alzheimer type dementia, J. Neurol. Sci., 62:159-170.

Fisher, R.S., Buchwald, N.A., Hull, C.D., and Levine, M.S., 1988, GABAergic basal forebrain neurons project to the neocortex: the localization of glutamic acid decarboxylase and choline acetyltransferase in feline corticopetal neurons, J. Comp. Neurol., 272:489-502.

Fisher, W., Gage, F.H., and Björklund, A., 1989, Degenerative changes in forebrain cholinergic nuclei correlate with cognitive impairments in aged rats, Euro. J. Neurosci., 1:34-45.

Fisone, G., Wu, C.F., Consolo, S., Nordström, O., Brynne, N., Bartfai, T., Melander, T., and Hökfelt, T., 1987, Galanin inhibits acetylcholine release in the ventral hippocampus of the rat: histochemical, autoradiographic, in vivo, and in vitro studies, Proc. Natl. Acad. Sci. USA, 84:7339-7343.

Flood, J.F., Hernandez, E.N., and Morley, J.E., 1987, Modulation of memory processing by neuropeptide Y, Brain Res., 421:280-290.

Flood, J.F., Baker, M.L., Hernandez, E.N., and Morley, J.E., 1989, Modulation of memory processing by neuropeptide Y varies with brain injection site, Brain Res., 503:73-82.

Fonnum, F., Storm-Mathisen, J., and Divac, I., 1981, Biochemical evidence for glutamate as neurotransmitter in corticostriatal and corticothalamic fibres in rat brain, Neurosci., 6:863-873.

Fonnum, F. and Walaas, I., 1978, The effect of intrahippocampal kainic acid injections and surgical lesions on neurotransmitters in hippocampus and septum, J. Neurochem., 31:1173-1181.

Foote, S.L., Bloom, F.E., and Aston-Jones, G., 1983, Nucleus locus coerleus: new evidence of anatomical and physiological specificity, Physiol. Rev., 63:844-914.

Forno, L.S., 1978, The locus coeruleus in Alzheimer's disease, J. Neuropath. Exp. Neurol., 37:614.

Francis, P.T., Carl, R., Pearson, A., Lowe, S.L., Neal, J.W., Stephens, P.H., Powell, T.P.S., and Bowen, D.M., 1987, The dementia of Alzheimer's disease; an update, J. Neurol. Neurosurg. Psychiatr., 50:242-243.

Freund, T.F. and Antal, M., 1988, GABA-containing neurons in the septum control inhibitory interneurons in the hippocampus, Nature, 366:170-173.

Fuller, T.A., Russchen, F.T., and Price, J.L., 1987, Sources of presumptive glutamatergic/aspartergic afferents to the rat ventral striatopallidal region, J. Comp. Neurol., 258:317-338.

Fulwiler, C.E. and Saper C.B., 1984, Subnuclear organization of the efferent connections of the parabrachial nucleus in the rat, Brain Res. Rev., 7:229-259.

Fuster, J.M., 1989, "The Prefrontal Cortex: Anatomy, Physiology, and Neuropsychology of the Frontal Lobe", Raven Press, New York.

Fuster, J.M., Bauer, R.H., and Jervey, J.P. 1982, Cellular discharge in the dorsolateral prefrontal cortex of the monkey in cognitive tasks, Exp. Neurol., 77:679-694.

Gabbott, P.L.A. and Somogyi, T., 1984, The single section Golgi-impregnation procedure: methodological description, J. Neurosci. Meth., 11:221-230.

Gaffan, D., and Harrison, S., 1987, Amygdalectomy and disconnection in visual learning for auditory secondary reinforcement by monkeys, J. Neurosci., 7:2285-2292.

Galey, D., Durkin, T., Sifakis, G., Kempf, E., and Jaffard, R., 1985,

Facilitation of spontaneous and learned spatial behaviours following 6-hydroxydopamine lesions of the lateral septum: a cholinergic hypothesis, Brain Res., 340:171-174.

Gaykema, R.P.A., Luiten, P.G.M., Nyakas, C., and Traber, J., 1990, Cortical projection patterns of the medial septum-diagonal band complex, J. Comp. Neurol., 293:103-124.

Gerfen, C.R., 1984, The neostriatal mosaic: compartmentalization of corticostriatal input and striatonigral output system, Nature, 311:461-464.

Gerfen, C.R., 1985, The neostriatal mosaic, I. Compartmental organization of projections from the striatum to the substantia nigra in the rat, J. Comp. Neurol., 236:454-476.

Gerfen, C.R., 1989, The neostriatal mosaic: a striatal patch-matrix organization is related to cortical lamination, Science, 246:385-388.

Gilad, G.M., Gilad, V.H., and Rabey, J.M., 1986, Dopaminergic modulation of the septo-hippocampal cholinergic system activity under stress, 1986, Life Sci., 39:2387-2393.

Giuffrida, R. and Rustioni, A., 1988, Glutamate and aspartate immunoreactivity in corticothalamic neurons of rat, in: "Cellular Thalamic Mechanisms", M. Bentivoglio and R. Spreafico, eds., Elsevier, Amsterdam, pp. 311-320.

Goldman-Rakic, P.S., 1987, Circuitry of primate prefrontal cortex and regulation of behavior by representational memory, in: "Handbook of Physiology: The Nervous System", Vol. V., Part 1, V.B. Mountcastle, F. Plum and S.R. Geiger, eds., American Physiol. Society, Maryland, pp. 373-417.

Goldman-Rakic, P.S. and Selemon, L.D., 1986, Topography of corticostriatal projections in nonhuman primates and implications for functional parcellation of the neostriatum, in: "Cerebral Cortex", (Vol. 5), E.G. Jones and A. Peters, eds., Plenum Press, New York, pp. 447-466.

Greene, R.W., Gerber, U., and McCarley, R.W., 1989, Cholinergic activation of medial pontine reticular formation neurons in vitro, Brain Res., 476:154-159.

Griffith, W.H., 1988, Membrane properties of cell types within guinea pig basal forebrain nuclei in vitro, J. Neurophysiol., 59:1590-1612.

Griffith, W.H. and Matthews, R.T., 1986, Electrophysiology of AChE positive neurons in basal forebrain slices, Neurosci. Lett., 71:169-174.

Groenewegen, H.J., Berendse, H.W., Meredith, G.E., Haber, S.N., Voorn, P., Wolters, J.G., and Lohman, A.H.M., 1990, Functional anatomy of the ventral, limbic system-innervated striatum, in: "The Mesolimbic Dopamine System: From Motivation to Action", P. Willner and J. Scheel-Krüger, eds., John Wiley and Sons, Ltd.

Groenewegen, H.J., Vermeulen-Van der Zee, E., te Kortshot, A., and Witter, M.P., 1987, Organization of the projections from the subiculum to the ventral striatum in the rat: a study using anterograde transport of Phaseolus vulgaris leucoagglutinin, Neurosci., 23:103-120.

Grove, E.A., 1988, Neural associations of the substantia innominata in the rat: afferent connections, J. Comp. Neurol., 277:315-346.

Grove, E.A., Domesick, V.B., and Nauta, W.J., 1986, Light microscopic evidence of striatal input to intrapallidal neurons of cholinergic cell group Ch4 in the rat: a study employing the anterograde tracer Phaseolus vulgaris leucoagglutinin (PHA-L), Brain Res., 367:379-384.

Guyenet, P.G. and Byrum, C.E., 1985, Comparative effects of sciatic nerve stimulation, blood pressure, and morphine on the activity of A5 and A6 pontine noradrenergic neurons, Brain Res., 327:191-201.

Haber, S.N., Groenewegen, H.J., Grove, E.A., and Nauta, W.J.H., 1985, Efferent connections of the ventral pallidum: evidence of a dual striato-pallidofugal pathway, J. Comp. Neurol., 235:322-335.

Haber, S.N., Lind, E., Klein, C., and Groenewegen, H.J., 1990, Topographic organization of the ventral striatal efferent projections in the *Rhesus* monkey: an anterograde tracing study, J. Comp. Neurol., 293:282-298.

Haber, S.N. and Nauta, W.J.H., 1983, Ramifications of the globus pallidus in the rat as indicated by patterns of immunohistochemistry, Neurosci., 9:245-260.

Hallanger, A.E., Price, S.D., Steininger, T., and Wainer, B.H., 1988, Mesopontine tegmental projections to the nucleus basalis of Meynert: an ultrastructural study, Soc. Neurosci. Abstr., 14:1184.

Hallanger, A.E. and Wainer, B.H., 1988, Ascending projections from the pedunculopontine tegmental nucleus and adjacent mesopontine tegmentum in the rat, J. Comp. Neurol., 274:483-515.

Haring, J.H. and Wang, R.Y., 1986, The identification of some sources of afferent input to the rat nucleus basalis magnocellularis by retrograde transport of horseradish peroxidase, Brain Res., 366:152-158.

Haroutunian, V., Kanof, P.D., and Davis, K.L., 1989, Interactions of forebrain cholinergic and somatostatinergic systems in the rat, Brain Res., 496:98-104.

Haroutunian, V., Kanof, P.D., Tsuboyama, G., and Davis, K.L., 1990, Restoration of cholinomimetic activity by clonidine in cholinergic plus noradrenergic lesioned rats, Brain Res., 507:261-266.

Haroutunian, V., Mantin, R., Campbell, G.A., Tsuboyama, G.K., and Davis, K.L., 1987, Cysteamine-induced depletion of central somatostatin-like immunoreactivity: effects on behavior, learning, memory and brain neurochemistry, Brain Res., 403:234-242.

Heilman, K.M., Watson, R.T., Valenstein, E., Goldberg, M.E., 1987, Attention; behavior and neural mechanisms, in: "Handbook of Physiology: The Nervous System", Vol. V., Part 1, V.B. Mountcastle, F. Plum, and Geiger, S.R., eds., American Physiol. Society, Maryland, pp. 461-481.

Heimer, L. and Alheid, G.F., 1991, Piecing together the puzzle of basal forebrain anatomy, in: "The Basal Forebrain: Anatomy to Function", T.C. Napier, P.W. Kalivas, and I. Hanin, eds., Plenum Press, New York (in press).

Heimer, L., de Olmos, J.S., Alheid, G.F., and Záborszky, L., 1991a, "Perestroika" in the basal forebrain: opening the borders between neurology and psychiatry, Progr. Brain Res., Vol 87, (in press).

Heimer, L., Zahm, D.S., Churchill, L., Kalivas, P.W., and Wohltmann, C., 1991b, Specificity in the projection patterns of accumbal core and shell in the rat, Neurosci., (in press).

Heimer, L. and Wilson, R.D., 1975, The subcortical projections of allocortex: similarities in the neural associations of the hippocampus, the piriform cortex and the neocortex, in: "Golgi Centennial Symposium Proceedings," M. Santini, ed., Raven Press, New York, pp. 177-193.

Herbert, H., Moga, M.M., and Saper, C.B., 1990, Connections of the parabrachial nucleus with the nucleus of the solitary tract and the medullary reticular formation in the rat, J.Comp. Neurol., 293:540-580.

Herkenham, M., 1978, The connections of the nucleus reuniens thalami: evidence for a direct thalamo-hippocampal pathway in the rat, J. Comp. Neurol., 177:589-610.

Herkenham, M., 1979, The afferent and efferent connections of the ventromedial thalamic nucleus in the rat, J. Comp. Neurol., 183:487-518.

Ichiyama, Y., Arai, H., Kosaka, K., and Izuka, R., 1986, Morphological and biochemical changes in the cholinergic and monaminergic systems in Alzheimer-type dementia, Acta Neuropathol., 70:112-116.

Ingham, C.A., Bolam, J.P., and Smith, A.D., 1988, GABA-immunoreactive synaptic boutons in the rat basal forebrain: comparison of neurons that project to the neocortex with pallidosubthalamic neurons, J. Comp. Neurol., 273:263-282.

Ingham, C.A., Bolam, J.P., Wainer, B.H., and Smith, A.D., 1985, A correlated light and electron microscopic study of identified cholinergic basal forebrain neurons that project to the cortex in the rat, J. Comp. Neurol., 239:176-192.

Inoue, M. Oomura, Y., Aou, S., Nishino, H., and Sikdar, S.K., 1985, Reward related neuronal activity in monkey dorsolateral prefrontal cortex during feeding behavior, Brain Res., 326:307-312.

Irle, E. and Markowitsch, H.J., 1986, Afferent connections of the substantia innominata/basal nucleus of Meynert in carnivores and primates, J. Hirnforsch., 27:343-367.

Jacobs, B.L., 1987, Brain monoaminergic unit activity in behaving animals, Prog. Psychobiol., Physiol., Psychol., 12:171-206.

Jacobs, B.L., Fornal, C.A., and Wilkinson, L.O., 1990, Neurophysiological and neurochemical studies of brain serotonergic neurons in behaving animals, Ann. NY. Acad. Sci., 600:260-271.

Jasper, H.H. and Tessier, J., 1971, Acetylcholine liberations from cerebral cortex during paradoxical (REM) sleep, Science, 172:601-602.

Jellinger, K., 1990, New developments in the pathology of Parkinson's disease, in: "Advances in Neurology", Vol. 53, M.B. Streifler, A.D. Kprczyn, E. Melamed and M.B.H. Youdim, eds., Raven Press, New York, pp. 1-16.

Jones, E.G., Burton, H., Saper, C.B., and Swanson, L.W., 1976, Midbrain, diencephalic, and cortical relationships of the basal nucleus of Meynert and associated structures in primates, J. Comp. Neurol., 167:385-420.

Jones, B.E. and Cuello, A.C., 1989, Afferents to the basal forebrain cholinergic cell area from the pontomesencephalic- catecholamine, serotonin, and acetylcholine-neurons, Neurosci., 31:37-61.

Jones, B. and Mishkin, M., 1972, Limbic lesions and the problem of stimulus-reinforcement associations, Exp. Neurol., 36:362-377.

Jones, B.E. and Moore, R.Y., 1977, Ascending projections of the locus coeruleus in the rat, II. Autoradiographic study, Brain Res., 127:23-53.

Jones, B.E. and Yang, T.Z., 1985, The efferent projections from the reticular formation and the locus coeruleus studied by anterograde and retrograde axonal transport in the rat, J. Comp. Neurol., 242:56-92.

Jones, E.G. and Powell, T.P.S., 1970, An experimental study of converging sensory pathways within the cerebral cortex of the monkey, Brain, 93:793-820.

Juliano, S.L., Ma, W., Bear, M.F., and Eslin, D., 1990, Cholinergic manipulation alters stimulus-evoked metabolic activity in cat somatosensory cortex, J. Comp. Neurol., 297:106-120.

Kafetzopoulos, E., Holzhauer, M.S., and Huston, J.P., 1986, Substance P injected into the region of the nucleus basalis magnocellularis facilitates performance of an inhibitory avoidance task, Psychopharm., 90:281-283.

Kanai, T., and Szerb, J.C., 1965, Mesencephalic reticular activating system and cortical acetylcholine output, Nature, 205:80-82.

Kelley, A.E. and Domesick, V.B., 1982, The distribution of the projection from the hippocampal formation to the nucleus accumbens in the rat: an anterograde and retrograde-horseradish peroxidase study, Neurosci., 7:2321-2335.

Kelley, A.E., Domesick, V.B., and Nauta, W.J.H., 1982, The amygdalostriatal projection in the rat: an anatomical study by anterograde and retrograde tracing methods, Neurosci., 7:615-630.

Kelley, A.E. and Stinus, L., 1984, The distribution of the projection from the parataenial nucleus of the thalamus to the nucleus accumbens in the rat: an autoradiographic study, Exp. Brain Res., 54:499-512.

Köhler, C., Chan-Palay, V., and Wu, J.Y., 1984, Septal neurons containing glutamic acid decarboxylase immunoreactivity project to the hippocampal region in the rat brain, Anat. Embryol., 169:41-44.

Kolb, B., 1984, Functions of the frontal cortex of the rat: a comparative review, Brain Res. Rev., 8:65-98.

Krayniak, P.F., Meibach, R.C., and Siegel, A., 1981, A projection from the entorhinal cortex to the nucleus accumbens in the rat, Brain Res., 209:427-431.

Krettek, J.E. and Price, J.L., 1978, Amygdaloid projections to subcortical structures within the basal forebrain and brainstem in the rat and cat, J. Comp. Neurol., 178:225-254.

Krieger, M.S., Conrad, L.C.A., and Pfaff, D.W., 1979, An autoradiographic study of the efferent connections of the ventromedial nucleus of the hypothalamus, J. Comp. Neurol. 183:785-816.

Lamour, Y., Dutar, P., and Jobert, A., 1984a, Septo-hippocampal and other medial septum-diagonal band neurons: electrophysiological and pharmacological properties, Brain Res., 309:227-239.

Lamour, Y., Dutar, P., Jobert, A., 1984b, Cortical projections of the nucleus of the diagonal band of Broca and of the substantia innominata in the rat: an anatomical study using the anterograde transport of a conjugate of wheat germ agglutinin and horseradish peroxidase, Neurosci., 12:395-408.

Lamour, Y., Dutar, P., Rascol, O., and Jobert, A., 1986, Basal forebrain neurons projecting to rat frontoparietal cortex: electrophysiological and pharmacological properties, Brain Res., 362:122-131.

Lasiter, P.S., Deems, D.A., and Garcia, J., 1985, Involvement of the anterior insular gustatory neocortex in taste-potentiated odor aversion learning, Physiol. Behav., 34:71-77.

LeDoux, J.E., 1987, Emotion, in: "Handbook of Physiology: The Nervous System", Vol. V., Part 1, V.B. Mountcastle, R. Plum, and S.R. Geiger, eds., American Physiol. Society, Maryland, pp. 419-459.

LeDoux, J.E., Ruggiero, D.A., and Reis, D.J., 1985, Projections to the subcortical forebrain from anatomically defined regions of the medial geniculate body in the rat, J. Comp. Neurol., 242:182-213.

LeDoux, J.E., Farb, C., and Ruggiero, D.A., 1990, Topographic organization of neurons in the acoustic thalamus that project to the amygdala, J. Neurosci., 10:1043-1054.

Leranth, C. and Frotscher, M., 1989, Organization of the septal region in the rat brain: cholinergic-GABAergic interconnections and the termination of hippocampo-septal fibers, J. Comp. Neurol., 289:304-314.

Levey, A.I., Hallanger, A.E., and Wainer, B.H., 1987, Cholinergic nucleus basalis neurons may influence the cortex via the thalamus, Neurosci. Lett., 74:7-13.

Levin, E.D., McGurk, S.R., Rose, J.E., and Butcher, L.L., 1990, Cholinergic-dopaminergic interactions in cognitive performance, Behav. Neural Biol., 54:271-299

Lindsley, D.B., Bowden, J.W., and Magoun, H.W., 1949, Effect upon the EEG of acute injury to the brain stem activating system, EEG Clin. Neurophysiol., 1:475-486.

Lindvall, O. and Björklund, A., 1979, Dopaminergic innervation of the globus pallidus by collaterals from nigrostriatal pathway, Brain Res., 172:169-173.

Lo Conte, G., Casamenti, F., Bigl, V., Milaneschi, E., and Pepeu, G., 1982, Effect of magnocellular forebrain nuclei lesions on acetylcholine output from the cerebral cortex, electrocorticogram and behaviour, Arch. Ital. Biol., pp. 176-188.

Luiten, P.G.M., Gaykema, R.P.A., Traber, J., and Spencer, D.G., 1987, Cortical projection patterns of magnocellular basal nucleus subdivisions as revealed by anterogradely transported Phaseolus vulgaris leucoaagglutinin, Brain Res., 413:229-250.

Ma, C., Hohmann, C., Coyle, J.T., and Juliano, S.L., 1989, Lesions of the basal forebrain alter stimulus-evoked metabolic ativity in mouse somatosensory cortex, J. Comp. Neurol., 288:414-427.

Macchi, G. and Bentivoglio, M., 1986, The thalamic intralaminar nuclei and the cerebral cortex, in: "Cerebral Cortex", (Vol. 5), E.G. Jones and A. Peters, eds., Plenum Press, New York, pp. 355-401.

Malthe-Sorenssen, D., Cheney, D.L., and Costa, E., 1978a, Modulation of acetylcholine metabolism in the hippocampal cholinergic pathway by intraseptally injected substance P, J. Pharmacol. Exp. Therap., 206:21-28.

Malthe-Sorenssen, D., Wood, P.L., Cheney, D.L., and Costa, E., 1978b, Modulation of the turnover rate of acetylcholine in rat brain by intraventricular injections of thyrotropin-releasing hormone, somatostatin, neurotensin and angiotensin II, J. Neurochem., 31:685-691.

Mann, D.M.A., 1988, Neuropathological and neurochemical aspects of Alzheimer's disease, in: "Handbook of Psychopharmacology", L.L. Iversen, S.D. Iversen, and S.H. Snyder, eds., Plenum, New York, pp. 1-56.

Mann, D.M.A. and Yates, P.O., 1983, Pathological basis for neurotransmitter changes in Parkinson's disease, Neuropathol. Appl. Neurobiol., 9:3-19.

Martinez-Murillo, R., Blasco, I., Alavrez, F.J., Villalba, R., Solano, M.L., Montero-Caballero, I., and Rodrigo, J., 1988a, Distribution of enkephalin-immunoreactive nerve fibers and terminals in the region of the nucleus basalis magnocellularis of the rat: a light and electron microscopic study, J. Neurocytol., 17:361-376.

Martinez-Murillo, R., Semenenko, F., and Cuello, A.C., 1988b, The origin of tyrosine hydroxylase immunoreactive fibers in the regions of the nucleus basalis magnocellularis of the rat, Brain Res., 451:227-236.

Martinez-Murillo, R., Villalba, R.M., and Rodrigo, J., 1990, Immunocytochemical localization of cholinergic terminals in the region of the nucleus basalis magnocellularis of the rat: a correlated light and electron microscopic study, Neurosci., 36:361-376.

Mason, S.T. and Fibiger, H.C., 1979, Possible behavioural function for noradrenaline-acetylcholine interaction in brain, Nature, 277:396-397.

McGeer, P.L., McGeer, E.G., and Suzuki, J.S., 1977, Aging and extrapyramidal function, Arch Neurol., 34:33-35.

McGeer, P.L., McGeer, E.G., Suzuki, J., Dolman, C.E., and Nagai, T., 1984, Aging, Alzheimer's disease, and the cholinergic system of the basal forebrain, Neurol., 34:741-745.

McGeorge, A.J. and Faull, R.L.M., 1989, The organization of the projection from the cerebral cortex to the striatum in the rat, Neurosci., 29:503-537.

McGinty, D., and Szymusiak, R., 1990, Keeping cool: a hypothesis about the mechanisms and functions of slow-wave sleep, TINS, 13:48-487.

McGurk, S.R., Levin, E.D., and Butcher, L.L., 1988, Cholinergic-dopaminergic interactions in radial-arm maze performance, Behav. Neural Biol., 49:234-239.

McGurk, S.R., Levin, E.D., and Butcher, L.L., 1989, Nicotinic-dopaminergic relationships and radial-arm maze performance in rats, Behav. Neural Biol., 52:78-86.

McKellar, S. and Loewy, A.D., 1982, Efferent projections of the A1 catecholamine cell group in the rat: an autoradiographic study, Brain Res., 241:11-29.

McKenna, T.M., Ashe, J.H., and Weinberger, N.M., 1989, Cholinergic modulation of frequency receptive fields in auditory cortex, I, Frequency-specific effects of muscarinic agonists, Synapse, 4:30-43.

McNaughton, N., Azmitia, E.C., Williams, J.H., Buchan, A., and Gray, J.A., 1980, Septal elicitation of hippocampal theta rhythm after localized de-afferentation of serotoninergic fibers, Brain Res., 200:259-269.

Melander, T., Hökfelt, T., and Rökaeus, A., 1986, Distribution of galanin-like immunoreactivity in rat central nervous system, J. Comp. Neurol., 248:475-517.

Mesulam, M.M., and Geula, C., 1988, Nucleus basalis (Ch4) and cortical cholinergic innervation in the human brain: observations based on the distribution of acetylcholinesterase and choline acetyltransferase. J. Comp. Neurol., 275:216-240.

Mesulam, M.M. and Mufson, E.J., 1984, Neural inputs into the nucleus basalis of the substantia innominata (Ch4) in the rhesus monkey, <u>Brain</u>, 107:253-274.

Mesulam, M.M., Mufson, E.J., Levey, A.I., and Wainer, B.H., 1983a, Cholinergic innervation of cortex by the basal forebrain: cytochemistry and cortical connections of the septal area, diagonal band nuclei, nucleus basalis (substantia innominata), and hypothalamus in the rhesus monkey, <u>J. Comp. Neurol.</u>, 214:170-197.

Mesulam, M.M., Mufson, E.J., Wainer, B.H., and Levey, A.I., 1983b, Central cholinergic pathways in the rat: an overview based on an alternative nomenclature (Ch1-Ch6), <u>Neurosci.</u>, 10:1185-1201.

Mesulam, M.M., Mufson, E.J., and Wainer, B.H., 1986a, Three-dimensional representation and cortical projection topography of the nucleus basalis (Ch4) in the macaque: concurrent demonstration of choline acetyltransferase and retrograde transport with a stabilized tetramethylbenzidine method for HRP, <u>Brain Res.</u>, 367:301-308.

Mesulam, M.M., Volicer, L., Marquis, J.K., Mufson, E.J., and Green, R.C., 1986b, Systematic regional differences in the cholinergic innervation of the primate cerebral cortex: distribution of enzyme activities and some behavioral implications, <u>Ann. Neurol.</u>, 19:144-151.

Metherate, R., and Weinberger, N.M., 1990, Cholinergic modulation of responses to single tones produces tone-specific receptive field alterations in cat auditory cortex, <u>Synapse</u>, 6:133-145.

Milner, T.A., Joh, T.H., Miller, R.J., and Pickel, V.M., 1984, Substance P, neurotensin, enkephalin, and catecholamine-synthesizing enzymes: light microscopic localizations compared with autoradiographic label in solitary efferents to the rat parabrachial region, <u>J. Comp. Neurol.</u>, 226:434-447.

Moga, M.M., Herbert, H., Hurley, K.M., Yasui, Y., Gray, T.S., and Saper, C.B., 1990, Organization of cortical, basal forebrain, and hypothalamic afferents to the parabrachial nucleus in the rat, <u>J. Comp. Neurol.</u>, 295:624-661.

Mogenson, G.J., Swanson, L.W., and Wu, M., 1983, Neural projections from nucleus accumbens to globus pallidus, substantia innominata, and lateral preoptic-lateral hypothalamic area: an anatomical and electrophysiological investigation in the rat, <u>J. Neurosci.</u>, 3:189-202.

Molliver, M.E., 1987, Serotonergic neuronal systems: what their anatomic organization tells us about function., <u>J. Clin. Psychopharm.</u>, 7:3S-23S.

Moore, S.D. and Guyenet, P.G., 1983, An electrophysiological study of the forebrain projection of nucleus commissuralis: preliminary identification of presumed A2 catecholaminergic neurons, <u>Brain Res.</u>, 263:211-222.

Morgan, D.G., May, P.C., and Finch, C.E., 1987, Dopamine and serotonin systems in human and rodent brain: effects of age and neurodegerative disease, <u>J. Am. Geriatr. Soc.</u>, 35:334-345.

Moroni, F., Peralta, E., Cheney, D.L., and Costa, E., 1978, On the regulation of GABA neurons in the caudatus, pallidus and nigra: effects of opioids and dopamine agonists, <u>J. Pharmacol. Exp. Ther.</u>, 208:190-194.

Moruzzi, G. and Magoun, H.W., 1949, Brain stem reticular formation and activation of the EEG, <u>Electroenceph. Clin. Neurophysiol.</u>, 1:455-473.

Mufson, E.J., Bothwell, M., Hersh, L.B., and Kordower, J.H., 1989, Nerve growth factor receptor immunoreactive profiles in the normal, aged human basal forebrain: colocalization with cholinergic neurons, <u>J. Comp. Neurol.</u>, 285:196-217.

Nagel, J.A. and Huston, J.P., 1988, Enhanced inhibitory avoidance learning produced by post-trial injections of substance P into the basal forebrain, <u>Behavl. Neural Biol.</u>, 49:374-385.

Nakajima, Y., Nakajima, S., Obata, K., Carlson, C.G., and Yamaguchi, K., 1985, Dissociated cell culture of cholinergic neurons from nucleus basalis of Meynert and other basal forebrain nuclei, <u>Proc. Natl. Acad. Sci. USA</u>, 82:6325-6329.

Nakajima, Y., Nakajima, S., and Inoue, M., 1988, Pertussis toxin-insensitive

G protein mediates substance P-induced inhibition of potassium channels in brain neurons, Proc. Natl. Acad. Sci. USA, 85:3643-3647.

Napier, T.C., and Potter, P.E., 1989, Dopamine in the ventral pallidum/substantia innominata: biochemical and electrophysiological studies, Neuropharmacology, 28:757-760.

Napier, T.C., Muench, M.B., and Maslowski, R.J., 1991, Is dopamine a neurotransmitter within the ventral pallidum/substantia innominata?, in: "Basal Forebrain: Anatomy to Function", T.C. Napier, P.W. Kaliwas, and I. Hanin, eds., Plenum Press, New York (in press).

Nauta, W.J.H. and Kuypers, H.G.J.M., 1958, Some ascending pathways in the brain stem reticular formation, in: "Reticular Formation of the Brain", H.H. Jasper, L.D. Proctor, R.S. Knighton, W.C. Noshay, and R.T. Costello, eds., Little, Brown, Boston, pp. 3-30.

Nauta, W.J.H., Smith G.P., Faull, R.L.M., and Domesick, V.B., 1978, Efferent connections and nigral afferents of the nucleus accumbens septi in the rat, Neurosci., 3:385-401.

Nicoll, R.A., 1988, The coupling of neurotransmitter receptors to ion channels in the brain, Science, 241:545-550.

Nilsson, O.G., Strecker, R.E., Daszuta, A., and Björklund, A., 1988, Combined cholinergic and serotonergic denervation of the forebrain produces severe deficits in a spatial learning task in the rat, Brain Res., 453:235-2456.

Nishijo, H., Ono T., and Nishino, H., 1988, Single neuron responses in amygdala of alert monkey during complex sensory stimulation with affective significance, J. Neurosci., 8:3570-3583.

Nitecka, L., and Frotscher, M., 1989, Organization and synaptic interconnections of GABAergic and cholinergic elements in the rat amygdaloid nuclei: single- and double-immunolableing studies, J. Comp. Neurol., 279:470-488.

Norgren, R., 1978, Projections from the nucleus of the solitary tract in the rat, Neurosci., 3:207-218.

Nyakas, C., Luiten, P.G.M., Spencer, D.G., and Traber, J., 1987, Detailed projection patterns of septal and diagonal band efferents to the hippocampus in the rat with emphasis on innervation of CA1 and dentate gyrus, Brain Res. Bull., 18:533-545.

Olton, D.S., Wenk, G.L., and Markowska, A.M., 1991, Basal forebrain, memory, attention, in: "Activation to Acquisition: Functional Aspects of the Basal Forebrain Cholinergic System", R. Richardson, ed., Birkhäuser, Boston (in press).

Palazzi, E., Fisone, G., Hökfelt, T., Bartfai, T., and Consolo, S., 1988, Galanin inhibits the muscarinic stimulation of phosphoinositide turnover in rat ventral hippocampus, Eur. J. Pharmacol., 148:479-480.

Palkovits, M., 1984, Distribution of neuropeptides in the central nervous system: a review of biochemical mapping studies, Progr. Neurobiol., 23:151-189.

Palkovits, M., Záborszky, L., Feminger, A., Mezey, E., Fekete, M.I.K., Herman, J.P., Kanyicska, B., and Szabo, D., 1980, Noradrenergic innervation of the rat hypothalamus: experimental biochemical and electron microscopic studies, Brain Res., 191:161-172.

Palmer, A.M., Francis, P.T., Bowen. D.M., Benton, J.S., Neary, D., Mann, D.M.A., and Snowden, J.S., 1987, Catecholaminergic neurons assessed ante-mortem in Alzheimer's disease, Brain Res., 414:365-375.

Pandya, D.N. and Yeterian, E.H., 1985, Architecture and connections of cortical association areas, in: "Cerebral Cortex", Vol. 4, A. Peters and E.G. Jones, eds., Plenum Press, New York, pp. 3-61.

Pare, D., Smith, Y., Parent, A., and Steriade, M., 1988, Projections of brainstem core cholinergic and non-cholinergic neurons of cat to intralaminar and reticular thalamic nuclei, Neurosci., 25:69-86.

Parent, A., 1990, Extrinsic connections of the basal ganglia, TINS, 13:254-258.

Penny, G.R., Afsharpour, S., and Kitai, S.T., 1986, The glutamate decarboxylase-leucine enkephalin-, methionine enkephalin- and substance P-immunoreactive neurons in the neostriatum of the rat and cat: evidence for partial population overlap, Neurosci, 17:1011-1045.

Petsche, H., Stumpf, C.H., and Gogolak, G., 1962, The significance of the rabbit's septum as a relay station between the midbrain and the hippocampus, I. The control of hippocampus arousal activity by the septum cells, Electroenceph. Clin. Neurophysiol., 14:202-211.

Peters, A., Palay, S.G., and Webster, H. de F., 1976, "The Fine Structure of the Nervous System: The Neurons and Supporting Cells", W.B. Saunders.

Pirch, J.H., Corbus, M.J., Rigdon, G.C., and Lyness, W.H., 1986, Generation of cortical event-related slow potentials in the rat involves nucleus basalis cholinergic innervation, Electroenceph. Clin. Neurophys., 63:464-475.

Pirch, J., Rigdon, G., Rucker, G., and Turco, K., 1991, Basal forebrain modulation of cortical cell activity during conditioning, in: "The Basal Forebrain: Anatomy to Function", T.C. Napier, P. Kaliwas and I. Hanin, eds., Plenum Press, New York (in press)

Phillipson, O.T. and Griffiths, A.C., 1985, The topographic order of inputs to the nucleus accumbens in the rat, Neurosci., 16:275-296.

Price, J.L. and Amaral, D.G., 1981, An autoradiographic study of the projections of the central nucleus of the monkey amygdala, J. Neurosci., 11:1242-1259.

Price, J.L., Russchen, F.T., and Amaral, D.G., 1987, The limbic region, II, The amygdaloid complex, in: "Handbook of Chemical Neuroanatomy, Vol. 5, Integrated Systems of the CNS, Part I", A. Björklund, T. Hökfelt, and L.W. Swanson, eds., Elsevier Science Publications, pp. 279-388.

Price, D.L., Whitehouse, P.J., and Struble, R.G., 1986, Cellular pathology in Alzheimer's and Parkinson's diseases, TINS, 9:29-33.

Rasmusson, D.D., and Dykes, R.W., 1988, Long-term enhancement of evoked potentials in cat somatosensory cortex produced by co-activation of the basal forebrain and cutaneous receptors, Exp. Brain Res., 70:276-286.

Ricardo, J.A., 1981, Efferent connections of the subthalamic region in the rat, II, The zona incerta, Brain Res., 214:43-60.

Ricardo, J.A. and Koh, E.T., 1978, Anatomical evidence of direct projections from the nucleus of the solitary tract to the hypothalamus, amygdala, and other forebrain structures in the rat, Brain Res., 153:1-26.

Richardson, R.T. and Delong, M.R., 1988, A reappraisal of the functions of the nucleus basalis of Meynert, TINS, 11:264-267.

Richardson, R.T., and DeLong, M.R., 1991, Functional implications of tonic and phasic activity changes in nucleus basalis neurons, in: "Activation to Acquisition: Functional Aspects of the Basal Forebrain Cholinergic System", R. Richardson, ed., Birkhäuser, Boston (in press).

Richardson, R.T., Mitchell, S.J., Baker, F.H., and DeLong, M.R., 1988, Responses of nucleus basalis of Meynert neurons in behaving monkeys in: "Cellular Mechanisms of Conditioning and Behavioral Plasticity", C.D. Woody, D.L. Alkon, and J.L. McGaugh, eds., Plenum Press, New York, pp. 161-173.

Riche, D., de Pommery, J., and Menetrey, D., 1990, Neuropeptides and catecholamines in efferent projections of the nuclei of the solitary tract in the rat, J. Comp. Neurol., 293:399-424.

Richter-Levin, G., and Segal, M., 1989, Spatial performance is severely impaired in rats with combined reduction of serotonergic and cholinergic transmission, Brain Res., 477:404-407.

Riekkinen, Jr., P., Sirviö, J., Miettinen, R., and Riekkinen, P., 1990, Interaction between raphe dorsalis and nucleus basalis magnocellularis in the regulation of high-voltage spindle activity in rat neocortex, Brain Res., 526:31-36.

Rigdon, G.C., and Pirch, J.H., 1986, Nucleus basalis involvement in conditioned neuronal responses in the rat frontal cortex, J. Neurosci., 6:2535-2542.

Robinson, S.E., 1986, Contribution of the dorsal noradrenergic bundle to the effect of amphetamine on acetylcholine turnover, Adv. Behav. Biol., 30:43-50.

Robinson, S.E., 1989, 6-Hydroxydopamine lesion of the ventral noradrenergic bundle blocks the effect of amphetamine on hippocampal acetylcholine, Brain Res., 397:181-184.

Robinson, S.E., Malthe-Sorenssen, D., Wood, P.L., and Commissiong, J., 1979, Dopaminergic control of the septal-hippocampal cholinergic pathway, J. Pharmacol. Exp. Therap., 208:476-479.

Rolls, E.T., 1989, Information processing in the taste system of primates, J. Exp. Biol., 146:141-164

Rosenkilde, C.E., Bauer, R.H., and Fuster, J.M., 1981, Single cell activity in ventral prefrontal cortex of behaving monkeys, Brain Res., 209:375-394.

Royce, G.J. and Mourey, R.J., 1985, Efferent connections of the centromedian and parafascicular thalamic nuclei: an autoradiographic investigation in the cat, J. Comp. Neurol., 235:277-300.

Ruggiero, D.A., Mraovitch, S., Granata, A.R., Anwar, M., and Reis, D.J., 1987, A role of insular cortex in cardiovascular function, J. Comp. Neurol., 257:189-207.

Russchen, F.T., Amaral, D.G., and Price, J.L., 1985a, The afferent connections of the substantia innominata in the monkey, Macaca fascicularis, J. Comp. Neurol., 242:1-27.

Russchen, F.T., Bakst, I., Amaral, D.G., and Price, J.L., 1985b, The amygdalostriatal projections in the monkey: an anterograde tracing study, Brain Res., 329:241-257.

Russchen, F.T. and Price, J.L., 1984, Amygdalostriatal projections in the rat; topographical organization and fiber morphology shown using the lectin PHA-L as as anterograde tracer, Neurosci. Lett., 47:15-22.

Rye, D.B., Wainer, B.H., Mesulam, M.-M., Mufson, E.J., and Saper, C.B., 1984, Cortical projections arising from the basal forebrain: a study of cholinergic and noncholinergic components combining retrograde tracing and immunohistochemical localization of choline acetyltransferase, Neurosci., 13:627-643.

Saper, C.B., 1982, Convergence of autonomic and limbic connections in the insular cortex in the rat, J. Comp. Neurol., 210:163-173.

Saper, C.B., 1984, Organization of cerebral cortical afferent systems in the rat, I. Magnocellular basal nucleus, J. Comp. Neurol., 222:313-342.

Saper, C.B., 1985, Organization of cerebral cortical afferent systems in the rat, II. Hypothalamocortical projections, J. Comp. Neurol., 237:21-46.

Saper, C.B., 1987, Diffuse cortical projection systems: anatomical organization and role in cortical function, in: "Handbook of Physiology: The Nervous System", Vol. V., Part 1, V.B., Mountcastle, F. Plum and S. Geiger, eds., Amer. Physiol. Soc., Bethesda, pp. 169-210.

Saper, C.B. and Chelimsky, T.C., 1984, A cytoarchitectonic and histochemical study of nucleus basalis and associated cell groups in the normal human brain, Neurosci., 13:1023-1037.

Saper, C.B. and Loewy, A.D., 1980, Efferent connections of the parabrachial nucleus in the rat, Brain Res., 197:291-317.

Saper, C.B., Swanson, L.W., and Cowan, W.M., 1976, The efferent connections of the ventromedial nucleus of the hypothalamus of the rat, J. Comp. Neurol., 169:409-442.

Saper, C.B., Swanson, L.W., and Cowan, W.M., 1978, The efferent connections of the anterior hypothalamus of the rat, cat, and monkey, J. Comp. Neurol., 182:575-600.

Saper, C.B., Swanson, L.W., and Cowan, W.M., 1979, An autoradiographic study of the efferent connections of the lateral hypothalamic area in the rat, J. Comp. Neurol., 183:689-706.

Saper, C.B., Wainer, B.H., and German, D.C., 1987, Axonal and transneuronal transport in the transmission of neurological disease: potential role in system degenerations, Neurosci., 23:389-398.

Sarter, M., Bruno, J.P., and Dudchenko, P., 1990, Activating the damaged basal forebrain cholinergic system: tonic stimulation versus signal amplification, <u>Psychopharmacol.</u>, 101:1-17.

Sarter, M., Schneider, H.H., and Stephans, D.N., 1988, Treatment strategies for senile dementia: antagonist ß-carbolines, <u>TINS</u>, 11:13-17.

Sato, H., Hata, Y., Hagihara, K., and Tsumoto, T., 1987, Effects of cholinergic depletion on neuron activities in the cat visual cortex, <u>J. Neurophysiol.</u>, 58:781-794.

Satoh, K. and Fibiger, H.C., 1986, Cholinergic neurons of the laterodorsal tegmental nucleus: efferent and afferent connections, <u>J. Comp. Neurol.</u>, 253:277-302.

Sawchenko, P.E., Arias, C., and Bittencourt, J.C., 1990, Inhibin ß, somatostatin, and enkephalin immunoreactivities coexist in caudal medullary neurons that project to the paraventricular nucleus of the hypothalamus, <u>J. Comp. Neurol.</u>, 291:269-280.

Scheibel, M.E. and Scheibel, A.B., 1958, Structural substrates for integrative patterns in the brain stem reticular core, <u>in</u>: "Reticular Formation of the Brain", H.H. Jasper, L.D. Proctor, R.S. Knighton, W.C. Noshay, and R.T. Costello, eds., Little, Brown, Boston, pp. 31-55.

Schwaber, J.S., Rogers, W.T., Satoh, K., and Fibiger, H.C., 1987, Distribution and organization of cholinergic neurons in the rat forebrain demonstrated by computer-aided data acquisition and three-dimensional reconstruction, <u>J. Comp. Neurol.</u>, 263:309-325.

Selemon, L.D. and Goldman-Rakic, P.S., 1985, Longitudinal topography and interdigitation of corticostriatal projections in the rhesus monkey, <u>J. Neurosci.</u>, 5:776-794.

Semba, K., 1991, The cholinergic basal forebrain: A critical role in cortical arousal, <u>in</u>: "The Basal Forebrain: Anatomy to Function", T.C. Napier, P. Kaliwas, and I. Hanin, eds., Plenum Press, New York (in press).

Semba, K. and Fibiger, H., 1989, Organization of central cholinergic systems, <u>Prog. Brain Res.</u>, 79:37-63.

Semba, K., Reiner, P.B., McGeer, E.G., and Fibiger, H., 1987, Morphology of cortically projecting basal forebrain neurons in the rat as revealed by intracellular iontophoresis of horseradish peroxidase, <u>Neurosci.</u>, 20:637-651.

Semba, K., Reiner, P.B., McGeer, E.G., and Fibiger, H.C., 1988, Brainstem afferents to the magnocellular basal forebrain studied by axonal transport, immunohistochemistry, and electrophysiology in the rat, <u>J. Comp. Neurol.</u>, 267:433-453.

Sesack, S.R., Deutch, A.Y., Roth, R.H., and Bunney, B.S., 1989, Topographical organization of the efferent projections of the medial prefrontal cortex in the rat: an anterograde tract-tracing study with *Phaseolus vulgaris* leucoagglutinin, <u>J. Comp. Neurol.</u>, 290:213-242.

Shipley, M.T. and Geinisman, Y., 1984, Anatomical evidence for convergence of olfactory, gustatory, and visceral afferent pathways in mouse cerebral cortex, <u>Brain Res. Bull.</u>, 12:221-226.

Shu, S.Y., Penny, G.R., and Peterson, G.M., 1988, The "marginal division": a new subdivision in the neostriatum of the rat, <u>J. Chem. Neuroanat.</u>, 1:147-163.

Sillito, A.M. and Kemp, J.A., 1983, Cholinergic modulation of the functional organization of the cat visual cortex, <u>Brain Res.</u>, 289:143-155.

Simerly, R.B. and Swanson, L.W., 1988, Projections of the medial preoptic nucleus: a *Phaseolus vulgaris* leucoagglutinin anterograde tract-tracing study in the rat, <u>J. Comp. Neurol.</u>, 270:209-242

Simon, L., LeMoal, M., and Calas, A., 1979, Efferents and afferents of the ventral tegmental-A10 region studied after local injection of [^{3}H]leucine and horseradish peroxidase, <u>Brain Res.</u>, 178:17-40.

Smith, D.A. and Bolam, J.P., 1990, The neural network of the basal ganglia as revealed by the study of synaptic connections of identified neurones, <u>TINS</u>, 13:259-265.

Sofroniew, M.V., Eckenstein, F., Thoenen, H., and Cuello, A.C., 1982,

Topography of choline acetyltransferase-containing neurons in the forebrain of the rat, Neurosci. Lett., 33:7-12.

Sofroniew, M.V., Pearson, R.C.A., and Powell, T.P.S., 1987, The cholinergic nuclei of the basal forebrain of the rat: normal structure, development and experimentally induced degeneration, Brain Res., 411:310-331.

Sofroniew, M.V., Priestly, J.V., Consolazione, A., Eckenstein, F., and Cuello, A.C., 1985, Cholinergic projections from the midbrain and pons to the thalamus in rat, identified by combined retrograde and choline acetyltransferase immunohistochemistry, Brain Res., 329:213-223.

Starzl, T.E., Taylor, C.W., and Magoun, H.W., 1951, Ascending conduction in reticular activating system, with special reference to the diencephalon, J. Neurophysiol., 14:461-477.

Staubli, U. and Huston, J.P., 1980, Facilitation of learning by post-trial injection of substance P into the medial septal nucleus, Behav. Brain Res., 1:245-255.

Steriade, M., 1970, Ascending control of thalamic and cortical responsiveness, Int. Rev. Neurobiol., 12:87-144.

Steriade, M. and Llinas, R.R., 1988, The functional states of the thalamus and the associated neuronal interplay, Physiol. Rev., 68:649-742.

Steriade, M. and McCarley, R.W., 1990, "Brainstem Control of Wakefulness and Sleep", Plenum Press, New York.

Steriade, M., Paré, D., Parent, A., and Smith Y., 1988, Projections of cholinergic and non-cholinergic neurons of the brainstem core to relay and associational thalmic nuclei in the cat and macaque monkey, Neurosci., 25:47-67.

Stewart, M. and Fox, S.E., 1990, Do septal neurons pace the hippocampal theta rhythm? TINS, 13:163.

Swanson, L.W., 1976, An autoradiographic study of the efferent connections of the preoptic region in the rat, J. Comp. Neurol., 167:227-256.

Swanson, L.W., 1987, The hypothalamus, in: "Handbook of Chemical Neuroanatomy: Integrated Systems of the CNS", (Part I, Vol. 5), A. Björklund, T. Hökfelt, and L.W. Swanson, eds., Elsevier, pp. 1-124.

Swanson, L.W. and Cowan, W.M., 1979, The connections of the septal region in the rat, J. Comp. Neurol., 186:621-656.

Swanson, L.W. and Köhler, C., 1986, Anatomical evidence for direct projections from the entorhinal area to the entire cortical mantle in the rat, J. Neurosci., 6:3010-3023.

Szerb, J.C., 1967, Cortical acetylcholine release and electroencephalographic arousal, J. Physiol., 192:329-343.

Szigethy, E., Leonard, K., and Beaudet, A., 1990, Ultrastructural localization of [^{125}I]neurotensin binding sites to cholinergic neurons of the rat nucleus basalis magnocellularis, Neurosci., 36:377-391.

Szymusiak, R. and McGinty, D., 1989, Sleep-waking discharge of basal forebrain projection neurons in cats, Brain Res. Bull., 22:423-430.

ter Horst, G.J. and Luiten, P.G.M., 1986, The projections of the dorsomedial hypothalamic nucleus in the rat, Brain Res. Bull., 16:231-248.

Thorpe, S.J., Rolls, E.T., and Maddison, S., 1983, The orbitofrontal cortex: neuronal activity in the behaving monkey, Exp. Brain Res., 49:93-115.

Tomlinson, B.E., Irving, D., and Blessed, G., 1981, Cell loss in the locus coeruleus in senile dementia of Alzheimer's type, J. Neurol. Sci., 49:419-428.

Turner, B.H., Mishkin, M., and Knapp, M., 1980, Organization of the amygdalopetal projections from modality-specific cortical association areas in the monkey, J. Comp. Neurol., 191:515-543.

van der Kooy, D., Koda, L.Y., McGinty, J.F., Gerfen, C.R., and Bloom, F.E., 1984, The organization of projections from the cortex, amygdala, and hypothalamus to the nucleus of the solitary tract in the rat, J. Comp. Neurol., 224:1-24.

Vanderwolf, C.H., 1988, Cerebral activity and behavior; control by central cholinergic and serotonergic systems, Int. Rev. Neurobiol., 30:225-330

Vanderwolf, C.H., Baker, G.B., and Dickson, C., 1990, Serotonergic control of cerebral activity and behavior: models of dementia, <u>Ann. NY. Acad. Sci.</u>, 600:366-383.

van Hoesen, G.W., 1981, The differential distribution, diversity and sprouting of cortical projections to the amygdala in the Rhesus monkey, <u>in</u>: "The Amygdaloid Complex", (INSERM Symposium No. 20), Y. Ben-Ari, ed., Elsevier, North Holland, pp. 77-104.

van Hoesen, G.W. and Pandya, D.N., 1975, Some connections of the entorhinal (area 28) and perirhinal (area 35) cortices of the rhesus monkey, I. Temporal lobe afferents, <u>Brain Res.</u>, 95:1-24.

van Wimersma Greidanus, T.B., Van Praag, M.C.G., Kalmann, R., Rinkel, G.J.E., Croiset, G., Hoeke, E.C., Van Egmond, M.A.H., and Fekete, M., 1982, Behavioral effects of neurotensin, <u>Ann. NY Acad. Sci.</u>, 400:319-329.

Vécsei, L., Király, C., Bollók, Nagy, A., Varga, J., Penke, B., and Telegdy, G., 1984, Comparative studies with somatostatin and cysteamine in different behavioral tests on rats, <u>Pharmacol. Biochem. Behav.</u>, 21:833-837.

Versteeg, D.H.G., Van der Gugten, J., de Jong, W., and Palkovits, M., 1976, Regional concentrations of noradrenaline and dopamine in the brain, <u>Brain Res.</u>, 113:563-574.

Vertes, R.P., 1977, Selective firing of rat pontine gigantocellular neurons during movement and REM sleep, <u>Brain Res.</u>, 128:146-152.

Vertes, R.P., 1979, Brain stem gigantocellular neurons: patterns of activity during behavior and sleep in the freely moving rat, <u>J. Neurophysiol.</u>, 42:214-228.

Vertes, R.P., 1981, An analysis of ascending brain stem systems involved in hippocampal synchronization and desynchronization, <u>J. Neurophysiol.</u>, 46:1140-1159.

Vertes, R.P., 1988, Brainstem afferents to the basal forebrain in the rat, <u>Neurosci.</u>, 24:907-935,

Vertes, R.P., 1990a, Fundamentals of brainstem anatomy: a behavioral perspective, <u>in</u>: "Brainstem Mechanisms of Behavior", W.R. Klemm and R.P. Vertes, eds., John Wiley and Sons, New York, pp. 33-103.

Vertes, R.P., 1990b, Brainstem mechanisms of slow-wave sleep and REM sleep, <u>in</u>: "Brainstem Mechanisms of Behavior", W.R. Klemm and R.P. Vertes, eds., John Wiley and Sons, New York, pp. 535-583.

Vertes, R.P. and Martin, G.F., 1988, Autoradiographic analysis of ascending projections from the pontine and mesencephalic reticular formation and the median raphe nucleus in the rat, <u>J. Comp. Neurol.</u>, 275:511-541.

Vertes, R.P., Martin, G.F., and Waltzer, R., 1986, An autoradiographic analysis of ascending projections from the medullary reticular formation in the rat, <u>Neurosci.</u>, 19:873-898.

Wainer, B.H., Bolam, J.P., Freund, T.F., Henderson, Z., Totterdell, S., and Smith, A.D., 1984, Cholinergic synapses in the rat brain: a correlated light and electron microscopic immunohistochemical study employing a monoclonal antibody against choline acetyltransferase, <u>Brain Res.</u>, 308:69-76.

Walaas, I. and Fonnum, F., 1979, The distribution and origin of glutamate decarboxylase and choline acetyltransferase in ventral pallidum and other basal forebrain regions, <u>Brain Res.</u>, 177:325-336.

Walker, L.C., Koliatsos, V.E., Kitt, C.A., Richardson, R.T., Rökaeus, Ä., and Price, D.L., 1989, Peptidergic neurons in the basal forebrain magnocellular complex of the rhesus monkey, <u>J. Comp. Neurol.</u>, 280, 272-282.

Wenk, G.L., 1984, Pharmacological manipulations of the substantia innominata-cortical cholinergic pathway, <u>Neurosci. Lett.</u>, 51:99-103.

Wilson, F.A.W. and Rolls, E.T., 1990a, Neuronal responses related to the novelty and familiarity of visual stimuli in the substantia innominata, diagonal band of Broca, and periventricular region of the primate basal forebrain, <u>Exp. Brain Res.</u>, 80:104-120.

Wilson, F.A.W. and Rolls, E.T., 1990b, Neuronal responses related to reinforcement in the primate basal forebrain, Brain Res., 509:213-231.

Witter, M.P., Groenewegen, H.J., Lopes da Silva, F.H., and Lohman, A.H.M., 1989, Functional organization of the extrinsic and intrinsic circuitry of the parahippocampal region, Progr. Neurobiol., 33:161-252.

Wood, P.L. and Cheney, D.L., 1979, The effect of muscarinic receptor blockers on the turnover rate of acetylcholine in various regions of the rat brain, Can. J. Physiol. Pharmacol., 57:404-411.

Wood, P.L. and McQuade, P., 1986, Substantia innominata-cortical cholinergic pathway: regulatory afferents, Adv. Behav. Biol., 30:999-1006.

Wood, P.L. and Richard, J., 1982, GABAergic regulation of the substantia innominata-cortical cholinergic pathway, Neuropharmacol., 21:969-972.

Wood, P.L., Cheney, D.L., and Costa, E., 1979, Modulation of the turnover rate of hippocampal acetylcholine by neuropeptides: possible site of action of α-melanocyte-stimulating hormone, adrenocorticotrophic hormone and somatostatin, J. Pharm. Exp. Ther., 209:97-103.

Woody, C.D., 1982, Acquisition of conditioned facial reflexes in the cat: cortical control of different facial movements, Fed. Proc., 41:2160-2168.

Woolf, N.J. and Butcher, L.L., 1986, Cholinergic systems in the rat brain, III. Projections from the pontomesencephalic tegmentum to the thalamus, tectum, basal ganglia, and basal forebrain, Brain Res. Bull., 16:603-637.

Woolf, N.J., Eckenstein, F., and Butcher, L.L., 1984, Cholinergic systems in the rat brain, I. Projections to the limbic telencephalon, Brain Res. Bull., 13:751-784.

Woolf, N.J., Hernit, M.C., and Butcher, L.L., 1986, Cholinergic and non-cholinergic projections from the rat basal forebrain revealed by combined choline acetyltransferase and Phaseolus vulgaris leucoagglutinin immunohistochemistry, Neurosci. Lett., 66:281-286.

Yau, W.M., Dorset, J.A., and Youther, M.L., 1986, Evidence for galanin as an inhibitory neuropeptide on myenteric cholinergic neurons in the guinea-pig small intestine, Neurosci. Lett., 3:305-308.

Yau, W.M., Lingle, P.F., and Youther, M.L., 1983, Modulation of cholinergic neurotransmitter release from myenteric plexus by somatostatin, Peptides, 4:49-53.

Záborszky, L., 1982, Afferent connections of the medial basal hypothalamus, Adv. Anat. Embryol. Cell Biol. 69:1-107.

Záborszky, L., 1989a, Afferent connections of the forebrain cholinergic projection neurons, with special reference to monoaminergic and peptidergic fibers, in: "Central Cholinergic Synaptic Transmission", M. Frotscher and U. Misgeld, eds., Birkhäuser, Basel, pp. 12-32.

Záborszky, L., 1989b, Peptidergic-cholinergic interactions in the basal forebrain, in: "Alzheimer's Disease: Advances in Basic Research and Therapies", R.J. Wurtman, S.H. Corkin, J.H. Growdon, and E. Ritter-Walker, eds., Proc. Fifth Meeting Int. Study Group on the Pharmacology of Memory Disorders Associated with Aging, CBSMCT, Cambridge, Massachusetts, pp. 521-528.

Záborszky, L. and Braun, A., 1988, Peptidergic afferents to forebrain cholinergic neurons, Soc. Neurosci. Abstr., 14:905.

Záborszky, L. and Cullinan, W.E., 1989, Hypothalamic axons terminate on forebrain cholinergic neurons: an ultrastructural double-labeling study using PHA-L tracing and ChAT immunocytochemistry, Brain Res., 479:177-184.

Záborszky, L. and Heimer, L., 1989, Combinations of tracer techniques, especially HRP and PHA-L, with transmitter identification for correlated light and electron microscopic studies, in: "Neuroanatomical Tract-Tracing Methods 2: Recent Progress", L. Heimer and L. Záborszky, eds., Plenum Press, New York, pp. 49-96

Záborszky, L. and Luine, V.N., 1987, Evidence for existence of monoaminergic-cholinergic interactions in the basal forebrain, <u>J. Cell.</u> <u>Biol.</u>, Suppl. 11D, 187.

Záborszky, L., Alheid, G.F., Alones, V., Oertel, W.H., Schmechel, D.E., and Heimer, L., 1982, Afferents of the ventral pallidum studied with a combined immunohistochemical-anterograde degeneration method, <u>Soc.</u> <u>Neurosci. Abstr.</u>, 8:218.

Záborszky, L., Carlsen, J., Brashear, H.R., and Heimer, L., 1986a, Cholinergic and GABAergic afferents to the olfactory bulb in the rat with special emphasis on the projection neurons in the nucleus of the horizontal limb of the diagonal band, <u>J. Comp. Neurol.</u>, 243:488-509.

Záborszky, L., Eckenstein, F., Leranth, Cs., Oertel, W., Schmechel, D., Alones, V., and Heimer, L., 1984b, Cholinergic cells of the ventral pallidum: a combined electron microscopic immunocytochemical, degeneration and HRP study, <u>Soc. Neurosci. Abst.</u>, 10:8.

Záborszky, L., Heimer, L., Eckenstein, F. and Leranth, C., 1986b, GABAergic input to cholinergic forebrain neurons: an ultrastructural study using retrograde tracing of HRP and double immunolabeling, <u>J. Comp. Neurol.</u>, 250:282-295.

Záborszky, L., Leranth, C., and Heimer, L., 1984a, Ultrastuctural evidence of amydalofugal axons terminating on cholinergic cells of the rostral forebrain, <u>Neurosci. Lett.</u>, 52:219-225.

Záborszky, L., Luine, V.N., Snavely, L., Heimer, L., 1986c, Biochemical changes in the cholinergic forebrain system following transection of the ascending brainstem fibers, <u>Soc. Neurosci. Abstr.</u>, 12:571.

Záborszky, L., Luine, V.N., Cullinan, W.E., and Heimer, L., 1991, Direct catecholaminergic-cholinergic interactions in the basal forebrain: morphological and biochemical studies, (submitted).

Zemlan, F.P., Behbehani, M.M., and Beckstead, R.M., 1984, Ascending and descending projections from nucleus reticularis magnocellularis and nucleus reticularis gigantocellularis: an autoradiographic and horseradish peroxidase study in the rat, <u>Brain Res.</u>, 292:207-220.

γ-AMINOBUTYRIC ACID AND μ-OPIOID RECEPTOR LOCALIZATION

AND ADAPTATION IN THE BASAL FOREBRAIN

Lynn Churchill, Andrea Bourdelais, Mark Austin[1],
Daniel S. Zahm[2] and Peter W. Kalivas

Department of Veterinary & Comparative Anatomy,
Pharmacology and Physiology, Washington State
University, Pullman, WA;
[1]National Institute of Mental Health, Bethesda, MD;
[2]Department of Anatomy and Neurobiology,
St. Louis School of Medicine, St. Louis, MO

INTRODUCTION

The projection from the nucleus accumbens to ventral
pallidum has been functionally implicated in the integration of
motivation and locomotion, since the limbic system and the
extrapyramidal motor system interconnect in this projection
(Mogenson et al., 1980). The nucleus accumbens receives
innervation from dopaminergic neurons in the ventral tegmental
area and substantia nigra (Fallon and Moore, 1978; Gerfen et
al., 1987) and projects topographically onto the ventral
pallidum (Conrad and Pfaff, 1976; Nauta et al., 1978; Mogenson
et al., 1983; Groenewegen and Russchen, 1984). The projection
from nucleus accumbens to ventral pallidum appears to contain γ-
aminobutyric acid (GABA) and enkephalin. Electrolytic lesions
of the nucleus accumbens significantly decreased glutamic acid
decarboxylase (GAD), the synthetic enzyme for GABA, in the
ventral pallidum (Walaas and Fonnum, 1979) and ibotenic acid
lesions of the nucleus accumbens significantly decreased
enkephalin-like immunoreactivity in the ventral pallidum
(Zaborszky et al., 1985). Electrophysiological studies revealed
that ventral pallidal neurons decreased their rate of discharge
in response to stimulation of the nucleus accumbens and that the
GABA antagonist, picrotoxin, increased the spontaneous firing
frequency of these neurons (Mogenson et al., 1983). Behavioral
studies indicated that GABA injections into the ventral pallidum
blocked the increase in locomotor activity induced by either
dopamine or opioid agonists (Jones and Mogenson, 1980; Mogenson
and Nielsen, 1983; Swerdlow et al., 1984; Austin and Kalivas,
1988; 1989). These biochemical, electrophysiological and
behavioral studies support the anatomical evidence that GABA and
enkephalin colocalize in terminals innervating ventral pallidal
neurons (Zahm et al., 1985).

The Basal Forebrain, Edited by T.C. Napier *et al.*
Plenum Press, New York, 1991

Since the behavioral or biochemical responses to pharmacological stimulation of the nucleus accumbens differ depending upon the part stimulated, anatomical and connectional differences between different regions of the nucleus accumbens have been suggested (Albert et al., 1989; Russell et al., 1989; Vaccarino and Rankin, 1989; Yoshikawa et al., 1989). In addition to rostrocaudal and mediolateral distinctions alluded to in the functional studies, core and shell subdistricts are also recognized in the nucleus accumbens (Zaborszky et al., 1985; Paxinos and Watson, 1986; Groenewegen et al., 1989; Voorn et al., 1989; Zahm and Heimer, 1990), while the ventral pallidum has been subdivided into dorsolateral and ventromedial compartments (Zahm and Heimer, 1988; Zahm, 1989). A specificity of the projection patterns of the nucleus accumbens core and shell to the dorsolateral and ventromedial compartments of the ventral pallidum and other downstream targets has been revealed by following iontophoresis of anterograde and retrograde tracers (Churchill et al., 1990; Heimer et al., in press).

Furthermore, discrete quinolinic acid lesions in specific compartments of the nucleus accumbens have revealed upregulation of GABA receptors on terminals within subdistricts of the nucleus accumbens and cell bodies within separate compartments of the ventral pallidum (Churchill et al., 1990). In contrast, μ-opioid receptor autoradiography, using Tyr-D-Ala-Gly-mePhe-Gly-OH(DAGO; a peptidase-resistant analog of enkephalin) revealed that μ-opioid receptors were localized on cell bodies in the nucleus accumbens but did not change in the ventral pallidum after destruction of the striatopallidal projection (Churchill et al., 1990). In this chapter, we compare these data and data following lesions in the ventral pallidum. Further evidence for GABA receptor localization and compartmentation in the ventral pallidum is presented based upon *in situ* hybridization histochemistry using oligonucleotide probes for the mRNAs of GAD and the $\alpha 1$ subunit of the $GABA_A$ receptor.

METHODS

The experimental procedures for quinolinic acid lesions in the nucleus accumbens, receptor autoradiography, morphometric analyses and Fluoro-Gold tracer histochemistry have been described previously (Churchill et al., 1990). For the ventral pallidal lesion, quinolinic acid (50-100 nmol/0.5 μl) or nicotinic acid (100-200 nmol/0.5 μl) as a control (Foster et al., 1983) was unilaterally injected into the ventral pallidum [A/P 8.7 mm; M/L 2.1 mm; D/V 1.8 mm; according to the atlas of Paxinos and Watson (1986)]. Two weeks after the injection, the rats were anesthetized with halothane, and their brains were removed and frozen in isopentane cooled with dry ice. Ten μm sections were cut on a cryostat and mounted on chrome alum-gel subbed slides for receptor binding studies. Experimental and control sections were simultaneously incubated for 30 min on ice in 50 nM [^{3}H]muscimol (Amersham, 12-23 Ci/mmol) after 3 preincubations for 5 min each on ice and washed with several 2 sec dips in 0.05 M Tris citrate buffer, pH 7. Non-specific binding was conducted in the presence of 100 μM GABA and did not show any labeling above film background levels. The sections were placed against [3H] sensitive Hyperfilm (Amersham). Binding density was quantified using a Nikon photometer with a 0.25 mm aperture at 3 spots within 3 sections of each brain as indicated previously (Churchill et al., 1990).

In situ hybridization experiments were performed as previously described (Young, 1989; Churchill et al., in press). Briefly, coronal sections (14 μM) that were thaw-mounted onto subbed slides were fixed in 4% paraformaldehyde, 0.02% diethyl pyrocarbonate in phosphate-buffered saline, rinsed in buffer without fixative and dried. The GAD oligonucleotide probe, provided by C. Gerfen, NIMH, was complementary to bases 704-751 of the cat glutamate decarboxylase mRNA (Kobayashi et al., 1987).The GABA$_A$ receptor α1 subunit probe, provided by S. Lolait and L.Mahan, NIMH, was complementary to bases 1048-1095 of the rat mRNA (Lolait et al., 1989a,b). Oligonucleotides were labeled with [35S]dATP to a specific activity of 4200-4600 Ci/mmol using terminal deoxynucleotidyl transferase and extracted with phenol (pH 8.0)/chloroform/isoamyl alcohol (50:49:1). The aqueous phase was extracted again with chloroform/isoamyl alcohol (49:1) and the oligonucleotide precipitated with 5M NaCl and 100% ethanol at -20°C. After centrifugation and rinsing with 70% ethanol, the probe was resuspended in 10 mM Tris HCl, 1 mM EDTA, pH 7.6, and 0.05 M dithiothreitol. Prior to hybridization, the probe was diluted to 300,000-500,000 dpm/40 μl with buffer containing 50% formamide, 600 mM NaCl, 80 mM Tris HCl (pH 7.5), 4 mM EDTA, 0.1% sodium pyrophosphate, 0.2% sodium dodecyl sulfate, 0.2 mg/ml heparin sulfate, 10% dextran sulfate, and 100 mM dithiothreitol. The sections were rinsed in 0.25% acetic anhydride, 0.1 M triethanolamine HCl in 0.9% NaCl (pH 8) for 10 min, and delipidated in a graded series of ethanol, chloroform and ethanol. Hybridization buffer was applied to tissue sections which were coverslipped with parafilm and incubated at 37°C for 16-20 hrs. Then the slides were rinsed thoroughly, dipped in NTB 3 emulsion and exposed for 5 weeks. The sections were developed, fixed and counterstained with cresyl violet.

RESULTS

Quinolinic acid lesions that destroyed the lateral core, medial shell and dorsomedial core of the nucleus accumbens, or the ventral pallidum, produced an increase in the refractive index of the unstained tissue, which allowed an evaluation of the extent of the lesion. Representative lesions are illustrated in Fig. 1. The spread of quinolinic acid was limited in the dorsoventral and mediolateral directions, allowing us to evaluate receptor changes in specific compartments of the nucleus accumbens. The lesion extends in the rostrocaudal plane as well (Churchill et al., 1990), but since the projection fascicles carry the toxin in a topographically-restricted manner, the lesions do not cross into other compartments at other levels of the accumbens. Since ventral pallidal neurons were more resistant to quinolinic acid than striatal neurons, the concentration necessary to destroy neurons in the ventral pallidum also resulted in lesions within the lateral nucleus accumbens rostrally.

[3H]Muscimol binding in the injection site increased 115-
127% compared to the contralateral side in the lateral nucleus
accumbens and dorsomedial core lesion, but did not change in the
medial shell lesion (Fig. 2). Differences were expressed as %
of the noninjected side, since no differences were observed
between the noninjected side and the nicotinic acid-injected
controls (Churchill et al., 1990). In contrast, [3H]muscimol
binding decreased significantly to 24-40% of the noninjected
side at the lesion site in the dorsolateral and ventromedial
compartments of the ventral pallidum, after a lesion that
destroyed all of the dorsolateral and most of the ventromedial
district. [125I]DAGO binding decreased to 60-70% of the
noninjected side in the nucleus accumbens lesions and to 46-60%
of the noninjected side in the dorsolateral and ventromedial
compartments of the ventral pallidal lesion (Fig. 3). No large
differences were observed in the decrements in [125I]DAGO binding
in the different compartments of the nucleus accumbens.

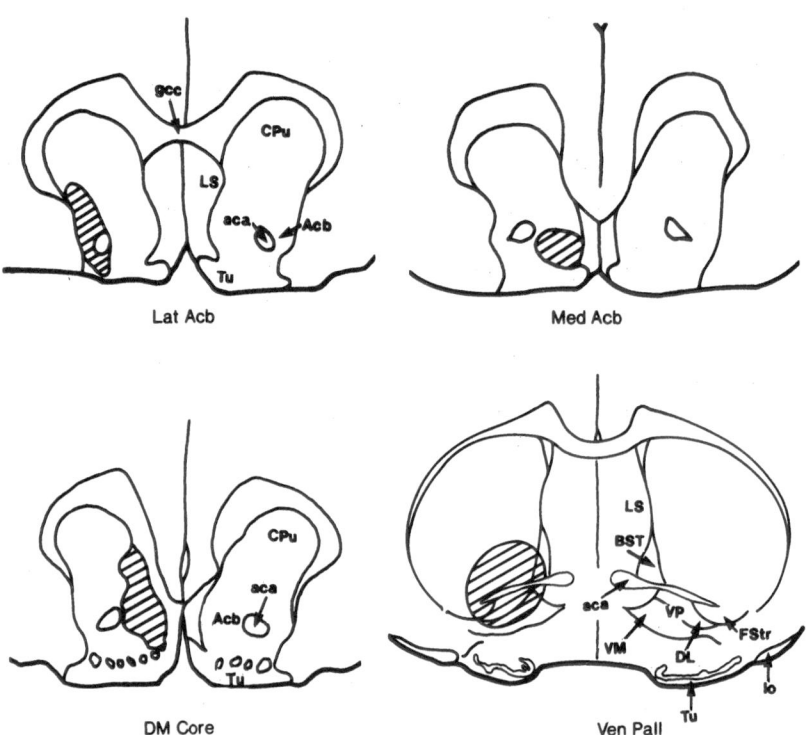

Fig. 1.   Drawings of representative lesions in each compartment
           of the nucleus accumbens and the ventral pallidum.
           Abbreviations: gcc, corpus callosum; LS, lateral
           septum; CPu, caudate-putamen; aca, anterior
           commissure; Acb, nucleus accumbens; Lat, lateral; Med,
           medial; DM, dorsomedial; Tu, olfactory tubercle; FStr,
           fundus striatum; BST, bed nucleus of stria terminalis;
           VP, Ven Pall, ventral pallidum; VM, ventromedial; DL,
           dorsolateral; lo, lateral olfactory tract.

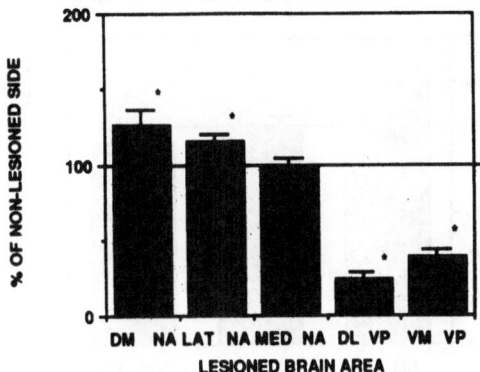

GABA-A RECEPTORS IN LESION SITE

Fig. 2.   A bar graph illustrating [^{3}H]muscimol binding to the
lesion site in the nucleus accumbens or ventral
pallidum as % of the noninjected side. Data are
expressed as mean ± S.E. with the asterisk indicating
$p < 0.05$ by the paired Student's *t* test.
Abbreviations: NA, nucleus accumbens; Med, medial;
Lat, lateral; others as in Fig. 1.

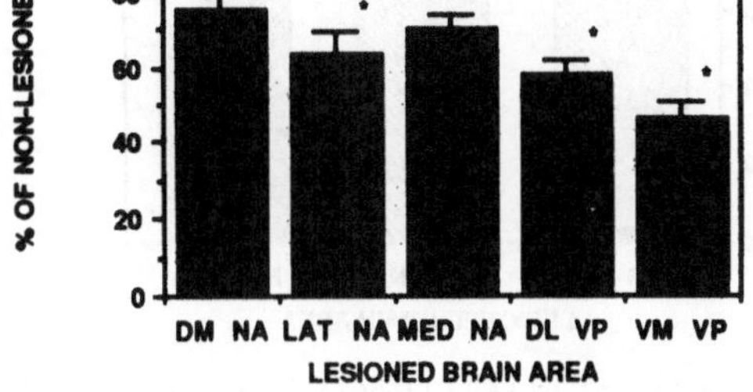

MU-OPIOID RECEPTORS IN LESION SITE

Fig. 3.   A bar graph illustrating [^{125}I]DAGO binding to the
lesion site in the nucleus accumbens or ventral
pallidum as % of the noninjected side. The expression
of the data and the abbreviations are the same as in
Fig. 2.

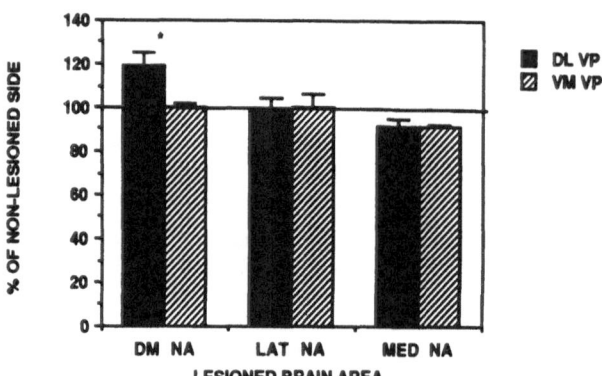

Fig. 4.   A bar graph of [³H]muscimol binding in the ventral
          pallidum after discrete quinolinic acid lesions in
          specific compartments of the nucleus accumbens. The
          expression of the data and abbreviations are the same
          as in Fig. 2.

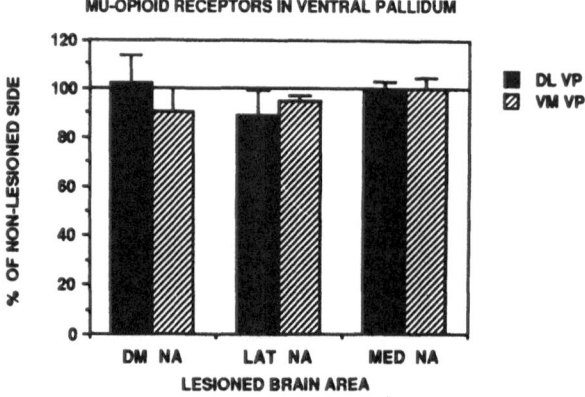

Fig. 5.   A bar graph of [¹²⁵I]DAGO binding in the ventral
          pallidum after discrete quinolinic acid lesions in
          specific compartments of the nucleus accumbens. The
          expression of the data and abbreviations are the same as
          in Fig. 2.

[³H]Muscimol binding did not change in the ventral pallidum after lesions in either the lateral core or medial shell of the nucleus accumbens (Fig. 4). In comparison, lesions in the dorsomedial core resulted in increases in [³H]muscimol binding to 119% of the noninjected side in the dorsolateral compartment of the ventral pallidum. Morphometric analysis of the lesions revealed a positive correlation coefficient for the linear regression analysis of the relationship between the size of the dorsomedial core lesion and the increasing [³H]muscimol binding (0.876; $p < 0.05$). No significant correlation was observed between the size of the lesion and [³H]muscimol binding when the lesion was in the medial shell ($r = 0.344$, $p > 0.05$) or lateral core ($r = 0.032$, $p > 0.05$). In contrast, [¹²⁵I]DAGO binding did not change in the ventral pallidum after lesions in any of the accumbal compartments (Fig. 5).

Iontophoresis of the retrograde tracer Fluoro-gold into the dorsolateral or ventromedial compartments of the ventral pallidum demonstrated that the dorsomedial core projects to the dorsolateral ventral pallidum, while the medial shell projects to the ventromedial compartment (Fig. 6). The iontophoretic injection sites were localized to either the dorsolateral or ventromedial compartment (Fig. 6A) and the retrogradely labeled cells at two different levels of the nucleus accumbens were localized to either the core or medial shell, respectively (Fig. 6B).

The oligonucleotide probes for GAD or the α1 subunit of the GABA$_A$ receptor labeled a large population of neurons in the dorsolateral compartment of the ventral pallidum and a smaller population of neurons in the ventromedial district (Fig. 7). In comparison, the expression of GAD mRNA in the nucleus accumbens was less evident in the ventral pallidum but still present (Fig. 8). The number of neurons labeled was greater in the shell than in the core. In contrast, neurons containing mRNA for the α1 subunit of the GABA receptor were absent in the nucleus accumbens. Labeled neurons were observed in the ventral pallidum and diagonal band of Broca in these same sections, indicating that the probe was working in these sections.

DISCUSSION

The data in this study support principles of organization that distinguish a core and shell region in the nucleus accumbens and a dorsolateral and ventromedial compartment in the ventral pallidum. Discrete quinolinic acid lesions in the lateral and dorsomedial core result in upregulation of GABA$_A$ receptors in the nucleus accumbens, whereas lesions in the medial shell do not, suggesting that the GABA$_A$ receptors are localized on terminals within the nucleus accumbens core. The lack of change in GABA$_A$ receptors in the medial shell may be due to a different mechanism for adaptation of the GABA$_A$ receptors on the terminals in the medial shell. *In situ* hybridization studies indicate that GABA$_A$ receptors are not on cell bodies in the nucleus accumbens since mRNA for the α1 subunit of the GABA$_A$ receptor was not observed in cells in either the core or shell of the nucleus accumbens. The lack of mRNA for the GABA$_A$ receptor contrasts with the presence of cells labeled by the oligonucleotide probe for GAD mRNA in both the core and shell of the nucleus accumbens (Chesselet et al., 1987). In contrast, many cells in the ventral pallidum contain α1 subunit mRNA for the GABA$_A$ receptor with the dorsolateral compartment of the ventral pallidum having the highest concentration of labeled cells.

Molecular cloning of subunit cDNAs for the GABA$_A$ receptor have revealed at least 4 classes of subunits (α, β, γ, and δ) with a number of variants in each class (Levitan et al., 1988; Khrestchatisky et al., 1989; Shivers et al., 1989; Lolait et al, 1989a,b; Malherbe et al., 1990). Reconstitution studies reveal that although each cloned subunit can form a homomeric channel activated by high concentrations of GABA (Levitan et al., 1988; Khrestchatisky et al., 1989; Shivers et al., 1989), coexpression of the α1, β and γ2 subunit is required to form a functional receptor that mimics the *in vivo* pharmacology (Schofield, 1989). *In situ* hybridization histochemistry at a low resolution revealed that within the basal forebrain, the α1, β2 and γ2 subunits were localized prominently in the ventral pallidum (Montpied et al., 1988; Sequier et al., 1988; Shivers et al., 1989; Malherbe et al., 1990); whereas the δ subunit was localized prominently in the nucleus accumbens. These findings suggest that the regulation of locomotor activity in the ventral pallidum by GABA$_A$ receptors that mimic the *in vivo* pharmacology results from an action postsynaptic to the GABAergic afferents rather than presynaptic regulation of GABA release from the accumbens-pallidal projection.

Discrete lesions in the dorsomedial core resulted in upregulation of GABA$_A$ receptors in the dorsolateral compartment of the ventral pallidum in a fashion similar to that observed in the globus pallidus after striatal lesions (Pan et al., 1983). In contrast, lesions in the lateral core did not upregulate GABA$_A$ receptors in the dorsolateral compartment. Fluoro-gold tracer histochemistry demonstrated that the dorsomedial and lateral core project specifically to the dorsolateral compartment of the ventral pallidum. Anterograde tracer histochemistry with *Phaseolus vulgaris* agglutinin (PHA-L), which has even greater advantages at demonstrating the terminal projection, further supports these observations (Heimer et al., in press; Zahm and Heimer, 1990). The PHA-L injection into ventral core labeled axons of passage and terminal puncta mainly in the dorsomedial compartment of the ventral pallidum with a slight distribution of label in the ventromedial compartment. In comparison, the PHA-L injection into the shell labeled exclusively the ventromedial compartment of the ventral pallidum. The differences in these results may be due to the differences in the size of the two lesions. Perhaps destruction of a small percentage of the GABAergic and enkephalin-positive boutons that surround the pallidal neurons stimulates the upregulation response, whereas destruction of a larger percentage of the projection terminals results in modification of other transmitter systems. An even larger lesion of the lateral core projection did not upregulate the GABA$_A$ receptors in dorsolateral ventral pallidum (Churchill et al., in press), supporting the conclusion that the destruction of this projection does not induce upregulation. Other explanations might be: (1) that the dorsomedial core projection is uniquely different from the lateral core projection in regulating the GABA$_A$ receptor; or (2) that the ventral pallidal neurons that receive innervation from the medial shell or lateral core are also innervated by other GABAergic afferents and may not respond to partial GABAergic denervation.

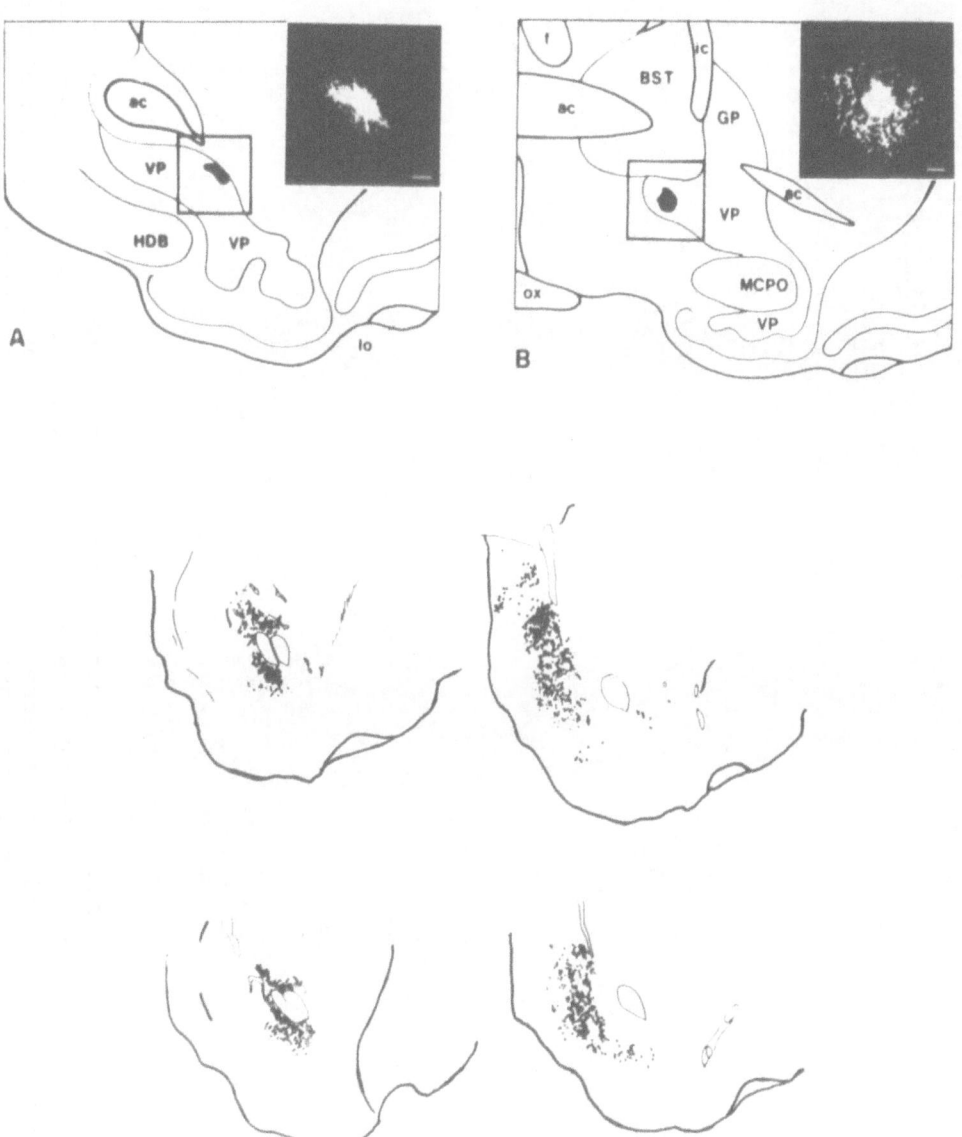

Fig. 6.   The iontophoretic injection sites for Fluoro-gold into the dorsolateral ventral pallidum (A) or the ventromedial compartment (B) and the maps of retrogradely labeled neurons at two different levels of the nucleus accumbens for each of the respective injections. This figure illustrates a portion of a larger study by Heimer et al., in press.

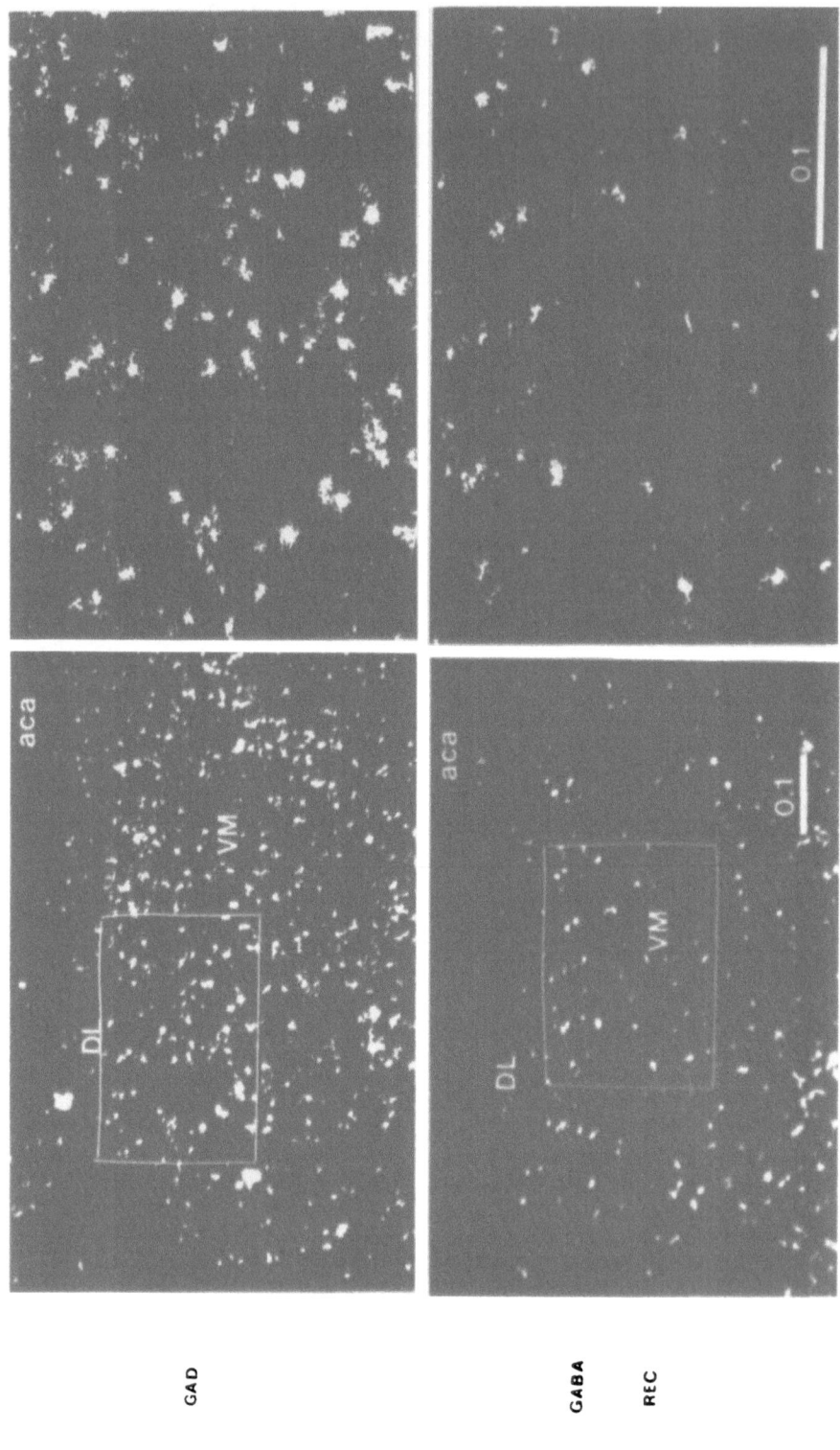

Fig. 7.   *In situ* hybridization of oligonucleotide probes for mRNAs of glutamic acid decarboxylase (GAD) or the α1 subunit of the GABA_A receptor (GABA Rec) in the ventral pallidum. The corresponding figures to the right are a higher power magnification of the box in each of the figures. Bar = 0.1 mm. Abbreviations: aca, anterior commissure; DL, dorsolateral; VM, ventromedial.

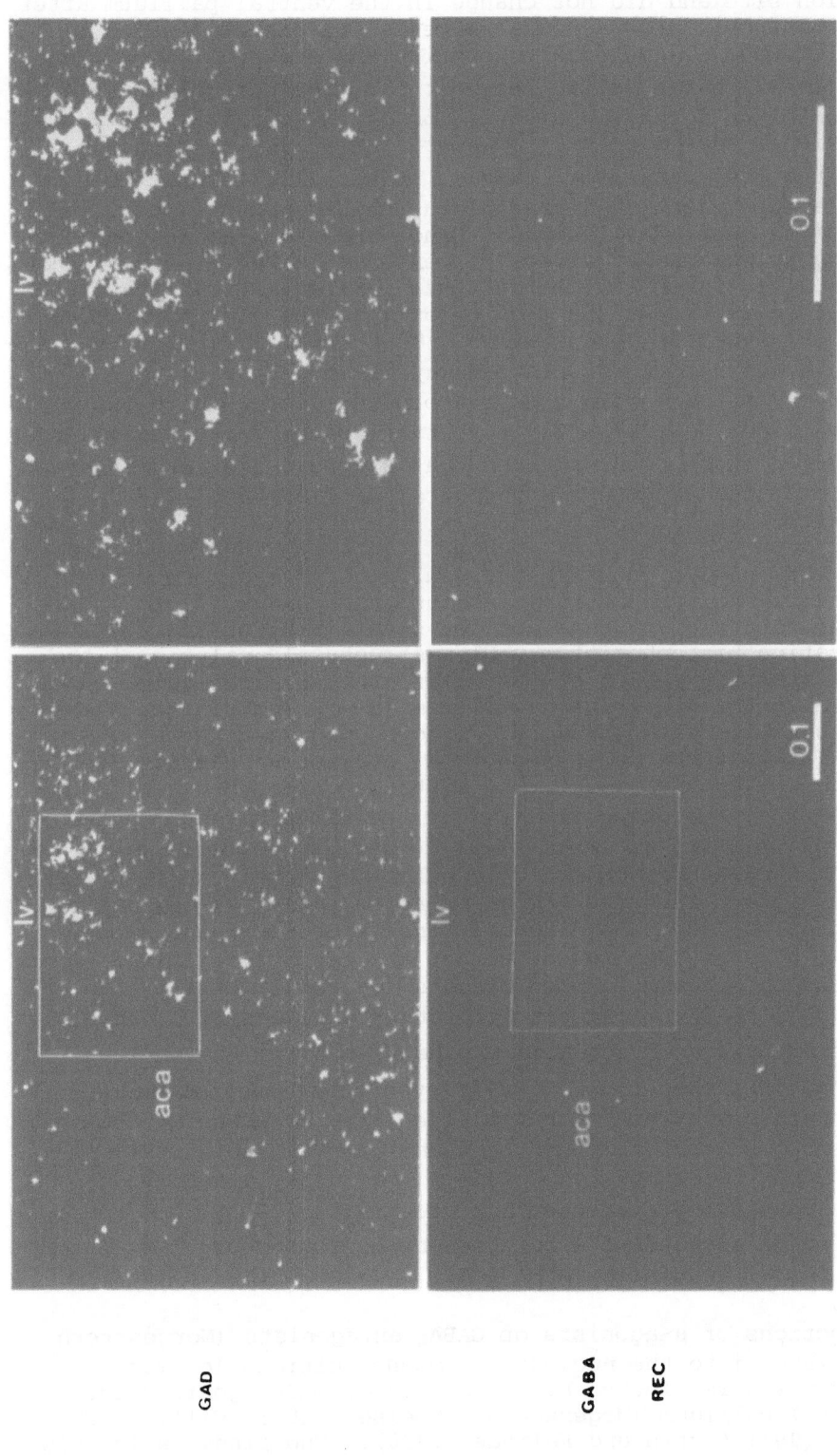

Fig. 8. *In situ* hybridization of probes for mRNAs of GAD and the α1 subunit of the GABA$_A$ receptor (GABA Rec) in the nucleus accumbens. The figures to the right are a higher magnification of the box in each figure. Bar = 0.1 mm. Abbreviations: lv, lateral ventricle; aca, anterior commissure. This figure represents part of a larger study by Churchill et al., in press.

In contrast to GABA$_A$ receptors, [^{125}I]DAGO binding to μ-opioid receptors decreased significantly in all of the compartments of the nucleus accumbens and ventral pallidum in the lesion site and did not change in the ventral pallidum after lesions destroying the accumbal enkephalinergic projection to this structure. These results confirm the observations of Pan et al. (1983) that μ-opioid receptors are not upregulated in the globus pallidus after the destruction of the striatal innervation 7 days earlier. However, these results do not confirm the observations of Waksman et al. (1987) that the μ-opioid receptor density in the globus pallidus decreased 7 and 21 days after striatal lesions. Since the striatal lesion in the study by Waksman et al. (1987) extends to the globus pallidus, the possibility exists that pallidal neurons have been destroyed by the spread of the lesion into the globus pallidus or by trans-synaptic degeneration. The possibility of trans-synaptic degeneration in this pathway is suggested by the observation that μ-opioid receptor binding decreased in the globus pallidus 8 weeks after striatal lesions (Abou-Khalil et al., 1984). These results argue that μ-opioid agonists injected into the nucleus accumbens and ventral pallidum act via a postsynaptic action.

The interpretation of these results depends on the capacity of quinolinic acid to destroy intrinsic neurons without affecting fibers of passage and the sensitivity of the *in situ* hybridization histochemistry. Ultrastructural analysis of the striatum after injection of 60 nmoles of quinolinic acid demonstrated that the presynaptic components were intact when the postsynaptic elements were destroyed (Schwarcz et al., 1983). The fact that some receptors remained unchanged after these lesions (Churchill et al., 1990) supports the specificity of the lesion. Negative results for the GABA$_A$ receptor probe for the α1 subunits in the nucleus accumbens cannot rule out the presence of GABA$_A$ receptors at or below the detection limit for this technique. The presence of GABA$_A$ receptors on neurons in the ventral pallidum and diagonal band of Broca in these same sections, however, demonstrates that the detection limit is sufficient to observe labeled neurons (Churchill et al., in press). Also GABA$_A$ receptors with other variants of α and β subunits as well as δ subunits may be present in the nucleus accumbens, (Shivers et al., 1989), even though the α1 and β2 subunit were not evident (Churchill et al., in press). However, the combination of both of these techniques to demonstrate the lack of GABA$_A$ receptors in neurons in the nucleus accumbens provides convincing evidence that presynaptic GABA$_A$ receptors containing α1 subunits are not present on the accumbal projection to the ventral pallidum.

Injections of μ-agonists or GABA$_A$ antagonists (Morgenstern et al., 1984) into the nucleus accumbens initiate locomotor activity which is blocked by injections of GABA agonists into the ventral pallidum (Mogenson and Nielsen, 1983; Williams and Herberg, 1987; Austin and Kalivas, 1990). The findings in this paper support the conclusion from behavioral experiments that μ-

(Pert and Sivit, 1977; Kalivas et al., 1983; Stinus et al., 1985) and ventral pallidum (Austin and Kalivas, 1990). These findings also show that GABA$_A$ receptors act postsynaptically in the ventral pallidum rather than by regulating presynaptic release of GABA. These conclusions are in contrast to the possible interpretation derived from two behavioral studies that suggest that GABA$_A$ agonists modulate presynaptic release of GABA in the ventral pallidum (Scheel-Kruger, 1984; Baud et al., 1988). In the study by Baud et al. (1988), higher doses of muscimol (50 ng compared with 2-10 ng in other studies) were used to initiate locomotor activity in the ventral pallidum. Swerdlow and Koob (1984) noted that at these higher doses, locomotor activity increased, whereas at the lower doses muscimol inhibited locomotor activity. Perhaps diffusion of the agonist to other nearby brain regions or a non-specific action of this drug is producing these results. In the Scheel-Kruger (1984) study, no numerical data are presented, so that the reliability of the finding is difficult to evaluate.

SUMMARY

In conclusion, GABA$_A$ receptors containing the α1 subunit are localized on postsynaptic neurons in the ventral pallidum, mainly in the dorsolateral compartment and on presynaptic terminals in the nucleus accumbens. μ-opioid receptors are localized on postsynaptic neurons in both the nucleus accumbens and ventral pallidum, and therefore may be regulating presynaptic release of enkephalin from the accumbens-pallidal projection. Discrete lesions in the dorsomedial core of the nucleus accumbens will upregulate GABA$_A$ receptors in the dorsolateral compartment of the nucleus accumbens in a fashion similar to the upregulation of GABA$_A$ receptors in the globus pallidus after striatal lesions. However, larger lesions of the lateral core projection to the dorsolateral compartment of the ventral pallidum do not upregulate the GABA$_A$ receptors, suggesting that the mechanisms for upregulation of GABA$_A$ receptors are specific to the dorsomedial core or a smaller lesion. The uniqueness of the compartments within the nucleus accumbens and the ventral pallidum are supported by these receptor and mRNA analyses.

REFERENCES

Abou-Khalil, B., Young, A. B., and Penney, J. B., 1984, Evidence for the presynaptic localization of opiate binding sites on striatal efferent fibers, Brain Res., 323:21-29.

Albert, D. J., Petrovic, D. M., and Walsh, M. L., 1989, Medial accumbens lesions attenuate testosterone-dependent aggression in male rats, Physiol. Behav., 46:625-631.

Austin, M. C., and Kalivas, P. W., 1988, The effect of cholinergic stimulation in the nucleus accumbens on locomotor behavior, Brain Res., 441:209-214.

Austin, M. C., and Kalivas, P. W., 1989, Blockade of enkephalinergic and GABAergic mediated locomotion in the nucleus accumbens by muscimol in the ventral pallidum, Jpn. J. Pharmacol., 50:487-490.

Austin, M. C., and Kalivas, P. W., 1990, Enkephalinergic and
    GABAergic modulation of motor activity in the ventral
    pallidum, _J. Pharmacol. Exp. Ther._, 252:1370-1377.

Baud, P., Mayo, W., LeMoal, M., and Simon, H., 1988, Locomotor
    hyperactivity in the rat after infusion of muscimol and [D-
    ala^2]met-enkephalin into the nucleus basalis
    magnocellularis. Possible interaction with cortical
    cholinergic projections, _Brain Res._, 452:203-211.

Chesselet, M.-F., Weiss, L., Wuenschell, C., Tobin, A. J., and
    Affolter, H.-U., 1987, Comparative distribution of mRNAs
    for glutamic acid decarboxylase, tyrosine hydroxylase and
    tachykinins in the basal ganglia: An _in situ_ hybridization
    study in the rodent brain, _J. Comp. Neurol._, 262:125-140.

Churchill, L., Dilts, R. P., and Kalivas, P. W., 1990, Changes
    in γ-aminobutyric acid, μ-opioid and neurotensin receptors
    in the accumbens-pallidal projection after discrete
    quinolinic acid lesions in the nucleus accumbens, _Brain
    Res._, 511:41-54.

Churchill, L., Bourdelais, A., Austin, M. C., Lolait, S.J.,
    Mahan, L.C., O'Carroll, A.-M., and Kalivas, P. W., in
    press, Lack of presynaptic GABA$_A$ receptors containing α1
    and β2 subunits on the projection from the nucleus
    accumbens to the ventral pallidum: Evidence derived from
    receptor autoradiography and _in situ_ hybridization,
    Synapse.

Conrad, L. C. A., and Pfaff, D. W., 1976, Autoradiographic
    tracing of nucleus accumbens efferents in the rat, _Brain
    Res._, 113:589-596.

Fallon, J. H., and Moore, R. Y., 1978, Catecholamine innervation
    of the basal forebrain. IV. Topography of the dopamine
    projection to the basal forebrain and neostriatum, _J. Comp.
    Neurol._, 180:545-580.

Foster, A. C., Collins, J. F., and Schwarcz, R., 1983, On the
    excitotoxic properties of quinolinic acid, 2,3-piperidine
    dicarboxylic acids and structurally related compounds,
    Neuropharmacology, 22:1331-1342.

Gerfen, C. R., Herkenham, M., and Thibault, J., 1987, The
    neostriatal mosaic: II. Patch- and matrix-directed
    mesostriatal dopaminergic and non-dopaminergic systems, _J.
    Neurosci._, 7:3915-3934.

Groenewegen, H. J., and Russchen, F. T., 1984, Organization of
    the efferent projections of the nucleus accumbens to
    pallidal, hypothalamic, and mesencephalic structures: A
    tracing and immunohistochemical study in the cat, _J. Comp.
    Neurol._, 223:347-367.

Groenewegen, H. J., Meredith, G. E., Berendse, H. W., Voorn, P.,
    and Wolters, J. G., 1989, The compartmental organization of
    the ventral striatum in the rat, _in_: "Neural Basis for
    Disorders in Movement," Grossman and Sambrook, eds., Libbey
    and Co., London, pp. 45-54.

Heimer, L., Zahm, D. S., Churchill, L., Kalivas, P. W., and
    Wohltman, C., 1990, Specificity in the projection pattern
    of accumbal core and shell, _Neuroscience_, in press.

Jones, D. L., and Mogenson, G. J., 1980, Nucleus accumbens to
    globus pallidus GABA projection: Electrophysiological and
    iontophoretic investigations, _Brain Res._, 188:93-105.

Kalivas, P. W., Widerlov, E., Stanley, D., Breese, G., and
    Prange, A. J. Jr., 1983, Enkephalin action on the
    mesolimbic system: A dopamine-dependent and a dopamine-
    independent increase in locomotor activity, J. Pharmacol.
    Exp. Ther., 227:229-237.

Kobayashi, Y., Kaufman, D.L. and Tobin, A.J., 1987, Glutamic
    acid decarboxylase cDNA: nucleotide sequence encoding an
    enzymatically active fusion protein, J. Neurosci., 7:2768-
    2772.

Khrestchatisky, M., MacLennan, A.J., Chiang, M.-Y., Xu, W.,
    Jackson, M.B., Brecha, N., Sternini, C., Olsen, R.W. and
    Tobin, A.J., 1989, A novel α subunit in rat brain GABA$_A$
    receptors, Neuron, 3:745-753.

Levitan, E.S., Schofield, P.R., Burt, D.R., Rhee, L.M., Wisden,
    W., Kohler, M., Fujita, N., Rodriquez, H.F., Stephenson,
    A., Darlison, M.G., Barnard, E.A. and Seeburg, P.H., 1988,
    Structural and functional basis for GABA$_A$ receptor
    heterogeneity, Nature 335:76-79.

Lolait, S.J., O'Carroll, A.-M., Kusano, K., Muller, J.M.,
    Brownstein, M.J. and Mahan, L.C., 1989a, Cloning and
    expression of a novel GABA$_A$ receptor, FEBS Letters,
    246:145-148.

Lolait, S.J., O'Carroll, A.-M., Kusano, K. and Mahan, L.C.,
    1989b, Pharmacological characterization and region specific
    expression in brain of β2 and β3 subunits of the rat GABA$_A$
    receptor, FEBS Letters, 258:17-21.

Malherbe, P., Sigel, E., Baur, R., Persohn, E., Richards, J.G.
    and Mohler, H., 1990, Functional expression and sites of
    gene transcription of a novel α subunit of the GABA$_A$
    receptor in rat brain, FEBS Letters 260:261-265.

Mogenson, G. J., and Nielsen, M. A., 1983, Evidence that an
    accumbens to subpallidal GABAergic projection contributes
    to locomotor activity, Brain Res. Bull., 11:309-314.

Mogenson, G. J., Jones, D. L., and Yim, C. Y., 1980, From
    motivation to action: Functional interface between the
    limbic system and the motor system, Prog. Neurobiol.,
    14:69-97.

Mogenson, G. J., Swanson, L. W., and Wu, M., 1983, Neural
    projections from nucleus accumbens to globus pallidus,
    substantia innominata, and lateral preoptic-lateral
    hypothalamic area: An anatomical and electrophysiological
    investigation in the rat, J. Neurosci., 3:189-202.

Montpied, P., Martin, B.M., Cottingham, S.L., Stubblefield,
    B.K., Ginns, E.I., and Paul, S.M., 1988, Regional
    distribution of the GABA$_A$/benzodiazepine receptor (α
    subunit) mRNA in rat brain, J. Neurochem., 51:1651-1654.

Morgenstern, R., Mende, T., Gold, R., Lemme, P., and Oelssner,
    W., 1984, Drug-induced modulation of locomotor
    hyperactivity induced by picrotoxin in nucleus accumbens,
    Pharmacol. Biochem. Behav., 21:501-506.

Nauta, W. J. H., Smith, G. P., Faull, R. L. M., and Domesick, V.
    B., 1978, Efferent connections and nigral afferents of the
    nucleus accumbens septi in the rat, Neuroscience, 3:385-
    401.

Pan, H. S., Frey, K. A., Young, A. B., and Penney, J. B. Jr., 1983, Changes in [^{3}H]muscimol binding in substantia nigra, entopeduncular nucleus, globus pallidus, and thalamus after striatal lesions as demonstrated by quantitative receptor autoradiography, J. Neurosci., 3:1189-1198.

Paxinos, G., and Watson, C., 1986, "The Rat Brain in Stereotaxic Coordinates," 2nd edition, Academic Press, Orlando, FL.

Pert, A., and Sivit, C., 1977, Neuroanatomical focus for morphine and enkephalin-induced hypermotility, Nature, 265:645-647.

Russell, V. A., Allin, R., Lamm, M. C. L., and Taljaard, J. J. F., 1989, Increased dopamine D$_2$ receptor-mediated inhibition of [^{14}C]acetylcholine release in the dorsomedial part of the nucleus accumbens, Neurochem. Res., 14:877-881.

Scheel-Kruger, J., 1984, On the role of GABA for striatal functions. Interaction between GABA and enkephalin in the pallidal systems, Neuropharmacology, 23:867-868.

Schofield, P.R., 1989, The GABA$_A$ receptor: molecular biology reveals a complex picture. Trends in Pharmaceut. Sci., 10: 476-478.

Schwarcz, R., Whetsell, W. O. Jr., and Mangano, R. M., 1983, Quinolinic acid: An endogenous metabolite that produces axon-sparing lesions in rat brain, Science, 219:316-318.

Sequier, J.M, Richards, J.G., Malherbe, P., Price, G.W., Mathews, S., and Mohler, H., 1988, Mapping of brain areas containing RNA homologous to cDNAs encoding the α and β subunits of the rat GABA$_A$ γ-aminobutyrate receptor, Proc. Natl. Acad. Sci., 85:7815-7819.

Shivers, B.D., Killisch, I., Sprengel, R., Sontheimer, H., Kohler, M., Schofield, P.R., and Seeburg, P.H., 1989, Two novel GABA$_A$ receptor subunits exist in distinct neuronal subpopulations, Neuron, 3:327-337.

Stinus, L., Winnock, M., and Kelley, A.E., 1985, Chronic neuroleptic treatment and mesolimbic dopamine denervation induce behavioral supersensitivity to opiates, Psychopharmacology, 85: 323-328.

Swerdlow, N. R., and Koob, G. F., 1984, The neural substrates of apomorphine-stimulated locomotor activity following denervation of the nucleus accumbens, Life Sci., 35:2537-2544.

Swerdlow, N. R., Swanson, L. W., and Koob, G. F., 1984, Substantia innominata: Critical link in the behavioral expression of mesolimbic dopamine stimulation in the rat, Neurosci. Lett., 50:19-24.

Vaccarino, F. J., and Rankin, J., 1989, Nucleus accumbens cholecystokinin (CCK) can either attenuate or potentiate amphetamine-induced locomotor activity: Evidence for rostral-caudal differences in accumbens CCK function, Behav. Neurosci., 103:831-836.

Voorn, P., Gerfen, C. R., and Groenewegen, H. J., 1989, Compartmental organization of the ventral striatum of the rat: Immunohistochemical distribution of enkephalin, substance P, dopamine and calcium binding protein, J. Comp. Neurol., 289:189-201.

Waksman, G., Hamel, E., Delay-Goyet, P., and Roques, B. P., 1987, Neutral endopeptidase-24.11, μ and δ-opioid receptors after selective brain lesions: An autoradiographic study, Brain Res., 436:205-216.

Walaas, I., and Fonnum, F., 1979, The distribution and origin of glutamate decarboxylase and choline acetyltransferase in ventral pallidum and other basal forebrain regions, <u>Brain Res</u>., 177:325-336.

Williams, S. F., and Herberg, L. J., 1987, Motivational vs. motor effects of striatal and pallidal GABAergic projections to subthalamic and entopeduncular nuclei, ventromedial thalamus and ventral globus pallidus, <u>Pharmacol. Biochem. Behav</u>., 26:49-55.

Yoshikawa, T., Fukamauchi, F., Shibuya, H., and Takahashi, R., 1989, Regional heterogeneity with the nucleus accumbens concerning the effects of dopaminergic agents on the content of cholecystokinin, <u>Neurochem. Int</u>., 14:467-469.

Young, W. S. III, 1989, *In situ* hybridization histochemical detection of neuropeptide mRNA using DNA and RNA probes, <u>Meth. in Enzymol.</u>, 168:702-710.

Zaborszky, L., Alheid, G. F., and Heimer, L., 1985, Mapping of transmitter-specific connections: Simultaneous demonstration of anterograde degeneration and changes in the immunostaining pattern induced by lesions, <u>J. Neurosci. Methods</u>, 14:255-266.

Zahm, D. S., 1989, The ventral striatopallidal parts of the basal ganglia in the rat. II. Compartmentation of ventral pallidal efferents. <u>Neuroscience</u>, 30:33-50.

Zahm, D. S., and Heimer, L., 1988, Ventral striatopallidal parts of the basal ganglia in the rat: I. Neurochemical compartmentation as reflected by the distributions of neurotensin and substance P immunoreactivity, <u>J. Comp. Neurol</u>., 272:516-535.

Zahm, D. S., and Heimer, L., 1990, Two transpallidal pathways originating in nucleus accumbens, <u>J. Comp. Neurol</u>., 302:437-446.

Zahm, D. S., Zaborszky, L., Alones, V. E., and Heimer, L., 1985, Evidence for the coexistence of glutamate decarboxylase and met-enkephalin immunoreactivities in axon terminals of rat ventral pallidum, <u>Brain Res</u>., 325:317-321.

CALCIUM-BINDING PROTEIN (CALBINDIN D-28K) IMMUNOREACTIVE NEURONS IN THE

BASAL FOREBRAIN OF THE MONKEY AND THE RAT: RELATIONSHIP WITH THE

CHOLINERGIC NEURONS

Howard T. Chang and Hui Kuo

Department of Anatomy and Neurobiology
The University of Tennessee, Memphis College of Medicine
875 Monroe Ave., Memphis, Tennessee 38163

INTRODUCTION

A previous study has shown that many neurons in the monkey nucleus basalis of Meynert (NBM) are immunoreactive for calbindin-D-28k (CaBP), a Vitamin D-dependent calcium binding protein (Celio and Norman, 1985). More recently, it has been shown that many cholinergic neurons (i.e., immunoreactive for choline acetyltransferase, ChAT) in the human NBM are immunoreactive for CaBP, and that these neurons are adversely affected in Alzheimer's patients (Ichimiya et al., 1989), thus suggesting that CaBP may play an important role in the normal functions of human NBM cholinergic neurons. However, whether cholinergic NBM neurons in other species also express CaBP immunoreactivity has remained unclear. It is also not clear whether all cholinergic neurons in the primate NBM are immunoreactive for CaBP. We report here that most, but not all, of the cholinergic NBM neurons in the Rhesus monkey are immunoreactive for CaBP. On the other hand, none of the rat cholinergic NBM neurons express CaBP immunoreactivity.

MATERIALS AND METHODS

Female young adult (4-9 yrs.) Rhesus monkey (Macaca mulatta) brains were obtained from normal donors used in pancreatic transplant experiments unrelated to this study. After exsanguination, the monkeys were perfused with a fixative solution consisted of 4% paraformaldehyde, 0.5% glutaraldehyde and 0.2% picric acid in 0.1 M phosphate buffer (pH 7.4). Normal adult (200 - 400 g) male and female Sprague-Dawley rats were also fixed by perfusion with the same fixative solution. Parasagittal, horizontal, or frontal 50 µm thick sections were cut on a Vibratome® and prepared for double-labeling immunocytochemistry as described previously (Chang, 1988). A rabbit antiserum raised against human placental choline acetyltransferase (ChAT) (Chemicon Inc.) and a mouse monoclonal antibody raised against CaBP (a gift from Dr. M.R. Celio) were used in double-labeling immunofluorescence reactions to compare the distribution of CaBP neurons with that of the ChAT neurons. The specificities of these primary antibodies have been described previously (Bruce et al., 1985; Celio et al., 1988; German et al., 1985). Fluorescene isothiocyanate (FITC) labeled donkey anti-mouse IgG and Texas Red® labeled donkey anti-rabbit IgG (both from Jackson Lab.) were used to detect CaBP and ChAT immunoreactive neurons, respectively. For method specificity control, adjacent sets of tissue sections were incubated for demonstration of other antigens using the same secondary antibodies, each displayed specific patterns of labeling as reported in previous studies. The immunofluorescently labeled neurons were examined on an Olympus BH-2 microscope and recorded on Kodak T-Max-400

The Basal Forebrain, Edited by T.C. Napier *et al.*
Plenum Press, New York, 1991

films. In order to capture both CaBP and ChAT immunoreactive neurons within the same black-&-white micrograph, some rat brain sections were further reacted with mouse peroxidase-anti-peroxidase (PAP) complex to label the FITC-labeled CaBP immunoreactive neurons with immunoperoxidase reaction products (Chang, '88) (Fig. 4).

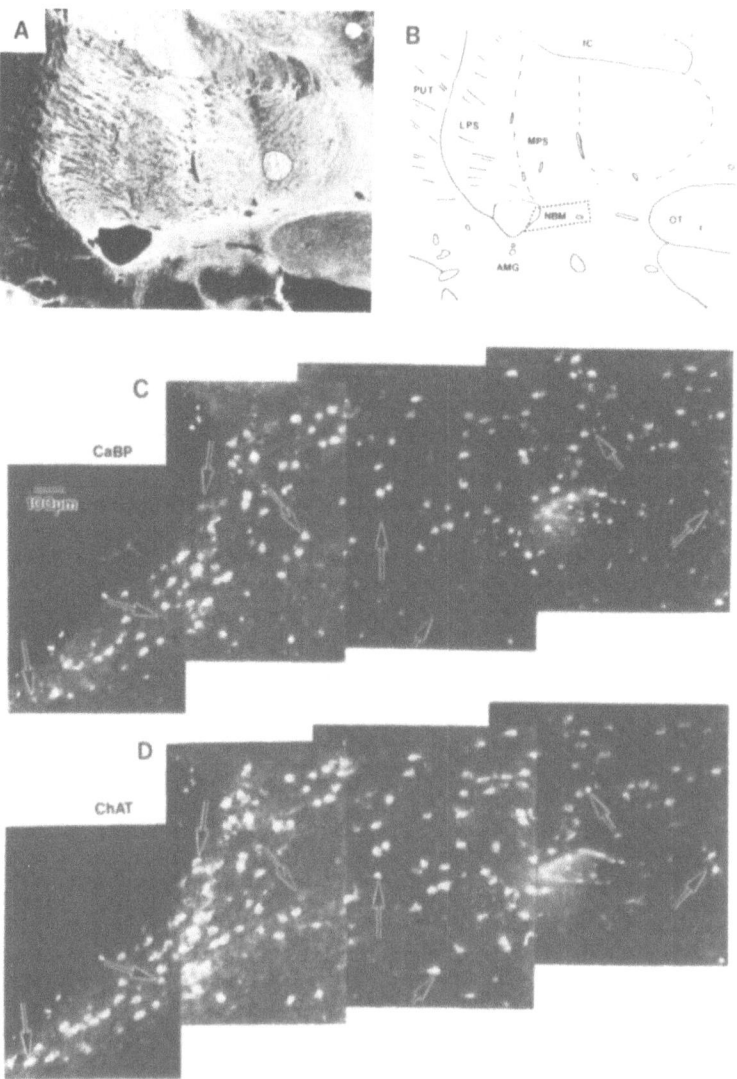

Fig. 1. A low magnification dark-field micrograph and a drawing showing the location of the monkey NBM in a frontal section are shown in **A** and **B**, respectively. AMG amygdala; IC internal capsule; LPS lateral pallidal segment; MPS medial pallidal segment; OT optic tract; PUT putamen. NBM neurons immunoreactive for CaBP and ChAT are shown in **C** and **D**, respectively. Arrows point to some of the ChAT neurons not immunoreactive for CaBP. A double-headed arrow points to a neuron immunoreactive only for CaBP. Note that most but not all of the ChAT neurons are immunoreactive for CaBP.

Somatic cross-sectional areas of labeled neurons were measured by tracing the enlarged negative images using a macroprojector (Documator, Aus Jena), and then fed into a Macintosh Plus® (Apple Computer) computer with a MacTablet® (Summagraphics) using the MacMeasure® program written by Wayne Rasband of the National Institute of Mental Health.

In this study, the NBM neurons in the monkey are defined as the cholinergic neurons intercalated within both the external and the internal medullary lamina, as well as the dense cluster of cholinergic neurons located in the substantia innominata ventral to the globus pallidus and extending rostrally under the anterior commissure (Mesulam et al., 1984). The rat NBM consists of the Ch4 group of cholinergic neurons (Mesulam et al., 1983) located in the sublenticular part of the substantia innominata, caudal and ventral to the globus pallidus and the ventral pallidum.

RESULTS

In agreement with previous observations made in primates (Celio and Norman, 1985; Ichimiya et al., 1989), many neurons located in the Rhesus monkey NBM are immunoreactive for CaBP. Analysis of the double-labeled materials indicated that virtually all of the CaBP immunoreactive neurons in the monkey NBM are cholinergic (i.e., immunoreactive for ChAT) (Fig. 1, a double-headed arrow points to an exception to this generalization). The converse, however, is not true: up to 20% of the cholinergic neurons in the monkey NBM are not immunoreactive for CaBP (arrows in Fig. 1C and D).

In contrast to the monkey, none of the ChAT immunoreactive neurons in the rat NBM are immunoreactive for CaBP (Fig. 2, 3, 4). Only a few CaBP immunoreactive neurons were found next to the ChAT immunoreactive neurons in the rat NBM. Relative to the majority of the ChAT neurons, the CaBP immunoreactive neurons seemed to be preferentially located in a more caudal, lateral, and ventral part of the rat substantia innominata (Fig. 2, 3, 4). More caudally, the number of CaBP immunoreactive neurons increased, and appeared to be contiguous with the CaBP immunoreactive neurons in the neighboring amygdaloid complex. A comparison of the somatic cross-sectional areas indicates that CaBP immunoreactive neurons are smaller than the ChAT immunoreactive neurons (Fig. 5). It is interesting to note that a band of axon terminals immunoreactive for CaBP is found in the caudal globus pallidus, the ventral pallidum and the sublenticular substantia innominata, partially overlapping the cholinergic neurons.

DISCUSSION

In order to investigate the role of CaBP in the function of cholinergic NBM neurons in other species, it is important to first determine the relationship between the CaBP immunoreactive neurons and the cholinergic neurons. Our present results not only confirm previous observations that many CaBP immunoreactive neurons are found in the monkey NBM (Celio and Norman, 1985), but also show that most of these CaBP immunoreactive neurons are indeed cholinergic neurons. This agrees well with the observation made in human NBM (Ichimiya et al., 1989) in which most cholinergic neurons are also immunoreactive for CaBP. In this study, most but not all of the monkey cholinergic NBM neurons showed CaBP immunoreactivity. Whether the lack of expression of CaBP immunoreactivity reflects a diseased state of the NBM cholinergic neurons as found in Alzheimer's patients (Ichimiya et al., 1989) remains to be determined.

Although previous single-labeling immunocytochemical studies (Celio, 1990) have reported the presence of CaBP immunoreactive neurons in the rat NBM, their cholinergic nature has remained unclear. The surprising finding in this study is that, in contrast to the monkey NBM, the rat NBM cholinergic neurons are not immunoreactive for CaBP. The rat CaBP immunoreactive neurons in the NBM region are fewer in number and smaller in size than the ChAT neurons. As these CaBP immunoreactive neurons appeared to be contiguous with those in the neighboring amygdaloid complex, it is likely that these CaBP immunoreactive neurons in the rat NBM region

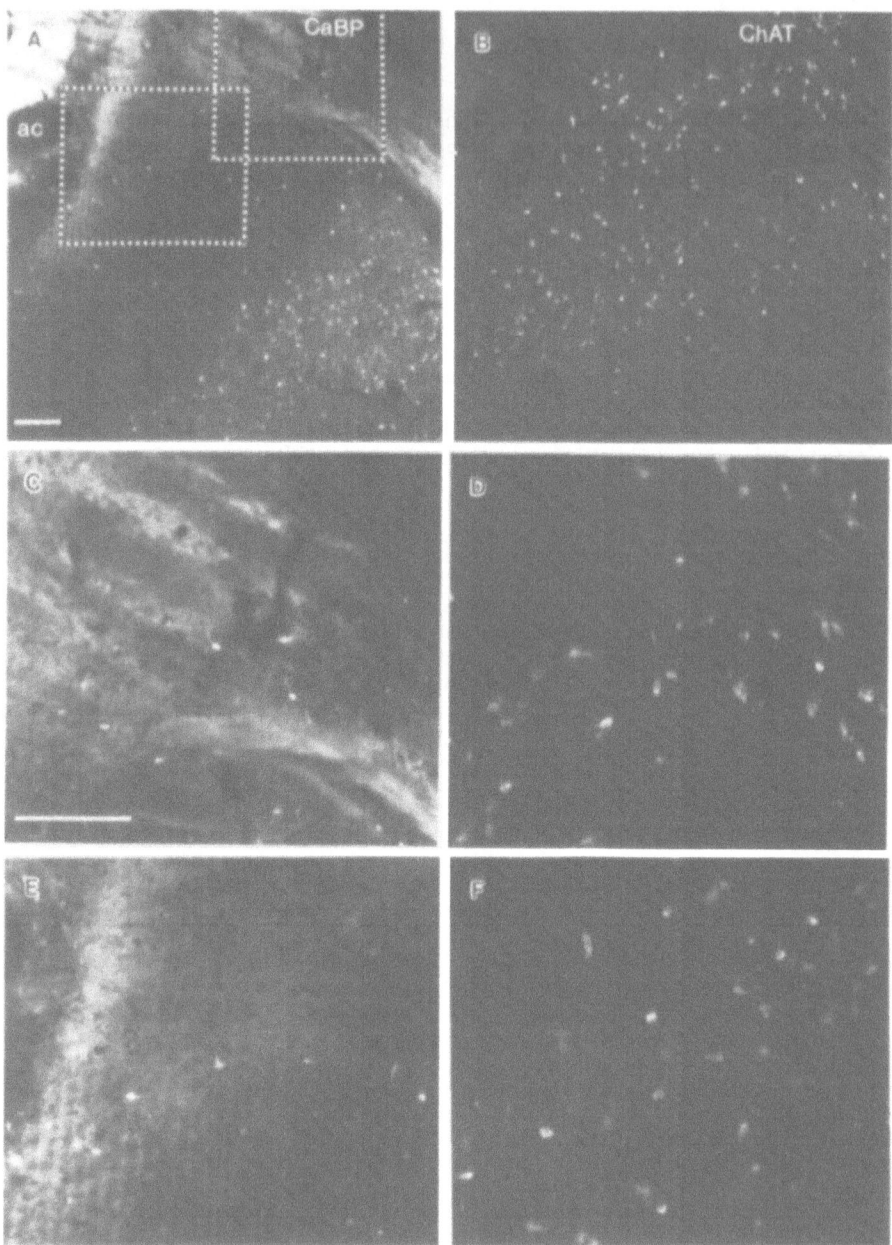

Fig. 2. In a parasagittal section of the rat NBM, neurons immunoreactive
for CaBP, and those immunoreactive for ChAT are shown in **A** and **B**,
respectively. ac anterior commissure. Boxed areas in **A** are
shown at higher magnifications in **C** and **E**. Identical areas
showing the ChAT neurons are shown in **D** and **F**, respectively.
Note that none of the ChAT neurons are immunoreactive for CaBP.
Both scale bars (in **A** and **C**) are 200 μm.

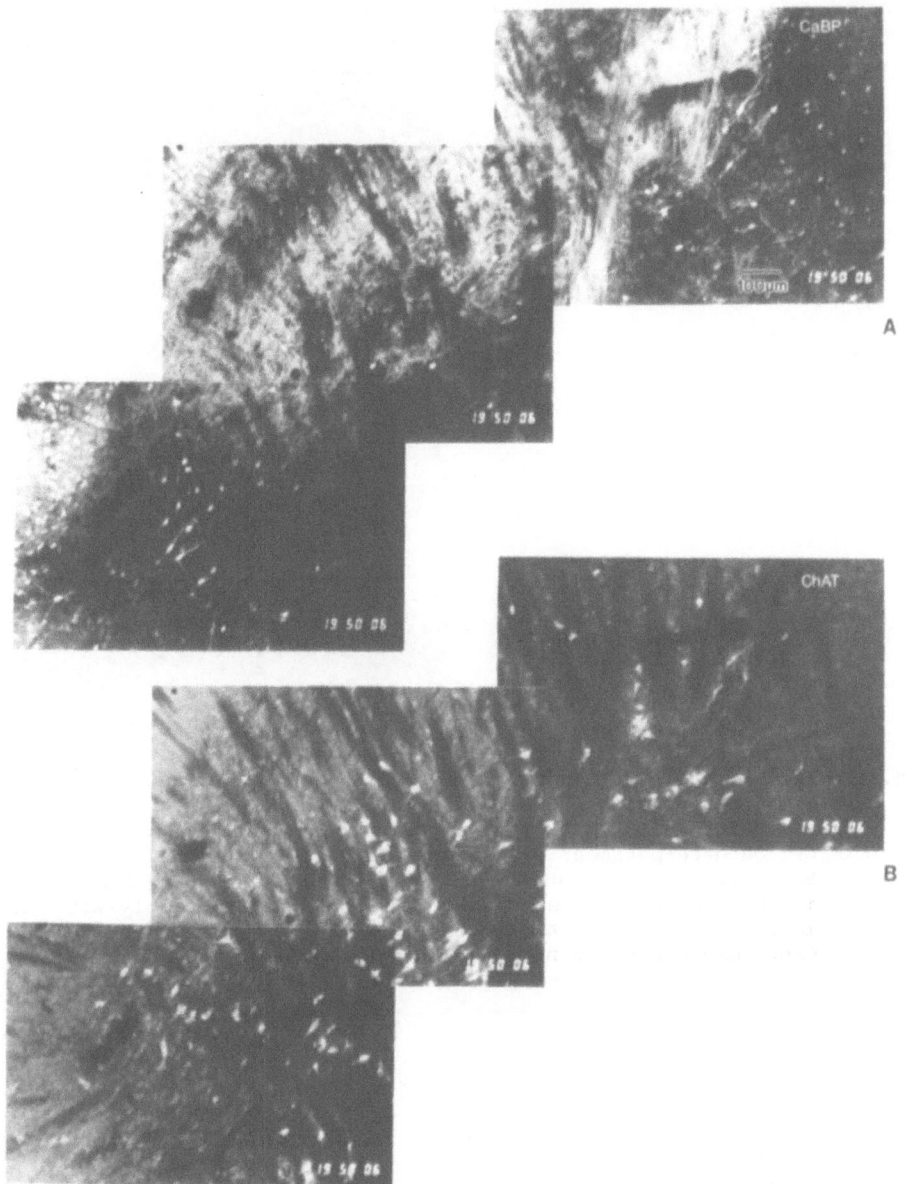

Fig. 3. Neurons immunoreactive for CaBP, and those immunoreactive for ChAT
are shown in **A** and **B**, respectively in a horizontal section of the
rat NBM region (Animal # 195006). Rostral and lateral directions
are to the top and to the left of the figure, respectively. Note
that the distribution of cholinergic neurons is partially
overlapped by CaBP immunoreactive axons and terminals.

actually represent neurons of the "extended amygdala" which bridges the bed
nucleus of the stria terminalis and the centro-medial amygdala through the
sublenticular substantia innominata (Alheid and Heimer, 1988). Results
from our preliminary studies indicate that none of the CaBP immunoreactive
neurons in the rat NBM region could be retrogradely labeled with tracers
deposited in the cerebral cortex.

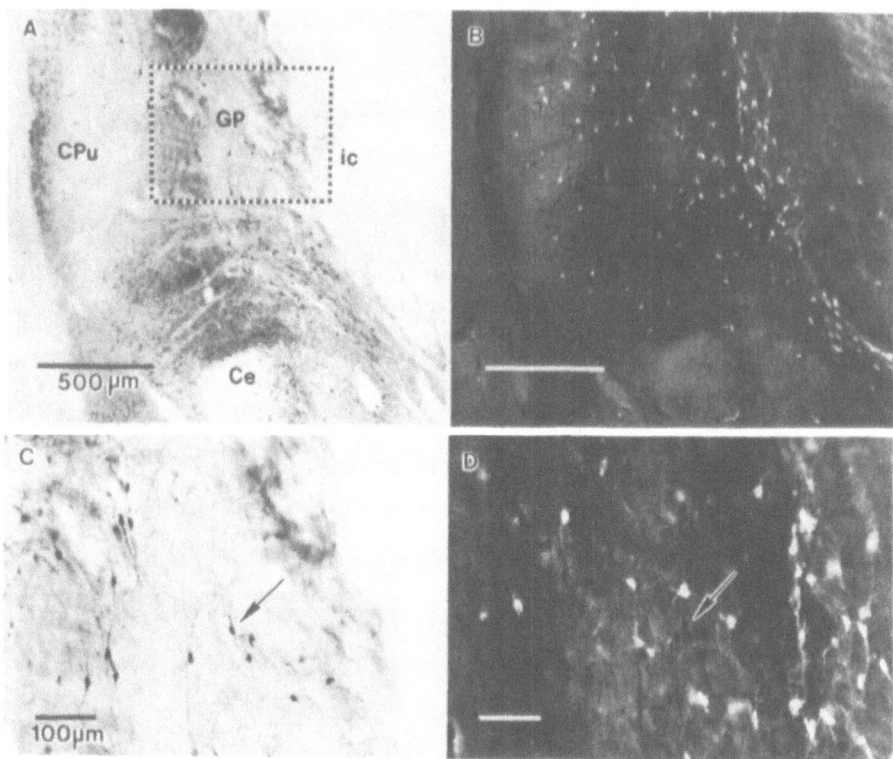

Fig. 4. In a frontal section of the rat basal forebrain, neurons
immunoreactive for CaBP, and those immunoreactive for ChAT are
shown in **A** and **B**, respectively. Ce central amygdaloid nucleus;
CPu caudate-putamen; GP globus pallidus; ic internal capsule.
Boxed area in **A** is shown in higher magnification in **C** and **D**. The
immunoperoxidase labeled CaBP neurons contrast sharply with the
Texas Red-labeled ChAT neurons in **D**. Arrow points to the same
CaBP immunoreactive neuron within both **C** and **D**.

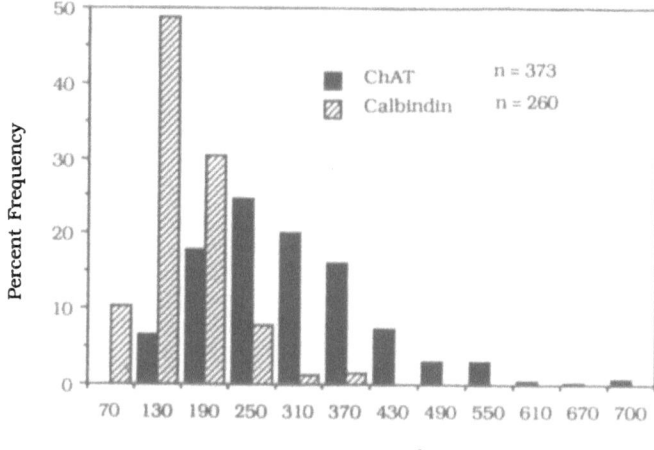

Fig. 5. Histogram showing the distribution of the somatic cross-sectional
areas of CaBP immunoreactive neurons and those of the ChAT
immunoreactive neurons in the rat NBM region.

In this study, CaBP immunoreactive axons and terminals are observed to partially overlap the cholinergic NBM neurons in the rat. Thus, the rat cholinergic NBM neurons, instead of being the origin of CaBP-immunoreactive fibers (as the primate cholinergic NBM neurons), may in fact be postsynaptic to axons containing CaBP immunoreactivity. These CaBP immunoreactive axons may correspond to the striopallidal fibers arising from the medium sized striatal projection neurons many of which have been shown to be CaBP immunoreactive (Gerfen et al., 1985), or from the CaBP immunoreactive neurons of the "extended amygdala" complex (Alheid and Heimer, 1988; Celio, 1990). The difference in the expression of CaBP immunoreactivity in the rat and monkey cholinergic NBM neurons suggests that Vitamin D-dependent calcium homeostasis may have different roles in the primate and the rodent basal forebrain cholinergic functions.

Acknowledgement

We thank Dr. Marco R. Celio for his generous gift of antibodies to CaBP, and also Kelly Bennie for her skillful technical assistance. This study was supported by USPHS Grant AG05944, Biomedical Research Support Grant RR05423, a grant from the Alzheimer's Disease and Related Disorders Association, and a predoctoral fellowship from the Neuroscience Center of Excellence of The University of Tennessee, Memphis.

REFERENCES

Alheid, G. F. and Heimer, L., 1988, New perspectives in basal forebrain organization of special relevance for neuropsychiatric disorders: The striatopallidal, amygdaloid and corticopetal components of substantia innominata, Neurosci. 27:1-39.

Bruce, G., Wainer, B. H. and Hersh, L. B., 1985, Immuno-affinity purification of human choline acetyltransferase; comparison of the brain and placental enzymes, J. Neurochem. 45:611-620.

Celio, M. R., 1990, Calbindin D-28k and parvalbumin in the rat nervous system, Neurosci. 35:375-475.

Celio, M. R., Baier, W., Schärer, L., De Viragh, P. A. and Gerday, C., 1988, Monoclonal antibodies directed against the calcium binding protein parvalbumin, Cell Calcium. 9:81-86.

Celio, M. R. and Norman, A. W., 1985, Nucleus basalis of Meynert neurons contain the vitamin D-induced calcium-binding protein (Calbindin-D-28K), Anat. Embryol. 173:143-148.

Chang, H. T., 1988, Dopamine - acetylcholine interaction in the striatum: A dual-labeling immunocytochemical study of tyrosine hydroxylase and choline acetyltransferase positive elements in the rat, Brain Res. Bull. 21:295-304.

Gerfen, C. R., Baimbridge, K. G. and Miller, J. J., 1985, The neostriatal mosaic: Compartmental distribution of calcium-binding protein and parvalbumin in the basal ganglia of the rat and monkey, Proc. Nat. Acad. Sci. USA. 82:8780-8784.

German, D. C., Bruce, G. and Hersh, L. B., 1985, Immunohistochemical staining of cholinergic neurons in the human brain using a polyclonal antibody to human choline acetyltransferase, Neurosci. Lett. 61:1-5.

Ichimiya, Y., Emson, P. C., Mountjoy, C. Q., Lawson, D. E. M. and Iizuka, R., 1989, Calbindin-immunoreactive cholinergic neurons in the nucleus basalis of Meynert in Alzheimer-type dementia, Brain Res. 499:402-406.

Mesulam, M.-M., Mufson, E. J., Wainer, B. H. and Levey, A. I., 1983,
    Central cholinergic pathways in the rat: An overview based on an
    alternative nomenclature (Ch1-Ch6), Neurosci. 10:1185-1201.

Mesulam, M.-M., Mufson, E. J., Levey, A. I. and Wainer, B. H., 1984, Atlas
    of cholinergic neurons in the forebrain and upper brainstem of the
    Macaque based on monoclonal choline acetyltransferase
    immunohistochemistry and acetylcholinesterase histochemistry,
    Neurosci. 12:669-686.

GLUTAMATE-LIKE IMMUNOREACTIVITY IS PRESENT WITHIN CHOLINERGIC
NEURONS OF THE LATERODORSAL TEGMENTAL AND PEDUNCULOPONTINE
NUCLEI

J.R. Clements[1], D.D. Toth, D.A. Highfield
and S.J. Grant

School of Life and Health Sciences and
Department of Psychology, University of
Delaware, Newark, DE 19716
[1]Current address: Dept. Vet. Anatomy, Texas
A&M University, College Station, TX 77843

INTRODUCTION

For the last decade the functional organization of
cholinergic neurons has dominated studies of the basal
forebrain. Cholinergic neurons in the brain, exclusive of motor
neurons and interneurons, are found in two spatially separate
groups (Armstrong et al., 1983, Mesulam et al., 1984). The
rostral group, located in the basal forebrain, has received
substantial attention because of its corticopedal projections
(Mesulam et al., 1984) and its' degeneration in Alzheimer's
disease (Coyle et al., 1983). The caudal group is found in the
laterodorsal tegmental nucleus (LDT) and pedunculopontine
nucleus (PPT) within the pontine tegmentum (Vincent et al.,
1983; Mesulam et al., 1984; Satoh and Fibiger, 1986), and is the
source of cholinergic innervation to the basal forebrain,
thalamus and brainstem (Sofroniew et al., 1985; Hallenger et
al., 1987; Maley et al., 1988; Rye et al., 1988; Jones, 1990).

There are clear parallels between the organization of these
cholinergic neurons and monoamine cell groups (Cooper, Bloom,
and Roth, 1986; Saper, 1987). Like the amine neurons,
cholinergic neurons form diffuse, long-axoned pathways which
arise from a restricted number of subcortical cell clusters. But
cholinergic and aminergic systems differ with respect to the
neurochemical homogeneity of these clusters. In general, all of
the neurons in nuclei such as the locus coeruleus, dorsal raphe
and substantia nigra (pars compacta) use a monoamine transmitter
and share similar inputs and outputs. In contrast, the rostral
cholinergic neurons are only one component of the basal
forebrain. One consequence of this is that the rostral
cholinergic neurons are distributed among a variety of non-
cholinergic neurons. It is now recognized, for example, that
the hippocampal and corticopedal projections from the basal
forebrain consist of GABAergic as well as cholinergic neurons
(Brashear et al., 1986; Fisher et al., 1988; Fruend and Antal,
1988).

Whether the mesopontine nuclei that contain the caudal cholinergic cell groups also contain non-cholinergic neurons remains unresolved. Neurons using amine neurotransmitters are found in the same pontine regions as the cholinergic neurons; but, at least in the rat, the amine containing neurons are spatially segregated from the cholinergic neurons in specific nuclei such as the locus coeruleus, raphe, and substantia nigra (Rye et al., 1987; Sutin and Jacobowitz, 1988). Nor do the LDT and PPT appear to contain GABAergic neurons (Kiosaka et al., 1988; Sutin and Jacobowitz, 1988). A variety of neuropeptides have been found in LDT and PPT neurons, as in the basal forebrain, including substance P, corticotropin releasing factor, atrial or brain natriuretic peptides, somatostatin, bombesin/gastrin, and opioid peptides (Matsuzaki et al., 1981; Vincent and Satoh, 1984; Crawley et al., 1985; Standaert et al., 1986; Vincent et al., 1986; Sutin and Jacobowitz, 1988). In the LDT and PPT however, peptides are generally co-localized with markers for cholinergic neurons (Crawley et al., 1985; Standaert et al., 1986; Vincent et al., 1986). Non-cholinergic peptide containing neurons are present in these regions, but again they are located in adjacent nuclei such as Barrington's nucleus and the dorsal and ventral tegmental nuclei of Gudden which are almost devoid of cholinergic neurons (Vincent and Satoh, 1984; Yamano and Tohyama, 1987; Sutin and Jacobowitz, 1988). Based on cytoarchitectonic, connectional, and neurochemical evidence, it has been further maintained that cholinergic neurons of the PPT can be differentiated from a medially adjacent non-cholinergic 'midbrain extrapyramidal area' (Rye et al., 1987).

Although direct evidence has been lacking, it has been suggested that excitatory amino acid neurotransmitters are present in these regions as well (Steriade and McCarley, 1990). We therefore decided to examine the relationship between brainstem cholinergic neurons and neurons exhibiting glutamate-like immunoreactivity. Because previous studies have shown that only cholinergic neurons in the pontine tegmentum (but interestingly not in the basal forebrain), stain deeply for Nicotinamide Adenine Dinucleotide Phosphate diaphorase (NADPH-d), counterstaining with NADPH-d histochemistry was used to compare distribution of cholinergic and glutamate-like immunoreactive (GLI) neurons within the same region (Vincent et al., 1983; Vincent et al., 1986). Portions of these studies have been previously reported (Clements and Grant, 1990a, b).

METHODS
Eight adult male Spraque Drawley rats were anesthetized with choral hydrate. Blood was cleared from the brain by perfusing 100 ml of warm 0.9% saline (100 ml) containing 200 I.U. of heparin and 0.2% sodium nitrite (Sigma Chemical Co., St. Louis, MO) through the ascending aorta. The brain from each animal was fixed with 500 ml of cold (4°C) 4% paraformaldehyde in 0.1M phosphate buffer (pH 7.4). After blocking, brains were post-fixed in 4% paraformaldehyde for 6 hours at 4°C. Brains were stored either in phosphate buffer prior to sectioning on a vibrating microtome, or were immersed in 30% sucrose overnight and then sectioned on a freezing microtome. Alternating sections from each brain were processed for glutamate-like immunoreactivity, NADPH-d histochemistry, and both glutamate immunohistochemistry and NADPH-d.

Sections for glutamate immunohistochemistry were incubated

in a 1:75,000 dilution of an anti-glutamate monoclonal antibody (Glu-2) in 10 mM phosphate buffered saline (PBS, pH 7.5) for 12 hours. This antibody has been described and characterized previously (McDonald et al., 1989). Following 3 five minute rinses in PBS, sections were incubated sequentially in a goat antimouse biotinylated secondary antibody for four hours and an avidin-biotin peroxidase complex for four hours. The secondary antibody and ABC complex were from a Vectastain Elite goat anti mouse kit (Vector Laboratories Inc., Burlingame, CA). Glutamate-like immunoreactivity in the tissue sections was visualized by incubating the sections, on ice, in a solution of 10 mg DAB and 0.01% hydrogen peroxide in 20 ml of 0.1M tris buffer. All immunocytochemistry was done at 4°C, and sections were thoroughly rinsed with PBS between each incubation. All tissue sections from both the immunocytochemical and histochemical protocols were rinsed in PBS and mounted on gelatin coated slides. The diaminobenzidene reaction product within the GLI neurons was intensified in a solution of reduced 1% osmium tetroxide for one hour. Control tissue sections, omitting either the primary or secondary antibody, or pre-adsorbing the primary antibody with the glutamate-BSA conjugate, were also processed.

Every second section was histochemically labeled for NADPH-d, and every third section was double-labeled for NADPH-d and glutamate-like immunoreactivity. The NADPH-d histochemical protocol used was that of Scherer-Singler, modified by the addition of 0.025% $MgCl_2$ (Scherer-Singler et al., 1983). Tissue sections that were double-labeled for both NADPH-d and glutamate-like immunoreactivity were first reacted histochemically to visualize NADPH-d reactivity. Following the histochemical protocol, tissue sections were labeled immunocytochemically with the anti-glutamate antibody, as described above.

RESULTS

GLI neurons were found to be distributed throughout the mesopontine tegmentum. Distinct populations of GLI neurons were present within the confines of both the LDT and PPT and in the following brainstem nuclei: the dorsal raphe, raphe pontis, median raphe; the trigeminal motor nucleus; and the reticulotegmental nucleus (Fig. 1). Comparison with an alternate section stained for NADPH-d clearly showed that GLI neurons occurred within the same general regions as the cholinergic neurons (Fig. 2). The numbers of both GLI and NADPH-d positive neurons in the LDT paralleled each other throughout the rostral-caudal extent of this nucleus (Fig. 3). GLI neurons within the LDT ranged in size from 9-22 μm in diameter and exhibited a variety of morphological types. Within individual GLI neurons, reaction product was not confined to the neuronal perikarya, but also extended into the neuronal processes. GLI neurons present in the region of the PPT extended from the cuneiform nucleus into the superior cerebellar peduncle (Fig 4A). These GLI neurons were somewhat smaller than those in the LDT, ranging in size from 8-15 μm in diameter.

In all control tissue sections, where either the primary or secondary antibody was omitted, or where the glutamate antibody had been pre-adsorbed with 48 ug/ml of a glutamate-BSA conjugate, immunostaining was eliminated (inset, Fig. 3).

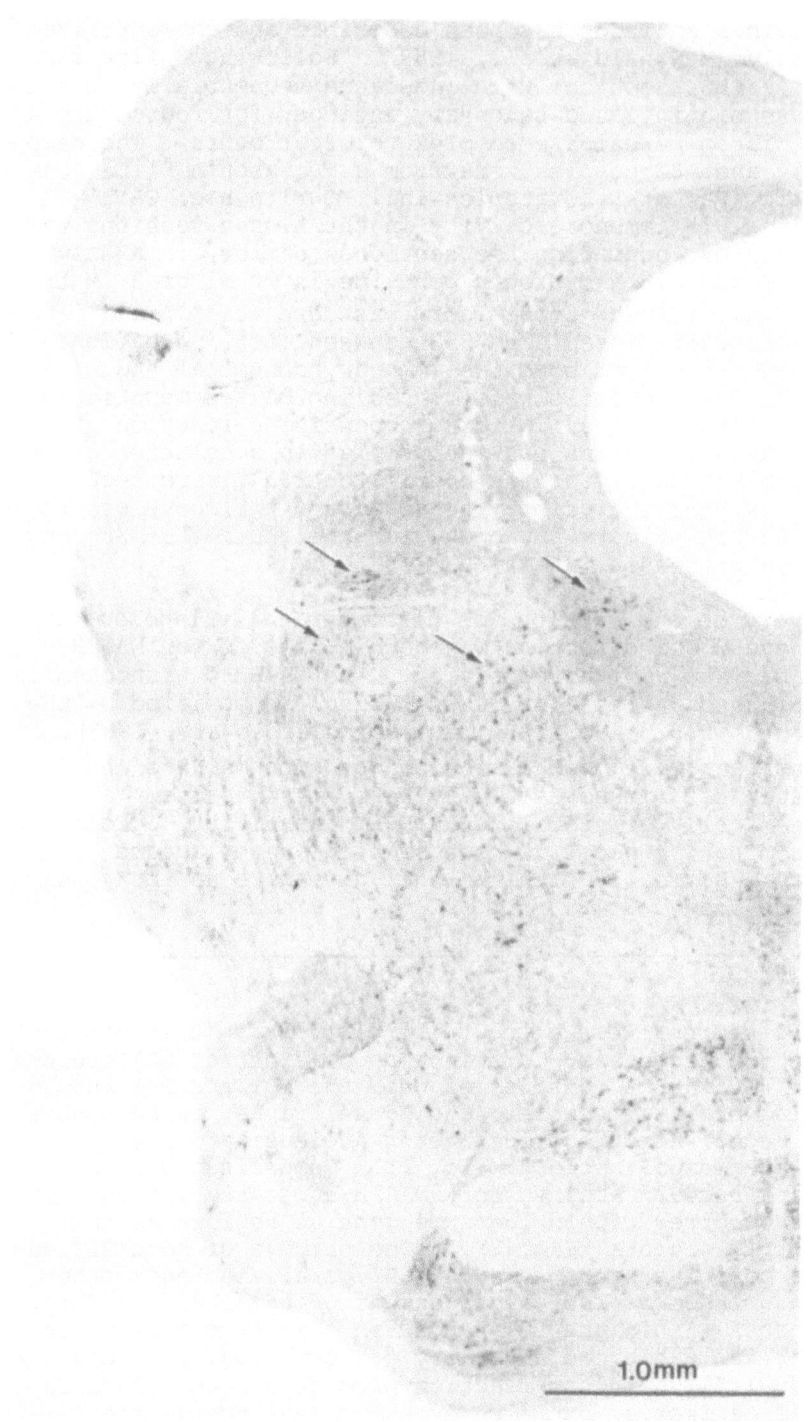

Fig. 1. Glutamate-like immunoreactivity in the midbrain of
the rat. Note the intensely-immunoreactive
populations of glutamate neurons (arrows) within the
LDT, PPT, and intermediate parabrachial region. Some
neurons in other brainstem nuclei, including the
raphe nuclei, the trigeminal motor nucleus and the
reticulotegmental nucleus also exhibited glutamate-
like immunoreactivity.

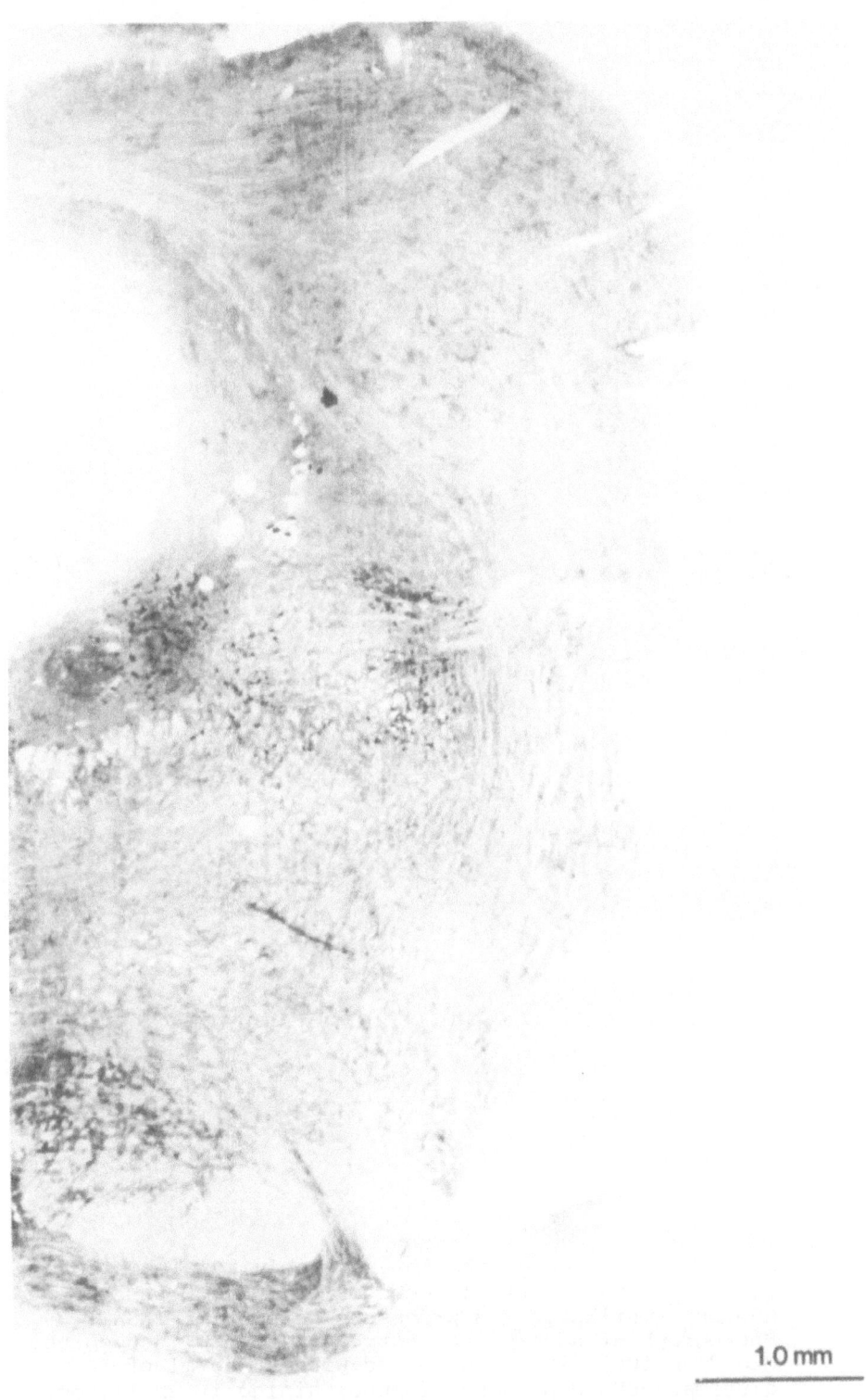

1.0 mm

Fig. 2.   In an alternate section histochemically stained
for NADPH-d, populations of cholinergic neurons are
present within the LDT and PPT.

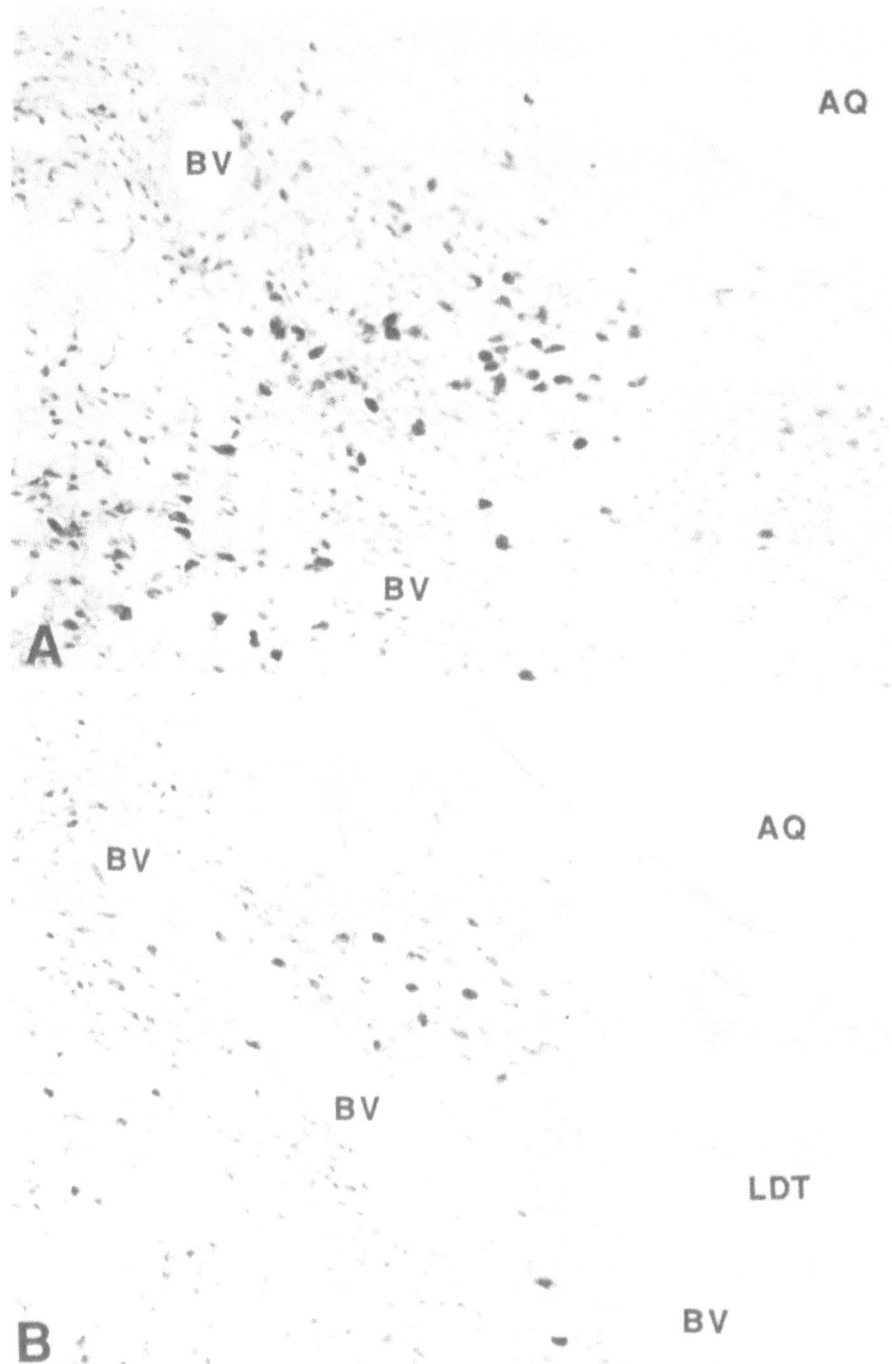

Fig. 3. A,B. Glutamate-like immunoreactivity within both caudal and more rostral sections of LDT. Both neuronal perikarya and processes were GLI. INSET: Photograph of adsorbtion control in the LDT. When the Glu-2 antibody is pre-adsorbed with 48 ug/ml of glutamate-BSA conjugate, immunostaining is abolished. AQ - aqueduct, BV - blood vessel.

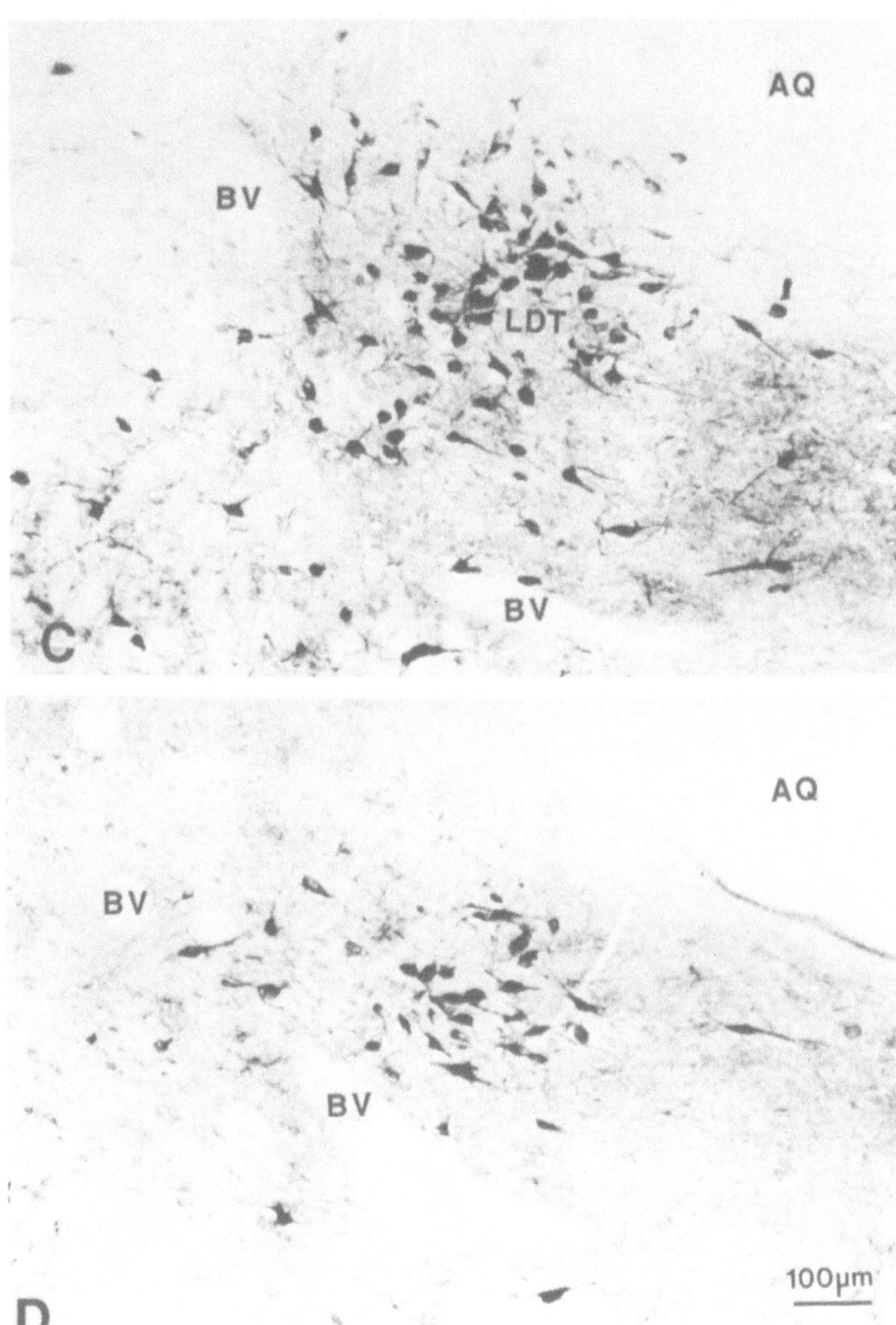

Fig. 3 con't. C,D. Note the similar distribution of
NADPH-d positive neurons in alternate sections. These
distribution patterns suggested that some neurons may
be both GLI and NADPH-d positive. Abbreviations
defined in Fig. 3A,B.

Fig. 4. The distribution of GLI neurons (A) within the PPT
also paralleled the distribution of NADPH-D reactive
neurons (B).

In tissue sections processed for NADPH-d and glutamate-like immunoreactivity, many neurons in the LDT and PPT were double-labeled with both chromagens (Fig. 5). However, some neurons were only GLI while some neurons were solely reactive for NADPH-d. There appeared to be no distinction between classes of neurons based on size or morphology. Individual GLI neurons were unipolar, bipolar and multipolar. At higher magnifications, it is apparent that glutamate-like immunoreactivity fills both the neuronal somata and processes, even in neurons reactive for NADPH-d.

DISCUSSION

The results indicate a widespread distribution of neurons exhibiting glutamate-like immunoreactivity throughout the pontine tegmentum. In particular, substantial numbers of GLI neurons were found among the cholinergic neurons within the LDT and PPT as indicated by deep staining for NADPH-d (Vincent et al., 1983). Glutamatergic neurons have been reported to be present in areas which also contain cholinergic neurons (Ottersen and Storm-Mathiesen, 1984), but to our knowledge, our data constitute the first anatomical evidence for a potential excitatory amino acid transmitter component to the LDT and PPT (Clements and Grant, 1990a,b). Furthermore, the double-labeled neurons described indicate that some neurons exhibit both NADPH-d reactivity and glutamate-like immunoreactivity. These findings raise the possibility that glutamate is a co-transmitter in a sub-population of brainstem cholinergic neurons.

The Glu-2 glutamate antibody that was used has been characterized (McDonald et al., 1989) and exhibits cross-reactivities of approximately 1% with GABA, glutamate, lysine, and taurine; and 3% with aspartate and cysteine sulfinate. This antibody exhibits no detectable large molecule reactivity on immunoblots, therefore the possibility that the immunostaining seen was due in part to glutamic acid decarboxylase (GAD) is unlikely. It is also unlikely that the Glu-2 antibody could recognize and detect n-acetylaspartyl glutamate (NAAG) in this tissue, since paraformaldehyde was used for fixation and NAAG does not have a primary amine available for fixation. Although some basal forebrain cholinergic neurons are immunoreactive for NAAG, NAAG immunoreactivity has not been observed in the LDT or PPT (Forloni et al., 1987), but has been reported to be present in almost all locus coeruleus neurons. However, we saw no GLI neurons within the locus coeruleus. It is therefore doubtful that the glutamate-like immunolabeling reported here was due to recognition of NAAG.

Although the development of immunocytochemical protocols useful for localizing glutamate is relatively recent, there is strong evidence that the antibody used here visualizes transmitter pools of glutamate. Anti-glutamate antibodies identify reliably nerve terminals characterized physiologically as excitatory (Ottersen and Storm-Mathiesen, 1984; Clements et al., 1986, 1987, 1990; Miller et al., 1988; Petrusz and Rustioni, 1989). The immunocytochemical demonstration of neuronal cell bodies whose processes are excitatory and that use glutamate as a transmitter is more problematic since glutamate present in the cell body may function in a metabolic capacity (Hertz et al., 1983). However, in previous studies done with this glutamate antibody, known excitatory neurons, such as the granule cells in the cerebellum and hippocampus, were shown to be glutamate-like immunoreactive (Clements et al., 1986, 1987,

1990). The glutamatergic neurons identified in both the LDT and PPT in this study exhibited strong glutamate immunolabeling when compared to other neurons in the neuropil. We believe that the high concentration of glutamate present in these neurons does indeed represent a transmitter pool of glutamate. Further studies examining the transmitter present in the terminal projections of these neurons are required to confirm this hypothesis.

Within the confines of the LDT and PPT there was no segregation of glutamatergic and cholinergic (NADPH-d positive) neurons. Thus, both basal forebrain cholinergic neurons (which are intermixed with GABAergic and other neurons) and brainstem cholinergic neurons, are located in neurochemically heterogenous regions (Brashear et al., 1986; Fisher et al., 1988). This is in contrast to the adjacent monoaminergic nuclei such as the noradrenergic locus coeruleus, the serotonergic raphe and dopaminergic substantia nigra, where almost all the neurons contain the same transmitter.

The close proximity of cholinergic and glutamatergic neurons in the LDT and PPT suggested that acetylcholine and glutamate were contained within the same neuron. Co-localization was confirmed in the double-labeling experiments. Although NADPH-d appears to be a selective marker for cholinergic neurons in the pontine tegmentum, it is conceivable that some neurons exhibiting NADPH-d staining and glutamate-like immunoreactvity were not cholinergic, or that the NADPH-d histochemical procedure resulted in a false positive immunostaining for glutamate. To eliminate this possibilty, sections from a limited number of cases were double-labeled with choline acetyltransferase, a positive marker for cholinergic neurons, and glutamate (Clements and Grant, 1990b). As with NADPH-d, neurons in the LDT and PPT were immunoreactive for both Glu-2 and ChA-T. The preliminary double-labeling immunocytochemical observations provide additional support for the hypothesis that glutamate and acetylcholine are co-localized in a sub-population of LDT and PPT neurons and function as co-transmitters.

Cholinergic neurons in the PPT and LDT are known to contain multiple peptide co-transmitters. Some cholinergic neurons have been shown to contain several peptides, and it has been suggested that all peptides in the LDT and PPT are co-localized with acetylcholine (Matsuzaki et al., 1981; Crawley et al., 1985; Standaert et al., 1986; Vincent et al., 1986; Sutin and Jacobowitz, 1988). Although co-localization of acetylcholine and glutamate has not been reported in the CNS, the presence of glutamatergic neurons in other brain regions that contain high concentrations of cholinergic cells has been reported (Ottersen and Storm-Mathiesen, 1984). One of these regions was the ventral pallidum which raises the possibility that there is a glutamatergic component to the basal forebrain cholinergic neurons. This possibility is supported by recent evidence showing that rat cerebral cortex cholinergic synaptosomes release glutamate upon depolarization (Docherty et al., 1987).

Although efferents from the PPT and LDT are known to have a non-cholinergic component, tract tracing experiments are now needed to confirm which projections may be glutamatergic (Sugimoto and Hattori, 1984; Satoh and Fibiger, 1986; Beninato

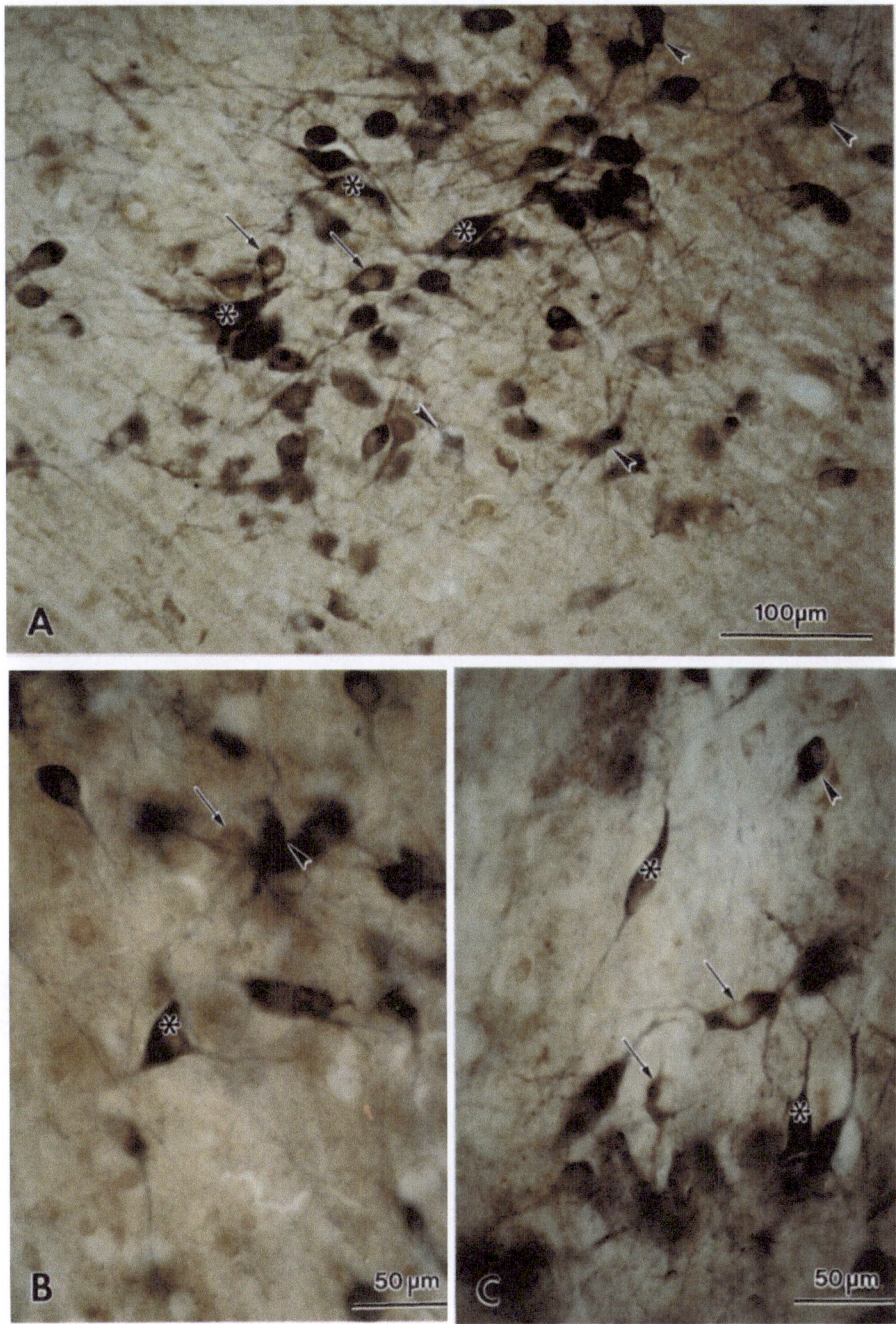

Fig. 5 A-C. Neurons in the LDT were double-labeled with
NADPH-d and glutamate-like immunoreactivity
(asterisks). Some neurons were single-labeled for
GLI (arrows) or NADPH-d (arrowheads). There appeared
to be no distinction between GLI or NADPH-d neurons
based on size or morphology. NADPH-d and glutamate-
like immunoreactivity were present in both neuronal
somata and processes.

and Spencer, 1987,1988; Goldsmith and Kooy, 1988; Rye et al., 1988; Semba et al., 1988, 1989; Gould et al., 1989). However, there is sufficient evidence available at this time to suggest that glutamate may be a transmitter in the basal forebrain as well as the substantia nigra.

First, it is known that there are reciprocal connections between the basal forebrain and the LDT and PPT, but it does not appear that there are direct connections between the cholinergic neurons of the two regions (Semba et al., 1988, 1989). Although the present results raise the posssibility that glutamatergic neurons could be synapsing on basal forebrain cholinergic neurons, further anatomical and physiological studies are required to resolve this question. It has already been shown that components of the excitatory response of thalamic neurons to LDT and PPT stimulation are not blocked by cholinergic antagonists (Kayama et al., 1986; Hu et al., 1989). One question that remains is whether synaptic activation following LDT and PPT stimulation of cortically projectiong, putative cholinergic basal forebrain neurons, is selectively blocked by cholinergic antagonists or by excitatory amino acid antagonists.

Second, the PPT lies within a region known to be connected reciprocally with the ventral striatal system (Lee et al., 1988; Rye et al., 1987; Saper and Loewy, 1982; Mogenson, this book). These projections as described by others in this book, are thought to mediate the locomotor effects of stimulant drugs. It is currently a matter of controversy whether the LDT and PPT cholinergic neurons are the targets of ventral pallidal projections. Based on physiological evidence it has been suggested that the cholinergic neurons are the anatomical substrate of the physiologically defined 'mesencephalic locomotor region' (Garcia-Rill et al., 1987). However, anatomical studies indicate that the extrapyramidal projections terminate in a region medial to the cholinergic neurons, designated as the 'midbrain extrapyramidal area' (Rye et al., 1987). Since GLI neurons were also found medial to cholinergic neurons of the PPT, it is possible that both the midbrain extrapyramidal area and the mesencephalic locomotor center are primarily glutamatergic. Confirmation of this hypothesis will require a detailed description of both the topographic distribution of GLI neurons and their efferent projections.

Finally, there is strong evidence that dopamine neurons in the substantia nigra receive both a cholinergic and an excitatory amino acid projection from the PPT and LDT. Tract tracing studies have established that nigral dopamine neurons receive a direct projection from the PPT and LDT which is only partially cholinergic (Beninato and Spencer, 1987,1988; Gould et al.,1989). Although additional experiments are needed to confirm that the glutamate immunoreactive neurons described here are the source of excitatory amino acid input to the nigra from the PPT and LDT, physiological studies also support a potential glutamatergic input to the nigra. Both cholinergic and excitatory amino acid antagonists can block synaptic activation of nigral neurons following stimulation of the PPT (Scarnati et al., 1986; Clarke et al., 1987). In studies employing antidromic activation, two neurophysiologically distinct neuronal populations in PPT which project to the nigra were identified and were suggested to correspond to the two neurochemical inputs (Scarnati et al., 1987). The physiological properties of these two populations (long vs. short conduction velocity, narrow vs. wide action potential widths) are consistent with *in vitro* intracellular studies of LDT and PPT neurons where two neuronal populations were distinguished on the basis of intrinsic membrane conductances (Leonard and Llinas, 1988; Wilcox et al., 1989). One population, characterized by potassium mediated long afterhyperpolarizations, was identified as cholinergic neurons by intracellular injection of Lucifer Yellow and double labeling with NADPH-d. The transmitter used by the second population, characterized by the presence of low threshold calcium currents, has not yet been determined. The present results raise the possiblity that these neurons are glutamatergic.

The recent interest in cholinergic systems has substantiated the earlier concept of an ascending cholinergic activating system proposed by Lewis and Shute (Vincent et al., 1986; Steriade and McCarley, 1990). Both the basal forebrain and the brainstem cholinergic neurons, along with the monoaminergic neurons, comprise a diffuse, projecting neuronal network now considered to be the substrates of the classical reticular activating system (Saper, 1987; Steriade and McCarley, 1990; Semba, this book; Richardson and DeLong, this book). Our data suggests that a brainstem excitatory amino acid system arising from the PPT and LDT also could contribute to the functions associated with this system, which include general regulation of arousal, locomotion, behavioral state, and sensory responsiveness. Glutamate neurotransmission also may be involved in the actions of drugs of abuse such as amphetamine and cocaine. Finally, these results raise the possibility that dysfunction of this glutamatergic system may contribute to neurodegenerative diseases associated with basal forebrain systems along with the cholinergic neurons in Alzheimer's disease and dopaminergic neurons in Parkinson's disease (Beninato and Spencer, 1987).

ACKNOWLEDGEMENTS

We thank Dr. J. Madl, Colorado State University, for providing the anti-glutamate antibody and glutamate-BSA conjugate, and Dr. B. Wainer for the AB-8 antibody for choline acetyltransferase. This work was supported by: NIH FIRST award DE08185 (JRC); the University of Delaware Research Foundation (JRC); NIMH award MH45610 (SJG); the State of Delaware, and ICI Pharmaceuticals (SJG).

REFERENCES

Armstrong, D.A., Saper, C.B., Levey, A.I., Wainer, B.H., and Terry, R.D., 1983, Distribution of cholinergic neurons in the rat brain demonstrated by the immuohistochemical localization of choline-acetyltransferase, J. Comp. Neurol., 200:53.

Beninato, M., and Spencer, R.F., 1987, Cholinergic projection to the rat substantia nigra from the pedunculopontine tegmental nucleus, Brain Res., 412:169.

Beninato, M., and Spencer, R.F., 1988, The cholinergic innervation of the rat substantia nigra: a light and electron microscopic immunohistochemical study, Exp. Brain Res., 72:178.

Brashear, H.R., Zaborszky, L., and Heimer, L., 1986, Distribution of gabaergic and cholinergic neurons in the rat diagonal band, Neurosci., 17:439.

Clarke, P.B.S., Hommer, D.W., Pert, A., and Skirboll, L.R., 1987, Innervation of substantia nigra neurons by cholinergic afferents from pedunculopontine nucleus in the rat: neuroanatomical and electrophysiological evidence, Neurosci., 23:1011.

Clements, J.R., and Grant, S.J., 1990a, Glutamate-like immunoreactivity in neurons of the laterodorsal tegmental and pedunculopontine nuclei, Neurosci. Lett., 120:70.

Clements, J.R., and Grant, S.J., 1990b, Glutamate and acetylcholine are colocalized in the laterodorsal tegmental and pedunculopontine nuclei., Neurosci. Abst., 16:1189.

Clements, J.R., Magnusson, K.R., and Beitz, A.J., 1990, Ultrastructural description of glutamate-, aspartate-, taurine- and glycine- like immunoreactive terminals from five rat brain regions, J.Electron Microscopic Tech., 15:49.

Clements, J.R., Magnusson, K.R., Mullett, M.A., and Beitz, A.J., 1987, An ultrastructural description of glutamate- and taurine-like immunoreactive elements in the rat hippocampus, Neurosci. Suppl., 22:S123.

Clements, J.R., Monaghan, P.L., Madl, J.E., Larson, A.A., and Beitz, A.J., 1986, An ultrastructural examination of glutamate- and aspartate-like immunoreactive cells and processes in the cerebellar cortex of the rat, Soc. Neurosci. Abstr., 12:462.

Cooper, J.R., Bloom, F.E., and Roth, R.H., 1986 , "The Biochemical Basis of Neuropharmacology, 5th ed.", Oxford Univ. Press, NY.

Coyle, J.T., Price, D.L., and DeLong, M.R., 1983, Alzheimer's disease: A disorder of cortical cholinergic innervation, Science, 219:1184.

Crawley, J.N., Olschowka, J.A., Diz, D.I., and Jacobowitz, D.M., 1985, Behavioral investigation of the coexistence of substance P, corticotropin releasing factor, and acetylcholinesterase in lateral dorsal tegmental neurons projecting to the medial frontal cortex of the rat, Peptides, 6:891.

Docherty, M., Bradford, H.F., and Wu, Y.-Y., 1987, Co-release of glutamate and aspartate from cholinergic and GABAergic synaptosomes, Nature, 330:64.

Fisher, R.S., Buchwald, N.A., Hull, C.D., and Levine, M.S., 1988, GABAergic basal forebrain neurons project to the neocortex: The localization of glutamic acid decarboxylase and choline acetyltransferase in feline corticipetal neurons, J. Comp. Neurol., 272:489.

Forloni, G., Grzanna, R., Blakely, R.D., and Coyle, J.T., 1987, Co-localization of N-acetyl-aspartyl-glutamate in central cholinergic, noradrenergic, and serotonergic neurons, Synapse, 1:455.

Fruend, T., and Antal, M., 1988, GABA containing neurons in the septum control inhibitory interneurons in the hippocampus, Nature, 336:170.

Garcia-Rill, E., Houser, C.R., Skinner, R.D., Smith, W., and Woodward, D.J., 1987, Locomotion inducing sites in the vicinity of the pedunculopontine nucleus, Brain Res. Bull., 18:731.

Goldsmith, M., and Kooy, D.V.D., 1988, Separate non-cholinergic descending projections and cholinergic ascending projections from the nucleus tegmenti pedunculopontinus, Brain Res., 445:386.

Gould, E., Woolf, N.J., and Butcher, L.L., 1989, Cholinergic projections to the substantia nigra from the pedunculopontine and lateral dorsal tegmental nuclei, Neurosci, 28:611.

Hallenger, A.E., Levey, A.I., Lee, H.J., Rye, D.B., and Wainer, B.H., 1987, The origins of cholinergic and other subcortical afferents to the thalamus in the rat, J. Comp. Neurol., 262:105.

Hertz, L., Kvamme, E., McGeer, E.G., and Schousboe, A., eds.,1983, "Glutamine, Glutamate and GABA in the Central Nervous System," Alan R. Liss, New York.

Hu, B., Steriade, M., and Deschenes, M., 1989, The effects of brainstem peribrachial stimulation on neurons of the lateral geniculate nucleus, Neurosci., 31:13.

Jones, B.E., 1990, Immunohistochemical study of choline acetyltransferase-immunoreactive processes and cells innervating the pontomedullary reticular formation in the rat, J. Comp. Neurol., 295:485.

Kayama, Y., Sumitomo, I., and Ogawa, T., 1986, Does the ascending cholinergic projection inhibit or excite neurons in the rat thalamic reticular nucleus? J. Neurophys., 56:1310.

Kiosaka, T., Tauchi, M., and Dahl, J.L., 1988, Cholinergic neurons containing GABA-like and/or glutamic acid decarboxylase-like immunoreactivities in various brain regions of the rat, Exp. Brain Res., 70:605.

Lee, H.J., Rye, D.B., Hallenger, A.E., Levey, A.I., and Wainer, B.H., 1988, Cholinergic vs. noncholinergic efferents from the mesopontine tegmentum to the extrapyramidal motor system nuclei., J.Comp. Neurol.,275:469.

Leonard, C.S., and Llinas, R., 1988, Electrophysiology of thalamic-projecting cholinergic brainstem neurons and their inhibition by ACh, Neurosci. Abstr., 14:297.

Lewis, P.R. and C.C.D. Shute., 1967. The cholinergic limbic system: Projection to hippocampal formation, medial cortex, nuclei of the ascending cholinergic reticular system, and the subfornical organ and supra-optic crest, Brain., 90: 521.

Maley, B.E., Frick, M.L., Levey, A.I., Wainer, B.H., and Elde, R.P., 1988, Immunohistochemistry of choline acetyltransferase in the guinea pig brain, Neurosci. Lett., 84:137.

Matsuzaki, T., Shiosaka, S., Inagaki, S., Sakanaka, M., Takatsuki, K., Takagi, H., Senba, E., Kawai, Y., and Tohyama, M., 1981, Distribution of neuropeptides in the dorsal pontine tegmental area of the rat, Cell. Molec. Biol., 27:499.

McDonald, A.J., Beitz, A.J., Larson, A.A., Kuriyama, R., Sellitto, C., and Madl, J.E., 1989, Co-localization of glutamate and tubulin in putative excitatory neurons of the hippocampus and amygdala: an immunohistochemical study using monoclonal antibodies, Neuroscience, 30:405.

Mesulam, M.-M., Mufson, E.J., Levey, A.I., and Wainer, B.H., 1984, Atlas of cholinergic neurons in the forebrain and upper brainstem of the macaque based on monoclonal choline acetyltransferase immunohistochemistry and acetylcholinesterase histochemistry, Neurosci., 12:669.

Miller, K.E., Clements, J.R., Larson, A.A., and Beitz, A.J., 1988, Organization of glutamate-like immunoreactivity in the rat superficial dorsal horn: light and electron microscopic observations, Synapse, 2:28.

Mogenson, G.J., and Yang, C.R., 1991, The contribution of basal forebrain to limbic motor integration and the mediation of motivation to action, in: "The Basal Forebrain: Anatomy to Function", T.C. Napier, P.W. Kalivas, and I. Hanin, eds., Plenum Press, New York, (in press).

Ottersen, O.P., and Storm-Mathiesen, J., 1984, Glutamate- and GABA- containing neurons in the mouse and rat brain, as demonstrated with a new immunocytochemical technique, J. Comp. Neurol., 229:374.

Petrusz, P., and Rustioni, A., 1989, Immunocytochemistry of excitatory amino acids in brain, in: "Techniques in Immunocytochemistry, Vol 4.", G.R. Bullock and P. Petrusz, ed., Academic Press, New York, p. 253.

Richardson, R.T., and DeLong, M.R., 1991, Electrophysiological studies of the functions of the nucleus basalis in primates, in: "The Basal Forebrain: Anatomy to Function", T.C. Napier, P.W. Kalivas, and I. Hanin, eds., Plenum Press, New York, (in press).

Rye, D.B., Lee, H.J., Saper, C.B., and Wainer, B.H., 1988, Medullary and spinal efferents of the pedunculopontine tegmental nucleus and adjacent mesopontine tegmentum in the rat, J. Comp. Neurol., 269:315.

Rye, D.B., Saper, C.B., Lee, H.J., and Wainer, B.H., 1987, Pedunculopontine tegmental nucleus of the rat: cytoarchitecture, cytochemistry, and some extrapyramidal connections of the mesopontine tegmentum, J. Comp. Neurol., 259:483.

Saper, C.B., 1987, Diffuse cortical projection systems: anatomical organization and role in cortical function, in: "Handbook of Physiology. Section 1: The Nervous System. Volume V. Higher Functions of the Brain, Part 2.", V.B. Mountcastle, F. Plum and S.R. Geiger, eds., American Physiological Society, Bethesda, Md, p. 169.

Saper, C.B., and Loewy, A.D., 1982, Projections of the pedunculopontine tegmental nucleus in the rat: evidence for additional extrapyramidal circuitry, Brain Res, 252:367.

Satoh, K., and Fibiger, H.C., 1986, Cholinergic neurons of the laterodorsal tegmental nucleus: Efferent and afferent connections, J. Comp. Neurol., 253:277.

Scarnati, E., Prioria, A., Campana, E., and Pacitti, C., 1986, A microiontophoretic study of the nature of the putative synaptic neurotransmitter involved in the pedunculopontine substantia nigra pars compacta excitatory pathway in the rat, Exp. Brain Res., 62:470.

Scarnati, E., Prioria, A., DiLoreto, S., and Pacitti, C., 1987, The reciprocal electrophysiological influence between the nucleus tegmenti pedunculopontinus and the substantia nigra in normal and decorticated rats, Brain Res., 423:116.

Scherer-Singler, U., Vincent, S.R., Kimura, H., and McGeer, E.G., 1983, Demonstration of a unique population of neurons with NADPH diaphorase histochemistry, J. Neurosci. Meth., 9:229.

Semba, K., Reiner, P.B., McGeer, E.G., and Fibiger, H.C., 1988, Brainstem afferents to the magnocellular basal forebrain studied by axonal transport, immunohistochemistry, and electrophysiology in the rat, J. Comp. Neurol, 267:433.

Semba, K., Reiner, P.B., McGeer, E.G., and Fibiger, H.C., 1989, Brainstem projecting neurons in the rat basal forebrain: Neurochemical, topographical and physiological distinctions from cortically projecting cholinergic neurons, Brain. Res. Bull., 22:501.

Semba, K., 1991, The cholinergic basal forebrain: A critical role in cortical arousal, in: "The Basal Forebrain: Anatomy to Function", T.C. Napier, P.W. Kalivas, and I. Hanin, eds., Plenum Press, New York, (in press).

Sofroniew, M.V., Priestly, J.V., Consolazione, A., Eckenstein, F., and Cuello, A.C., 1985, Cholinergic projections from the midbrain and pons to the thalamus in the rat, identified by combined retrograde tracing and choline acetyltransferase immunohistochemistry, Brain Res., 329:213.

Standaert, D.G., Saper, C.B., Rye, D.B., and Wainer, B.H., 1986, Colocalization of atriopeptin like immunoreactivity with choline acetyltransferase and substance P like immunoreactivity in the pedunculopontine and laterodorsal tegmental nuclei in the rat, Brain Res., 382:163.

Steriade, M., and McCarley, R. W., 1990, "Brainstem Control of Wakefulness and Sleep", Plenum Press, New York.

Sugimoto, T., and Hattori, T., 1984, Organization and efferent projections of nucleus tegmenti pedunculopontinus pars compacta with special reference to its cholinergic aspects, Neurosci., 11:931.

Sutin, E.L., and Jacobowitz, D.M., 1988, Immunocytochemical localization of peptides and other neurochemicals in the rat laterodorsal tegmental nucleus and adjacent area, J. Comp. Neurol, 270:243.

Vincent, S.R., and Satoh, K., 1984, Corticotropin releasing factor (CRF) immunoreactivity in the dorsolateral pontine tegmentum: further studies on the micturition reflex system, Brain Res., 308:387.

Vincent, S.R., Satoh, K., Armstrong, D.M., and Fibiger, H.C., 1983, NADPH- diaphorase: A selective histochemical marker for the cholinergic neurons of the pontine reticular formation, Neurosci. Lett., 43:31.

Vincent, S.R., Satoh, K., Armstrong, D.M., Panula, P., Vale, W., and Fibiger, H.C., 1986, Neuropeptides and NADPH-diaphorase activity in the ascending cholinergic reticular system of the rat, Neurosci, 17:167.

Wilcox, K.S., Grant, S.J., Burkhardt, B., and Christoph, G.R., 1989, Electrophysiological properties of lateral dorsal tegmental neurons in vitro, Brain Res. Bull., 22:557.

Yamano, M., and Tohyama, M., 1987, Afferent and efferent enkephalinergic systems of the tegmental nuclie of Gudden in the rat: an immunocytochemical study, Brain Res., 408:22.

ELECTROPHYSIOLOGIC CHARACTERISTICS OF BASAL FOREBRAIN NEURONS

IN VITRO

William H. Griffith[1], Joan A. Sim[1,2] and Robert T. Matthews[3]

Departments of Medical Pharmacology & Toxicology[1] and Anatomy[3],College of Medicine, Texas A&M University, College Station, TX; 77843

INTRODUCTION AND BACKGROUND

The medial septum (MS) and nucleus of the diagonal band (nDB) are nuclei of the basal forebrain that contain both cholinergic and non-cholinergic neurons. Since the work of Green and Arduini (1954), it has been established that the MS/nDB form a crucial link between the brainstem reticular formation and the hippocampus. In order to understand the function of the basal forebrain component of this pathway, much research has been directed toward categorizing MS/nDB cell types physiologically, primarily with extracellular unit recording techniques in vivo. Experiments in anesthetized or curarized animals have distinguished three types of MS/nDB neurons based on spontaneous firing patterns (Apostol and Creutzfeldt, 1974; Lamour et al., 1984; Segal, 1974; Vinogradova et al., 1987). One neuron type fires in an irregular, single spike pattern. A second type of neuron fires in bursts of 2 to 10 spikes and the burst frequency is synchronized with hippocampal slow rhythmic activity (theta rhythms). These neurons may switch under certain conditions between irregular and burst-type firing patterns. In addition, antidromically activated septo-hippocampal neurons are more likely to be bursting neurons (50%) than are unidentified MS/nDB neurons (30%; Lamour et al., 1984). A third type of neuron reported by a few laboratories has a pacemaker-like firing pattern that does not convert to other patterns (Apostol and Creutzfeldt, 1974; Lamour et al., 1984; Vinogradova et al., 1987).

Besides a neuron classification based on firing pattern, investigators have classified MS/nDB neurons as to conduction velocity, firing rate, efferent target areas, response to peripheral stimuli, response to stimulation of afferent input, and response to iontophoresed drugs (Assaf and Miller, 1978; Bassant et al., 1988; Brazhnik and Vinogradova, 1988; Disturnal et al., 1985; Dutar et al., 1985; Hubbard et al., 1979; McLennan and

---

[2] permanent address: Department of Pharmacology, University College London, London WC1E 6BT, U.K.

The Basal Forebrain, Edited by T.C. Napier *et al.*
Plenum Press, New York, 1991

Miller, 1974; Segal, 1976; Stewart and Fox, 1989; Vinogradova et al., 1980). In general, different in vivo classification schemes do not correlate well with one another, suggesting that additional approaches are needed for identifying cell types, such as defining physiological properties of neurochemically identified neurons. The purpose of the present study is to review the intrinsic membrane properties of MS/nDB neurons that could contribute to their repetitive firing activity, specifically those properties that could control either rhythmic burst activity or regular, pacemaker firing. Evidence is also reviewed for the cholinergic nature of one cell type. We will attempt to correlate our in vitro studies with existing in vivo data to better understand the function of the basal forebrain.

DESCRIPTION OF CELL TYPES

We have described three principle categories of neurons in vitro in the MS/nDB (Griffith and Matthews, 1986; Griffith, 1988). These include: first, cells that fire slowly in response to membrane depolarization where the rate is limited by the development of a slow afterhyperpolarization (AHP); second, neurons that fire repetitively once threshold is reached; and third, cells that fire in a burst-pattern. Each cell type is shown in Fig. 1 and will be discussed in more detail below.

1. Slow-afterhyperpolarization (S-AHP) cell

An example of an S-AHP cell is shown in Fig. 1 (left column). A single action potential is usually evoked during a 300 ms pulse of threshold depolarization. The spike is followed by a post-spike AHP of approximately 200-700 ms duration, the signature for this cell type. If a spike is generated by a short (3 ms) current step or antidromic stimulation, the AHP often decays as a single exponential function with a time constant of decay of 150-200 ms. Other S-AHP cells at similar membrane potentials, however, may exhibit an AHP decay with a second component lasting 2-5 sec (Fig. 2).

The conductance responsible for the slow-AHP has been shown to be a calcium (Ca^{2+})-activated potassium (K^+) conductance (Griffith, 1988). Addition of either cadmium (Cd^{2+}) or zero-Ca^{2+}/high-Mg^{2+} solutions reversibly blocked the AHP. The K^+-dependence was established with the predicted shift in the reversal potential of the AHP after addition of high external K^+ concentrations. Likewise, the conductance responsible for the long-duration AHP, i.e. second component, in S-AHP cells is a Ca^{2+}-activated K^+ conducatance (unpublished observation) but needs to be further characterized.

The source of the Ca^{2+} responsible for activating the K^+ conductance is most probably Ca^{2+} entry during the action potential. S-AHP cells have a mean spike duration of approximately 1.4 ms (measured at one-fourth amplitude base to peak) with a prominent shoulder on the falling phase of the spike (Fig. 2). Addition of Cd^{2+} shortens the action potential duration and reduces this repolarizing shoulder (Griffith, 1988). A high-threshold Ca^{2+} current is a likely candidate to mediate this Ca^{2+} entry during the spike since depolarizations positive to -60 mV are most effective in generating the repolarizing "hump", and maintained membrane potentials of -60 mV are most effective for

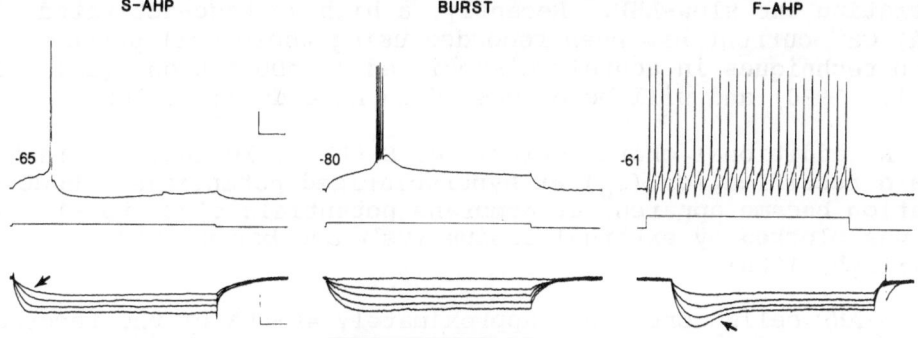

Fig. 1. Identified cell types within the MS/nDB. Each column shows a different cell type. Records in the top row show action potentials in response to depolarizing current pulses. The membrane potentials are indicated to the left of each record. Lower records show voltage responses to hyperpolarizing current steps. Note the fast inward rectification in the S-AHP cell as well as the time-dependent rectification in the F-AHP cell (arrows). See text for further discussion. Calibration: 20 mV, 1 nA and 40 ms.

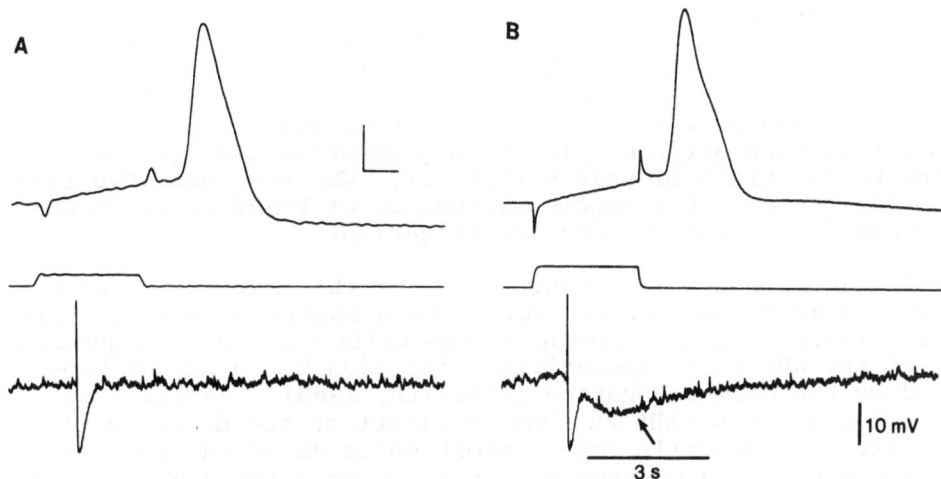

Fig. 2. Example of two S-AHP cells. A. A single action potential (top) is generated by a 3 ms current pulse (middle). The spike is followed by a slow-AHP of 600 ms duration that is more clearly seen in the chart record (bottom). Note the slow time scale for the chart records (calibration is the same as B). Calibration for spikes in A and B: 20 mV, 1 ms, 1 nA; membrane potential for both cells was -60 mV. B. Using the same protocol as A, a different S-AHP displays a two component AHP. In this cell a slow-AHP, similar to that in A, is followed by a long-AHP of many seconds duration (arrow).

generating the slow-AHP. Recently, a high voltage-activated (HVA) Ca^{2+}-current has been recorded using whole-cell patch-clamp techniques in acutely dissociated MS/nDB neurons (Griffith et al., 1990) and will be discussed in more detail below.

An additional characteristic of S-AHP cells includes fast inward rectification (I_{IR}) at hyperpolarized potentials. Rectification became apparent at membrane potentials close to -100 mV and was blocked by external cesium (Cs^+) and barium (Ba^{2+}) (Griffith, 1988).

S-AHP cells constitute approximately 40-45% of the recorded neurons in vitro. Importantly, these cells are presumed to be cholinergic, based on a double-labelling technique of intra-cellular injection of Lucifer Yellow paired with acetylcholin-esterase (AChE) histochemistry (Griffith and Matthews, 1986).

2. Fast-afterhyperpolarization (F-AHP) cell

F-AHP cells are characterized by a short duration (50-100 ms) post-spike AHP which follows a very fast (5 ms) spike undershoot. These two components have previously been considered together (Griffith and Matthews, 1986; Griffith, 1988) but undoubtedly reflect different conductances. Only the 50-100 ms AHP (termed fast-AHP) will be reviewed.

The short duration of the fast-AHP may allow repetitive firing in F-AHP cells. In response to a threshold depolarization, this repetitive discharge rate is in vivid contrast to the firing pattern in S-AHP cells (Fig. 1). However, the firing patterns observed in F-AHP cells are not stereotyped, and two different patterns have been observed. First, cell discharge throughout the duration of stimulation with little spike frequency adaptation or, second, cells that exhibit repetitive spiking associated with varying degrees of frequency adaptation. This latter type often display a long-duration AHP (2-5 sec) following the train of spikes (Fig. 3). Whether these two firing patterns reflect a subclassification of F-AHP cells is unknown and is currently under investigation.

The ionic mechanism responsible for the fast-AHP is also a Ca^{2+}-activated K^+ conductance but of much shorter duration. Elevated extracellular K^+ concentrations shift the reversal potential of the AHP in a depolarizing direction by an amount predicted by the Nernst equation (Griffith, 1988). Addition of Cd^{2+} blocks the fast-AHP but has no effect on the duration of the spike. F-AHP cells have a short spike duration (approximately 0.6 ms) with no shoulder on the repolarizing phase. The contrasting spike durations of F-AHP and S-AHP cells are another illustration of their different intrinsic properties.

A final characteristic of F-AHP cells that distinguish them from other cell types is the nature of membrane rectification at hyperpolarized potentials. F-AHP cells show a time-dependent inward rectification similar to that generated by the Q-current described in hippocampal pyramidal cells (Halliwell and Adams, 1982). Rectification in F-AHP cells can be observed at membrane potentials more negative than approximately -80 mV (see Fig. 1, F-AHP cell) and often result in rebound excitation following the stimulus. External Cs^+ but not Ba^{2+} blocked inward rectification (Griffith, 1988).

F-AHP cells comprise 50-55% of the recorded cells and may represent more than one cell type or possibly several chemically-distinct neuron subclasses. F-AHP cells did not stain positively for AChE and are, therefore, not cholinergic.

3. Burst-firing cells

A small percentage of cells (10-15%) fire bursts of action potentials reminiscent of the burst-firing pattern described previously in thalamic neurons (Llinas and Jahnsen, 1982; Jahnsen and Llinas, 1984). Bursts consist of sodium-dependent action potentials (3-10) riding on a low-threshold, Ca^{2+}-dependent depolarization (Fig. 1). In MS/nDB cells, the size of the burst is controlled by the duration of the low-threshold depolarization, although the complete mechanism(s) responsible for burst generation is not known. For example, the activation and inactivation characteristics of the low-threshold depolarization as well as other inward and outward conductances need to be identified in these neurons.

Recently, we have described a low voltage-activated (LVA) Ca^{2+} current in acutely dissociated MS/nDB neurons (Griffith et al., 1990). Using the whole-cell patch-clamp configuration, with interfering conductances reduced, we demonstrated that the LVA Ca^{2+}-current was activated at potentials positive to -75 mV and displayed significant voltage-dependent inactivation at positive potentials. The divalent cationic blockers Cd^{2+} and nickel reduced the LVA current. These patch-clamp data demonstrate a Ca^{2+}-dependent LVA current that could generate the low-threshold depolarization and contribute to burst-firing in these cells.

It should be mentioned that some variation in burst-firing has been observed in MS/nDB cells. In addition to the pattern described above, some neurons exhibit a burst of spikes at the beginning of a depolarizing current step followed by a much reduced rate of spiking; or cessation of spiking altogether. As will be discussed below, the generation of burst-firing is voltage-dependent (see also Fig. 4) and a number of additional conductances may contribute to bursting in these cells.

The spike duration of burst-firing cells ranges between the values for S-AHP and F-AHP neurons. Likewise, a characteristic form of inward rectification is not a robust feature of these cells. As illustrated in Fig. 1, neither I_{IR} nor I_Q was consistently observed in these cells. The neurochemical identity of burst-firing cells is not known at this time.

In summary, we have described cell types recorded in vitro in the MS/nDB of guinea pig (see Griffith and Matthews, 1986; Griffith, 1988 for additional details). Qualitatively similar results have been observed in the MS/nDB of rat (Sim, unpublished observation; Matthews and Lee, unpublished observation), mouse (Matthews, unpublished observation), as well as rat nucleus basalis neurons (Sim, unpublished observation). Our description of cell types differs somewhat from an early description by Segal (1986) where cells in the MS of rat were described as spontaneous high-resistance and quiescent low-resistance types. Burst-firing neurons have been reported in the guinea pig septum by Alvarez de Toledo et al., (1988) who showed that action potential bursts had both Na^+ and Ca^{2+} components and were followed by a post-burst AHP. These authors suggested that

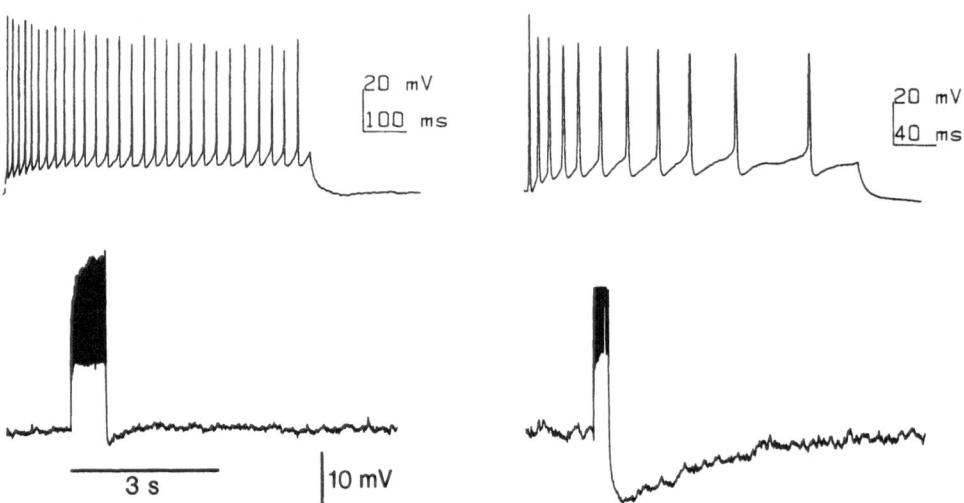

Fig. 3.   Example of two repetitively-firing cells.   A.   Voltage
traces in response to a 300 pA current pulse.   Top
record shows individual action potentials while the
bottom trace shows a chart record of the response at a
higher gain and slower time scale.   Note the absence
of a post-train AHP.   B.   A similiar protocol in
another cell shows pronounced spike-frequency adaptation (top) and the presence of a long-duration post-
train AHP (bottom).   Membrane potential for both
cells, -65 mV.

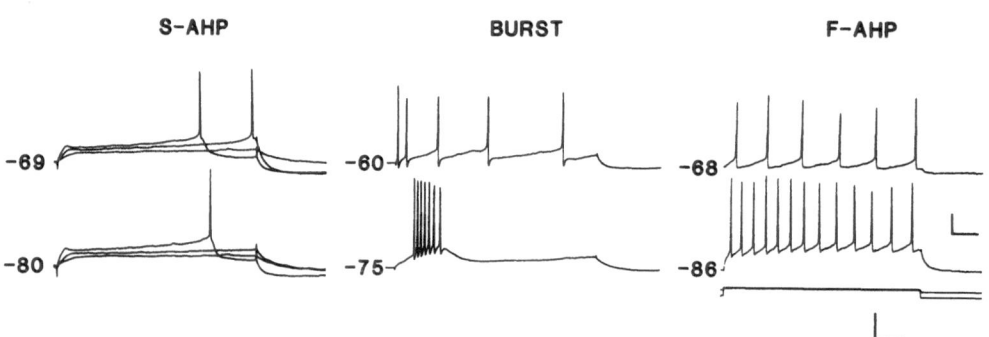

Fig. 4.   Voltage-dependent firing in different cell types.
Three different cells are shown at two membrane
potentials (to the left of the records).   The S-AHP
cell fired a single spike at both potentials, whereas
the F-AHP cell fired repetitively (at different rates)
at each potential.   Only the burst-firing cell changed
the pattern of firing - from a burst mode at hyperpolarized levels to repetitive mode at depolarized
potentials.   Calibration: top, 20 mV and 40 ms;
bottom, 1 nA and 40 ms.

a post-burst AHP along with other membrane properties which
limit repetitive firing may contribute to a sustained rhythmic
burst firing in some cells.

VOLTAGE-DEPENDENT CHANGES IN ACTION POTENTIAL DISCHARGE

In the previous section, several factors controlling repetitive firing were described for each cell type. In addition,
firing patterns can also be governed by the resting potential of
the cell. Voltage-dependent changes in spike discharge are prominent features of MS/nDB neurons and are most clearly seen in
burst-firing cells (Fig. 4, middle column). At positive membrane potentials (-60 mV), a substantial fraction of the LVA
Ca^{2+}-current is inactivated and therefore, a current stimulus
generates repetitive spiking. On the other hand, membrane
hyperpolarization (-75 mV) removes this inactivation and generates a different firing mode. Under these conditions, a depolarizing pulse is capable of generating a burst complex.
Similar voltage-dependent modes of firing have been described in
thalamic bursting neurons (Llinas and Jahnsen, 1982; Jahnsen and
Llinas, 1984).

The low percentage of bursting-neurons (10-15%) that we
record in vitro is in contrast to the approximately 50% of
rhythmically bursting septohippocampal cells recorded in vivo
(Lamour et al., 1984). One possible explanation for these differences is that both S-AHP and F-AHP cells have the intrinsic
capabilities to convert into a burst-firing mode with appropriate changes in membrane potential. However, this explanation
does not seem to be sufficient. As shown in Fig. 4, neither an
S-AHP nor an F-AHP cell generated a burst-firing pattern when
hyperpolarized prior to depolarizing current steps. Differences
in the number of spikes (during the 300 ms depolarization)
occurred at different membrane potentials, but the firing
pattern did not change, i.e., single spikes for S-AHP cells or
repetitive firing in F-AHP cells. Therefore, the intrinsic
properties of the identified cells alone are not enough to
explain the high percentage of rhythmic bursts recorded in vivo.
Alternatively, a complex synaptic circuitry of incoming afferents along with intraseptal synapses must also contribute to
rhythmic burst-firing in vivo.

CONTROL OF CELL DISCHARGE IN S-AHP CELLS: INTRINSIC PROPERTIES

In addition to the burst-firing pattern described above,
the rhythmic discharge of single spikes in S-AHP cells is also
dependent on membrane potential. Although most S-AHP cells are
quiescent at rest (-65 mV to -70 mV), a slow regular firing
pattern of single action potentials can be established with
sufficient membrane depolarization (Fig. 5). Once a threshold
level is attained, a regularity persists. This pattern can be
maintained for many minutes, suggesting a contribution of
intrinsic membrane properties to the rhythm. One hypothesis
proposed for a similar rhythm in other CNS neurons is a balance
between inward (depolarizing) Ca^{2+} conductances and one or more
outward (hyperpolarizing) K^+ conductances (Williams et., 1984;
Burlhis and Aghajanian, 1987). A number of conductances already
described in S-AHP cells could generate this regular firing
pattern. These conductances include, first, a high-threshold

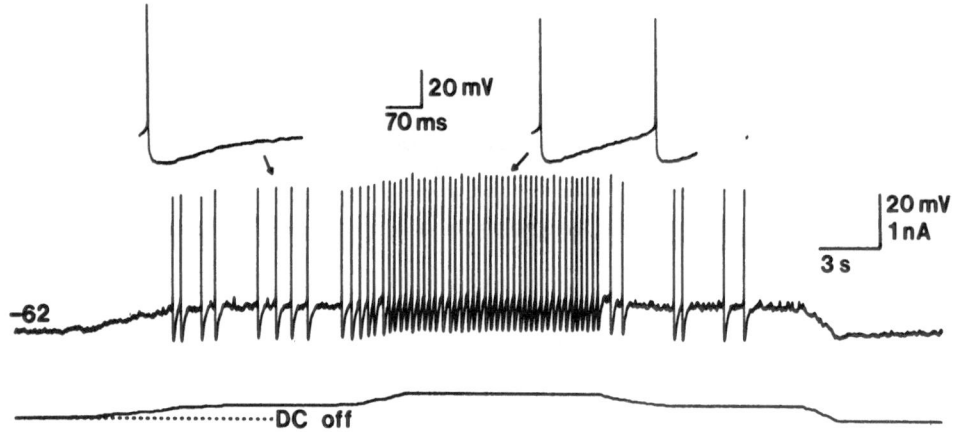

Fig. 5.   Sustained firing in an S-AHP cell. A normally quiescent cell can be induced to fire action potentials
when depolarized by dc current (lower records). Once
a stable level is maintained the cell fires rhythmically for the duration of the applied current. Individual action potentials are shown above at a faster
time base (modified from Griffith and Matthews, 1986).

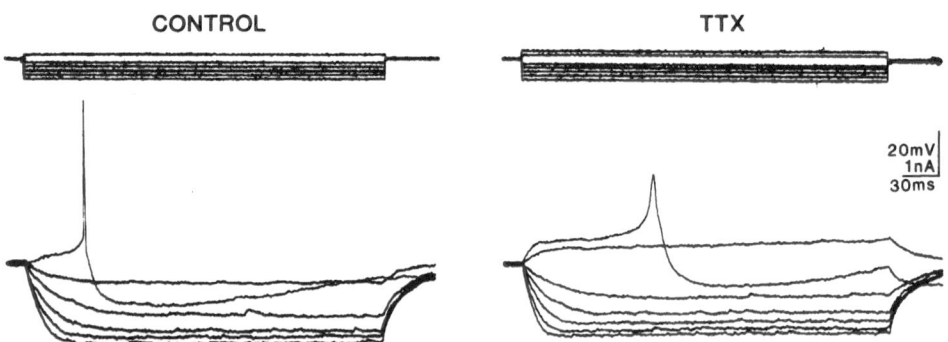

Fig. 6.   Tetrodotoxin (TTX)-insensitive spikes in an S-AHP
cell. In control, voltage records are shown in
response to hyperpolarizing and depolarizing current
pulses (top). A single action potential is generated
with depolarizing current. After the addition of TTX
(0.5 μM) a smaller, TTX-insensitive component of the
spike is seen. Note the slow-AHP that follows the
spike. Membrane potential was -61 mV.

Ca^{2+} spike, second, a Ca^{2+}-activated K$^+$ conductance that follows the spike, and third, a transient hyperpolarizing conductance that controls repolarization. A possible scenario could consist of a non-inactivating Ca^{2+} conductance which slowly depolarizes the cell until threshold is reached for spike generation. During the spike, sufficient Ca^{2+} enters to activate a repolarizing Ca^{2+}-activated K$^+$ hyperpolarization that approaches -75 to -80 mV. With decay of the K$^+$ conductance the membrane potential will slowly depolarize again with the rate controlled, in part, by the transient hyperpolarizing conductance. Briefly reviewed below is evidence for each of these three conductances in S-AHP cells.

1. High-threshold Ca^{2+} spike

Figure 6 shows a tetrodotoxin (TTX)-insensitive spike in an S-AHP cell. Addition of Cd^{2+} or zero-Ca^{2+}/high Mg^{2+} solutions blocked the small action potential suggesting a Ca^{2+}-sensitivity of the response (data not shown). The Ca^{2+}-spike was considered high-threshold since it was generated at potentials more positive than -60 mV after the cell was maintained for many minutes at these depolarized levels. Consistent with the observation of a high-threshold Ca^{2+}-spike, is the HVA Ca^{2+}-current recorded in our patch-clamp studies (Griffith et al., 1990).

2. Ca^{2+}-activated K$^+$ conductance (slow-AHP)

The K$^+$ conductance responsible for the slow-AHP has been described previously, including it's Ca^{2+}- and K$^+$-dependencies (Griffith, 1988). An example of a slow-AHP following either a mixed Na$^+$/Ca^{2+} action potential or a pure Ca^{2+}-dependent spike is shown in Fig. 6. The large hyperpolarizing conductance after the spike is sufficient to slow cell firing to 5-10 Hz even with substantial depolarizing current applied (Fig. 5 and unpublished observation). This is in contrast to F-AHP cells that can fire at frequencies of 100-200 Hz for many seconds with equivalent depolarizations (unpublished observation).

The importance of the slow-AHP in controlling firing rate has been confirmed in other experiments where the slow-AHP is selectively blocked by the bee venom toxin apamin. In 30 nM apamin, S-AHP cells fired much more rapidly in response to a depolarizing current when compared to pre-apamin control (Matthews and Lee, 1990).

3. Transient outward current (A-current)

A transient outward current, similar to that described previously (Connor and Stevens, 1971; Nehr, 1971) has been reported in all MS/nDB cell types (Griffith and Sim, submitted). The activation and inactivation characteristics of this current along with it's pharmacological sensitivity to 4-aminopyridine (4-AP) is quite similar to A-currents described in many different neurons (Rogawski, 1985). This conductance could contribute to the rate of repolarization following membrane hyperpolarization. This is observed under current-clamp, where a series of hyperpolarizing current steps reveal an undershoot or "droop" of the membrane potential upon repolarization (Fig. 7A, arrow). 4-AP (100 μM) effectively blocked this response and changed the rate of repolarization (Fig. 7B).

Using the single electrode voltage-clamp, the underlying current responsible for this transient hyperpolarization is shown in Fig. 7C. A clamp protocol was used such that an A-current was generated at -40 mV following a conditioning hyperpolarizing pulse. Voltage jumps as short as 10-20 ms could generate the current. A description of the voltage-dependency and pharmacology of this A-current has been performed (Griffith and Sim, submitted).

A second, more slowly inactivated K^+ conductance also sensitive to 4-AP is present in S-AHP cells (unpublished observation). An example of the slowly inactivating conductance is shown in Fig. 4 (S-AHP cell). Following depolarizing current steps from negative potentials, a transient hyperpolarization is observed at the beginning of the pulse. The slow rate of rise of the membrane potential to reach spike threshold can be controlled, in part, by this second 4-AP sensitive current. Similar results have been suggested for the influence of 4-AP sensitive currents in controlling firing rates of hippocampal pyramidal cells (Storm, 1988).

In summary, the three above mentioned conductances, i.e. a high-threshold Ca^{2+} spike, a Ca^{2+}-activated K^+ conductance and a transient A-current are present in S-AHP cells and provide a starting point to explain the slow, regular firing-pattern in S-AHP cells. Additional synaptic conductances, as well as inward rectification undoubtedly contribute to controlling firing rate, and a challenge in the future will be to discern more detailed mechanisms to describe this firing pattern.

SUMMARY AND CONCLUSIONS

Our data show that different cell types recorded in vitro can be identified by their intrinsic membrane properties. One type of neuron, namely S-AHP cells, have the ability to fire single action potentials in a rhythmic fashion following sufficient membrane depolarization. The rate is apparently controlled by several voltage-dependent conductances. S-AHP cells are normally quiescent at their resting potentials but will discharge once threshold is reached (-55 to -60 mV). Importantly, S-AHP (or F-AHP) cells will not convert into burst-firing neurons merely with changes in membrane potential. On the other hand, burst-firing cells have the ability to switch to a repetitive-firing pattern following membrane depolarization. All of these data provide a first step in an understanding of the firing rates of basal forebrain neurons, however, our results must be consolidated with existing in vivo studies for a more general understanding of basal forebrain function.

Comparing our data to an in vivo preparation of the MS/nDB with synaptic afferents surgically removed may be one approach to correlating in vitro and in vivo studies. Vinogradova et al. (1980) used single unit recording techniques in unanesthetized chronic rabbits and compared the firing rates of cells before and after deafferentation. These authors reported a preservation of burst-firing neurons (25% of the cells) after deafferentation but with a significant reduction in the mean frequency of bursts. In addition a higher percentage of regularly firing cells also occurred following deafferentation (Vinogradova et al., 1980). It is interesting to speculate that these regularly

firing cells may correspond to S-AHP cells in our in vitro studies, and some of the burst-firing units may correspond to the
burst-firing cells we record in slices.   Nevertheless, the in
vivo data strongly suggests that endogenous regular spiking as
well as rhythmic burst capabilities are present in some MS/nDB
cells, however, the firing rates of most MS/nDB neurons are

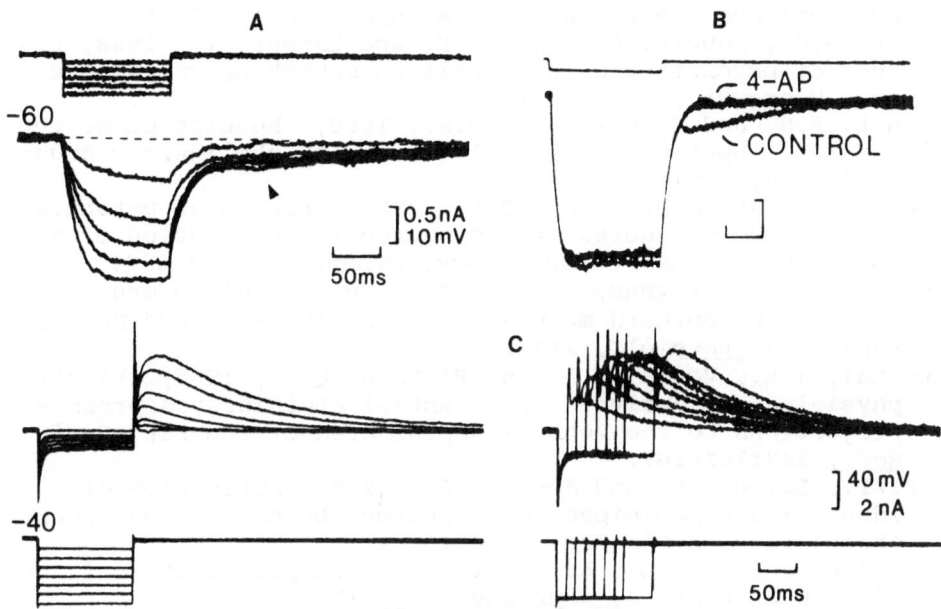

Fig. 7.   Delayed hyperpolarization and A-current. A. From -60
mV, increasingly larger current pulses (top) generate
membrane hyperpolarizations that display a delayed hyperpolarization upon repolarization (arrow). B. In
another cell, a similar protocol was used to show the
delayed hyperpolarization was blocked by 100 μM 4-
aminopyridine (4-AP). C. Voltage-clamp records of
outward currents generated following hyperpolarizing
jumps from -40 mV. To the right, time-dependency of
the outward current is shown. TTX (0.5 μM) was present in C.

strongly influenced by synaptic afferents (see also Vinogradova
et al., 1980; 1987). The endogenous activity in vivo can be explained, in part, by the intrinsic properties elucidated in our
in vitro studies.   How the synaptic afferents control MS/nDB
circuitry and integrative output is premature to speculate without a more thorough understanding of the synaptic mechanisms involved. It is possible that future in vitro studies will help
define these mechanisms and again contribute to an understanding
of basal forebrain function.

Acknowledgements: This work was supported by PHS Grant AG07805.

REFERENCES

Alvarez de Toledo, G. and Lopez-Barneo, J., 1988, Ionic basis of the differential neuron activity of guinea-pig septal nucleus studied in vitro, J. Physiol., 396:399-415.

Apostol, G. and Creutzfeldt, O.D., 1974, Crosscorrelation between the activity of septal units and hippocampal EEG during arousal, Brain Research, 67:65-75.

Assaf, S.Y. and Miller, J.J., 1978, The role of a raphe serotonin system in the control of septal unit activity and hippocampal desynchronization, Neuroscience, 3:539-550.

Bassant, M.H., Jobert, A., Dutar, P. and Lamour, Y., 1988, Effect of psychotropic on identified septohippocampal neurons, Neurosci., 27:911-920.

Brazhnik, E.S. and Vinogradova, O.S., 1988, Modulation of the afferent input to the septal neurons by cholinergic drugs, Brain Res., 451:1-12.

Burlhis, T.M. and Aghajanian, G.K., 1987, Pacemaker potentials of serotonergic dorsal raphe neurons: contribution of a low-threshold Ca^{2+} conducatance, Synapse, 1:582-588.

Connor, J.A. and Stevens, C.F., 1971, Voltage clamp studies of a transient outward membrane current in gastropod neural sonata, J. Physiol., 213:21-30.

Disturnal, J.E., Veale, W.L. and Pittman, Q.J., 1985, Electrophysiological analysis of potential arginine vasopressin projections to the ventral septal area of the rat, Brain Res., 342:162-167.

Dutar, P., Lamour, Y. and Jobert, A., 1985, Activation of identified septo-hippocampal neurons by noxious peripheral stimulation, Brain Res., 328:15-21.

Green, J.D. and Arduini, A., 1954, Hippocampal electrical activity in arousal, J. Neurophysiol., 17:533-557.

Griffith, W.H. and Matthews, R.T., 1986, Electrophysiology of AChE-positive neurons in basal forebrain slices, Neurosci. Lett., 71:169-174.

Griffith, W.H., 1988, Membrane properties of cell types within guinea pig basal forebrain nuclei in vitro, J. Neurophysiol., 59:1590-1612.

Griffith, W.H., Taylor, L. and Davis, M.J., 1990, Whole-cell calcium currents in acutely dissociated medial septum/ diagonal band neurons, Neurosci. Abstr., in press.

Griffith, W.H., Sim, J.A., Comparison of 4-aminopyridine and tetrahydroaminoacridine on basal forebrain neurons. (submitted).

Halliwell, J.V. and Adams, P.R., 1982, Voltage-clamp analysis of muscarinic excitation in hippocampal neurons. Brain Res., 250:71-92.

Hubbard, J.I., Mills, R.G. and Sirett, N.E., 1979, Responses in the diagonal band of Broca evoked by stimulation of the fornix in the cat, J. Physiol., 292:233-249.

Jahnsen, H. and Llinas, R., 1984, Electrophysiological properties of guinea-pig thalamic neurons: an in vitro study, J. Physiol., 349:205-226.

Lamour, Y., Dutar, P. and Jobert, A., 1984, Septo-hippocampal and other medial septum-diagonal band neurons: electrophysiological and pharmacological properties, Brain Res., 309:227-239.

Llinas, R. and Jahnsen, H., 1982, Electrophysiology of mammalian thalamic neurons in vitro, Nature, 297:406-408.

Matthews, R.T. and Lee, W.L., 1990, Effects of apamin on cho-
    linergic and noncholinergic medial septal/diagonal band
    (MS/DB) neurons of the guinea pig in vitro, Neurosci.
    Abstr., in press.
McLennan, H. and Miller, J.J., 1974, The hippocampal control of
    neuronal discharges in the septum of the rat, J. Physiol.,
    237:607-624.
Nehr, E., 1971, Two fast transient current components during
    voltage-clamp on snail neurons, J. Gen. Physiol., 58:36-53.
Rogawski, M.A., 1985, The A-current: how ubiguitous a feature
    of excitable cells is it?, TINS, 8:214-219.
Segal, M., 1974 Responses of septal nuclei neurons to microion-
    tophoretically administered putative neurotransmitters,
    Life Sci., 14:1345-1351.
Segal, M., 1976, Brain stem afferents to the rat medial septum,
    J. Physiol. 261:617-631.
Segal, M., 1986, Properties of rat medial septal neurons re-
    corded in vitro, J. Physiol., 379:309-330.
Stewart, M. and Fox, S.E., 1989, Two populations of rhythmi-
    cally bursting neurons in rat medial septum are revealed by
    atropine, J. Neurophysiol., 61:982-993.
Storm, J.F., 1988, Temporal integration by a slowly inactivat-
    ing K^+ current in hippocampal neurons, Nature, 336:379-381.
Vinogradova, O.S., Brazhnik, E.S., Karanov, A.M. and Zhadina,
    S.D., 1980, Neuronal activity of the septum following
    various types of deafferentation, Brain Res., 187:353-368.
Vinogradova, O.S., Zhadina, S.D. and Brazhnik, E.S., 1987, Back-
    ground activity pattern of guinea pig septal neurons in
    vitro, Neurophysiol. (Russia), 19:427-433.
Williams, J.T., North, R.A., Shefner, S.A., Nishi, S. and Egan,
    T.M., 1984, Membrane properties of rat locus coeruleus
    neurons, Neurosci., 13:137-156.

SUBSTANCE P EXCITES CULTURED CHOLINERGIC NEURONS IN THE BASAL FOREBRAIN

Yasuko Nakajima*, Peter R. Stanfield#,
Kazuhiko Yamaguchi= and Shigehiro Nakajima+

Department of Biological Sciences
Purdue University
West Lafayette, IN

INTRODUCTION

With the introduction of choline acetyltransferase immunocytochemistry, it has recently been shown that there are nuclei containing cholinergic neurons in the basal forebrain, which innervate wide areas of the cerebral cortex and hippocampus. These basal forebrain nuclei include the nucleus basalis of Meynert, the medial septal nucleus and the diagonal band nuclei. These cholinergic neurons in the basal forebrain nuclei severely degenerate in patients with Alzheimer's disease (Coyle et al., 1983; Terry and Katzman, 1983). Despite their clinical importance, physiological and pharmacological properties of these cholinergic neurons are not well understood.

Substance P, a tachykinin, seems to play an important role in the modulation of the cholinergic neurons in the basal forebrain. Bolam et al. (1986) and Beach et al. (1987) found that the cholinergic neurons are innervated by substance P containing nerves. In patients with Alzheimer's disease, substance P receptors in the basal forebrain (Dietl et al., 1986) and substance P in the cerebral cortex (Beal and Mazurek, 1987; Crystal and Davies, 1982) are reduced as compared with other peptides. In animal behavior studies, facilitation of learning has been reported by injection of substance P into the nucleus basalis (Kafetzopoulos et al., 1986), as well as by the peripheral administration of substance P (Tomaz and Huston, 1986). Substance P is known to pass the blood brain barrier (Banks and Kastin, 1985). Therefore, administration of substance P may have a therapeutic effect in patients with Alzheimer's disease.

In 1985 we developed primary dissociated cholinergic neuron cultures from the basal forebrain nuclei of the rat and found that substance P excites these cholinergic neurons (Nakajima et al., 1985). Subsequently, we found that this excitation by substance P is due to the suppression of

---

*Present address: Department of Anatomy and Cell Biology, University of
  Illinois College of Medicine at Chicago, Chicago, IL. 60612
#Present address: Department of Physiology, University of Leicester,
  Leicester,LE1 7RH. England.
=Present address: National Institute for Physiological Sciences, Myodaiji,
  Okazaki,444 Japan.
+Present address: Department of Pharmacology, University of Illinois
  College of Medicine at Chicago, Chicago, IL. 60612

inwardly rectifying K-channels (Stanfield et al., 1985; Yamaguchi et al., 1990). This action of substance P on the K-channels is mediated through a pertussis toxin-insensitive G protein (Nakajima et al., 1988). Here we will describe the biophysical properties of the K-channels that are modulated by substance P and discuss the signal transduction mechanism of this substance P action.

PRIMARY CELL CULTURES OF CHOLINERGIC NUCLEI IN THE BASAL FOREBRAIN

We have established a technique of making primary cell cultures of brain nuclei of postnatal rats and mice (Nakajima et al., 1985; Masuko et al., 1986) and have succeeded in making cholinergic neuron cultures of the nucleus basalis of Meynert separately from other cholinergic nuclei. We have also cultured the medial septal nucleus and diagonal band nuclei together. The salient features of our technique are to make brain slices with a vibratome (Lancer 1000) and to remove the brain nucleus tissue (for example, the nucleus basalis) under direct visual observation with a dissecting microscope. These brain tissues were dissociated and cultured on a feeder layer of astrocytes. Figure 1 is an example of brain slices treated with acetylcholinesterase (AChE) histochemistry, showing the nucleus basalis, the distinctly dark areas at the ventral medial aspect of the globus pallidus. Recently we have improved our previously reported method (Nakajima et al., 1985) by using papain instead of trypsin for dissociation and by supplementing the culture medium with 5% rat serum and 10% horse serum instead of 10% fetal bovine serum and 10% horse serum (Nakajima et al., 1988). These modifications yield healthier cultures.

The cholinergic nature of the cultured neurons was determined with choline acetyltransferase (ChAT) immunocytochemistry. The cholinergic neurons were large with a soma diameter of 20-25 μm. If we counted only large neurons (20 μm or larger in diameter), about 75% were cholinergic. Therefore, we did electrophysiological experiments on large neurons. In order to confirm that the neurons were indeed cholinergic, in some

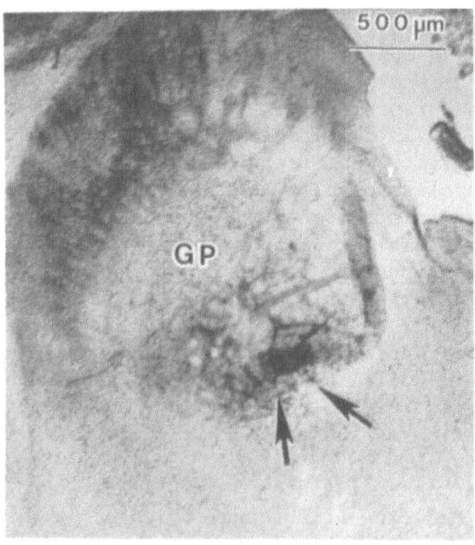

Fig. 1. Coronal vibratome section (150 μm thick) of the forebrain of a 2-day old newborn rat. The brain slice was treated with AChE histochemistry. The nucleus basalis (arrows) is a dark area positive to the AChE reaction, located at the ventromedial aspect of the globus pallidus (GP).

experiments we treated the neurons with AChE histochemistry after physiological recording. Levey et al. (1983) reported that at least in the basal forebrain, neurons with strong AChE positivity are cholinergic since all of them also showed ChAT positivity. We found that 12 out of 13 recorded cells produced AChE positivity (Nakajima et al., 1985). Figure 2A illustrates cholinergic neurons showing ChAT immunoreactivity, and Figure 2B shows a neuron strongly positive to the AChE reaction.

Cultured cholinergic neurons did not usually show spontaneous firing of action potentials, but steady depolarizations produced trains of action potentials, which adapted relatively quickly (Fig. 3A and B) (Nakajima et al., 1985). This fast spike adaptation of cultured cholinergic neurons is in sharp contrast to the absence of adaptation in potentials of cultured noradrenergic neurons from the locus coeruleus upon steady depolarization (Masuko et al., 1986). However, cholinergic neurons occasionally (5 of 61 cells) produced spontaneous spike activity (Fig. 3C and D) (Nakajima et al., 1985). Four cells among these rare five cells appeared to be contacted by other neurons, and three out of these five cells possessed small spontaneous potential changes resembling postsynaptic potentials. Possibly, the activity of these neurons was driven by presynaptic elements.

ELECTROPHYSIOLOGICAL STUDIES OF SUBSTANCE P EFFECTS

Cultured cholinergic neurons from the nucleus basalis of Meynert were used. Experiments were performed with the whole cell-patch clamp method (Hamill et al., 1981). During the experiments cultures were superfused with an oxygenated Krebs solution containing 146 mM NaCl, 5 mM KCl, 2.4 mM CaCl$_2$, 1.3 mM MgCl$_2$, 11 mM D-glucose, 5 mM HEPES/NaOH buffer and 1 μM tetrodotoxin (pH 7.4). The standard pipette (internal) solution contained 120 mM K-aspartate, 40 mM NaCl, 3 mM MgCl$_2$, 0.5 mM EGTA, 0.25 mM CaCl$_2$, 2 mM Na$_2$ATP, 5 mM HEPES/KOH buffer and ~ 6 mM KOH (pH 7.2). Substance P was applied by pressure ejection from a glass pipette placed about 10-20 μm from the cell body. The bath temperature was 29-34°C.

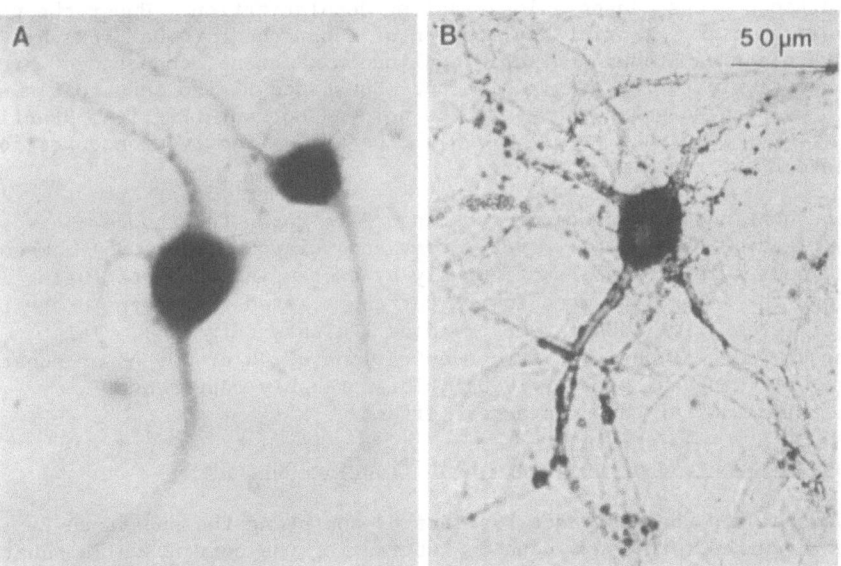

Fig. 2. Dissociated cultured cholinergic neurons from the nucleus basalis of the rat. (A) Neurons showing immunoreactivity to ChAT. Cultured for 21 days. (B) A neuron showing a strong histochemical reaction to AChE. Cultured for 23 days.

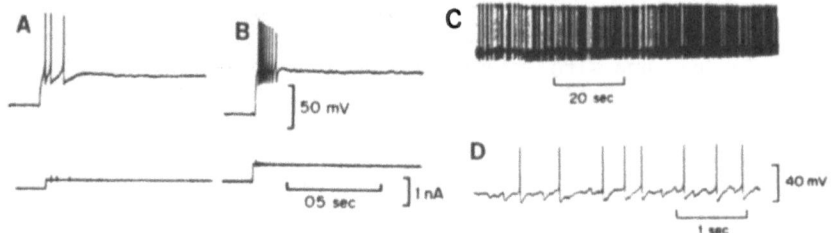

Fig. 3. (A and B) Trains of action potentials (upper traces) elicited by
stepwise constant currents (lower traces) in cultured neurons
from medial septal and diagonal band nuclei. The whole-cell
version of patch clamp using the current-clamp mode was used.
Resting potential was -89 mV. Reproduced from Nakajima et al.
(1985). (C and D) Spontaneous firing recorded from a neuron
cultured from the nucleus basalis of Meynert. The quasi-steady
potential between the spikes was about -63 mV. An intracellular
microelectrode was used to record the potential (See Masuko et
al., 1986, for the method used). The neurons were cultured for
32 days and recorded at 32°C.

Substance P Application Produced a Slow Inward Current and a Decrease in Membrane Conductance. Continuous Presence of Substance P produced Desensitization

Substance P is known to excite certain neurons (Dun and Karczmar, 1979;
Nicoll et al., 1980; Otsuka et al., 1982), and Adams et al. (1983) reported
that substance P produces depolarization in sympathetic neurons by
inhibiting the M-current. Application of substance P on our cultured
cholinergic neurons caused a long-lasting depolarization. Under the whole-
cell voltage clamp, a brief application of substance P (0.06-3 μM) by
pressure ejection produced a rapid decline in membrane conductance which
recovered slowly to its original value. Substance P also induced a small
(70-100 pA) inward-going change in the holding current (Fig. 4A; Stanfield
et al., 1985; Yamaguchi et al., 1990). The half-effective concentration of
substance P was roughly 40 nM.

The continuous presence of substance P is known to produce
desensitization (Lee et al., 1982, Dryer and Chiappinelli, 1985). When we
applied substance P (3 μM) continuously by exchanging the superfusing
solution, the conductance and inward current started to return to their
original levels with a half-time of about 1 minute (Fig. 5A). This
phenomenon was also observed at concentrations of 30 or 300 nM of substance
P (Fig. 5B; Yamaguchi et al., 1990). This probably represents
"desensitization" of the substance P effect.

Substance P Suppressed the Inward Rectification Current

We analyzed the substance P effect by computing the substance P-
sensitive current: this was done by subtracting the current while substance
P was effective from the current before applying substance P (control
current). The substance P-sensitive current plotted against voltage showed
a quite strong inward rectification (Fig. 4D) revealing the reversal
potential, approximately coinciding with E_K (the potassium equilibrium
potential) (Fig. 4D). Thus, it is very likely that the substance P-

sensitive current occurs mainly because a potassium conductance is inhibited (Stanfield et al., 1985; Yamaguchi et al., 1990). We have calculated the substance P-sensitive chord conductance from the current data. Although there was considerable scatter in our data, the relation between the conductance and voltage could be fitted by an equation of the Boltzmann type. Changes in the external potassium concentration shifted this relation along the voltage axis approximately in parallel with the shift in the potassium equilibrium potential (Stanfield, et al., 1985; Yamaguchi et al., 1990). This means that the activation of substance P-sensitive channels cannot be determined only by voltage, but is a function of the driving force

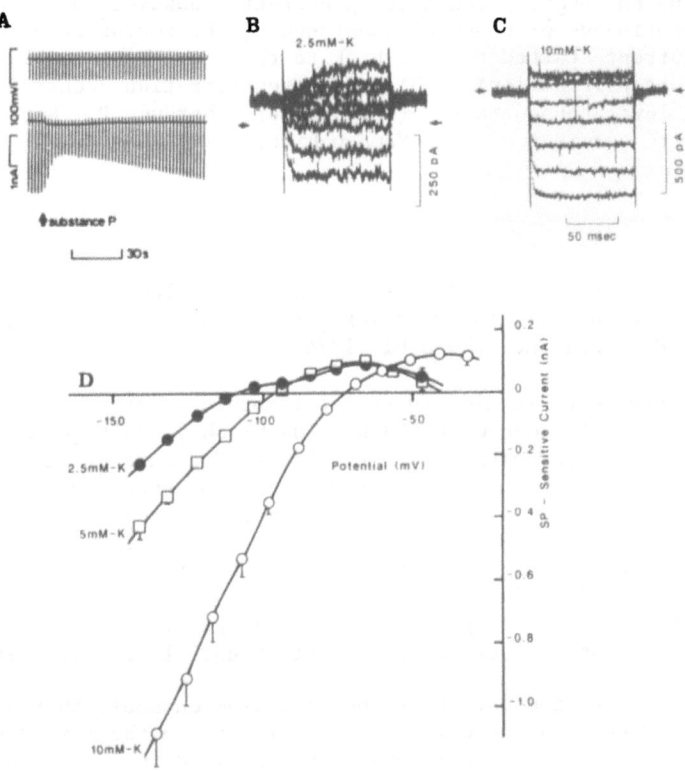

Fig. 4. Substance P effects on nucleus basalis neurons. (A) Application of 3 μM substance P (arrow) produced an inward current concomitant with a decrease in current under voltage-clamp. Intermittent voltage pulses were applied: each pulse is a square-wave depolarization (20 mV, 100 msec) followed by an interval (100 msec) and a hyperpolarization (60 mV, 100 msec). (B) Substance P-sensitive currents in 2.5 mM K solution were computed by subtracting the record during substance P action from the control record. Holding potential was -84 mV. Potential level for each record was (from the uppermost): -56, -66, -75, -94, -103, -113, -122, and -132 mV. The arrows indicate the zero level of substance P current. (C) Substance P-sensitive currents in 10 mM K solution. The holding potential was -69 mV; the potential level for each record was (from uppermost): -41, -51, -60, -79, -88, -98, -107, and -117 mV. (D) Mean values of substance P-sensitive currents plotted against potential for three $[K^+]_0$ solutions measured at 10-15 msec. Vertical bars, 1 s.e.m.. (A), reproduced, with permission, from Stanfield et al. (1985). (B) to (D), reproduced, with permission, from Yamaguchi et al. (1990).

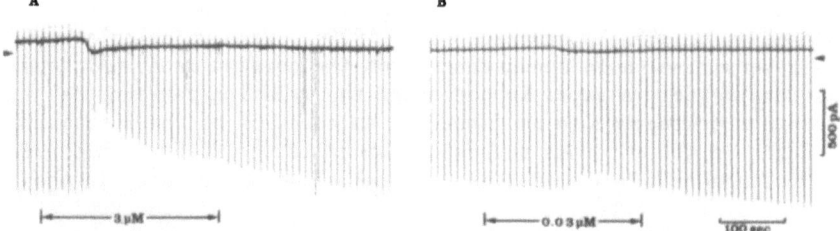

Fig. 5. (A) Application of substance P (3 μM) by exchanging the
superfusing solution induced a decrease in membrane conductance
and an inward-going holding current. However, despite the
continuous presence of substance P, the conductance and inward
current started to come back to their original levels
(desensitization). (B) This desensitization occurred even with
a lower concentration (0.03 μM) of substance P. Reproduced,
with permission, from Yamaguchi et al. (1990).

for potassium ions, namely, (E-E$_K$), in agreement with the situation in
skeletal muscle (Hodgkin and Horowicz, 1959; Leech and Stanfield, 1981) or
in oocytes (Hagiwara and Takahashi, 1974).

The inward rectification in skeletal muscle or oocytes is known to be
blocked by Cs$^+$ at low concentrations. The block is voltage-dependent, with
hyperpolarization producing a stronger block (Hagiwara et al., 1976; Gay
and Stanfield, 1977). In our cultured cells, the substance P-sensitive
current was blocked by Cs$^+$ at 0.1 mM in a voltage-dependent manner with an
equivalent valence for Cs$^+$ of 1.9 (Yamaguchi et al., 1990). This suggests
that Cs$^+$ blockage occurs at a multi-ion pore (Hille and Schwarz, 1978).
Ba^{2+} also blocked the substance P-sensitive current; in this block we
observed virtually no voltage-dependency (Yamaguchi et al., 1990). Rb$^+$ also
blocked the substance P-sensitive current (Stanfield et al., 1985).

All these data indicate that the potassium channels that are modulated
by substance P have characteristics very similar to the inwardly rectifying
channels first found by Katz in 1949, and analyzed in detail in skeletal
muscle or oocytes by Stanfield et al. (1981) and by Hagiwara (1983).
Previously, in cardiac muscle an inward rectifier was implicated as the
conductance which is responsible for the muscarine-induced hyperpolarization
(Sakmann et al., 1983). Also, in Aplysia neurons, an inwardly rectifying
current was implicated for the effect of serotonin (Benson and Levitan,
1983). In vertebrate neurons Stanfield et al. (1985) were the first to show
that the inwardly rectifying potassium channels serve as the effector for
the modulatory influence by a transmitter substance.

Substance P-Sensitive Current Did Not Show Time-Dependent Inactivation,
Whereas the Background Current Showed Time-Dependent Inactivation

In skeletal muscle and oocytes, hyperpolarization activates the
inwardly rectifying current, which soon starts to decay (inactivation)
(Adrian and Freygang, 1962; Almers, 1972; Ohmori, 1978; Stanfield et al.,
1981; Hagiwara, 1983). The time course of the substance P-sensitive current
shows that no inactivation occurs even at large hyperpolarizing potentials
(Fig. 4B and C). On the other hand, the background current (i.e., the
current remaining after the effect of substance P) showed an inactivation
upon hyperpolarization, suggesting that the normal membrane of the
cholinergic neuron has two types of potassium channels at the resting level:

one which is modulated by substance P (modifiable K-channels) and the other which is resistant to modulation (non-modifiable K-channels) (Yamaguchi et al., 1990).

INTRACELLULAR SIGNAL TRANSDUCTION OF THE SUBSTANCE P EFFECT

Nakajima et al. (1988) found that when the cholinergic neuron was pre-loaded with a non-hydrolyzable GTP analogue, GTPγS (guanosine 5'-[γ-thio]triphosphate) or Gpp(NH)p (5'-guanylyl imidodiphosphate), application of substance P produces an almost irreversible inhibition of the inward rectifying current, suggesting that the substance P effect is mediated by a GTP-binding protein (G protein). Furthermore, the substance P effect on the inward rectifier was pertussis toxin-insensitive (Nakajima et al., 1988). Therefore, the G protein involved is not G_i or G_o or transducin (G_t). In his study on sympathetic neurons, Pfaffinger (1988) also reported that a pertussis toxin-insensitive G protein is involved in the modulation of the M-current by muscarine or by t-LHRH.

There are several kinds of pertussis toxin-insensitive G proteins such as G_s (a stimulatory G protein), $G_{(x\ or\ z)}$ (G_x and G_z are the same G protein but named differently; Matsuoka et al., 1988; Fong et al., 1988), or low molecular weight GTP-binding proteins such as rap1-b (Siess et al., 1990) and p21ras. Among them G_s is unlikely to mediate the substance P effect because: (1) neither elevating nor lowering the level of intracellular cyclic AMP influenced the effect of substance P (Nakajima et al., 1988); and (2) cholera toxin, which activates G_s through ADP ribosylation (Cassel and Selinger, 1977) did not change the substance P action (unpublished data, Kozasa, Nakajima and Nakajima). Presently, we do not know what kind of G protein mediates the substance P effect on the inward rectification.

Using our technique of culturing brain nuclei, Masuko et al. (1986) succeeded in obtaining cultured noradrenergic neurons from the locus coeruleus. Subsequently, Inoue et al. (1988) studied the effects of somatostatin on the noradrenergic neurons. They found that somatostatin enhances the activity of inwardly rectifying potassium channels. The properties of this somatostatin-induced potassium conductance have a striking similarity to those of the potassium conductance inhibited by substance P (for somatostatin effects on other preparations, see Mihara et al., 1987; Pennefather et al., 1988; Yamashita et al., 1988). Thus, it appears that substance P and somatostatin are acting on the same channels, one reducing their activity, the other enhancing it. Furthermore, Inoue et al. (1988) found that a G protein is involved in the signal transduction of the somatostatin-induced enhancement of the potassium conductance in locus coeruleus neurons. However, the somatostatin effect, unlike the effect of substance P, was abolished completely by treating the neurons with pertussis toxin.

In the experiments so far described in this chapter, locus coeruleus neurons were used for the study of somatostatin effect, whereas nucleus basalis neurons were used for analyzing the substance P effect. However, we have preliminary evidence that substance P also produces excitation in locus coeruleus neurons, and the excitation seems to be brought about, at least partly, by a reduction in the activity of inwardly rectifying potassium channels (Masuko et al., 1986; Nakajima et al., 1988). Thus, it is very plausible that in the locus coeruleus neuron, the same potassium channel is regulated by somatostatin and substance P in an opposite manner.

Figure 6 is a hypothetical scheme describing the situation for the locus coeruleus neuron. In the center there is an inwardly rectifying potassium channel which could be termed as a modifiable potassium channel:

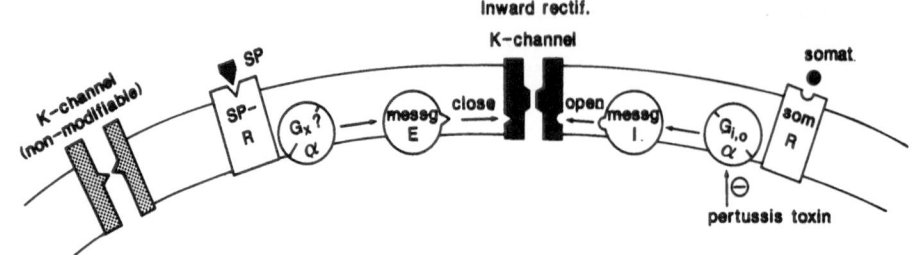

Fig. 6. A hypothetical scheme which is consistent with our results of
substance P and somatostatin experiments (Nakajima et al., 1988;
Inoue et al., 1988; Yamaguchi et al., 1990). SP (substance P)
attaches to the SP-R (substance P receptor). This activates a G
protein (this could be $G_{(x\ or\ z)}$, which is pertussis toxin-
<u>insensitive</u>). Then α-subunit is separated, and activates messg
E (messenger E), which binds to an (internal) receptor site of
the K-channel, causing closing of the channel. Messenger E may
not exist, and α-subunit may act directly on the K-channel. The
somatostatin part is almost the mirror image of the substance P
part. However, the G protein is pertussis toxin-<u>sensitive</u>, and
probably represents G_i or G_o. Somato = somatostatin. Som R =
somatostatin receptor. Messg I = hypothetical messenger, which
could be α_i or α_o subunit itself. To the far left, there is a K-
channel, which is not modified by the messengers. This
constitutes the main background conductance.

its activity is enhanced by somatostatin through a pertussis toxin-sensitive
G protein (G_i or G_o) or decreased by substance P <u>via</u> a pertussis toxin-
insensitive G protein ($G_{(x\ or\ z)}$?). It is still an open question whether or
not an excitatory or inhibitory messenger (messenger E or I in Fig. 6) is
the α-subunit of the respective G protein. In addition to this K-
conductance, there are other K-channels that are resistant to the modulatory
influences. They will remain active after most of the modifiable K-channels
are inhibited by substance P. We have shown that in the case of the
modifiable K-channels, a sudden hyperpolarizing step produces an inward
current which does not inactivate (Fig. 4B and C), whereas in the case of
the non-modifiable K-channels, hyperpolarizations do show inactivation
(Yamaguchi et al., 1990). Future experiments will tell us whether the
hypothetical scheme in Figure 6 is correct or needs modification.

ACKNOWLEDGEMENT

We thank Ms Linda Johnston and Miss Alyse Kondrat for editing the
manuscript. This research was supported by a NIH grant AG06098, an
Alzheimer's Disease and Related Disorders Association Grant to Yasuko
Nakajima, and by the Wellcome Trust to Peter R. Stanfield.

REFERENCES

Adams, P. R., Brown, D. A., and Jones, S. W., 1983, Substance P inhibits the
    M-current in bullfrog sympathetic neurones, <u>Br. J. Pharmac.</u>,
    79:330-333.
Adrian, R. H., and Freygang, W. H., 1962, The potassium and chloride
    conductance of frog muscle membrane, <u>J. Physiol.</u>, 163:61-103.

Almers, W., 1972, Potassium conductance changes in skeletal muscle and the potassium concentration in the transverse tubules, J. Physiol., 225:33-56.

Banks, W. A., and Kastin, A. J., 1985, Peptides and the blood-brain barrier: Lipophilicity as a predictor of permeability, Br. Res. Bull., 15:287-292.

Beach, T. G., Tago, H., and McGeer, E. G., 1987, Light microscopic evidence for a substance P-containing innervation of the human nucleus basalis of Meynert, Brain Res., 408:251-257.

Beal, M. F., and Mazurek, M. F., 1987, Substance P-like immunoreactivity is reduced in Alzheimer's disease cerebral cortex, Neurology, 37:1205-1209.

Benson, J. A., and Levitan, I. B., 1983, Serotonin increases an anomalously rectifying K+ current in the Aplysia neuron R15, Proc. Natl. Acad. Sci., U. S. A., 80:3522-3525.

Bolam, J. P., Ingham, C. A., Izzo, P. N., Levey, A. I., Rye, D. B., Smith, A. D., and Wainer, B. H., 1986, Substance P-containing terminals in synaptic contact with cholinergic neurons in the neostriatum and basal forebrain: A double immunocytochemical study in the rat, Brain Res., 397:279-289.

Cassel, D. and Selinger, Z., 1977, Mechanism of adenylate cyclase activation by cholera toxin: Inhibition of GTP hydrolysis at the regulatory site. Proc. Natl. Acad. Sci., U. S. A., 74:3307-3311.

Coyle, J. T., Price, D. L., and DeLong, M. R., 1983, Alzheimer's disease: A disorder of cortical cholinergic innervation, Science, 219:1184-1190.

Crystal, H. A., and Davies, P., 1982, Cortical substance P-like immunoreactivity in cases of Alzheimer's disease and senile dementia of the Alzheimer type, J. Neurochem., 38:1781-1784.

Dietl, M., Probst, A., and Palacios, J. M., 1986, Mapping of substance P receptor sites in the human brain: High densities in the substantia innominata and effect of senile dementia, Soc. Neurosci. Abstr., 12:831.

Dryer, S. E., and Chiappinelli, V. A., 1985, Properties of choroid and ciliary neurons in the avian ciliary ganglion and evidence for substance P as a neurotransmitter, J. Neurosci., 5:2654-2661.

Dun, N. J., and Karczmar, A. G., 1979, Actions of substance P on sympathetic neurons, Neuropharmacology., 18:215-218.

Fong, H. K. W., Yoshimoto, K. K., Eversole-Cire, P., and Simon, M. I., 1988, Identification of a GTP-binding protein α subunit that lacks an apparent ADP-ribosylation site for pertussis toxin, Proc. Natl. Acad. Sci., U. S. A., 85:3066-3070.

Gay, L. A., and Stanfield, P. R., 1977, Cs+ causes a voltage-dependent block of inward K currents in resting skeletal muscle fibres, Nature, 267:169-170.

Hagiwara, S., 1983, "Membrane potential-dependent ion channels in cell membrane. Phylogenetic and developmental approaches", Raven Press, New York.

Hagiwara, S., Miyazaki, S., and Rosenthal, N. P., 1976, Potassium current and the effect of cesium on this current during anomalous rectification of the egg cell membrane of a starfish, J. Gen. Physiol., 67:621-638.

Hagiwara, S., and Takahashi, K., 1974, The anomalous rectification and cation selectivity of the membrane of a starfish egg cell, J. Membrane Biol., 18:61-80.

Hamill, O. P., Marty, A., Neher, E., Sakmann, B., and Sigworth, F. J., 1981, Improved patch-clamp techniques for high-resolution current recording from cells and cell-free membrane patches, Pflüger's Arch., 391:85-100.

Hille, B., and Schwarz, W., 1978, Potassium channels as multi-ion single-file pores, J. Gen. Physiol., 72:409-442.

Hodgkin, A. L., and Horowicz, P., 1959, The influence of potassium and chloride ions on the membrane potential of single muscle fibres, J. Physiol., 148:127-160.

Inoue, M., Nakajima S., and Nakajima, Y., 1988, Somatostatin induces an inward rectification in rat locus coeruleus neurones through a pertussis toxin-sensitive mechanism, J. Physiol., 407:177-198.

Kafetzopoulos, E., Holzhäuer, M.-S., and Huston, J. P., 1986, Substance P injected into the region of the nucleus basalis magnocellularis facilitates performance of an inhibitory avoidance task, Psychopharmacology, 90:281-283.

Katz, B., 1949, Les constantes électriques de la membrane du muscle, Arch. Sci. Physiol., 3:285-299.

Lee, C. M., Iversen, L. L., Hanley, M. R. and Sandberg, B. E. B., 1982, The possible existence of multiple receptors for substance P, Naunyn-Schmiedeberg's Arch. Pharm., 318:281-287.

Leech, C. A., and Stanfield, P. R., 1981, Inward rectification in frog skeletal muscle fibres and its dependence on membrane potential and external potassium, J. Physiol., 319:295-309.

Levey, A. I., Wainer, B. H., Mufson, E. J., and M.-M. Mesulam, 1983, Co-localization of acetylcholinesterase and choline acetyltransferase in the rat cerebrum, Neuroscience, 9:9-22.

Masuko, S., Nakajima, Y., Nakajima, S., and Yamaguchi, K., 1986, Noradrenergic neurons from the locus coeruleus in dissociated cell culture: Culture methods, morphology and electrophysiology, J. Neurosci., 6:3229-3241.

Matsuoka, M., Itoh, H., Kozasa, T., and Kaziro, Y., 1988, Sequence analysis of cDNA and genomic DNA for a putative pertussis toxin-insensitive guanine nucleotide-binding regulatory protein α subunit, Proc. Natl. Acad. Sci., U. S. A, 85:5384-5388.

Mihara, S., North, R. A., and Surprenant, A., 1987. Somatostatin increases an inwardly rectifying potassium conductance in guinea-pig submucous plexus neurones. J. Physiol., 390:335-355.

Nakajima, Y., Nakajima, S., and Inoue, M., 1988, Pertussis toxin-insensitive G protein mediates substance P-induced inhibition of potassium channels in brain neurons, Proc. Natl. Acad. Sci., U. S. A., 85:3643-3647.

Nakajima, Y., Nakajima, S., Obata, K., Carlson, C. G., and Yamaguchi, K., 1985, Dissociated cell culture of cholinergic neurons from nucleus basalis of Meynert and other basal forebrain nuclei, Proc. Natl. Acad. Sci., U. S. A. 82:6325-6329.

Nicoll, R. A., Schenker, C., and Leeman, S. E., 1980, Substance P as a transmitter candidate, Ann. Rev. Neurosci., 3:227-268.

Ohmori, H., 1978, Inactivation kinetics and steady-state current noise in the anomalous rectifier of tunicate egg cell membranes, J. Physiol., 281:77-99.

Otsuka, M., Konishi, S., Yanagisawa, M., Tsunoo, A., and Akagi, H., 1982, Role of substance P as a sensory transmitter in spinal cord and sympathetic ganglia, in: "Ciba Foundation Symposium, Vol. 91: Substance P in the Nervous System", Pitman, London, pp. 13-34.

Pennefather, P. S., Heisler, S. and MacDonald, J. F., 1988, A potassium conductance contributes to the action of somatostatin-14 to suppress ACTH secretion, Brain. Res., 444:346-350.

Pfaffinger, P., 1988, Muscarine and t-LHRH suppress M-current by activating an IAP-insensitive G-protein, J. Neurosci., 8:3343-3353.

Sakmann, B., Noma, A., and Trautwein, W., 1983, Acetylcholine activation of single muscarinic K^+ channels in isolated pacemaker cells of the mammalian heart, Nature, 303:250-253.

Siess, W., Winegar, D. A., and Lapetina, E. G., 1990, Rap1-b is phosphorylated by protein kinase A in intact human platelets, Biochem. Biophys. Res. Commun., 170: 994-950.

Stanfield, P. R., Nakajima, Y., and Yamaguchi, K., 1985, Substance P raises neuronal membrane excitability by reducing inward rectification, Nature, 315:498-501.

Stanfield, P. R., Standen, N. B., Leech, C. A., and Ashcroft, F. M., 1981, Inward rectification in skeletal muscle fibres, in: "Adv. Physiol. Sci., Vol. 5: Molecular and cellular aspects of muscle function", Varger, E., Köver, A., Kovács, T., and Kovács, L., eds., Pergamon Press, New York, pp. 247-262.

Terry, R. D., and Katzman, R., 1983, Senile dementia of the Alzheimer type, Ann. Neurol., 14:497-506.

Tomaz, C., and Huston, J. P., 1986, Facilitation of conditioned inhibitory avoidance by post-trial peripheral injection of substance P, Pharmacol. Biochem. Behav., 25:469-472.

Yamaguchi, K., Nakajima, Y., Nakajima, S., and Stanfield, P. R., 1990, Modulation of inwardly rectifying channels by substance P in cholinergic neurones from rat brain in culture, J. Physiol., 426:499-520.

Yamashita, N., Shibuya, N., and Ogata, E., 1988, Requirement of GTP on somatostatin-induced K^+ current in human pituitary tumor cells, Proc. Natl. Acad. Sci., U. S. A., 85:4924-4928.

NEUROMODULATORY ACTIONS OF DOPAMINE AND CHOLECYSTOKININ IN

THE VENTRAL STRIATUM

Conrad Chi-Yiu Yim, Lisa Sheehy and Gordon Mogenson

Departments of Clinical Neurological Sciences and Physiology,
University of Western Ontario, London, Ontario, Canada, N6A 5A5

Electrophysiological investigations of the cellular actions of
dopamine had focused in the past on its direct action on cell bodies of
the postsynaptic neuron, assuming dopamine to be a neuromediating transmitter. Recent evidence suggests, however, that while dopamine undoubtedly has actions on the cell bodies of postsynaptic neurons, it may also
have an important presynaptic neuromodulatory action on non-dopaminergic
inputs to the ventral striatum (see Yim and Mogenson, 1986). There is
also evidence that peptides such as cholecystokinin which coexists with
dopamine in a subpopulation of the mesolimbic dopamine neurons may in
turn modulate the neuromodulatory action of dopamine.

The term neuromodulation does not have a very clear and precise
definition. Some investigators (Kupfermann, 1979; Kaczmarek and Levitan,
1987) consider it to be an action produced by a neurotransmitter that
does not involve direct membrane depolarization or hyperpolarization but
results in the modification of the postsynaptic effects brought about by
another neurotransmitter. The action of dopamine does not fit this
definition of neuromodulation entirely since there are reports of direct
postsynaptic effects of dopamine on neurons in the striatum as well as in
other brain areas (Kitai et al., 1976; Herrling and Hull, 1980; Bernardi
et al., 1978; Mercuri et al., 1985). However, some of the actions of
dopamine in the ventral striatum are still best described as neuromodulatory. For some years, research in our laboratory had been directed
towards investigating the functional role of dopamine in the nucleus
accumbens, and we have found in electrophysiological as well as
behavioural experiments evidence that dopamine at a titrated level of
activity modulates amygdala and hippocampal inputs to the accumbens
without producing direct effects on its own (Yim and Mogenson, 1982; Yang
and Mogenson, 1984; Yim and Mogenson, 1986; Yim and Mogenson, 1988; Yim
and Mogenson, 1989). Similar findings have since been reported by other
investigators interested in the actions of dopamine in the ventral as
well as dorsal striatum (Hirata et al., 1984; Bergstrom and Walters,
1984; Ferron et al., 1984; Abercrombie & Jacobs, 1985; Vives and
Mogenson, 1986; Thierry et al., 1988).

The Basal Forebrain, Edited by T.C. Napier *et al.*
Plenum Press, New York, 1991

ELECTROPHYSIOLOGICAL INVESTIGATIONS OF THE NEUROMODULATORY ACTION OF DOPAMINE IN THE NUCLEUS ACCUMBENS

Evidence of a neuromodulatory function for dopamine in the nucleus accumbens came initially from electrophysiological experiments. Single unit recordings made from the nucleus accumbens showed that accumbens neurons are strongly excited by amygdala (Yim and Mogenson, 1982) or hippocampal stimulation (Yang and Mogenson, 1984). These excitatory responses are attenuated by dopamine, either released endogenously by stimulation of the ventral tegmental area (VTA) or applied exogenously by microiontophoresis. The amount of dopamine (in terms of current used to stimulate the VTA or iontophoretic current used to deliver the drug) required to bring about an attenuation of the excitatory response had no or little effect on the spontaneous activity of the neuron, suggesting a neuromodulatory action. It is worth noting, however, that higher doses of dopamine also produced an inhibition of spontaneous activity as reported by other investigators, indicating that dopamine may have both neuromediating and neuromodulatory functions.

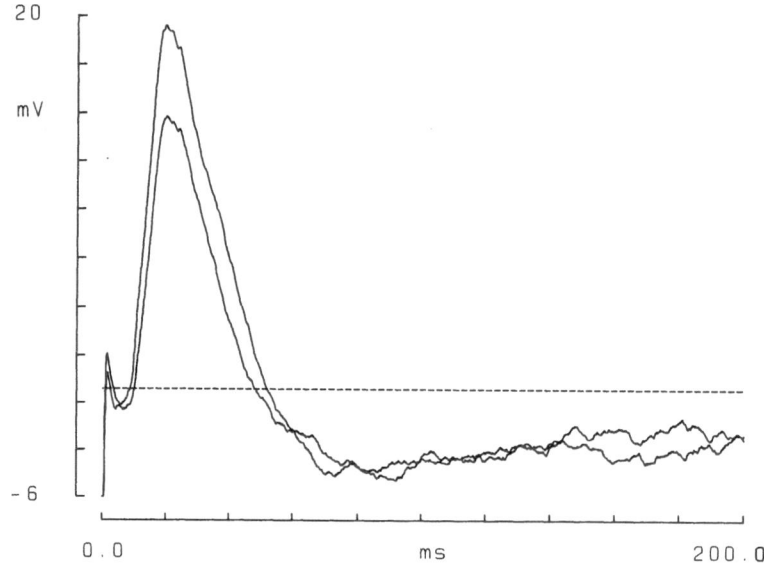

Fig. 1 EPSP-IPSP sequence recorded intracellularly from an accumbens neuron in vivo following amygdala stimulation. The larger amplitude EPSP was the control and the smaller amplitude EPSP was the attenuated response when amygdala stimulation was preceeded by VTA stimulations.

The observation on the neuromodulatory action of dopamine in extracellular single unit recordings was later extended to intracellularly recorded evoked potentials (Yim and Mogenson 1986, 1988). Intracellular recording was made from accumbens neurons of rats in vivo. Electrical stimulation of the amygdala or hippocampus was shown to produce EPSP-IPSP sequences in the majority of accumbens neurons, corresponding to the excitatory-inhibitory sequence observed in single unit recording experiments. It was also found that VTA stimulation produced a small membrane depolarization but as much as 60% reduction of the amplitude of the evoked EPSP (but not the IPSP) from amygdala or hippocampal stimulations (Figure 1). Iontophoretically applied dopamine produced similar results. These results indicate that the membrane depolarization and attenuation of the EPSP were not causally related and that dopamine probably acted at two different sites to bring about somatic membrane depolarization and attenuation of the evoked EPSP.

Evidence from the intracellular recording experiments suggests that dopamine appears to have a presynaptic neuromodulatory action in the ventral striatum in addition to a neuromediating action. The suggestion is consistent with pharmacological observations that dopamine attenuates the release of glutamate from cortical afferents innervating the caudate nucleus. While there is no direct observation to indicate that dopamine attenuates the release of glutamate from amygdala terminals, we have shown that dopamine affects the terminal excitability of hippocampal to accumbens afferents demonstrating directly that dopamine acts presynaptically in the nucleus accumbens (Yang and Mogenson, 1986).

Behavioural Evidence in Support of a Neuromodulatory Action for Dopamine

In view of the electrophysiological evidence of a neuromodulatory action for dopamine in the accumbens, it is of considerable interest to investigate whether or not a complementary behavioural role can be demonstrated. The amygdala and hippocampus are implicated in various adaptive behaviours and limbic-motor integration (Mogenson, 1977, 1987). More specifically, it has been shown that electrical stimulation of the amygdala produces arrest of locomotion (Ursin and Kaada, 1960) and lesions of the amygdala has been shown to increase exploration in the rat (White and Weingarten, 1976; Robinson, 1963). Since the amygdala to accumbens projection, investigated in electrophysiological experiments, likely mediates in part the behavioural responses elicited from amygdala stimulation, the hypolocomotor effect produced by amygdala stimulation was used as a behavioural model in investigating the neuromodulatory action of dopamine.

In a series of experiments to test the hypothesis that dopamine may modify amygdala induced reduction in locomotor activity, rats were implanted with injection cannula into the basal lateral nucleus of the amygdala and the nucleus accumbens. A steady baseline of locomotor activity level was induced by placing the animal in an activity box measuring 0.75 m by 0.75 m and divided into three chambers by partitions with communicating holes in them (see Yim and Mogenson, 1989). Animals when placed in the apparatus keep exploring the chambers in turn and exhibit a high level of spontaneous activity even on repeated trials. Their activity level was measured by the number of interruptions of four infrared photobeams which cris-crossed the activity box.

Bilateral injection of N-methyl-D-aspartic acid (NMDA, 0.2 or 0.4 μg) into the basolateral amygdala produced a dose dependent suppression of locomotor activity. This hypolocomotor effect of NMDA stimulation of the amygdala was reversed when dopamine was administered bilaterally to the nucleus accumbens (Yim and Mogenson, 1989). Since the release of dopamine into the accumbens is known to be associated with the production of hyperlocomotor activity, the suppression of locomotor activity could have simply been a summation of a stimulatory (dopamine injection) and an inhibitory (NMDA injection) action. However, the experiments demonstrated that a dose of dopamine as low as 3.5 μg was sufficient to produce complete reversal of the amygdala induced hypolocomotor activity. This dose of dopamine had no hyperlocomotor activity effect by itself and is much lower than doses used in similar experiments to stimulate hyperlocomotor activity in the rat.

Following the demonstration by Pijnenburg and co-workers that administration of dopamine to the nucleus accumbens increased locomotor activity in the rats, other investigators have subsequently confirmed that release of dopamine in the nucleus accumbens is associated with locomotion. What is not clear, however, is whether or not dopamine produced the hyperactivity through its neuromediating action or neuro-modulatory action in the accumbens. The behavioural experiment described above suggest that neuromodulatory action of dopamine in the accumbens may play an important role in the initiation of locomotor behaviour.

COMODULATORY ACTION BETWEEN DOPAMINE AND CHOLECYSTOKININ

The identification of the octapeptide cholecystokinin (CCK) in the VTA and the nucleus accumbens as well as the possible coexistence of CCK with dopamine in a subpopulation of the mesolimbic dopamine projection prompted pharmacological and behavioural experiments to investigate the functional role of this peptide in the nucleus accumbens. Experimental evidence that has been accumulated suggest that this octapeptide may indeed have a significant functional role in the ventral part of the striatum (Crawley et al., 1985; Meyer and Krauss, 1983; Schneider et al., 1983). Cholecystokinin has thus been shown to inhibit dopamine release in the nucleus accumbens (Fuxe et al., 1980; Markstein and Hokfelt, 1984; Voigt and Wang, 1984; Voigt et al., 1986; Lane et al., 1986; Phillips, et al., 1988), possibly acting as a negative feedback control at high rate of release dopamine (Phillips, et al., 1988; Wang, 1988). There is also evidence of dopamine inhibiting the release of CCK although its signifi-cance is not known (Meyer and Krauss, 1983; Martin et al., 1986; Hut-chison et al., 1986; Altar and Boyar, 1989). In electrophysiological studies, cholecystokinin has been shown to be a powerful excitatory agent on midbrain dopaminergic neurons (Chiodo and Bunney, 1983; Skirboll et al., 1981) and other CNS neurons (Ishibashi et al., 1979; Morin et al., 1983; Dodd and Kelly, 1981; Phillis and Kirkpatrick, 1980) and indeed the sustained depolarization (depolarization block) produced by CCK has been postulated as a mechanism of inhibiting dopamine release (Phillips et al., 1987).

In addition to its neuromediating actions and its inhibiting effect on dopamine release, there is also evidence that CCK may have other neuromodulating properties. Cholecystokinin has been shown to modulate the release of acetylcholine in the striatum (Arneric and Reis, 1986), to attenuate hippocampal evoked population spike in the accumbens and to potentiate the actions of dopamine and acetylcholine (DeFrance et al., 1984). It has also been shown to modulate dopamine binding (Murphy and Schuster, 1982) and may thereby modulate the action of dopamine on dopaminergic as well as non-dopaminergic synapses in the striatum.

Electrophysiological Investigations of Cholecystokinin-Dopamine Interac-tion

While there is ample evidence to suggest that CCK modulates dopamine release in the nucleus accumbens, little is known about its possible synaptic action in this nucleus. Wang and Hu (1986, 1988) have shown specifically that CCK has a primarily excitatory action on accumbens neurons and that dopamine antagonizes this excitatory action, and vice versa. Since the evidence was derived from iontophoretic applications of the putative transmitters, it appeared that the interaction observed by these investigators took place primarily at the postsynaptic somatic region. In view of the evidence that dopamine has a neuromodulatory

172

action on presynaptic non-dopaminergic terminals in the accumbens, and that CCK may be co-released with dopamine in this nucleus, we extended our previous electrophysiological experiments to investigate the possibility of a co-modulatory action between CCK and dopamine.

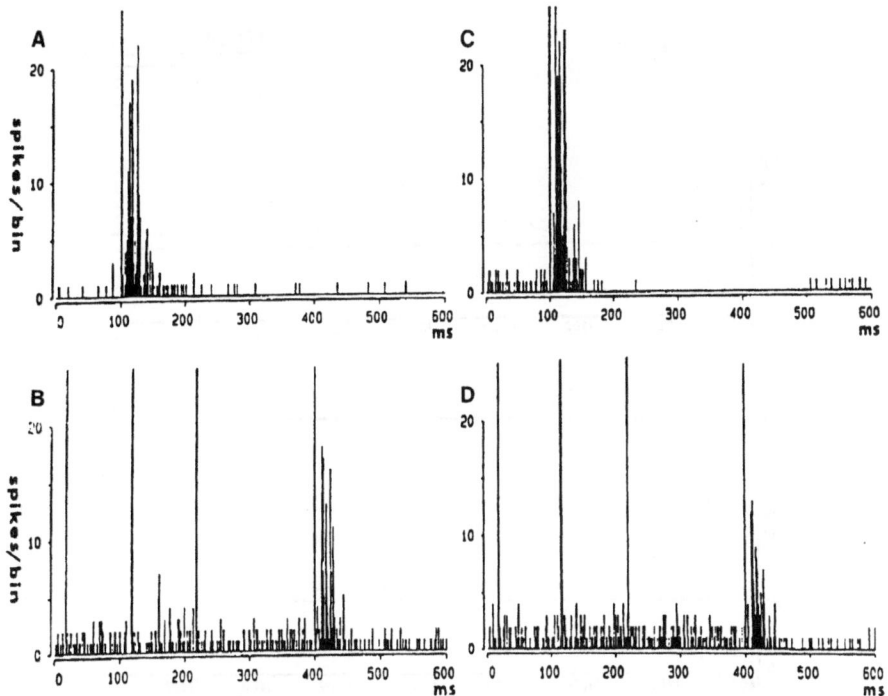

Fig. 2. Peristimulus time histogram to show excitatory responses of an accumbens neuron to electrical stimulation of the amygdala and its interaction with VTA stimulation and proglumide. Panels A and B show the attenuating effect of superimposed VTA stimulation on the response of the neuron to amygdala stimulation. Response was reduced by 31% from 242 evoked spikes in control (A) to 167 evoked spikes when VTA stimulation was superimposed (B). Panels C and D show the enhancement of the effect of VTA stimulation by iontophoretically applied proglumide. Panel C is the control after proglumide (not significantly different from that before proglumide application) and panel D shows the enhanced attenuating effect of VTA stimulation (67% reduction from 255 evoked spikes to 84 evoked spikes).

Single unit recordings were made from dorsal medial nucleus accumbens where cholecystokinin and dopamine were shown to co-exist (Hokfelt et al., 1980a; 1980b). The same experimental paradigm used in earlier studies to test the modulatory action of dopamine in the accumbens was employed. Responses of accumbens neurons to electrical stimulation of the amygdala and the effect of superimposed VTA stimulation were tested. As reported in earlier studies, accumbens neurons are consistently excited by amygdala stimulation. Stimulation of VTA at 10 Hz prior to amygdala stimulation produced significant attenuation of the response, as

shown in figure 2. The stimulating current to VTA was then adjusted to
produce a submaximal attenuation of the excitatory response to serve as a
control and the same sequence of tests were repeated following iontophoretic application of proglumide, a specific antagonist of cholecystokinin (Chiodo, et al., 1987; Freeman and Chiodo, 1988). Results show
that proglumide enhanced the effect of VTA stimulation by an average of
102% in approximately one half (n=56) of accumbens neurons tested (Fig.
2).

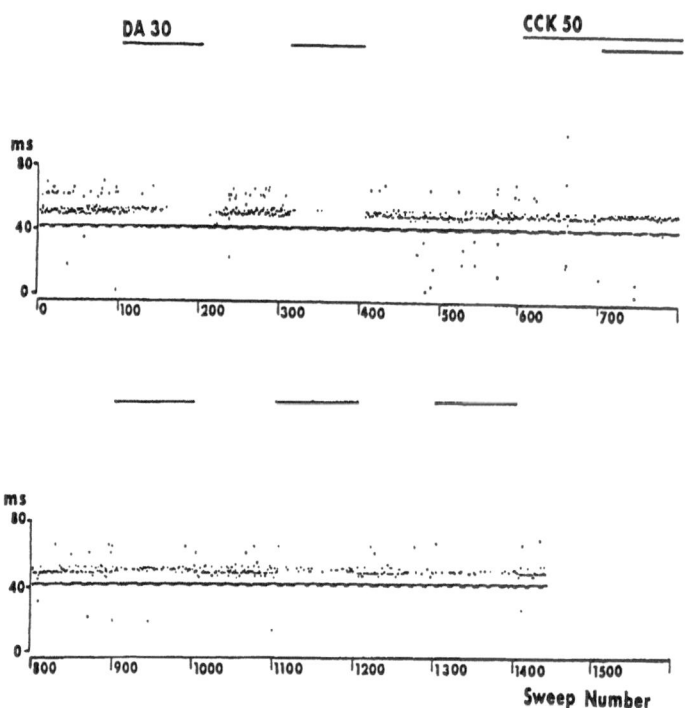

Fig. 3 Vertical rastor plot to show the attenuating effect of dopamine
on the excitatory response of an accumbens neuron to amygdala stimulation and the antagonistic effect of CCK. Horizontal bars above the
rastor plot show periods during which either dopamine or CCK was applied
iontophoretically. See text for details.

These results can be interpreted along with pharmacological data to
suggest that CCK was released together with dopamine during VTA stimulation, and that CCK inhibits the release of dopamine. By blocking the
feedback inhibition of CCK with proglumide, dopamine release was
increased and its modulatory action was enhanced and a greater attenuation of the excitatory response was observed. On the other hand, it is
also possible that tonic release of CCK interacted with dopamine at the
postsynaptic site, antagonized the action of dopamine and blocked its
suppression of the excitatory response produced by amygdala stimulation.
By preventing this antagonistic action of CCK, the effect of endogenously
released dopamine was enhanced. A second series of experiments suggests
that the latter possibility may be present in the accumbens.

In a second series of experiment, dopamine and CCK were applied
iontophoretically instead of being released endogenously. Figure 3 is an
example of the typical results obtained. The response of an accumbens
neuron to amygdala stimulation is plotted on a vertical raster plot in
which unit activities following a trigger to the computer are plotted on
the vertical Y-axis and successive sweeps are displayed along the X-axis.

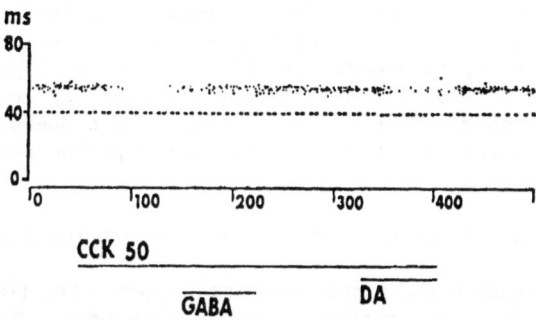

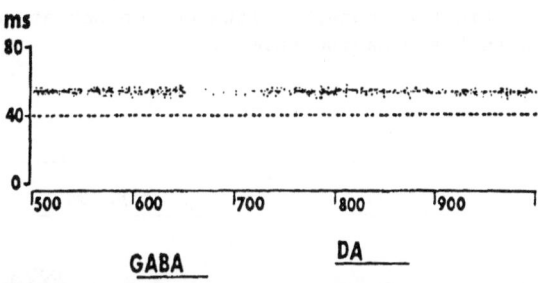

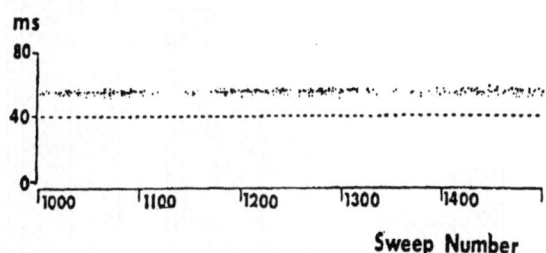

Fig. 4 Vertical rastor plot to show specific interaction between CCK and dopamine. Note that inhibitory response produced by GABA was not antagonized by CCK while that produced by dopamine was.

A stimulus pulse was delivered to the amygdala 40 ms after the start of the sweep and responses of the accumbens neuron to the stimulation were recorded as a series of dots plotted along the y axis. The first 100 sweeps of the raster plot show the control excitatory response of the accumbens neuron to amygdala stimulation. Beginning at sweep 100, dopamine was iontophoretically applied and it produced a significant suppression of the excitatory response of the accumbens neuron to amygdala stimulation. Concurrent application of CCK at 50 nA (sweeps 600-800) did not produce an enhancement of the response but blocked the attenuating effect of dopamine. The antagonism was specific. Inhibition produced by GABA titrated to mimic the effect of dopamine was not reversed by CCK (Fig. 4). Since both CCK and dopamine were applied iontophoretically, the effect was likely not due to pharmacological down regulation of the release of endogenous dopamine. The antagonistic effect of CCK appears likely due to an interaction at the post-synaptic site. Earlier experiments suggest that dopamine modulates the excitatory response of accumbens neurons to amygdala stimulation via presynaptic dopamine receptors on non-dopaminergic projection to the accumbens, the present result suggest a possible interaction between dopamine and CCK at these receptor sites.

Behavioural Investigations of Cholecystokinin-Dopamine Interaction

The electrophysiological observations suggest that CCK may function as an endogenous dopamine antagonist in the accumbens. To determine whether or not these observations have behavioural relevance, two series of experiments investigating possible interaction between dopamine and CCK were conducted in the behaving animal.

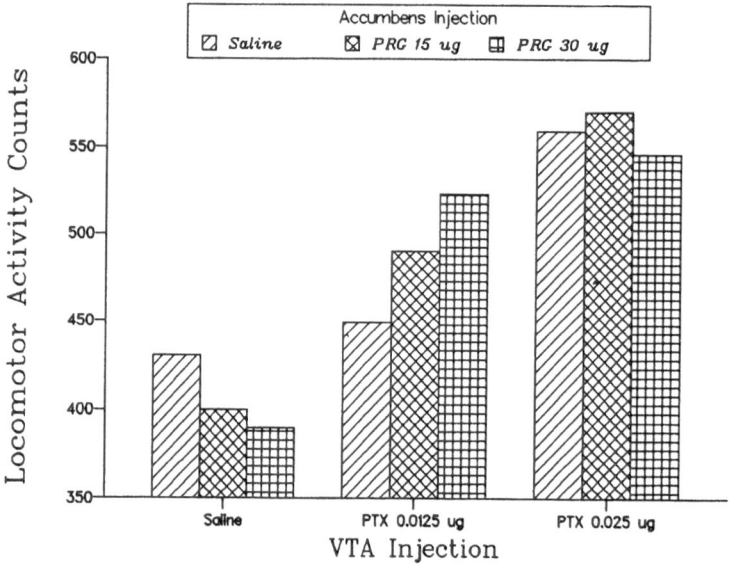

Fig. 5 Histogram to show effect of proglumide on picrotoxin induced hyperlocomotor activity. At moderate level of hyperlocomotion (0.0125 μg picrotoxin), proglumide produced a significant enhancement but at higher levels of activity, it produced an opposite effect. Activity produced by 0.025 μg of picrotoxin is at a level where the effect of proglumide begins to reverse.

In one series of experiments, hyperlocomotor activity was induced by injection of picrotoxin into the VTA. Previous experiments have demonstrated that the injection of picrotoxin into the VTA is reliable in eliciting hyperlocomotor activity in rats (Pijnenburg et al., 1973). The increase in locomotor activity was shown to be due to the release of dopamine in the nucleus accumbens. The hypothesis tested in this series of experiments is that CCK is co-released with dopamine in the accumbens following picrotoxin injection into the VTA, and if endogenous CCK is a functional antagonist of dopamine, then by blocking the effect of CCK with proglumide, the hyperlocomotor activity would be enhanced. Results from these experiments showed that proglumide had a biphasic effect on dopamine stimulated locomotor activity depending on the initial activity level. With a low dose of picrotoxin (0.0125 μg), locomotor activity was increased by an average of 6% compared to control, concurrent injection of proglumide into the accumbens further enhanced the locomotor activity by 20% in a dose dependent manner. But at higher doses of picrotoxin (0.025 or 0.05 μg), injection of proglumide into the accumbens produced an opposite effect and significantly attenuated the hyperlocomotor effect of picrotoxin. This biphasic action was puzzling. The enhancing effect of proglumide at low level of induced hyperlocomotor activity confirms the initial hypothesis and is consistent with electrophysiological observations as discussed above. The attenuating effect of proglumide at a high hyperactivity level is consistent with the observation of Crawley and co-workers (1985) but the possibility that proglumide itself had non-specific blocking effects cannot be ruled out.

TABLE I

Interaction between dopamine and/or CCK injection into the nucleus accumbens and NMDA injection into the amygdala on locomotor activity of the rats.

Activity counts are mean $\pm$ S.E.M.   n=36

Amygdala Injection	Accumbens Injection			
	Saline	DA 7 μg	CCK 40 ng	DA/CCK
Saline	595$\pm$32	704$\pm$45	561$\pm$25	725$\pm$41
NMDA 0.4 μg	387$\pm$37	544$\pm$42	435$\pm$31	465$\pm$39
Percent changes compared to saline/saline control				
Saline	--.--%	+18.32%	-5.71%	+21.85%
NMDA 0.4 μg	-34.96%	-8.57%	-26.89%	-21.85%

In another series of experiments, the same behavioural paradigm used to investigate the neuromodulatory action of dopamine was modified to test the co-modulatory action of cholecystokinin. Thirty six animals were tested with injections of saline or NMDA into the amygdala together with injection of either saline, dopamine, CCK or dopamine+CCK into the accumbens. Their spontaneous activities in the partitioned activity box following the different treatments are shown in Table 1. With saline injection into the amygdala, concurrent injection of 7 μg of dopamine into the accumbens produced a small increase in activity, consistent with repeated observations from previous experiments. Injection of 40 ng of

cholecystokinin into the accumbens did not produce any significant change
in the activity level whereas injection of a combination of CCK and
dopamine produced an enhancement of the activity level that is slightly
higher than that produced by dopamine alone. NMDA injection into the
amygdala produced a reliable suppression of locomotor activity which was
reversed by simultaneous injection of dopamine into the accumbens. Both
CCK and combination of CCK and dopamine produced a small but not significant reversal of the suppression. These results are in general consistent with the hypothesis that CCK acts as an antagonist to the action of
dopamine. Thus, the ability of dopamine to reverse the effect of NMDA
was blocked by CCK, an observation that is consistent with
electrophysiological data. What was unexpected, however, was that the
hyperactivity produced by dopamine injection alone was not blocked but
instead was slightly enhanced by CCK. This observation is consistent
with that reported by Crawley et al. (1985) but the contrasting effect
seen at different levels of activity is again puzzling. One possible
explanation is that CCK has different interaction with dopamine at
different synaptic sites. The ability of dopamine to reverse the effect
of NMDA likely occur at a presynaptic site on afferents from the
amygdala. Electrophysiological data show that CCK modulates this action
of dopamine. On the other hand, the hyperactivity produced by dopamine
itself is probably mediated by a direct action on the soma of output
neurons in the accumbens. It is plausible that CCK has a different
action here. Validity of this hypothesis remains to be investigated.

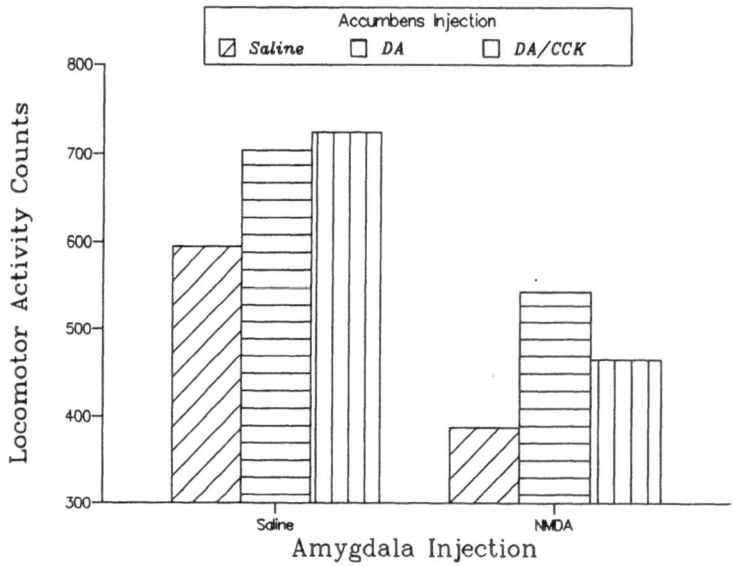

Fig. 6 NMDA injection into the amygdala suppresses locomotor
activity. Concomitant injection of dopamine into the accumbens reversed the suppression but this effect is in turn
partly blocked by CCK.

Conclusion

    Our earlier studies demonstrated that a significant functional role
of dopamine in the accumbens lies in its modulatory interaction with
other afferents to this nucleus. Extending those studies to its co-
transmitter CCK, we have shown that some of the actions of CCK are similarly modulatory in nature. More specifically, CCK appears to be an
endogenous functional antagonist of dopamine. At low doses, CCK does not

have significant direct effects of its own but blocks the action of dopamine. Unlike neuroleptics, however, this antagonistic effect of CCK cannot be due to competition at dopamine receptor sites since it was blocked by proglumide, a specific antagonist for CCK. There is some evidence that the interaction occurs between CCK receptor and the D_2 dopamine receptor subtype (Liang, personal communication), but the detailed mechanism remains to be investigated.

While we are encouraged by being able to demonstrate in behavioural experiments some of the neuromodulatory actions of CCK observed in electrophysiological experiments, results from those experiments must be considered empirical and preliminary. The biphasic effect observed in the experiments with proglumide as well as the differential effects observed with CCK injection into the accumbens depending on the treatment given to the amygdala further compound the contrasting observations reported in the literature. Crawley and her co-workers reported co-operative effects between dopamine and CCK (1985, 1988), but other studies have reported functional antagonisms (Katsuura and Itoh, 1982; Cohen et al., 1982; Van Ree et al., 1983; Takeda et al., 1986). Two recent reports, however, suggest possible explanation for the contrasting observations. Weiss and co-workers, (1988), showed that CCK potentiated amphetamine induced stereotypy but antagonized locomotor activity. Vaccarino and Vaccarino, (1989), reported that proglumide antagonized self stimulation when it was injected into the caudal area of the accumbens, but mildly facilitated the same behaviour when it was injected into the rostral regions.

Our approach to investigating the action of dopamine and CCK in the accumbens has been unique and has demonstrated rather interesting properties of the two putative transmitters. Both neurotransmitters show neuromediating and neuromodulatory actions in the same synaptic system. How the two transmitters and their respective neuromediating and neuromodulatory actions interplay, and which of the two actions is the more physiologically significant one remain the subject of an interesting research objective.

ACKNOWLEDGEMENT

Funding was supported by separate research grants to C. Yim and G. J. Mogenson from the Medical Research Council of Canada. We thank M. Wu, B. Shen and H. Y. Chow for technical assistance.

REFERENCES

Abercrombie, E.D., and Jacobs, B.L., 1985, Dopaminergic modulation of sensory responses of striatal neurons: single unit studies, Brain Res., 358:27-33.

Altar, C.A., and Boyar, W.C., 1989, Brain CCK-8 receptors mediate the suppression of dopamine release by cholecystokinin, Brain Res., 483:321-326.

Arneric, S.P., and Reis, D.J., 1986, Somatostatin and cholecystokinin octapeptide differentially modulate the release of acetylcholine from caudate nucleus but not cerebral cortex: role of dopamine receptor activation, Brain Res., 374:153-161.

Bergstrom, D.A., and Walters, J.R., 1984, Dopamine attenuates the effects of GABA on single unit activity in the globus pallidus, Brain Res., 310:23-33.

Bernardi, G., Marciani, M.G., Morocutti, C., Pavone, F., and Stanzione, P., 1978, The action of dopamine on rat caudate neurones intracellularly recorded, Neurosci.Lett., 8:235-240.

Chiodo, L.A., and Bunney, B.S., 1983, Proglumide: Selective anatagonism of excitatory effects of cholecystokinin in central nervous system, Science, 219:1449-1451.

Chiodo, L.A., Freeman, A.S., and Bunney, B.S., 1987, Electrophysiological studies on the specificity of the cholecystokinin antagonist pro-glumide, Brain Res., 410:205-211.

Cohen, S.L., Knight, M., Tamminga, C.A., and Chase, T.N., 1982, Chole-cystokinin effects on conditioned avoidance behaviour, stereotypy and catalepsy, Eur.J.Pharmacol., 83:213-222.

Crawley, J.N., 1988, Modulation of mesolimbic dopaminergic behaviors by cholecystokinin, Ann.N.Y.Acad.Sci., 537:380-396.

Crawley, J.N., Stivers, J.A., Blumstein, L.K., and Paul, S.M., 1985, Cholecystokinin potentiates dopamine-mediated behaviors: evidence for modulation specific to a site of coexistence, J.Neurosci., 5:1972-1983.

DeFrance, J.F., Sikes, R.W., and Chronister, R.B., 1984, Effects of CCK-8 in the nucleus accumbens, Peptides, 5:1-6.

Dodd, J., and Kelly, J.S., 1981, The actions of cholecystokinin and related peptides on pyramidal neurons of the mammalian hippocampus, Brain Res., 205:337-350.

Ferron, A., Thierry, A.M., Le-Douarin, C., and Glowinski, J., 1984, Inhibitory influence of the mesocortical dopaminergic system on spontaneous activity or excitatory response induced from the thalamic mediodorsal nucleus in the rat medial prefrontal cortex, Brain Res., 302:257-265.

Freeman, A.S., and Chiodo, L.A., 1988, Electrophysiological effects of cholecystokinin octapeptide on identified rat nigrostriatal dopa-minergic neurons, Brain Res., 439:266-274.

Fuxe, K., Andersson, K., Locatelli, V., Agnati, L.F., Hokfelt, T., Skirboll, L., and Mutt, V., 1980, Cholecystokinin peptides produce marked reduction of dopamine turnover in discrete areas in the rat brain following intraventricular injection, Eur.J.Pharmacol., 67:325-331.

Herrling, P.L., and Hull, C.D., 1980, Iontophoretically applied dopamine depolarizes and hyperpolarizes the membrane of cat caudate neurons, Brain Res., 192:441-462.

Hirata, K., Yim, C.Y., and Mogenson, G.J., 1984, Excitatory input from sensory motor cortex to neostriatum and its modification by condi-tioning stimulation of the substantia nigra, Brain Res., 321:1-8.

Hokfelt, T., Rehfeld, J., Skirboll, L., Ivemark, B., Goldstein, M., and Marley, K., 1980, Evidence for coexistence of dopamine and CCK in mesolimbic neurons, Nature, 285:476-478.

Hokfelt, T., Skirboll, L., Rehfeld, J., Goldstein, M., Marley, K., and Dann, O., 1980, A subpopulation of mesencephalic dopamine neurons projecting to limbic areas contains a cholecystokinin-like peptide: evidence from immunohistochemistry combined with retrograde tracing, Neuroscience, 5:2093-2124.

Hutchison, J.B., Strupish, J., and Nahorski, S.R., 1986, Release of endogenous dopamine and cholecystokinin from rat striatal slices: effects of amphetamine and dopamine antagonists, Brain Res., 370:310-314.

Ishibashi, S., Oomura, Y., Okajima, T., and Shibata, S., 1979, Chole-cystokinin, motilin and secretin effects on the central nervous system, Physiol.Behav., 23:401-403.

Kaczmarek, L.K. and Levitan, I.B. 1987, Neuromodulation, Oxford Univer-sity Press, New York.

Katsuura, G., and Itoh, S., 1982, Sedative action of cholecystokinin octapeptide on behavioral excitation by thyrotropin releasing hor-mone and methamphetamine in the rat, Jpn.J.Physiol., 32:83-91.

Kitai, S.T., Sugimori, M., and Kocsis, J.D., 1976, Excitatory nature of dopamine in the nigro-caudate pathway, Brain Res., 24:351-363.

Kupfermann, I., 1979, Modulatory actions of neurotransmitters, Ann. Rev. Neurosci., 2:447-465.

Lane, R.F., Blaha, C.D., and Phillips, A.G., 1986, In vivo electrochemical analysis of cholecystokinin-induced inhibition of dopamine release in the nucleus accumbens, Brain Res., 397:200-204.

Markstein, R., and Hokfelt, T., 1984, Effect of cholecystokinin-octapeptide on dopamine release from slices of cat caudate nucleus, J.Neurosci., 4:570-575.

Martin, J.R., Beinfeld, M.C., and Wang, R.Y., 1986, Modulation of cholecystokinin release from posterior nucleus accumbens by D-2 dopamine receptor, Brain Res., 397:253-258.

Mercuri, N., Bernardi, G., Calabresi, P., Cotugno, A., Levi, G., and Stanzione, P., 1985, Dopamine decreases cell excitability in rat striatal neurons by pre- and postsynaptic mechanisms, Brain Res., 358:110-121.

Meyer, D.K., and Krauss, J., 1983, Dopamine modulates cholecystokinin release in neostriatum, Nature, 301:338-340.

Mogenson, G.J., 1977, The Neurobiology of Behavior: An Introduction, Erlbaum.,Hillsdale., 1:

Mogenson, G.J., 1987, Limbic-Motor Integration. in: "Progress in Psychobiology and Physiological Psychology," A.N.Epstein, ed., Academic Press Inc., New York, p. 117-170.

Morin, M.P., De Marchi, P., Champagnat, J., Vanderhaeghen, J.J., Rossier, J., and Denavit-Saubie, M., 1983, Inhibitory effect of cholecystokinin octapeptide on neurons in the nucleus tractus solitarius, Brain Res., 265:333-338.

Murphy, R.B., and Schuster, D.I., 1982, Modulation of -dopamine binding by cholecystokinin octapeptide (CCK-8), Peptides, 3:539-543.

Phillips, A.G., Blaha, C.D., Fibiger, H.C., and Lane, R.F., 1988, Interactions between mesolimbic dopamine neurons, cholecystokinin, and neurotensin: evidence using in vivo voltammetry, Ann.N.Y.Acad.Sci., 537:347-361.

Phillips, A.G., Jakubovic, A., and Fibiger, H.C., 1987, Increased in vivo tyrosine hydroxylase activity in rat telencephalon produced by self-stimulation of the ventral tegmental area, Brain Res., 402:109-116.

Phillis, J.W., and Kirpatrick, J.R., 1980, The actions of motilin, cholecystokinin, somatostatin, vasoactive interstinal peptide, and other peptides on rat cerebral cortical neurons, Can.J.Physiol., 58:612-623.

Pijnenburg, A.J.J., Woodruff, G.N., and Van Rossum, J.M., 1973, Ergometrine-induced locomotor activity following intracerebral injection into the nucleus accumbens, Brain Res., 59:289-302.

Robinson, E., 1963, Effect of amygdalectomy on fear-motivated behaviour of rats, J.Comp.Physiol.Psychol., 56:814-820.

Schneider, L.H., Alpert, J.E., and Iversen, S.D., 1983, CCK-8 modulation of mesolimbic dopamine: antagonism of amphetamine-stimulated behaviors, Peptides, 4:749-753.

Skirboll, L.R., Grace, A.A., Hommer, D.W., Rehfeld, J., Goldstein, M., Hokfelt, T., and Bunney, B.S., 1981, Peptide-monoamine coexistence: studies of the actions of cholecystokinin-like peptides on the electrical activity of midbrain dopamine neurons, Neuroscience, 6:2111-2124.

Takeda, Y., Kamiya, Y., Honda, K., Takano, Y., and Kamiya, H., 1986, Effect of injection of CCK-8 into the nucleus caudatus on the behavior of rats, Jpn.J.Pharmacol., 40:569-575.

Thierry, A.M., Mantz, J., Milla, C., and Glowinski, J., 1988, Influence of the mesocortical/prefrontal dopamine neurons on their target cells, Ann.N.Y.Acad.Sci., 537:101-111.

Ursin, H., and Kaada, B.R., 1960, Subcortical structures mediating the attention response induced by amygdala stimulation, Exp.Neurol., 2:109-122.

Vaccarino, F.J., and Vaccarino, A.L., 1989, Antagonism of cholecystokinin function in the rostral and caudal nucleus accumbens: differential effects on brain stimulation reward, Neurosci.Lett., 97:151-156.

Van Ree, J.M., Gaffori, O., and De Wied, D., 1983, In rats the behavioral profile of CCK-8-related peptides resembles that of antipsychotic agents, Eur.J.Pharmacol., 93:65-78.

Vives, F., and Mogenson, G.J., 1986, Electrophysiological study of the effects of D1 and D2 dopamine antagonists on the interaction of converging inputs from the sensory-motor cortex and substantia nigra neurons in the rat, Neuroscience, 17:349-359.

Voigt, M., Wang, R.Y., and Westfall, T.C., 1986, Cholecystokinin octa-peptides alter the release of endogenous dopamine from the rat nucleus accumbens in vitro, J.Pharmacol.Exp.Ther., 237:147-153.

Voigt, M.M., and Wang, R.Y., 1984, In vivo release of dopamine in the nucleus accumbens of the rat: modulation by cholecystokinin, Brain Res., 296:189-193.

Wang, R.Y., 1988, Cholecystokinin, dopamine, and schizophrenia: recent progress and current problems, Ann.N.Y.Acad.Sci., 537:362-379.

Wang, R.Y., and Hu, X.T., 1986, Does cholecystokinin potentiate dopamine action in the nucleus accumbens, Brain Res., 380:363-367.

Weiss, F., Tanzer, D.J., and Ettenberg, A., 1988, Opposite actions of CCK-8 on amphetamine-induced hyperlocomotion and stereotypy following intracerebroventricular and intra-accumbens injections in rats, Pharmacol.Biochem.Behav., 30:309-317.

White, N., and Weingarten, H., 1976, Effects of amygdaloid lesions on exploration by rats, Physiol.Behav., 17:73-79.

Yang, C.R., and Mogenson, G.J., 1984, Electrophysiological responses of neurones in the nucleus accumbens to hippocampal stimulation and the attenuation of the excitatory responses by the mesolimbic dopa-minergic system, Brain Res., 324:69-84.

Yang, C.R., and Mogenson, G.J., 1986, Dopamine enhances terminal excit-ability of hippocampal- accumbens neurons via D2 receptor: role of dopamine in presynaptic inhibition, J.Neurosci., 6:2470-2478.

Yim, C.Y., and Mogenson, G.J., 1982, Response of nucleus accumbens neurons to amygdala stimulation and its modification by dopamine, Brain Res., 239:401-415.

Yim, C.Y., and Mogenson, G.J., 1986, Mesolimbic dopamine projection modulates amygdala-evoked EPSP in nucleus accumbens neurons: an in vivo study, Brain Res., 369:347-352.

Yim, C.Y., and Mogenson, G.J., 1988, Neuromodulatory action of dopamine in the nucleus accumbens: an in vivo intracellular study, Neuro-science, 26:403-415.

Yim, C.Y., and Mogenson, G.J., 1989, Low doses of accumbens dopamine modulates amygdala suppression of spontaneous exploratory activity in rats, Brain Res., 477:202-210.

IS DOPAMINE A NEUROTRANSMITTER WITHIN THE VENTRAL PALLIDUM/

SUBSTANTIA INNOMINATA?

T. Celeste Napier, Mary Beth Muench
Renata J. Maslowski and George Battaglia

Department of Pharmacology
Loyola University Chicago
Stritch School of Medicine
Maywood, Illinois

INTRODUCTION

In a classic treatise titled "Criteria for Identification of a Central Nervous System Transmitter" (1966), R. Werman provided a basis by which a chemical found in the brain could be classified as involved in the communication from one nerve cell to another. As listed in Werman's paper, the criteria are:

"The Criterion of the Inactivating Enzyme.
The Criterion of the Presence of the Transmitter.
The Criterion of Collectability of the Transmitter.
The Criterion of the Synthesizing Enzyme.
The Criterion of the Presence of Precursors.
The Criterion of a Specific Release Mechanism.
The Criterion of Identical Actions.
The Criterion of Pharmacological Identity."

These concepts were expanded by Barchas and colleagues (1978) to include: The criterion of presence of specific receptors.

These essays provide a standard for the identification and characterization of the various putative brain transmitters. This approach also is applicable to determine whether a known transmitter is neurally active in a discrete region of the brain.

The basal forebrain region focused upon in this chapter is the infracommissural extension of the external segment of the dorsal globus pallidus, i.e. the ventral pallidum (VP), and its caudal extension, the sublenticular substantia innominata (SI). The neurotransmitter, dopamine (DA), is discussed with regard to possible influences in the VP/SI. Following the criteria proposed by Werman (1966) and Barchas et al. (1978), the multidisciplinary approaches that helped establish DA as a neurotransmitter within the VP/SI will be presented. Subsequently, a discussion regarding the functional consequences of DA's influence on VP/SI neuronal activity will be provided.

The Basal Forebrain, Edited by T.C. Napier *et al.*
Plenum Press, New York, 1991

CRITERIA FULFILLED BY DOPAMINE WITHIN THE VENTRAL PALLIDUM

The Criterion of the Synthesizing Enzyme

Non-peptide neurotransmitters often are synthesized in the terminal endings of presynaptic neurons which allows for the local control of turnover rates necessary for rapid neurotransmission. Thus, the presence of a synthetic enzyme in a particular brain region can be a presynaptic indicator for the neurotransmitter. Tyrosine hydroxylase is the rate limiting synthetic enzyme for DA and other catecholamines, and tyrosine hydroxylase immunocytochemistry (TH) has revealed positively stained varicose fibers within several regions of the basal forebrain, including the VP/SI (Martinez-Murillo et al., 1988; Jones and Cuello, 1989; Zaborszky, 1989). Using electronmicroscopic procedures, (Zaborszky, 1989) verified that the TH positive varicosities make synaptic contacts within the VP/SI. The TH distribution is not the same as that for DA-*beta* hydroxylase, a marker for epinephrine/norepinephrine (Zaborszky et al., 1991), suggesting that some TH positive terminals are dopaminergic.

The Criterion of the Presence of the Transmitter

Anatomically, the presence of DA within the basal forebrain has been detected using two approaches. Lindvall and Bjorklund (1974), using the glyoxylic acid fluorescent histochemical method demonstrated intense fluorescence in several forebrain regions. The authors also described fibers of the medial forebrain bundle system/nigrostriatal pathway that send collaterals within the *ansa lenticularis*, including the region now referred to as the SI. The intensity was considerably less for the ventral structures of the forebrain than for dorsal and rostral locations. More recently, Voorn and colleagues (1986), using specific antibodies against DA, reported that the VP is "relatively sparsely innervated by DA fibers" which arise from the medial forebrain bundle. These fibers ramify into varicosities, supporting the work with TH that implicates the VP/SI as a termination site for DA fibers.

The presence of DA within VP tissue homogenates has now also been demonstrated using high performance liquid chromatographic separation with electrochemical detection (Table 1). These biochemical assays confirm the anatomical conclusion that innervation of this region by DA is relatively sparse compared to the striatum.

Another method employed by anatomists to implicate DA as a neurotransmitter within the basal forebrain, utilizes anterograde and retrograde tracers to determine if the midbrain regions containing dopaminergic somata project to the basal forebrain. Horseradish peroxidase (HRP; with and without wheat germ agglutinin) microinjected into the VP/SI retrogradely labels cells in the ventral tegmental area, the substantia nigra (Fallon and Moore, 1978; Russchen et al., 1985; Haring and Wang, 1986; Grove 1988; Martinez-Murillo et al., 1988; Semba et al., 1988; Jones and Cuello, 1989), and the retrorubral field and zona incerta of the substantia nigra (Deutch et al., 1988; Jones and Cuello, 1989). Furthermore, the HRP-positive midbrain neurons often are counterstained for TH (Deutch et al., 1988; Martinez-Murillo et al., 1988; Semba et al., 1988; Jones and Cuello, 1989; Zaborszky, 1989), suggesting that the neurons are catecholinergic, and because of their anatomical location, most likely dopaminergic. A complementary approach, using anterograde labeling of the VP/SI following microinjection of *Phaseolus vulgaris* leuco-agglutinin in the ventral tegmental area and substantia nigra (Grove, 1988; Zaborszky, 1989) substantiates the origin

Table 1. Monoamine Concentration in Various Forebrain Regions and the
Effect of 6-hydroxydopamine Treatment

TISSUE	DA	DOPAC	5HT	5HIAA
Striatum				
Control	357.0 ± 19	57.4 ± 3.6	11.6 ± 0.7	15.8 ± 1.1
6-OHDA	28.1 ± 11.4***	3.1 ± 1.4***	9.7 ± 0.8	15.5 ± 1.0
Globus Pallidus				
Control	127.0 ± 16	28.7 ± 5.1	39.7 ± 4.3	67.9 ± 5.7
6-OHDA	8.6 ± 2.2***	4.0 ± 0.9***	45.9 ± 2.0	60.6 ± 4.9
Ventral Pallidum				
Control	44.2 ± 10.2	12.4 ± 2.8	83.6 ± 5.2	74.4 ± 10.0
6-OHDA	12.8 ± 1.8**	4.9 ± 0.8*	78.7 ± 6.5	60.5 ± 6.4

*$p < 0.05$, ** $p < 0.01$, ***$p < 0.001$, compared with the respective control,
Student's t-test. Mean ± S.E.M. in pmoles/mg protein. Eight micrograms
of 6-hydroxydopamine (6-OHDA) was infused into the substantia nigra pars
compacta. Saline was similarly infused into the nigra of control rats.
One week later, the animals were sacrificed, the rostral striatum was
dissected out and the GP and the VP/SI were punched out of 500 micron
sections of tissue. Tissue content of DA and a metabolite
dihydroxyphenyl acetic acid (DOPAC), as well as serotonin (5HT) and its
major metabolite 5-hydroxyindoleacetic acid (5HIAA) were determined
simultaneously using HPLC and electrochemical detection. At least 11
tissues were assayed in each group. DA and DOPAC were decreased in each
tissue while 5HT and 5HIAA remained intact (from Napier and Potter,
1989).

of the DA projection to the basal forebrain. Furthermore, disruption of
this system by microinjections of the neurotoxin 6-hydroxydopamine,
greatly reduces the levels of DA and its metabolites in VP/SI tissue
(Table 1; Geula and Slevin, 1989). Thus, biochemical and anatomical
approaches agree that a dopaminergic projection arising from the midbrain
terminates within basal forebrain regions. This projection likely
reflects collateralization of the massive ascending system to the
striatum and nucleus accumbens.

The Criterion of the Inactivating Enzyme

"This is the criterion which states that the agent suspected to be
a transmitter must have an inactivating enzyme at or adjacent to the site
of its action" (Werman, 1966). This criterion is concerned with
termination of action of DA after it is released into the synapse; a feat
accomplished by an active uptake system to remove the neurotransmitter
from the cleft, and metabolic breakdown of the DA molecule into inactive
products. Because uptake sites for DA are located primarily on
dopaminergic axons and terminals, the detection of these sites
substantiates anatomical and biochemical evidences for the presence of

Table 2. Dopamine Uptake Sites in Forebrain Tissue Homogenates

REGION	% STRIATUM
Globus Pallidus	42 ± 8
Ventral Pallidum	23 ± 11

DA uptake sites were labeled using 1nM ^{3}H-GBR12935 specific binding was defined using 1μM mazindol. The density of ^{3}H-GBR12935-labelled sites in striatum was 819 fmol/mg protein from 3 determinations per region. Data are expressed as the mean percent of striatal binding ± S.E.M.

dopaminergic fibers. A marker of the high affinity uptake site for DA is GBR12935. We have determined that VP/SI tissue homogenates demonstrate ^{3}H-GBR12935 binding at a level that was 28% of striatal concentrations (Table 2). Even though the enzymes known to metabolize DA have not been identified in basal forebrain tissue, DOPAC, a product of these enzymes are found in significant concentrations in the VP/SI (Table 1; Ccula and Slevin, 1989). These data indicate that the processes necessary for the termination of DA's action is present in the basal forebrain.

The Criterion of the Presence of Specific Receptors

Current dogma contends that there are two major DA receptors, D1 and D2, with opposing effects on signal transduction. Binding sites indicative of D1 and D2 DA receptors have been visualized for the VP/SI autoradiographically (Gehlert and Wamsley, 1985; Dawson et al., 1986; Contreras et al., 1987; Beckstead et al., 1988; Besson et al., 1988; Camps et al., 1989a and b; Cortes et al., 1989; Richfield et al., 1989). We have substantiated these observations and quantified the concentration of D1 and D2 DA receptors using tissue homogenates. ^{3}H-SCH23390-labeled D1 and ^{3}H-spiperone-labeled D2 DA receptors were detected in VP/SI tissue. Using saturating concentrations of the radioligands, D1 and D2 receptor densities in VP/SI and GP represented approximately 30% of the striatum (Tables 3 and 4, respectively). Thus, both receptor subtypes are present within the VP/SI and, consistent with the concentration of DA in this tissue, the receptor concentration is considerably less than that observed for the striatum.

The Criterion of Identical Actions

This criterion requires that the suspected transmitter, when applied exogenously, "mimics the action of the natural transmitter" (Werman, 1966). However, to accomplish this task it would be necessary to know what is the action of DA and for dopaminoceptive brain regions, considerable debate still exists with regard to the nature of DA-induced responses. For example, electrical stimulation of dopaminergic somata causes release of terminal DA and evokes excitatory as well as inhibitory responses in dopaminoceptive brain regions, including the nucleus accumbens (see Mogenson, Jones and Yim, 1980), and striatum (Connor, 1970). Preliminary studies in our laboratory suggest that a similar scenario may exist for the VP/SI; however, inhibition was the more predominant response evoked (Fig. 1). It is not known if stimulation of the ventral tegmental area also influences these cells.

Table 3. D1 Receptor Binding in Forebrain Tissue Homogenates

REGION	D1 RECEPTORS (fmoles/mg protein)	% STRIATUM
Striatum (n=9)	477 ± 35	100 ± 7
Globus Pallidus (n=9)	203 ± 22	43 ± 5
Ventral Pallidum (n=9)	157 ± 28	33 ± 6

Radioligand binding data are expressed as the mean ± S.E.M.
N, number of determinations per region. D1 receptors were labeled
using a saturating concentration of ^{3}H-SCH23390 (2.8nM) in the
presence of 40nM ketanserin to preclude binding to $5HT_2$ serotonin
receptors. Specific binding was defined by 1µM (+)-butaclamol.

Table 4. D2 Receptor Binding in Forebrain Tissue Homogenates

REGION	D2 RECEPTORS (fmoles/mg protein)	% STRIATUM
Striatum (n=21)	213 ± 16	100 ± 8
Globus Pallidus (n=13)	43. ± 5	20 ± 2
Ventral Pallidum (n=12)	81 ± 15	38 ± 7

Radioligand binding data expressed as the mean ± S.E.M.
N, number of determinations per region. D2 receptors were
labeled using a saturating concentration of ^{3}H-spiperone
(1.3nM) in the presence of 40nM ketanserin to preclude binding
to $5HT_2$ serotonin and α adrenergic receptors.

    Like evoked responses, local application of DA using
microiontophoresis can result in either excitation and inhibition in the
nucleus accumbens (e.g., Woodruff et al., 1976), striatum (e.g., Bloom et
al., 1965), and globus pallidus (e.g., York, 1970). Equally complex, of
the VP/SI neurons that responded to microiontophoretic applications of DA
(43 of 102 neurons tested), 31% were suppressed (Figs. 3 and 4) and 12%
were excited (Napier et al., 1991). Even though much remains to be
studied with regard to dopaminergic influences on VP/SI neurons, these
experiments allow for the following conclusions: (1) Locally applied DA
effectively alters VP/SI neuronal activity. (2) Only a portion of the
encountered VP/SI neurons are sensitive to DA; an anticipated phenomenon
given the relatively low concentration of DA and its receptors in VP/SI
tissue.

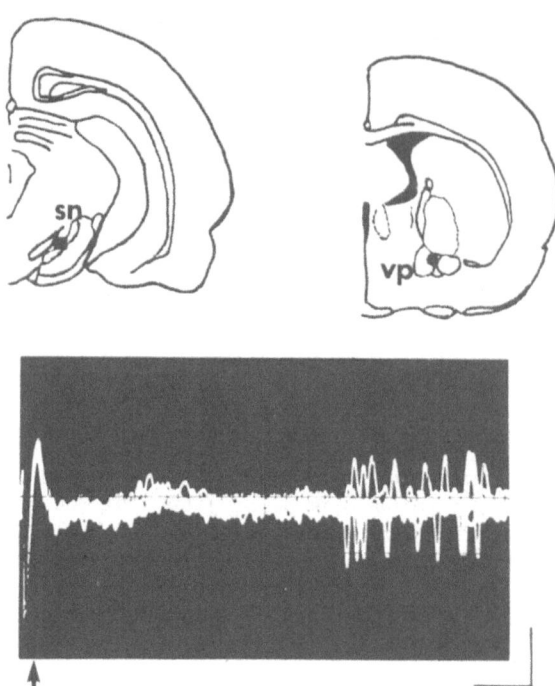

Fig. 1. Inhibition of VP/SI neuronal activity produced by nigral
stimulation. The left stereotaxic map shows the tip location of
the bipolar stimulating electrode within the substantia nigra
zona compacta. The right map illustrates the location of the cell
whose activity is shown in the oscilloscope photograph. The
photograph depicts 10 superimposed sweeps recorded from a single
VP cell. The calibration bar represents 60 μV vertically and 5
ms horizontally. The arrow reveals the stimulation artifact from
single, monopolar, 100 μA, 0.3 ms pulses applied to the
stimulation electrode.

   There are influences that could contribute to the observed response
with both of these electrophysiologic approaches. The stimulation-evoked
responses could reflect electrical excitation of non-dopaminergic
processes (either mono- or polysynaptically) as well as the dopaminergic
inputs. Microiontophoretic applications of a substance bathes the local
milieu around the recorded neuron and can affect presynaptic terminals
located within the diffusional distance of the ejected drug. Thus, some
of the effects observed with microiontophoretically applied DA could
reflect alterations in the release of a secondary neurotransmitter.
Given these confounding influences on the observed response to either
endogenously released or exogenously applied DA, it is very difficult to
determine if the two produce similar effects within any of the
dopaminoceptive brain regions, including the basal forebrain. However,
these electrophysiologic studies do provide evidence that DA is capable
of producing a physiological response by VP/SI neurons, and provide a
functional endorsement of anatomical and biochemical evidence for DA
neurotransmission in this region.

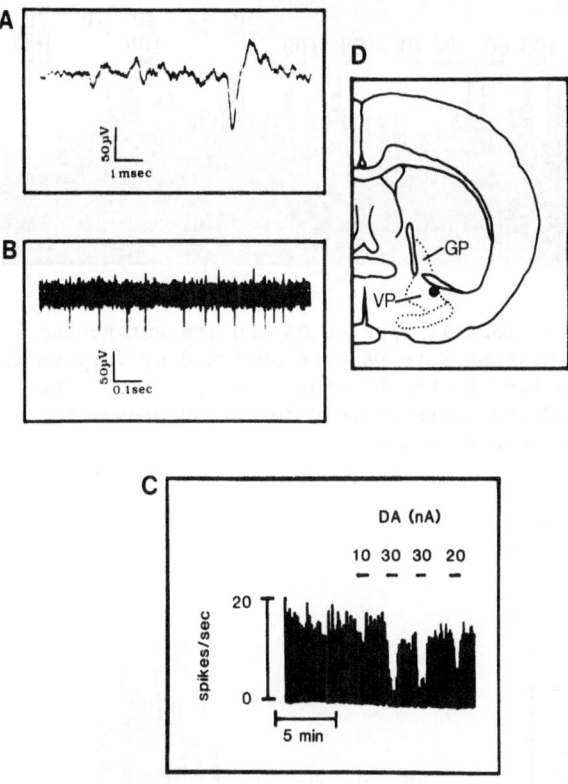

Fig. 2. Representative action potential configuration (A), firing pattern
(B), and frequency histogram (C), of a cell located within the VP
(D), of a chloral hydrate anesthetized rat. GP=globus pallidus.
DA ejected from the pipette using currents as indicated by the
numbers above the histogram (in nonoamperes [nA]).

The ability of antagonists to attenuate the DA-induced response was
tested using an iontophoretic current that produced approximately half
the maximal response obtained by the "current-response" protocol
illustrated in Fig. 3. Iontophoretically applied haloperidol was able to
antagonize the rate changes observed with DA (Fig. 3).

Because significant binding for both the D1 and D2 receptors
detected for VP/SI tissue, and DA does not discriminate the subtypes, it
was of interest to determine if VP/SI neurons were affected by
independent activation of either subtype. Systemic administration of the
D1 agonist, SKF38393 produced a dose dependent increase in firing in 17
of 25 cells tested (Fig. 4) (Maslowski and Napier, 1991). In contrast,
the D2 agonist, quipirole increased firing in only 3 of 12 cells tested,
but suppressed rates in 6 neurons (Fig. 5) (Maslowski and Napier, 1991).
These results demonstrate that VP/SI neurons respond to receptor-specific
agonists, and that they may serve opposing roles with regard to neuronal
activity in this region.

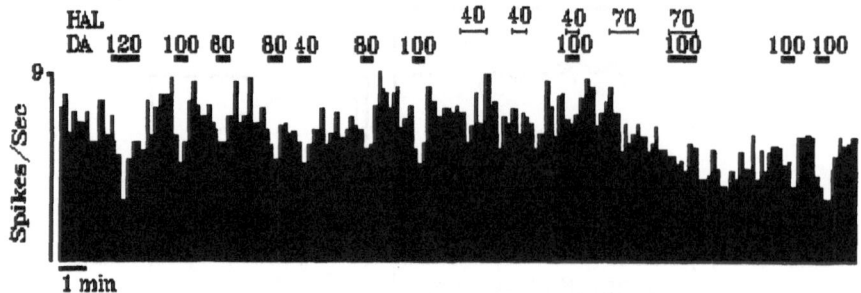

Fig. 3. Effects of locally applied DA and its antagonism by haloperidol (HAL). Responses to DA were analyzed by varying the current level applied to the DA-containing pipette. The current level that produced approximately 30-50% suppression was co-iontophoresed with HAL.

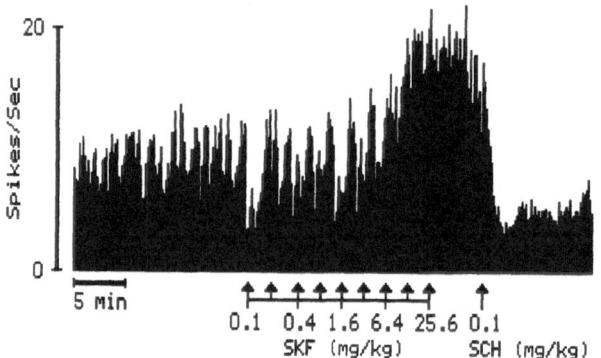

Fig. 4. Typical example of VP/SI responding to intravenous administration of the D1 receptor agonist, SKF23393 (SKF), in chloral hydrate anesthetized rats. The D1 antagonist, SCH23390 (SCH), reversed the agonist-induced excitation:

FUNCTIONAL RELEVANCE

Sufficient evidence now exists for the inclusion of the VP/SI into the category of dopaminoceptive brain regions. The functional relevance becomes apparent when the gamut of behaviors attributed to the VP/SI is considered.

Mogenson and colleagues have investigated extensively the contribution that the VP/SI provides to motor functions mediated by the nucleus accumbens - VP/SI - mesencephalic locomotor region system (for review see Mogenson and Yang, this book). The nucleus accumbens receives

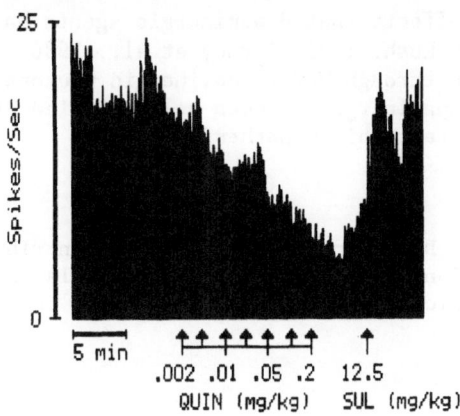

Fig. 5. VP/SI rate suppression mediated by quinpirole (QUIN) acting at D2
receptors. Sulpiride (SUL), the D2 antagonist, reversed this
effect.

significant dopaminergic inputs from the ventral tegmental area, and the
VP/SI has been shown to be influenced indirectly by this dopaminergic
communication to the nucleus accumbens (Yang and Mogenson, 1989). The
present paper demonstrates that, like the GP, a monosynaptic dopaminergic
input to the VP/SI provides an additional site of modulation for dopamine
in this limbic motor pathway.

Behavioral and electrophysiological studies suggest that VP/SI
brain regions play an important role in positive reinforcement (DeLong,
1971; Linseman, 1974; Mora et al., 1976; Rolls et al., 1979, 1980;
Richardson and DeLong, 1986; Huston et al., 1987; Hubner and Koob, 1990;
Wilson and Rolls 1990; for reviews see Richardson and DeLong, and Wilson,
this book). Since DA systems have long been known for their involvement
in reward phenomenon (for review see Wise, 1980), an understanding of the
contribution of VP/SI DA neurotransmission to this process may provide
new insights into reward mechanisms.

Finally, VP/SI contains a massive, cortically directed cholinergic
projection (for review see McGeer et al., 1986) that likely influences
cognitive functioning (for review see Olton et al., and Fibiger et al.,
this book). Zaborszky (1989) observed that terminals immunoreactive for
TH make monosynaptic contacts with VP/SI dendritic processes that are
immunoreactive for choline acetyltransferase, the cholinergic synthetic
enzyme. This provides the possibility that DA may regulate
monosynaptically the VP/SI cholinergic efferents. Electrophysiologic
characteristics of neurons recorded in the present study are similar to
those observed for VP/SI neurons activated antidromically by stimulation
of cholinergic terminal regions in the cortex (Aston-Jones et al., 1985;
Lamour et al., 1986; Reiner et al., 1987). Thus, some of the VP/SI
neurons with DA-induced responses in the present study may be
cholinergic. Additional support for this possibility comes from studies
of conditioned slow potential shifts recorded from the cortical surface
that are generated by VP/SI cholinergic neurons (Pirch et al., 1986, for
review see Pirch et al., this book). These event-related slow
potentials, reflective of associative or cognitive processes, are
enhanced by haloperidol (Pirch and Corbus, 1983) and suppressed by
amphetamine (Pirch 1977a, 1977b, 1980; Pirch et al., 1981a, 1981b),

suggesting that the effects that dopaminergic agents have on cognitive functions (Beatty and Rush, 1983; Carnoy et al., 1986; McGurk et al., 1988) may be mediated through VP/SI cholinergic neurons. Further experimentation on dopaminergic effects on identified VP/SI cholinergic neurons is needed to test this hypothesis.

ACKNOWLEDGEMENTS

The authors thank Ms. Dowon An for her valuable contributions to the presented studies. Funding provided by BRSG, Loyola University Chicago Stritch School of Medicine and USPHS grant MH45180 to TCN.

REFERENCES

Aston-Jones G., Shaver R., and Dinan T.G., 1985, Nucleus basalis neurons exhibit axonal branching with decreased impulse conduction velocity in rat cerebrocortex, Brain Res,. 325:271.

Barchas, J.D., Akil, H., Elliott, G.R., Holman, R.B., and Watson, S.J., 1978, Behavioral neurochemistry: Neuroregulators and behavioral states, Science, 200:964.

Beatty W.W., and Rush J.R., 1983, Spatial working memory in rats: Effects of monoaminergic antagonists, Pharmacol. Biochem. Behav., 18:7.

Beckstead R.M., 1988, Association of dopamine D1 and D2 receptors with specific cellular elements in the basal ganglia of the cat: The uneven topography of dopamine receptors in the striatum is determined by intrinsic striatal cells, not nigrostriatal axons, Neuroscience, 27:851.

Beckstead R.M., Wooten G.F., and Trugman J.M., 1988, Distribution of D1 and D2 dopamine receptors in the basal ganglia of the cat determined by quantitative autoradiography, J. Compar. Neurol., 268:131.

Besson M.-J., Graybiel A.M., and Nastuk M.A., 1988, [^{3}H]SCH 23390 binding to D1 dopamine receptors in the basal ganglia of the cat and primate: Delineation of striosomal compartments and pallidal and nigral subdivisions, Neuroscience, 26:101.

Bloom, F.E., Costa, E., and Salmoiraghi, G.C., 1965, Anesthesia and the responsiveness of individual neurons of the caudate nucleus of the cat to acetylcholine, norepinephrine and dopamine administered by microelectrophoresis, J. Pharm. Exp. Therap., 150:244.

Camps M., Cortes R., Gueye B., Probst A., and Palacios J.M., 1989a Dopamine receptors in human brain: Autoradiographic distribution of D2 sites, Neuroscience, 28:275.

Camps M., Kelly P.H., and Palacios, J.M., 1989b, Autoradiographic localization of dopamine D1 and D2 receptors in the brain of several mammalian species, J. Neural. Transm., 80:105.

Carnoy P., Ravard S., Wemerman B., Soubrie P.H, Simon P., 1986, Behavioral deficits induced by low doses of apomorphine in rats: Evidence for a motivational and cognitive dysfunction which discriminates among neuroleptic drugs, Pharmacol. Biochem. Behav., 25:503.

Connor, J.D., 1970, Caudate nucleus neurones: Correlation of the effects of substantia nigra stimulation with iontophoretic dopamine, J. Physiol., 208:691.

Contreras P.C., Quirion R., Gehlert D.R., Contreras M.L., and O'Donohue T.L., 1987, Autoradiographic distribution of non-dopaminergic binding sites labeled by [^{3}H] haloperidol in rat brain, Neurosci. Lett., 75:133.

Cortes R., Gueye B., Pazos A., Probst A., and Palacios J.M., 1989, Dopamine receptors in human brain: Autoradiographic distribution of D1 sites, Neuroscience, 23:263.

Dawson T.M., Barone P., Sidhu A., Wamsley J.K., and Chase T.N., 1986, Quantitative autoradiographic localization of D-1 dopamine receptors in the rat brain: Use of the iodinated ligand [125I]SCH23390. Neurosci. Lett., 68:261.

DeLong M.R., 1971, Activity of pallidal neurons during movement, J. Neurophysiol., 34:414.

Deutch A.Y., Goldstein M., Baldino F., Roth R.H., 1988, Telencephalic projections of the A8 dopamine cell group, in: "The Mesocorticolimbic Dopamine System, Annals of the New York Academy of Sciences", Vol. 537, P.W. Kalivas, and C.B. Nemeroff, eds., New York: The New York Academy of Sciences, p 27.

Fallon J.H., Moore R.Y., 1978, Catecholamine innervation of the basal forebrain IV. Topography of the dopamine projection to the basal forebrain and neostriatum, J. Comp. Neurol., 180:545.

Fibiger, H.C., Damsma, G. and Day, J.C. 1991, Behavioral pharmacology and biochemistry of central cholinergic neurotransmission, in: "The Basal Forebrain: Anatomy to Function: Advances in Experimental Medicine and Biology", T.C. Napier, P.W. Kalivas, I. Hanin, eds, New York, Plenum Publishing Corporation, in press.

Gehlert D.R., and Wamsley J.K., 1985, Dopamine receptors in the rat brain: Quantitative autoradiographic localization using [³H]sulpiride. Neurochem. Int., 7:717.

Geula C., and Slevin J.T., 1989, Substantia nigra 6-hydroxydopamine lesions alter dopaminergic synaptic markers in the nucleus basalis magnocellularis and striatum of rats, Synapse, 4:248.

Grove E.A., 1988, Neural associations of the substantia innominata in the rat: afferent connections, J. Comp. Neurol., 277:315.

Haring J.H., and Wang R.Y., 1986, The identification of some sources of afferent input to the rat nucleus basalis magnocellularis by retrograde transport of horseradish peroxidase, Brain Res., 366:152.

Hubner C.B., and Koob G.F., 1990, The ventral pallidum plays a role in mediating cocaine and heroin self-administration in the rat. Brain Res. 508:20.

Huston J.P., Kiefer S., Buscher W., and Monoz C., 1987, Lateralized functional relationship between the preoptic area and lateral hypothalamic reinforcement, Brain Res. 436:1.

Jones B.E., and Cuello A.C., 1989, Afferents to the basal forebrain cholinergic cell area from pontomesencephalic-catecholamine, serotonin, and acetylcholine-neurons, Neuroscience, 31:37.

Jones D.L., and Mogenson G.J., 1980, Nucleus accumbens to globus pallidus GABA projection subserving ambulatory activity, Amer. J. Physiol., 238:R65-R69.

Lamour Y., Dutar P., Rascol O., and Jobert A., 1986, Basal forebrain neurons projecting to the rat frontoparietal cortex: Electrophysiological and pharmacological properties, Brain Res. 362:122.

Lindvall, O., and Bjorklund, A., 1974, The organization of the ascending catecholamine neuron systems in the rat brain, ACTA Physiol. Scand., Supp. 412:1.

Lindvall O., and Bjorklund A., 1979, Dopaminergic innervation of the globus pallidus by collaterals from the nigrostriatal pathway, Brain Res., 172:169.

Linseman M.A., 1974, Inhibitory unit activity of the ventral forebrain during both appetitive and aversive Pavlovian conditioning, Brain Res. 80:146.

Martinez-Murillo R., Semenenko F., and Cuello A.C., 1988, The origin of tyrosine hydroxylase-immunoreactive fibers in the regions of the nucleus basalis magnocellularis of the rat, Brain Res., 451:227.

Maslowski, R.J., and Napier, T.C., 1991, Dopamine D1 and D2 agonists induce opposite changes in the firing rate of ventral pallidal neurons, Eur. J. Pharmacol., in press.

McGeer P.L., McGeer E.G., Kimura H., and Peng J.-F., 1986, Cholinergic neurons and cholinergic projections in the mammalian CNS, in: "Dynamics of Cholinergic Function: Advances in Behavioral Biology", Vol 30, I. Hanin, ed., Plenum Press: New York p 11.

McGurk S.R., Levin E.D. and Butcher L.L., 1988, Cholinergic-dopaminergic interactions in radial-arm maze performance, Behav. Neural Biol., 49:234.

Mogenson G.J., Jones, D.L., and Yim, C.Y., 1980, From motivation to action: functional interface between the limbic system and the motor system, Prog. Neurobiol. 14:69.

Mogenson G.J., and Yang C.R., 1991, The contribution of basal forebrain to limbic-motor integration and the mediation of motivation to action, in: "The Basal Forebrain: Anatomy to Function: Advances in Experimental Medicine and Biology", T.C. Napier, P.W. Kalivas, I. Hanin, eds, New York, Plenum Publishing Corporation, in press.

Mora F., Rolls E.T., and Burton M.J., 1976, Modulation during learning of the responses of neurons in the lateral hypothalamus to the sight of food, Exp. Neurol. 53:508.

Napier T.C., and Potter P.P., 1989, Dopamine in the rat ventral pallidum/substantia innominata: Biochemical and electrophysiological studies, Neuropharmacology, 28:757.

Napier, T.C., Simson, P.E. and Givens, B.S., 1991, Dopamine electrophysiology of ventral pallidal/substantia innominata neurons: Comparison with the dorsal globus pallidus. J. Pharmacol. Exp. Therap. in press.

Olton D., Markowska A., Voytko M.L., Givens B., Gorman L., and Wenk G., 1990, Basal forebrain cholinergic system: A functional analysis, in: "The Basal Forebrain: Anatomy to Function: Advances in Experimental Medicine and Biology", T.C. Napier, P.W. Kalivas, I. Hanin eds, New York: Plenum Publishing Corporation, in press.

Pirch J.H., 1977a, Effects of amphetamine and chlorpromazine on brain slow potentials in the rat, Pharmacol. Res. Comm., 9:669.

Pirch J.H., 1977b, Amphetamine effects on brain slow potentials associated with discrimination in the rat, Pharmacol. Biochem. Behav. 6:697.

Pirch J.H., 1980, Effects of dextroamphetamine on event-related potentials in rat cortex during a reaction time task, Neuropharmacology 19:365.

Pirch, J.H. and Corbus, M.J., 1983, Haloperidol antagonism of amphetamine-induced effects on event-related slow potentiate from rat cortex. Int. J. Neurosci., 18:137.

Pirch J.H., Corbus M.J., and Napier T.C., 1981a, Auditory cue preceding intracranial stimulation induces event-related potential in rat frontal cortex: Alterations by amphetamine, Brain Res. Bull., 7:799.

Pirch J.H., Corbus M.J., Rigdon G.C., and Lyness W.H., 1986, Generation of cortical event-related slow potentials in the rat involves nucleus basalis cholinergic innervation, Electroencephalogr. Clin. Neurophysiol., 63:464.

Pirch J.H., Napier T.C., and Corbus M.J., 1981b, Brain stimulation as a cue for event-related potentials in rat cortex: Amphetamine effects, Int. J. Neurosci., 15:217.

Pirch, J.H., Rigdon, G.C., Rucker, H.K. and Turco, K, 1991, Basal forebrain modulation of cortical cell activity during conditioning, in: "The Basal Forebrain: Anatomy to Function: Advances in Experimental Medicine and Biology", T.C. Napier, P.W. Kalivas, I. Hanin, eds, New York, Plenum Publishing Corporation, in press.

Reiner P.B., Semba K., Fibiger H.C., and McGeer E.G., 1987, Physiological evidence for subpopulations of cortically projecting basal forebrain neurons in the anesthetized rat, Neuroscience, 20:629.

Richardson R.T., and DeLong M.R., 1986, Nucleus basalis of Meynert neuronal activity during a delayed response task in monkey, Brain Res. 399:364.

Richardson R.T., and DeLong, M.R., 1991, Electrophysiological studies of the functions of the nucleus basalis in primates, in: "The Basal Forebrain: Anatomy to Function: Advances in Experimental Medicine and Biology", T.C. Napier, P.W. Kalivas, I. Hanin, eds., New York, Plenum Publishing Corporation, in press.

Richfield E.K., Penney J.B., and Young A.B., 1989, Anatomical and affinity state comparisons between dopamine D1 and D2 receptors in the rat central nervous system, Neuroscience, 30:767.

Rolls E.T., Burton M.J., and Mora F., 1980, Neurophysiological analysis of brain-stimulation reward in the monkey, Brain Res., 194:339.

Rolls E.T., Sanghera M.K., Roper-Hall A., 1979, The latency of activation of neurones in the lateral hypothalamus and substantia innominata during feeding in the monkey, Brain Res., 164:12.

Russchen F.T., Amaral D.G., and Price J.L., 1985, The afferent connections of the substantia innominata in the monkey, Macaca fascicularis, J. Comp. Neurol., 242:1.

Semba K., Reiner P.B., McGeer E.G., and Fibiger H.C., 1988, Brainstem afferents to the magnocellular basal forebrain studied by axonal transport, immunohistochemistry, and electrophysiology in the rat, J. Comp. Neurol., 267:433.

Voorn P., Jorritsma-Byham B., Van Dijk C., and Buijs R.M., 1986, The dopaminergic innervation of the ventral striatum in the rat: A light- and electron-microscopical study with antibodies against dopamine, J. Comp. Neurol., 251:84.

Werman, R., 1966, A review - Criteria for identification of a central nervous system transmitter. Comp. Biochem. Physiol., 18:745.

Wilson F.A.W., and Rolls E.T., 1990, Neuronal responses related to reinforcement in the primate basal forebrain, Brain Res., 509:213.

Wilson, F.A.W., 1991, The relationship between learning, memory and neuronal responses in the primate basal forebrain, in: The Basal Forebrain: Anatomy to Function: Advances in Experimental Medicine and Biology", T.C. Napier, P.W. Kalivas, I. Hanin, eds. New York, Plenum Publishing Corporation, in press.

Wise R.A., 1980, The dopamine synapse and the notion of 'pleasure centers' in the brain, Trends in Neurosci., 3:91.

Woodruff, G.N., McCarthy P.S., and Walker R.J., 1976, Studies on the pharmacology of neurons in the nucleus accumbens of the rat, Brain Res., 11:233.

Yang C.R., and Mogenson G.J., 1989, Ventral pallidal neuronal responses to dopamine receptor stimulation in the nucleus accumbens, Brain Res., 489:237.

York D.H., 1970, Possible dopaminergic pathway from substantia nigra to putamen, Brain Res., 20:233.

Zaborszky L., 1989, Afferent connections of the forebrain cholinergic projection neurons, with special reference to monoaminergic and peptidergic fibers, in: "Central Cholinergic Synaptic Transmission" M. Frotscher, U. Misgeld, eds., Basel Switzerland: Birkhauser Verlag. p. 12.

Zaborszky, L., Luine, V.N., Cullinan, W.E., Allen, D.L., and Heimer, L., 1991, Direct catecholaminergic-cholinergic interactions in the basal forebrain: Morphological and biochemical studies, (submitted).

THE CHOLINERGIC BASAL FOREBRAIN:

A CRITICAL ROLE IN CORTICAL AROUSAL

Kazue Semba

Department of Anatomy
Dalhousie University
Halifax, N.S. B3H 4H7
Canada

INTRODUCTION

Acetylcholine (ACh) has long been implicated in the regulation of arousal or wakefulness. However, the anatomical basis for this regulation had been missing because relatively little was known about the organization of central cholinergic pathways. During the last decade, however, specific immuno-histochemical markers became available, and by using these markers central cholinergic neurons have been mapped and their projections delineated (see Semba and Fibiger, 1989, for review). It is now well established that there are two major cholinergic projection systems in the CNS: cholinergic neurons in the basal forebrain project widely to the cerebral cortex, and those in the mesopontine tegmentum project heavily to the thalamus. Armed with these anatomical findings, researchers of behavioral state have begun to investigate the role of specific populations of central cholinergic neurons in the regulation of waking and sleep. One important conclusion which has emerged from such recent studies is that cholinergic neurons in the basal forebrain have a crucial role in cortical arousal. In the present paper, both anatomical and physiological evidence supporting this notion is discussed, and clues are explored as to how the activity of basal forebrain cholinergic neurons is regulated during different behavioral states.

The term arousal requires a cautionary note here, because its meaning varies. Arousal has often been defined in terms of electroencephalogram (EEG) or behavior, and EEG arousal is usually correlated with behavioral arousal. However, there are instances of dissociation. For example, during rapid eye movement (REM) sleep, cortical EEG is desynchronized or acti-vated despite concurrent muscle atonia. Dissociation can also be induced pharmacologically; animals injected with atropine appear to be behaviorally active, however, with synchronized or slow cortical EEG (Stumpf, 1965; Longo, 1966). Although mechanisms to integrate cortical and behavioral arousal are unclear, there is good evidence that activation of cortical EEG is associated with facilitation, in particular, enhancement of stimulus-specific responses in the cortex (see below). On this

The Basal Forebrain, Edited by T.C. Napier *et al.*
Plenum Press, New York, 1991

basis and because of this functional implication, in the present paper cortical arousal is defined in terms of EEG activation, i.e., low voltage fast activity or desynchronization.

THE ORIGINAL CONCEPT OF THE ASCENDING RETICULAR ACTIVATING SYSTEM BY MORRUZI AND MAGOUN

The impetus for early studies of sleep and arousal was provided by the concept of the ascending reticular activating system, which was proposed by Morruzi and Magoun in 1949. This concept, which predominated this area of research in the 1950's and 1960's, was based on data obtained with EEG recordings from the cerebral cortex combined with electrical stimulation of the brainstem in the encéphale isolé preparation of the cat. When the concept was put forward, its functional implication, i.e., that the cerebral cortex is activated during EEG desynchronization, was entirely hypothetical, and experimental evidence for facilitation in the cortex during brainstem stimulation was not provided until the late 1950's (see Steriade and McCarley, 1990).

In the original concept of the ascending reticular activating system proposed by Morruzi and Magoun, no anatomical substrate was suggested. In 1967, however, Shute and Lewis (1967; Lewis and Shute, 1967), using acetylcholinesterase histochemistry combined with lesions, mapped cholinergic pathways in the brain, and proposed that the ascending cholinergic pathways from the brainstem and the basal telencephalon are the anatomical substrate of the ascending reticular activating system. Because dopaminergic neurons in the substantia nigra also contain this enzyme marker, their projection to the striatum was included as a third component of the cholinergic activating system. However, the landmark studies by Shute and Lewis set a framework for subsequent research to identify the anatomical substrate for cortical arousal.

EARLY PHARMACOLOGICAL EVIDENCE SUGGESTING CHOLINERGIC INVOLVEMENT IN CORTICAL EEG ACTIVATION

Pharmacological evidence accumulated during the 1950's and 1960's indicated that cortical EEG desynchronization is mediated by ACh. ACh and cholinesterase inhibitors induce cortical EEG desynchronization, and these effects are antagonized by atropine (see Stumpf, 1965; Longo, 1966 for reviews). More direct evidence for the involvement of ACh was obtained by measuring ACh release from the cortex during reticular stimulation as well as during different behavioral states in cat and rabbit. In a pioneer study, Kanai and Szerb (1965) measured cortical ACh release in anesthetized cats with the cup method and using the dorsal muscle of the leech for bioassay of ACh. These authors demonstrated that mesencephalic reticular stimulation increases the rate of ACh release from the somatosensory and parietal cortices, to about 5-6 times baseline, and simultaneously produces cortical EEG desynchronization (Fig. 1). The EEG response, but not ACh release, was blocked by atropine, indicating that it is a postsynaptic action of ACh that is crucial for cortical EEG activation.

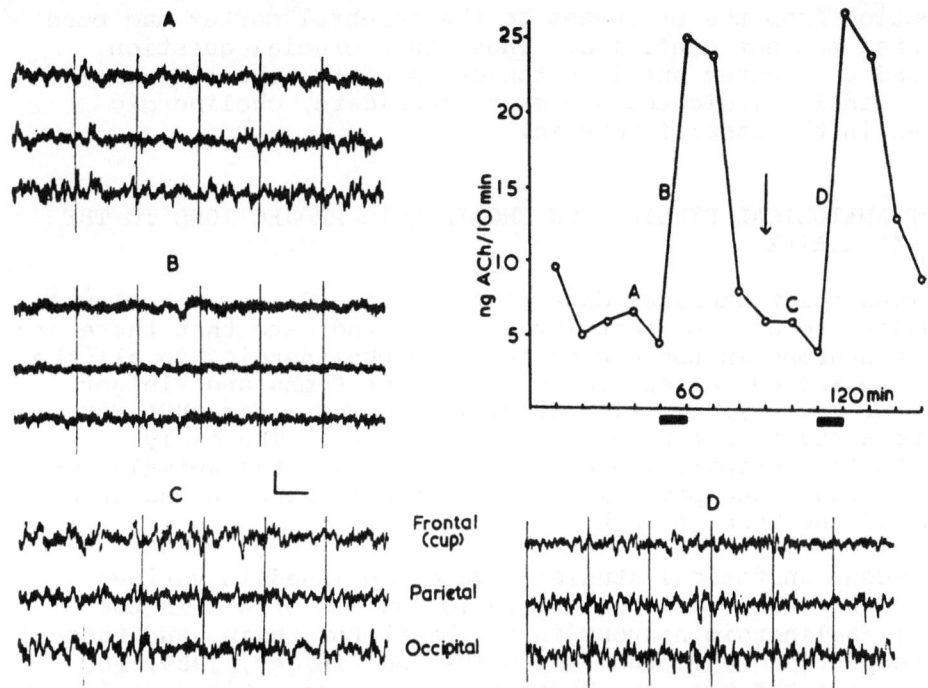

Fig. 1. Cortical EEG patterns (A-D) and ACh output from the
cortical surface during the electrical stimulation
(thick lines) of the midbrain reticular formation in
halothane-anesthetized cats. Note that the stimulation
induced activation, i.e., low voltage fast activity, in
the EEG, and elevated ACh output. Atropine (1 mg/kg,
i.v., arrow) blocked the EEG activation, whereas the
elevated ACh output was unaffected, which suggests that
the EEG activation is a postsynaptic effect of ACh.
(Reprinted from Szerb and Kanai, 1965, by courtesy of
the authors and permission of Macmillan Magazines Ltd.)

Additional evidence for the involvement of ACh in cortical
arousal came from a study by Collier and Mitchell (1967), who
measured ACh release from the visual cortex in freely moving
rabbits, also with the cortical cup method. Typically, the rate
of ACh release increased upon recovery from thiopental anesthesia, remained at a high level when the animal was behaviorally
active, decreased during inactive periods, and was further
reduced following an overdose with pentobarbital, falling to
undetectable levels at death. Cortical ACh release was also
measured during REM sleep in cats by Jasper and Tessier (1971).
The rate of ACh release was almost twice as high during waking
and REM sleep compared with slow wave sleep. Cortical EEG is
mostly desynchronized during waking as well as REM sleep,
whereas it is synchronized (high voltage slow activity) during
slow wave sleep. Thus, the state-dependent ACh release is
consistent with the view that an increase in ACh release is
associated with cortical EEG activation.

While these studies strongly supported the notion that
cortical ACh release increases during cortical arousal, the
question remained as to the source of ACh collected from the
cerebral cortex. The presence of a direct cholinergic

projection from the brainstem to the cerebral cortex had been suspected but not confirmed. Thus, this crucial question remained unanswered until anatomical studies since the early 1980's finally indicated one major candidate, cholinergic neurons in the basal forebrain.

RECENT ANATOMICAL FINDINGS ON CHOLINERGIC PROJECTIONS TO THE CEREBRAL CORTEX

Immunohistochemical data so far obtained with monoclonal antibodies to choline acetyltransferase indicate that there are no interneurons in the cortex that are cholinergic, in all the species examined, except for the rat (see Semba and Fibiger, 1989, for review). In situ hybridization data for mRNA for choline acetyltransferase have not appeared. The early investigators primarily used cats as experimental animals, which would preclude the possibility of cortical interneurons as a source of the released ACh.

Recent anatomical studies have established in various species that there are two extrinsic sources of ACh in the cortex: cholinergic neurons in the basal forebrain, and those in the mesopontine tegmentum (see Semba and Fibiger, 1989, for review). Of the two, the basal forebrain projection provides by far the more extensive cholinergic input to the cerebral cortex in that it innervates all regions of the cerebral cortex, in a topographic manner. In contrast, cholinergic neurons in the mesopontine tegmentum (the pedunculopontine and laterodorsal tegmental nuclei) appear to innervate selectively the medial prefrontal cortex, at least in the rat (Vincent et al., 1983; Satoh and Fibiger, 1986; Hallanger and Wainer, 1988); no comparable data for other species are available at present. Thus, cholinergic neurons in the basal forebrain would be the most likely candidates as the source of ACh released in the cortex during EEG activation, and these neurons also appear to serve as a crucial link to mediate the effects of reticular stimulation for inducing cortical EEG activation.

In addition to direct actions, cholinergic basal forebrain neurons may regulate cortical EEG via the thalamus. Anatomical studies during the past few years have indicated that a sub-population of cholinergic basal forebrain neurons projects to the reticular thalamic nucleus (Hallanger et al., 1987; Levey et al., 1987; Steriade et al., 1987). The reticular thalamic nucleus has widespread and yet topographically organized projections to all the other thalamic nuclei, and is believed to be the key structure in the generation of thalamocortical rhythmical activity or spindles (Steriade and Deschênes, 1984). Furthermore, at least some cholinergic basal forebrain neurons project to both the reticular thalamic nucleus and the cerebral cortex (Jourdain et al., 1989). Thus, cholinergic neurons in the basal forebrain may influence cortical neuronal activity both directly, to induce cortical EEG desynchronization, and indirectly via the thalamus, to block rhythmical thalamocortical oscillations. However, at present the effects of ACh on reticular thalamic neurons are somewhat controversial (Kayama et al., 1986; McCormick and Prince, 1986; Hu et al., 1989). Moreover, in addition to a cholinergic component, the basal forebrain projection to the reticular thalamic nucleus contains a GABAergic component as well (Jourdain et al., 1989; Porter and

Asanuma, 1989). The functional significance of cholinergic vs. GABAergic inputs from the basal forebrain to the reticular thalamic nucleus in the generation of thalamocortical rhythmical activity remains to be investigated.

THE CHOLINERGIC BASAL FOREBRAIN AND CORTICAL EEG PATTERNS

Septohippocampal System

In the scheme of the continuous distribution of cholinergic neurons in the basal forebrain (Schwaber et al., 1987; Semba and Fibiger, 1988), the septohippocampal projection is considered to be part of the entire cholinergic projection system from the basal forebrain to the cerebral cortex. It is one of the best studied cholinergic systems in the brain (Nicoll, 1985), and the role of ACh in the activation of hippocampal EEG has been the subject of intensive investigation since the 1950's. The pattern of EEG during activation in the hippocampus, however, is distinctly different from that in the neocortex, in that unlike the cortical EEG the hippocampal EEG displays a rhythmical pattern when activated, often called the hippocampal theta activity or the rhythmical slow activity (Vanderwolf, 1969; Vanderwolf and Robinson, 1981).

Since the first description of rhythmical slow waves in the hippocampus in 1938 by Jung and Kornmüller, various psychological concepts including attention, information processing, and response inhibition, to name a few, were put forward to explain the function of the hippocampal theta activity (see Bennett, 1971 for review). However, the Zeitgeist of psychology in the 1960's was in favor of descriptive terminology of behavior, as opposed to inferred psychological processes (e.g., Skinner, 1974). In this spirit of behaviorism, Vanderwolf (1969) carefully observed the correlation between hippocampal EEG patterns and behavior, in terms of movement and posture, in freely moving rats, and concluded that the hippocampal theta activity is most closely correlated with "voluntary" (designated later as type 1) movements including walking, rearing, and head movements. In contrast, "automatic" (or type 2) movements such as grooming and alert immobility were only sometimes accompanied by theta activity. A further understanding of behavioral correlates of the hippocampal rhythmical activity was obtained from pharmacological studies indicating that the hippocampal theta activity actually consists of two types, one sensitive and the other resistant to atropine. Both atropine-sensitive and atropine-resistant theta activities are present during type 1 behavior, whereas only atropine-sensitive theta activity is present during type 2 behavior (Kramis et al., 1975; Vanderwolf, 1988). Similar pharmacological differentiation has been described with neocortical EEG activation (Kramis et al., 1975), and the atropine-resistant components of hippocampal theta activity and cortical desynchronization have been suggested to be dependent on a serotonergic input (Vanderwolf and Baker, 1986; Vanderwolf, 1988).

Recently, Stewart and Fox (1989) demonstrated that pharmacological differentiation is also present with respect to what is thought to be the pacemaker for hippocampal theta activity *in vivo*, namely, rhythmically bursting neurons in the medial septum. In view of recent anatomical evidence that the septohippocampal pathway contains both cholinergic and GABAergic components (Köhler et al., 1984), Stewart and Fox (1990) have proposed the hypothesis that bursting medial septal neurons that are atropine-sensitive are cholinergic, whereas those that are atropine-resistant are GABAergic. Because GABAergic septo-hippocampal axons terminate on GABAergic interneurons in the hippocampus (Freund and Antal, 1988), Stewart and Fox (1990) further suggested that atropine-sensitive medial septal neurons terminate on both GABAergic interneurons and pyramidal neurons, whereas atropine-resistant cells terminate on inhibitory interneurons. Another corollary of the hypothesis by Stewart and Fox might be that a serotonergic input directly paces atropine-resistant, but not atropine-sensitive, septal neurons. These possibilities remain to be investigated both physiologically and anatomically.

Nucleus Basalis-Neocortex System

The role of the nucleus basalis in the cortical EEG activation has been studied by lesions and stimulation of the nucleus basalis, and, again, there is strong evidence that the activation of the cholinergic projection from the basal forebrain to the neocortex is crucial for cortical EEG desynchronization. Following ibotenic acid lesions of the nucleus basalis, which reduced acetylcholinesterase staining of corresponding cortical areas, Buzsáki et al. (1988) have reported that the cortical EEG recorded from cholinergically deafferentated areas of the cortex slowed down as indicated by power spectral analyses. Similar results have been reported by Stewart et al. (1984). In contrast, electrical stimulation of the nucleus basalis increased cortical ACh release, and also produced desynchronization of cortical EEG in the rat (Casamenti et al., 1986). Increases in slow components of cortical EEG have been reported to occur in advanced stages of Alzheimer's disease involving massive degeneration of basal forebrain cholinergic and other neurons (e.g., Coben et al., 1985).

In contrast to nucleus basalis lesions, following selective lesions of the thalamic reticular nucleus or extensive damage to virtually the entire thalamus, the desynchronized pattern of cortical EEG recovered in 1-2 weeks, whereas thalamocortical spindle activity never returned (Stewart et al., 1984; Buzsáki et al., 1988; Vanderwolf and Stewart, 1988).

In combination, the above studies suggest that cholinergic basal forebrain neurons play a critical role in the activation of cortical EEG. Furthermore, although the thalamus, and in particular, the intralaminar nucleus of the thalamus has been implicated in cortical arousal (e.g., Steriade, 1981), its principal function appears to be related to the generation of spindles (Andersen and Andersson, 1968; Steriade and Deschênes, 1984), and its role in neocortical activation or desynchronization may be less critical than has previously been assumed.

EEG AND BEHAVIORAL CORRELATES OF SINGLE UNIT ACTIVITIES FROM BASAL FOREBRAIN NEURONS

There have been two lines of research related to cortical arousal that uses single unit recordings in behaving animals. One comes from the tradition of behavioral state research, commonly conducted in cats or rats, with a focus on the pattern of neuronal firing during different behavioral states. The other is concerned with neuronal correlates of reward and movement control, with the primary focus on the specificity of activity during visual discrimination tasks in primates.

Chronic single unit recordings during sleep-waking cycles have been made from neurons in the basal forebrain regions which variably overlap with the cholinergic basal nuclear complex (Detári et al., 1984, 1987; Szymusiak and McGinty, 1986; Buzsáki et al., 1988). However, the fact that cholinergic neurons are intermixed with non-cholinergic neurons in all regions of the magnocellular basal forebrain makes these results less conclusive with respect to the transmitter content and projection of recorded neurons. To circumvent this problem, at least in part, a test for antidromic activation from the cortex is often used (Reiner et al., 1987). This capitalizes on the fact that cortically projecting neurons in nucleus basalis magnocellularis regions are 80-90% cholinergic in the rat (Rye et al., 1984). Adopting this strategy, Detári and Vanderwolf (1987) examined the correlation between cortical EEG patterns and firing rates of single units from the substantia innominata and nucleus basalis regions that were antidromically driven from the frontal cortex in urethane-anesthetized rats. Most of these cortically projecting neurons were strongly activated during spontaneous or peripherally elicited cortical EEG desynchronization. In contrast, many neurons which were not driven from the cortex increased their firing rates during high-amplitude slow wave activity in the cortical EEG.

Data considerably different from the above study have been reported by Szymusiak and McGinty (1989), who identified cortically projecting basal forebrain neurons by antidromic stimulation from the anterior cingulate bundle and the external capsule in freely moving cats. In this study, the majority of the neurons fired at a higher rate during slow wave sleep than during waking or REM sleep, although there was also a population of neurons that were active predominantly in waking. The reason for the larger population of slow wave sleep-active cells in this study compared with the study by Detári and Vanderwolf (1987) is not clear. However, it may be related to the species difference in the transmitter content of cortically projecting neurons: unlike in the rat, many cortically projecting neurons in the basal forebrain of the cat have been reported to be GABAergic (Fisher et al., 1988).

In primates, single units have been recorded in the nucleus basalis during a variety of discrimination tasks, and the presence of neurons which fire specifically in relation to reward has been documented (see Richardson and DeLong, and Wilson, this book). Cholinergic neurons in the nucleus basalis have been suggested to modulate conditioning-related cortical neuronal activity in the rat (Pirch et al., this book). Recently, nucleus basalis neurons that fire in relation to aversive aspect of stimuli have also been reported (Richardson and DeLong, this book). Similarly, neuronal responses related to the novelty and familiarity of visual stimuli have been described (Wilson and Rolls, 1990). To incorporate these new findings, it has been suggested that these basal forebrain neurons respond to arousing quality of aversive and appetitive stimuli (Richardson and DeLong, this book). Interestingly, the responses of the neurons described by Richardson and DeLong were apparently not homogeneous; some neurons increased, while others decreased, their firing rate in response to "arousing" stimuli. It was not known whether these neurons projected to the cortex.

The single unit studies discussed above are consistent with the view that basal forebrain neurons are involved in cortical activation. However, there appear to be two opposite patterns of activity in relation to cortical activation or arousal: either an increase or a decrease in firing rate during arousal. Interestingly, the view of the basal forebrain as a hypnotic, rather than arousal, center was expressed as early as in the 1960's (Sterman and Clemente, 1962; McGinty and Sterman, 1968), and continues to be influential to date (e.g., Szymusiak and McGinty, 1989). This apparent paradox may be related to the fact that not all cortically projecting neurons are cholinergic (Rye et al., 1984). Also, because an antidromicity test to confirm cortical projection was not always conducted in the previous studies, local interneurons might have been included in the recorded cell populations. It is interesting, in this regard, that Detári et al. (1990) have found negative correlations in the firing rate of a pair of adjacent neurons recorded simultaneously in the basal forebrain of anesthetized cats. The problem might be solved by identifying the transmitter content of single units which have been characterized for their firing characteristics in relation to different EEG patterns. However, this would involve intracellular labelling of chronically recorded neurons, followed by immunohistochemistry, which would seem to be a rather formidable challenge at the present time.

FACILITATION OF STIMULUS-SPECIFIC RESPONSES OF CORTICAL NEURONS BY ACH

One well known effect of ACh in the cerebral cortex is the facilitation of stimulus-specific responses. For example, without significantly increasing background activity, ACh increased responses of visual cortical neurons to optimal visual stimuli (Sillito and Kemp, 1983), as well as to electrical stimulation of the lateral geniculate nucleus; this latter effect was blocked by atropine (Sato et al., 1987). Similar enhancement of signal to noise ratio by ACh was seen in the response of cortical somatosensory neurons to tactile stimuli (Donoghue and Carroll, 1987; Spehlmann et al., 1971) or iontophoretically applied glutamate (Metherate et al., 1987).

Furthermore, in the cholinergically depleted visual cortex following unilateral kainate lesions of the nucleus basalis, neuronal responses to receptive field stimuli were reduced, although response specificity was not altered (Sato et al., 1987). The response magnitude was restored to control levels obtained from the intact side, by ionophoretic administration of ACh. Similarly, pairing of peripheral stimuli with nucleus basalis stimulation induced long-term (up to several hours) enhancement of somatosensory evoked potentials (Rasmusson and Dykes, 1988). A key demonstration to relate these ACh effects in the cortex to behavioral states comes from the study by Livingstone and Hubel (1981), in which the transition from slow wave sleep to waking was shown to parallel increases in the magnitude of responses of visual cortical neurons, as well as enhancement of receptive field specificity.

These findings provide strong support for the view that ACh
has facilitatory effects on the processing of sensory information in the cortex, and that this facilitation normally occurs
during EEG desynchronization and waking.

ASCENDING BRAINSTEM INPUTS TO THE CHOLINERGIC BASAL FOREBRAIN

The evidence so far discussed above suggests that basal
forebrain cholinergic neurons become active during cortical EEG
activation and increase ACh release in the cortex, facilitating
sensory responses of cortical neurons. However, it should be
recalled that electrical stimulation of the brainstem reticular
formation induces cortical EEG desynchronization and elevates
ACh release from the cortex. This means that although the
reticular stimulation cannot directly induce cortical ACh
release, it can do so indirectly, by activating inputs to basal
forebrain cholinergic neurons projecting to the cortex, and the
same scenario may account for the "arousing" effects of peripheral stimuli.

Reportedly, the most effective stimulation sites for
inducing cortical EEG desynchronization are the area of the
midbrain reticular formation just ventrolateral to the central
gray (e.g., Starzl et al., 1951; Fig. 2). Although the presence
of monoaminergic pathways in the brainstem was unknown at the
time of early studies by Morruzi, Magoun, and their associates,
the rich body of information obtained thereafter on ascending
monoaminergic as well as cholinergic pathways allows us now to
examine whether these transmitter-specific pathways were
activated during reticular brainstem stimulation. Comparisons
of the courses of these ascending pathways and the effective
stimulation sites for cortical arousal indeed clearly indicate
that the effective stimulation sites fall right on the major
ascending fiber bundles of the brainstem monoaminergic (Fig. 2A)
and cholinergic projections (Fig. 2B). Obviously, this raises
the possibility that the activity of cholinergic basal forebrain
neurons is regulated or modulated by monoaminergic and cholinergic inputs from the brainstem.

The effects of activating at least some of these brainstem
inputs to cholinergic basal forebrain regions have been studied
with extracellular recordings from cortically projecting neurons
in anesthetized rats (Semba et al., 1988b). Electrical stimulation at or near the pedunculopontine tegmental nucleus and the
dorsal raphe nucleus induced either an increase or a decrease in
firing rate in 18-21% of the cortically projecting neurons so
tested (Semba et al., 1988b; Fig. 3).

The anatomical basis for brainstem regulation of cholinergic basal forebrain neurons has been provided by recent
studies combining retrograde tracing and immunohistochemistry at
the light microscopic level (Semba et al., 1988b; Jones and
Cuello, 1989). Afferents from the brainstem which terminate in
cholinergic basal forebrain regions include: a serotonergic
afferent from the median and dorsal raphe nuclei, a noradrenergic input from the locus ceruleus, and a cholinergic afferent
from the mesopontine tegmentum (Fig. 4). These three inputs are
directed to all levels of the cholinergic basal forebrain
complex. In addition, dopaminergic neurons in the ventral
tegmental area and the substantia nigra pars compacta, as well

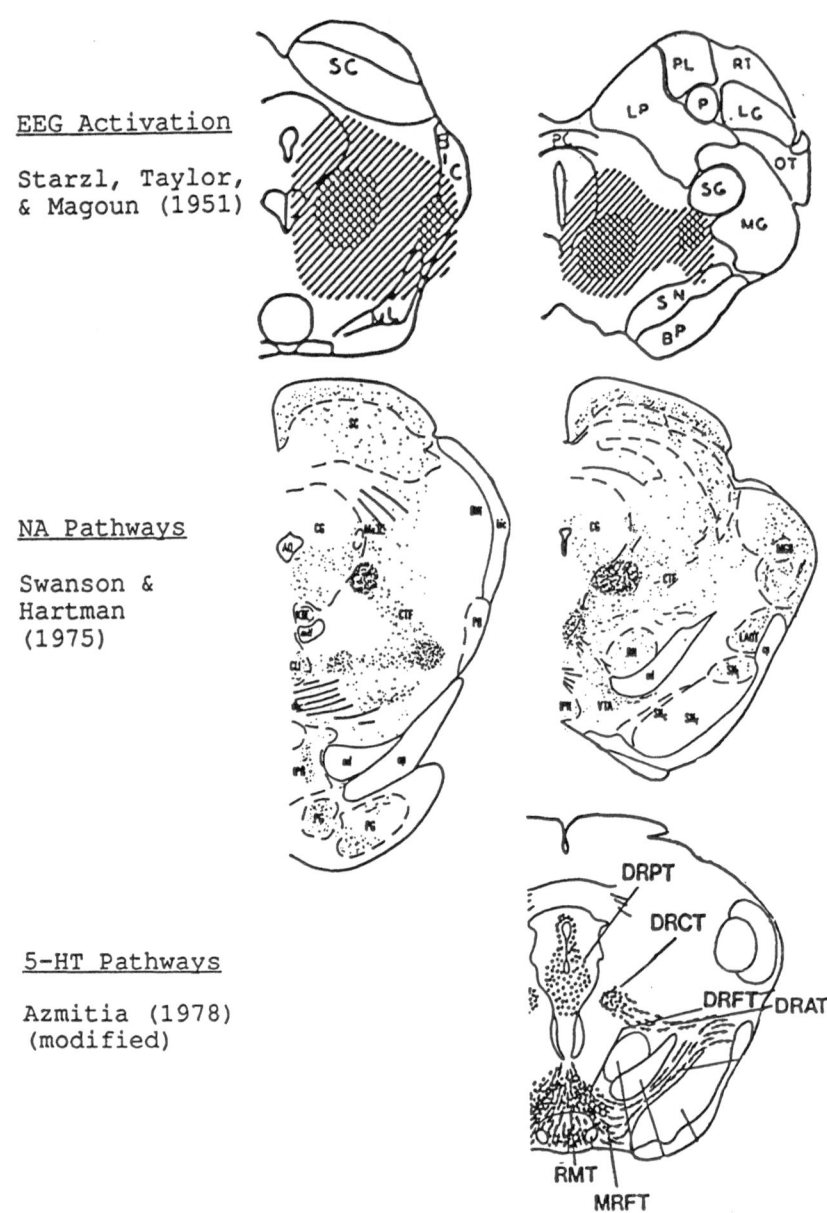

EEG Activation

Starzl, Taylor,
& Magoun (1951)

NA Pathways

Swanson &
Hartman
(1975)

5-HT Pathways

Azmitia (1978)
(modified)

Fig. 2A.   Comparisons between the effective stimulation sites
for inducing cortical EEG activation in the cat
(Starzl et al., 1951), and the courses in the upper
brainstem of ascending noradrenergic (NA; Swanson and
Hartman, 1975) and serotonergic (5-HT; modified from
Azmitia, 1978) pathways in the rat. Note that the
stimulation sites for cortical activation fall onto
the positions of major ascending noradrenergic and
serotonergic fiber bundles. (Reprinted or modified
from the original papers indicated by courtesy of the
authors and permission of The American Physiological
Society and Wiley-Liss, a division of John Wiley and
Sons, Inc.)

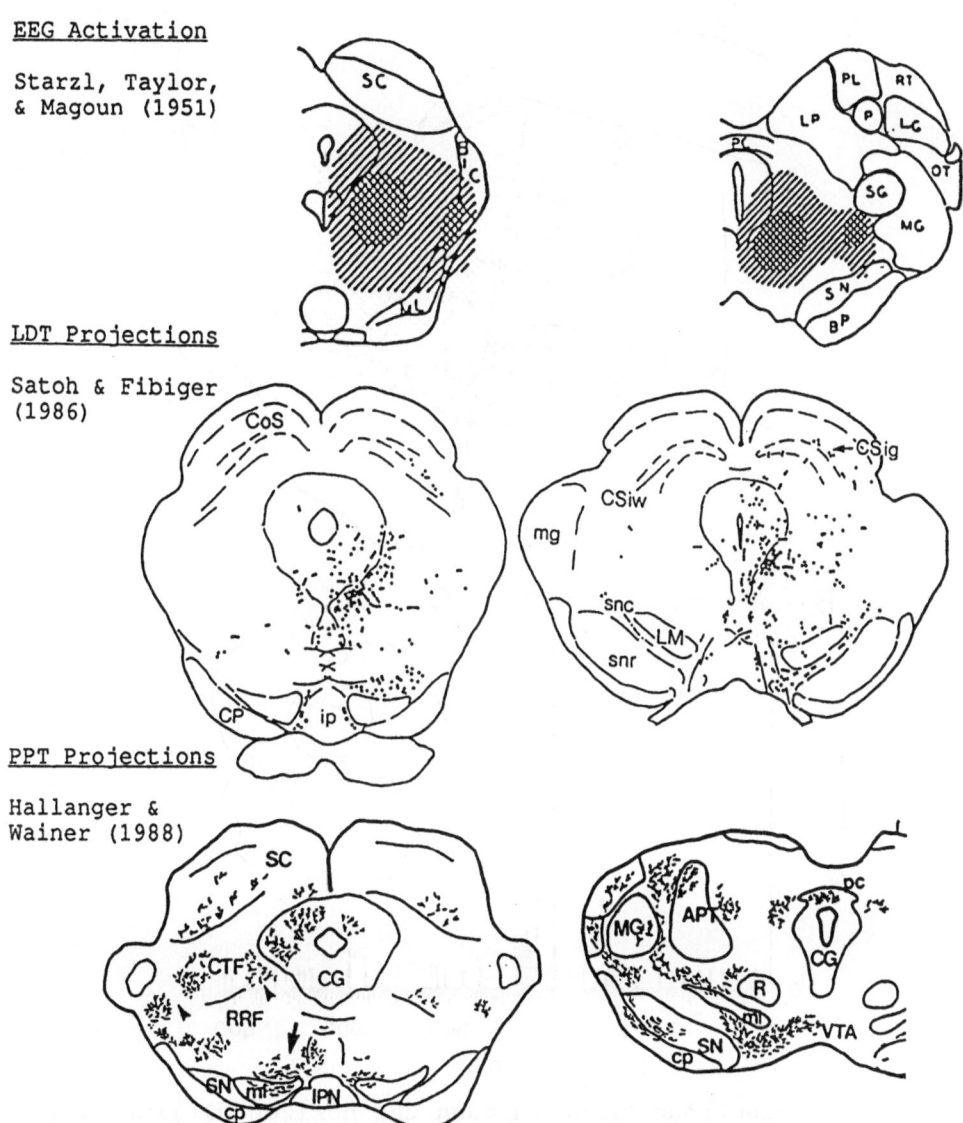

Starzl, Taylor,
& Magoun (1951)

LDT Projections

Satoh & Fibiger
(1986)

PPT Projections

Hallanger &
Wainer (1988)

Fig. 2B.   Comparisons between the upper brainstem sites
effective in inducing cortical EEG desynchronization
(Starzl et al., 1951), and the courses of ascending,
presumably cholinergic projections from the laterodorsal (LDT; Satoh and Fibiger, 1985) and the pedunculopontine tegmental nuclei (PPT; Hallanger and Wainer,
1988).   Note the correlation between the stimulation
sites and the position of major fiber bundles of these
projections.   (Reprinted or modified from the original
papers indicated courtesy of the authors and permission of The American Physiological Society and Wiley-
Liss, a division of John Wiley and Sons, Inc.)

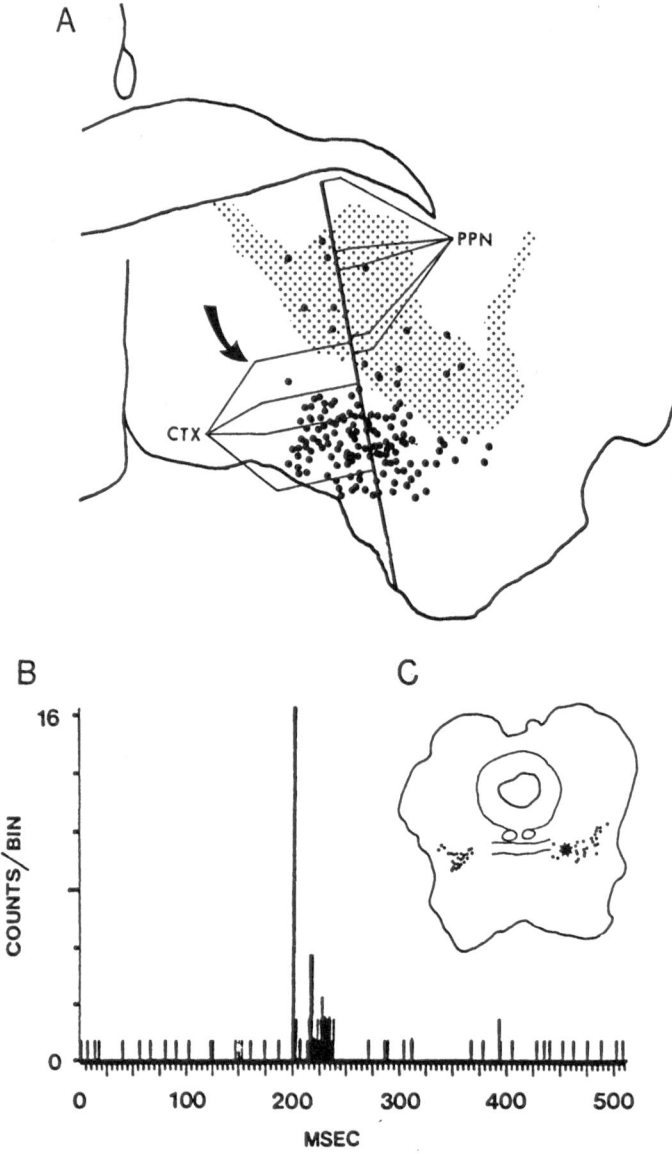

Fig. 3. A: An electrode track through the horizontal limb of the
diagonal band reconstructed from sections which
contained the electrode track and were stained for
cholinergic neurons (dots) and substance P (stippling).
Locations of cells driven antidromically from the cortex
(CTX) are mostly within the cluster of cholinergic
neurons, whereas those antidromic to pedunculopontine
tegmental nucleus (PPN) stimulation are located more
dorsally, in the ventral pallidum as delineated by
substance P neuropil (see also Semba et al., 1989).
B: A peristimulus time histogram of orthodromic
responses of the cortically projecting cell indicated by
an arrow in A. This cell responded with an initial
increase followed by a decrease in firing rate to the
electrical stimulation at the PPN. C: Stimulation site
in the PPN as confirmed by the presence of cholinergic
neurons. (From Semba et al., 1988b.)

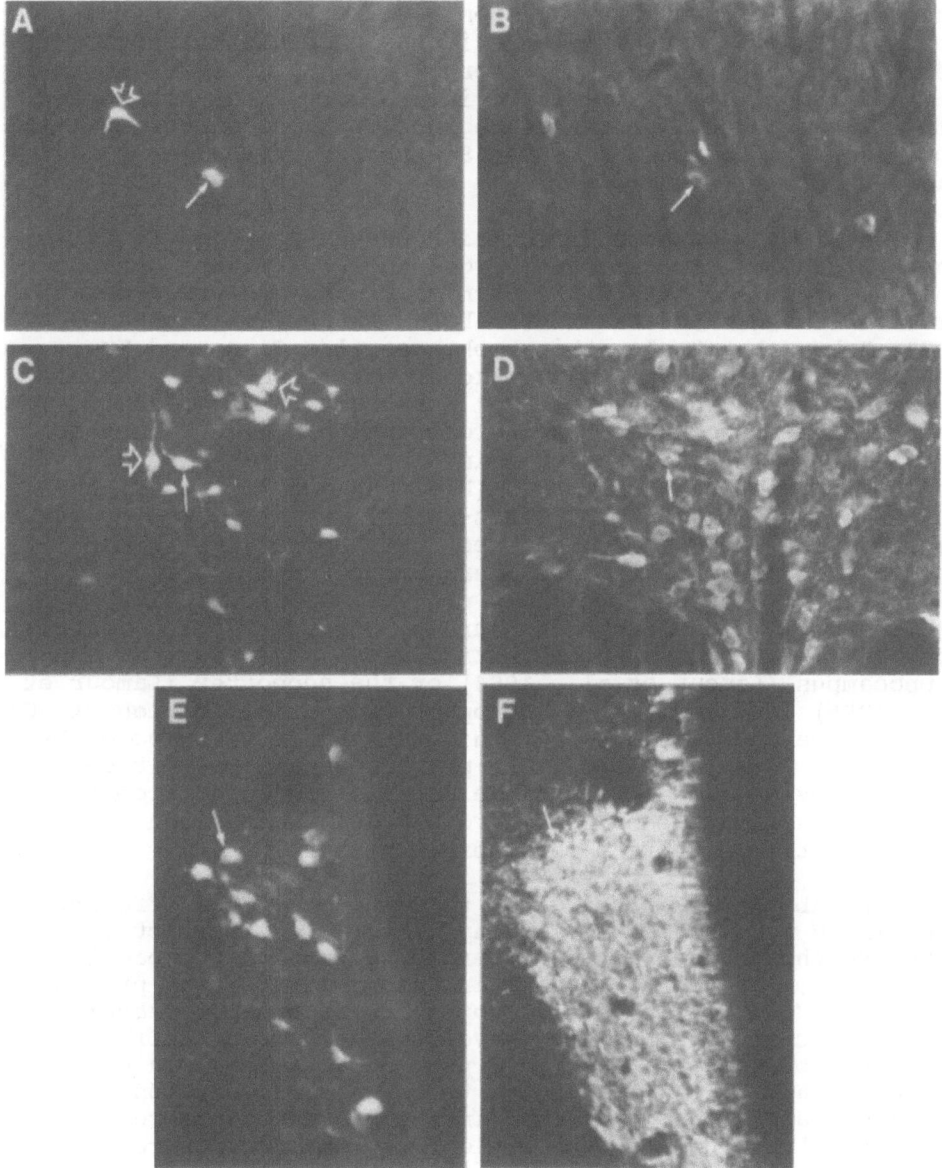

Fig. 4. Retrograde labelling in the pedunculopontine tegmental
nucleus (A), dorsal raphe nucleus (C), and locus
ceruleus (E) following injection of the fluorescent
tracer fluorogold into the horizontal limb of the
diagonal band-magnocellular preoptic area. Some of
these retrogradely labelled neurons are immunoreactive
for choline acetyltransferase (B), serotonin (D), or
tyrosine hydroxylase (F), suggesting that cholinergic,
serotonergic, and noradrenergic neurons in respective
brainstem areas project to the basal forebrain region.
(From Semba et al., 1988b.)

as adrenergic neurons in the ventrolateral medulla innervate many, if not all, regions of the cholinergic basal forebrain (Semba et al., 1988b). Iontophoretic application of dopamine has been reported to inhibit spontaneously active neurons in the ventral pallidum-substantia innominata region (Napier and Potter, 1989). There is also evidence that histaminergic neurons in the hypothalamic tuberomammillary nucleus, another aminergic cell group implicated in arousal (Vanni-Mercier et al., 1984), innervate rostral regions of the cholinergic basal forebrain (Panula et al., 1984; Steinbusch and Mulder, 1984; Watanabe et al., 1984).

What types of basal forebrain neurons are the targets of these monoaminergic and cholinergic inputs from the brainstem? Noradrenergic and adrenergic terminals make synaptic contacts with cholinergic neurons in the basal forebrain (Zaborszky, 1989; Zaborszky et al., this book; see also Chang and Kuo, 1989). These terminals also make synaptic contacts with non-cholinergic neurons in the basal forebrain (Chang, 1989). In contrast to these catecholaminergic inputs, the cholinergic input appears to be directed primarily to non-cholinergic neurons in the basal forebrain (Bialowas and Frotscher, 1987; Hallanger et al., 1988; Martinez-Murillo et al., 1990). There is evidence, however, to indicate that indirectly via non-cholinergic neurons, the cholinergic input has a strong excitatory effect on cortically projecting, presumably cholinergic, basal forebrain neurons. In anesthetized rats, most basal forebrain neurons antidromically driven from the hippocampus (Lamour et al., 1984) or the neocortex (Lamour et al., 1986) were excited by iontophoretic administration of ACh and cholinergic agonists, and this activation was blocked by atropine. These anatomical and physiological observations suggest the possibility that the cholinergic input from the brainstem activates non-cholinergic basal forebrain neurons, which, in turn, activate cholinergic neurons.

The above findings suggest that the brainstem afferents containing monoamines and ACh may directly or indirectly regulate the activity of cholinergic neurons in the basal forebrain. Functional studies in this regard are at present limited. However, intrinsic properties of basal forebrain cholinergic neurons *in vitro* have been described (Griffith et al., this book), and this would provide a basis on which to conduct pharmacological studies to characterize actions of various transmitters contained in the brainstem afferents. Another intriguing approach would be to measure ACh release from the cortex *in vivo* with high performance liquid chromatography (HPLC) during administration of the afferent transmitters in the cholinergic basal forebrain, or during stimulation of monoaminergic and cholinergic sites in the brainstem, with or without antagonists delivered to the basal forebrain. These approaches would provide us with functional data as to how the activity of cholinergic basal forebrain neurons is regulated and modulated by transmitter-specific inputs from the brainstem.

CONCLUSIONS

The studies reviewed above provide fairly comprehensive evidence for a critical role of ACh in cortical EEG arousal. During cortical EEG activation or desynchronization, cholinergic neurons in the basal forebrain appear to become more active, resulting in an increase in ACh release in the cortex. During EEG activation, sensory processing in the cortex is facilitated, and similar enhancement of stimulus-specific responses are seen with local administration of ACh in the cortex. In addition to such direct activation of cortical neurons, cholinergic basal forebrain neurons may enhance cortical arousal by inhibiting thalamocortical spindles associated with slow wave sleep, through their projection to the reticular thalamic nucleus. Thus, the cholinergic basal forebrain projection to the cortex appears to be a key ascending pathway responsible for cortical EEG activation or arousal. Considering some evidence to suggest additional roles played by non-cholinergic neurons, however, the role of basal forebrain cholinergic neurons in cortical arousal probably should not be regarded as an exclusive one.

What turns on cholinergic neurons in the basal forebrain during arousal to release ACh at their terminals in the cortex? Normally, the activation of EEG occurs over wide regions of the cortex, rather than locally. Because individual cholinergic basal forebrain neurons appear to have relatively small terminal fields, such global activation would require the activation of practically all cholinergic neurons in the basal forebrain. For this purpose, afferents from the brainstem in general would be ideal because unlike the afferents from the forebrain (e.g., Semba et al., 1988a; Semba and Fibiger, in preparation), brainstem afferents are directed bilaterally (with ipsilateral dominance), and tend to innervate all levels of the cholinergic basal forebrain (see above). In particular, a number of monoaminergic and cholinergic brainstem afferents, as discussed above, would be strong candidates for this role, because electrical stimulation of the midbrain reticular formation, which is known to induce both ACh release and EEG activation in the cortex, would activate at least some of these afferent pathways.

Currently available data with chronic single unit record-ings would allow us to speculate on the interplay of some pertinent neuronal populations during different behavioral states. Both noradrenergic and serotonergic neurons in the brainstem are active during waking, less active during slow wave sleep, and fall silent during REM sleep (see reviews by Jacobs, 1987; McGinty and Szymusiak, 1988). In contrast, the cholinergic neurons in the mesopontine tegmentum are active during both waking and REM sleep but less so during slow wave sleep (El Mansari et al., 1989; Steriade et al., 1990). Because the cortical EEG is desynchronized in both waking and REM sleep, it would appear that the cholinergic input plays a major role. However, in a cortical ACh release study by Kanai and Szerb (1965), systemic atropine, which would block not only cholin-ergic transmission in the cortex but also the postsynaptic effects of the ascending cholinergic input to the basal fore-brain, did not interfere with the increased ACh release from the cortex (while blocking EEG activation) in response to midbrain stimulation (Kanai and Szerb, 1965). Thus, it is possible that the activation of cholinergic basal forebrain neurons is dependent on not a single but multiple ascending systems from the brainstem. Furthermore, it is conceivable that in addition to the general arousal of the cortex presumably set by these brainstem afferents, intense activation of local cortical regions may also occur, and for such local activation, other afferents may be involved; for example, those from the forebrain are more selectively directed to subpopulations of cholinergic basal forebrain neurons. Such general arrangement would allow for both general and more selective activation of different cortical regions.

In conclusion, there is considerable anatomical and physiological evidence to suggest that the cholinergic projection from the basal forebrain to the cortex represents a major ascending system critically involved in cortical arousal, in particular that associated with the activated patterns of cortical EEG. The activation of this major cholinergic ascending system in the forebrain, in turn, appears to be directly and indirectly regulated by aminergic and cholinergic inputs from the brainstem, as well as other inputs. In this general scheme, the basal forebrain cholinergic projection to the cortex can be viewed as a critical and integral component of the global ascending activating system originating from the brainstem. Cortical arousal is normally coupled with behavioral arousal, except for during REM sleep or under certain pharmacological conditions. Information regarding the brainstem inputs to cholinergic basal forebrain neurons may hold a key to a better understanding of the mechanisms for integrating cortical and behavioral arousal.

ACKNOWLEDGEMENTS

The author wishes to thank Peter Reiner for insightful and stimulating discussions and comments, and John Szerb and Doug Rasmusson for a critical reading of an early version. Generous offers of the use of the figures reproduced in this paper by respective authors and publishers of the original articles are gratefully acknowledged. Supported by the Alzheimer Society of Canada. The author is a Medical Research Council Scholar.

REFERENCES

Andersen, P., and Andersson, S. A., 1965, "Physiological Basis of the Alpha Rhythm", Appleton-Century-Crofts, New York.

Asanuma, C., 1989, Axonal arborizations of a magnocellular basal nucleus input and their relation to the neurons in the thalamic reticular nucleus of rats, Proc. Nat. Acad. Sci. U.S.A., 86:4746-4750.

Azmitia, E. C., 1978, The serotonin-producing neurons of the midbrain median and dorsal raphe nuclei, in: "Handbook of Psychopharmacology, Vol. 9, Chemical Pathways in the Brain", L. L. Iversen. S. D., Iversen, and S. H. Snyder, eds., Plenum, New York, pp. 233-314.

Bennett, T. L., 1971, Hippocampal theta activity and behavior-a review, Commun. Behav. Biol., 6:37-48.

Bialowas, J., and Frotscher, M., 1987, Choline acetyltransferase-immunoreactive neurons and terminals in the rat septal complex: a combined light and electron microscopic study, J. Comp. Neurol., 259:298-307.

Buzsáki, G., Bickford, R. G., Ponomareff, G., Thal, L. J., Mandel, R., and Gage, F. H., 1988, Nucleus basalis and thalamic control of neocortical activity in the freely moving rat, J. Neurosci., 8:4007-4026.

Casamenti, F., Deffenu, G., Abbamondi, A. L., and Pepeu, G., 1986, Changes in cortical acetylcholine output induced by modulation of the nucleus basalis, Brain Res. Bull., 16:689-695.

Chang, H. T., 1989, Noradrenergic innervation of the substantia innominata: a light and electron microscopic analysis of dopamine ß-decarboxylase immunoreactive elements in the rat, Exp. Neurol., 104-101-112.

Chang, H. T., and Kuo, H., 1989, Adrenergic innervation of the substantia innominata: co-localization of phenylethanol-amine N-methyltransferase and tyrosine hydroxylase immunoreactivities within the same axons, Brain Res., 503:350-353.

Coben, L. A., Danziger, W., and Storandt, M., 1985, A longitudinal EEG study of mild senile dementia of Alzheimer type: changes at 1 year and at 2.5 years, Electroenceph. Clin. Neurophysiol., 61: 101-112.

Collier, B., and Mitchell, J. F., 1967, The central release of acetylcholine during consciousness and after brain lesions, J. Physiol. (Lond.), 188:83-98.

Detári, L., Juhasz, G., and Kukorelli, T., 1984, Firing properties of cat basal forebrain neurones during sleep-wakefulness cycle, Electroenceph. Clin. Neurophysiol., 58:362-368.

Detári, L., Juhasz, G., and Kukorelli, T., 1987, Neuronal firing in the pallidal region: firing patterns during sleep-wakefulness cycle in cats, Electroenceph. Clin. Neurophysiol., 67:159-166.

Detári, L., and Vanderwolf, C. H., 1987, Activity of identified cortically projecting and other basal forebrain neurones during large slow waves and cortical activation in anaesthetized rats, Brain Res., 437:1-8.

Detári, L., Vanderwolf, C. H., and Kukorelli, T., 1990, Inhibitory connections in the basal forebrain: a possible explanation for the ambiguous role of BFA in the regulation of sleep and wakefulness, in: "The Diencephalon and Sleep", M. Mancia and G. Marini, eds., Raven, New York, pp. 355-359.

Donoghue, J. P., and Carroll, K. L., 1987, Cholinergic modulation of sensory responses in rat primary somatic sensory cortex, Brain Res., 408:367-371.

El Mansari, M., Sakai, K., and Jouvet, M., 1989, Unitary characteristics of presumptive cholinergic tegmental neurons during the sleep-waking cycle in freely moving cats, Exp. Brain Res., 76:519-529.

Fisher, R. S., Buchwald, N. A., Hull, C. D., and Levine, M. S., 1988, GABAergic basal forebrain neurons project to the neocortex: the localization of glutamic acid decarboxylase and choline acetyltransferase in feline corticopetal neurons, J. Comp. Neurol., 272:489-502.

Freund, T. F., and Antal, M., 1988, GABA-containing neurons in the septum control inhibitory interneurons in the hippocampus, Nature, 336:170-163.

Griffith, W. H., Sim, J. A., and Matthews, R. T., 1991, Electrophysiologic characteristics of basal forebrain neurons in vitro, in "The Basal Forebrain: Anatomy to Function", C. T. Napier, P. W. Kalivas, and I. Hanin, eds, Plenum, New York.

Hallanger, A. E., Levey, A. I., Lee, H. J., Rye, D. B., and Wainer, B. H., 1987, The origin of cholinergic and other subcortical afferents to the thalamus in the rat, <u>J. Comp. Neurol.</u>, 262:105-124.

Hallanger, A. E., Price, S. D., Steininger, T., and Wainer, B. H., 1988, Mesopontine tegmental projections to the nucleus basalis of Meynert: an ultrastructural study, <u>Soc. Neurosci. Abstr.</u>, 14:118

Hallanger, A. E., and Wainer, B. H., 1988, Ascending projections from the pedunculopontine tegmental nucleus and the adjacent mesopontine tegmentum in the rat, <u>J. Comp. Neurol.</u>, 274:483-515.

Hu, B., Steriade, M., and Deschênes, M., 1989, The effects of brainstem peribrachial stimulation on perigeniculate neurons: the blockade of spindle waves, <u>Neurosci.</u>, 31:1-12.

Jacobs, B. L., 1987, Brain monoaminergic unit activity in behaving animals, <u>Prog. Psychobiol. Physiol. Psychol.</u>, 12:171-206.

Jasper, H.H., and Tessier, J., 1971, Acetylcholine liberation from cerebral cortex during paradoxical (REM) sleep, <u>Science</u>, 172:601-602.

Jones, B. E., and Cuello, A. C., 1989, Afferents to the basal forebrain cholinergic cell area from pontomesencephalic-catecholamine, serotonin, and acetylcholine-neurons, <u>Neurosci.</u>, 31:37-61.

Jourdain, A., Semba, K., and Fibiger, H. C., 1989, Basal forebrain and mesopontine tegmental projections to the reticular thalamic nucleus: an axonal collateralization and immunohistochemical study in the rat, <u>Brain Res.</u>, 505:55-65.

Jung, R., and Kornmüller, A. E., 1938, Eine Methodik der Ableitung lokalisierter Potentialschwankungen aus sucorticalen Hirngebieten, <u>Arch. Psychiat.</u>, 109:1-30.

Kanai, T., and Szerb, J. C., 1965, Mesencephalic reticular activating system and cortical acetylcholine output, <u>Nature</u>, 205:80-82.

Kayama, Y., Sumitomo, I., and Ogawa, T., 1986, Does the ascending cholinergic projection inhibit or excite neurons in the rat thalamic reticular nucleus? <u>J. Neurophysiol.</u>, 56:1310-1320.

Köhler, C., Chan-Palay, V., and Jang-Yen, W., 1984, Septal neurons containing glutamic acid decarboxylase immunoreactivity project to the hippocampal region in the rat brain, <u>Anat. Embryol.</u>, 169:41-44.

Kramis, R., Vanderwolf, C. H., and Bland, B. H., 1975, Two types of hippocampal rhythmical slow activity in both the rabbit and the rat: Relations to behavior and effects of atropine, diethyl ether, urethane and pentobarbital. <u>Exp. Neurol.</u>, 49:58-85.

Lamour, Y., Dutar, P., and Jobert, A., 1984, Septo-hippocampal and other medial septum-diagonal band neurons: electrophysiological and pharmacological properties, <u>Brain Res.</u>, 309:227-239.

Lamour, Y., Dutar, P., and Rascol, O., and Jobert, A., 1986, Basal forebrain neurons projecting to the rat frontoparietal cortex: electrophysiological and pharmacological properties, <u>Brain Res</u>., 362:122-132.

Levey, A. I., Hallanger, A., and Wainer, B. H., 1987, Cholinergic nucleus basalis neurons may influence the cortex via the thalamus, <u>Neurosci. Lett</u>., 74:7-13.

Lewis, P. R., and Shute, C. C. D., 1967, The cholinergic limbic system: projections to hippocampal formation, medial cortex, nuclei of the ascending cholinergic reticular system, and the subfornical organ and supra-optic crest, Brain, 90:521-540.

Livingstone, M. S., and Hubel, D. H., 1981, Effects of sleep and arousal on the processing of visual information in the cat, Nature, 291:554-561.

Longo, V. G., 1966, Behavioral and electroencephalographic effects of atropine and related compounds, Pharmacol. Rev., 18:965-996.

Martinez-Murillo, R., Villalba, R. M., and Rodrigo, J., 1990, Immunocytochemical localization of cholinergic terminals in the region of the nucleus basalis magnocellularis of the rat: a correlated light and electron microscopic study, Neurosci., 36:361-376.

McCormick, D. A., and Prince, D. A., 1986, Acetylcholine induces burst firing in thalamic reticular neurons by activating a potassium conductance, Nature, 319:402-405.

McGinty, D. J., and Sterman, M. B., 1968, Sleep suppression after basal forebrain lesions in the cat, Science, 160:1253-1255.

McGinty, D., and Szymusiak, R., 1988, Neuronal unit activity patterns in behaving animals: brainstem and limbic system, Ann. Rev. Psychol., 39:135-168.

Metherate, R., Tremblay, N., and Dykes, R. W., 1987, Acetylcholine permits long-term enhancement of neuronal responsiveness in cat primary somatosensory cortex, Neurosci., 22:75-81.

Morruzi, G., and Magoun, H. W., 1949, Brainstem reticular formation and activation of the EEG, Electroenceph. Clin. Neurophysiol., 1:455-473.

Napier, T. C., and Potter, P. E., 1989, Dopamine in the rat ventral pallidum/substantia innominata: biochemical and electrophysiological studies, Neuropharmacol., 28:757-760.

Nicoll, R. A., 1985, The septo-hippocampal projection: a model cholinergic pathway, TINS, December, 533-536.

Panula, P., Yang, H.-Y. T., and Costa, E., 1984, Histamine-containing neurons in the rat hypothalamus, Proc. Natl. Acad. Sci. U.S.A., 81:2572-2576.

Pirch, J., Rigdon, G., Rucker, H., and Turco, K., 1991, Basal forebrain modulation of cortical cell activity during conditioning, in "The Basal Forebrain: Anatomy to Function", C. T. Napier, P. W. Kalivas, and I. Hanin, eds, Plenum, New York.

Porter, L. L., and Asanuma, C., 1989, Ultrastructural and immunohistochemical observations on a projection from the magnocellular basal forebrain in rats, Soc. Neurosci. Abstr., 15:289.

Rasmusson, D. D., and Dykes, R. W., 1988, Long-term enhancement of evoked potentials in cat somatosensory cortex produced by co-activation of the basal forebrain and cutaneous receptors, Exp. Brain Res., 70:276-286.

Reiner, P. B., Semba, K., Fibiger, H. C., and McGeer, E. G., 1987, Physiological evidence for subpopulations of cortically projecting basal forebrain neurons in the anesthetized rat, Neurosci., 20:629-636.

Richardson, R. T., and DeLong, M. R., 1991, Electrical studies of the function of the nucleus basalis in primates, in "The Basal Forebrain: Anatomy to Function", C. T. Napier, P. W.

Kalivas, and I. Hanin, eds, Plenum, New York.

Rolls, E. T., Canghera, M. K., and Roper-Hall, A., 1979, The latency of activation of neurones in the lateral hypothalamus and substantia innominata during feeding in the monkey, Brain Res., 164:121-135.

Rye, D. B., Wainer, B. H., Mesulam, M. -M., Mufson, E. J., and Saper, C. B., 1984, Cortical Projections arising from the basal forebrain: a study of cholinergic and noncholinergic components employing combined retrograde tracing and immunohistochemical localization of choline acetyl-transferase, Neurosci., 13:627-643.

Sato, H., Hata, Y., Hagihara, K., and Tsumoto, T., 1987, Effects of cholinergic depletion on neuron activities in the cat visual cortex, J. Neurophysiol., 58:781-794.

Satoh, K., and Fibiger, H. C., 1986, Cholinergic neurons of the laterodorsal tegmental nucleus: Efferent and afferent connections, J. Comp Neurol., 253:277-302.

Schwaber, J. S., Rogers, W. T., Satoh, K., and Fibiger, H. C., 1987, Distribution and organization of cholinergic neurons in the rat forebrain demonstrated by computer-aided data acquisition and three-dimensional reconstruction, J. Comp. Neurol., 263:309-325.

Semba, K., and Fibiger, H. C., 1988, Time of origin of cholinergic neurons in the rat basal forebrain, J. Comp. Neurol., 269:87-95.

Semba, K., and Fibiger, H. C., 1989, Organization of central cholinergic systems, Prog. Brain Res., 79:37-63.

Semba, K., and Fibiger, C. H., Forebrain afferents to the magnocellular basal forebrain of the rat, in preparation.

Semba, K., Reiner, P. B., McGeer, E. G., and Fibiger, H. C., 1988a, Non-cholinergic basal forebrain neurons project to the contralateral basal forebrain in the rat, Neurosci. Lett., 84:23-28.

Semba, K., Reiner, P. B., McGeer, E. G., and Fibiger, H. C., 1988b, Brainstem afferents to the magnocellular basal forebrain studied by axonal transport, immunohisto-chemistry, and electrophysiology in the rat, J. Comp. Neurol., 267:433-453.

Semba, K., Reiner, P. B., McGeer, E. G., and Fibiger, H. C., 1989, Brainstem projecting neurons in the rat basal forebrain: neurochemical, topographical, and physiological distinctions from cortically projecting cholinergic neurons, Brain Res. Bull., 22:501-509.

Shute, C. C. D., and Lewis, P. R., 1967, The ascending cholinergic reticular system: neocortical, olfactory and subcortical projections, Brain., 90:497-520.

Sillito, A. M., and Kemp, J. A., 1983, Cholinergic modulation of the functional organization of the cat visual cortex, Brain Res., 289:143-155.

Skinner, B. F., 1974, "About Behavioralism", Alfred A. Knopf, New York.

Spehlmann, R., Daniels, J. C., and Smathers, C. C., Jr., 1971, Acetylcholine and the synaptic transmission of specific impulses to the visual cortex, Brain, 94:125-138.

Starzl, T.E., Taylor, C.W., and Magoun, H.W., 1951, Ascending conduction in reticular activating system, with special reference to the diencephalon, J. Neurophysiol., 14:461-477.

Steinbusch, H. W. M., and Mulder, A. H., 1984, Immunohisto-
    chemical localization of histamine neurons and mast cells
    in the rat brain, in: "Handbook of Chemical Neuroanatomy,
    Vol. 3: Classical Transmitters and Transmitter Receptors in
    the CNS, Part II", A. Björklund, T., Hökfelt, and M. J.
    Kuhar, eds., Elsevier, Amsterdam, pp. 126-140.
Steriade, M., 1981, Mechanisms underlying cortical activation:
    neuronal organization and properties of the midbrain
    reticular core and intralaminar thalamic nuclei, in: "Brain
    Mechanisms and Perceptual Awareness", O. Pompeiano, and C.
    Ajmone Marsan, eds., Raven, New York, pp. 327-377.
Steriade, M., Datta, S., Paré, D., Oakson, G., and Curró Dossi,
    R., 1990, Neuronal activities in brain-stem cholinergic
    nuclei related to tonic activation processes in
    thalamocortical systems, J. Neurosci., 10:2541-2559.
Steriade, M. and Deschênes, M., 1984, The thalamus as a neuronal
    oscillator, Brain Res. Rev., 8:1-63.
Steriade, M., and McCarley, R. W., 1990, "Brainstem Control of
    Wakefulness and Sleep", Plenum, New York.
Steriade, M., Parent, A., Paré, D., and Smith, Y., 1987,
    Cholinergic and non-cholinergic neurons of the cat basal
    forebrain project to the reticular and mediodorsal thalamic
    nuclei, Brain Res., 408:372-376.
Sterman, M. B., and Clemente, C. D., 1962, Forebrain inhibitory
    mechanisms: sleep patterns induced by basal forebrain
    stimulation in the behaving cat, Exp. Neurol., 6:103-117.
Stewart, D. J., MacFabe, D. F., and Vanderwolf, C. H., 1984,
    Cholinergic activation of electrocorticogram: role of the
    substantia innominata and effects of atropine and
    quinuclidinyl benzilate, Brain Res., 322:219-232.
Stewart, M., and Fox, S. E., 1989, Two populations of
    rhythmically bursting neurons in rat medial septum are
    revealed by atropine, J. Neurophysiol., 61:982-993.
Stewart, M., and Fox, S. E., 1990, Do septal neurons pace the
    hippocampal theta rhythm? TINS, 13:163-168.
Stumpf, C., 1965, Drug action on the electrical activity of the
    hippocampus, Int. Rev. Neurobiol., 7:77-138.
Swanson, L. W., and Hartman, B. K., 1975, The central adrenergic
    system. An immunofluorescence study of the location of
    cell bodies and their efferent connections in the rat
    utilizing dopamine-ß-hydroxylase as a marker, J. Comp.
    Neurol., 163:467-506.
Szymusiak, R., and McGinty, D., 1986, Sleep-related neuronal
    discharge in the basal forebrain of cats, Brain Res.,
    370:82-92.
Szymusiak, R., and McGinty, D., 1989, Sleep-waking discharges of
    basal forebrain projection neurons in cats, Brain Res.
    Bull., 22:423-430.
Vanderwolf, C. H., 1969, Hippocampal electrical activity and
    voluntary movement in the rat, Electroenceph. Clin.
    Neurophysiol., 26:407-418.
Vanderwolf, C. H., 1988, Cerebral activity and behavior: control
    by central cholinergic and serotonergic systems, Int. Rev.
    Neurobiol., 30, 225-340.
Vanderwolf, C.H., and Baker, G. B., 1986, Evidence that
    serotonin mediates non-cholinergic neocortical low voltage
    fast activity, non-cholinergic hippocampal rhythmical slow
    activity and contributes to intelligent behavior, Brain
    Res., 374:342-356.

Vanderwolf, C. H., and Robinson, T. E., 1981, Reticulo-cortical activity and behavior: A critique of the arousal theory and a new synthesis, <u>Behav. Brain Sci</u>., 4:459-514.

Vanderwolf, C. H., and Stewart, D. J., 1988, Thalamic control of neocortical activation: a critical re-evaluation, <u>Brain Res. Bull</u>., 20:529-638.

Vanni-Mercier, G., Sakai, K., and Jouvet, M., 1984, "Waking-state specific" neurons in the caudal hypothalamus of the cat, <u>C. R. Acad. Sci</u>., 298:195-220.

Vincent, S. R., Satoh, K., Armstrong, D. M., and Fibiger, H.C., 1983, Substance P in the ascending cholinergic reticular system, <u>Nature</u>, 306:688-691.

Wilson, F. A. W., 1991, The relationship between learning, memory and neuronal responses in the primate basal forebrain, <u>in</u> "The Basal Forebrain: Anatomy to Function", C. T. Napier, P. W. Kalivas, and I. Hanin, eds, Plenum, New York.

Wilson, F. A. W., and Rolls, E. T., 1990, Neuronal responses related to the novelty and familiarity of visual stimuli in the substantia innominata, diagonal band of Broca and periventricular region of the primate basal forebrain, <u>Exp. Brain Res</u>., 80:104-120.

Watanabe, T., Taguchi, Y., Shiosaka, S., Tanaka, J., Kubota, H., Terano, T., Tohyama, M., and Wada, H., 1984, Distribution of the histaminergic neuron system in the central nervous system of rats: a fluorescent immunohistochemical analysis with histidine decarboxylase as a marker, <u>Brain Res</u>., 295:13-25.

Zaborszky, L., 1989, Afferent connections of the forebrain cholinergic projection neurons, with special reference to monoaminergic and peptidergic fibers, <u>in</u> "Central Cholinergic Synaptic Transmission", M. Frotscher, and U. Misgeld, eds, Birkhauser, Basel, pp. 12-32.

Zaborszky, L., Cullinan, W. E., and Braun, A., 1991, Afferents to basal forebrain cholinergic projection neurons: an update, <u>in</u> "The Basal Forebrain: Anatomy to Function", C. T. Napier, P. W. Kalivas, and I. Hanin, eds, Plenum, New York.

BASAL FOREBRAIN MODULATION OF CORTICAL CELL

ACTIVITY DURING CONDITIONING

James Pirch, Greg Rigdon, Hubert Rucker and Kathy Turco

Department of Pharmacology
Texas Tech University Health Sciences Center
Lubbock, TX 79430

INTRODUCTION

Several investigators have found that basal forebrain neurons in the area of the nucleus basalis and substantia innominata respond to cues that signal the availability of reinforcement (DeLong, 1971; Mitchell et al., 1987; Richardson and DeLong, 1986, 1988; Rolls et al., 1979; Travis and Sparks, 1968; Wilson and Rolls, 1990a, 1990b). The reinforcement-related or conditioning-related responses of such basal forebrain neurons do not depend upon the sensory modality of the signal cue (Wilson and Rolls, 1990a). It is also known that cortical neurons respond to conditioned stimuli applied during similar behavioral paradigms (Aou et al., 1983; Boyd et al., 1982; Fuster et al., 1982; Kojima and Goldman-Rakic, 1982; Mauritz and Wise, 1986; Peterson, 1986; Watanabe, 1990), suggesting a possible relationship between cortical neuron responses and activity of basal forebrain neurons.

Basal forebrain cholinergic neurons may play an important role in modulation of cortical responses to sensory stimuli. Supporting evidence comes primarily from experiments examining the effects of cholinergic agonists or antagonists on responses of primary sensory cortex neurons. Effects consistent with a modulatory influence of acetylcholine have been observed for neurons in visual cortex (Greuel et al., 1988; Sillito and Kemp, 1983), somatosensory cortex (Donoghue and Carroll, 1987; Metherate et al., 1988a, 1988b) and auditory cortex (Ashe et al., 1989; McKenna et al., 1988, 1989; Metherate and Weinberger, 1989). In some cases, an enhancement of responsiveness was prolonged for a considerable time after the application of acetylcholine in combination with sensory stimulation.

Evidence that cholinergic mechanisms may be involved in cortical neuron plasticity associated with conditioning has been provided by the studies of Woody and his co-workers comparing changes in excitability of pyramidal tract neurons induced by eye blink conditioning and the effects of iontophoretically applied acetylcholine (Swartz and Woody, 1979, 1984; Woody, 1982; Woody et al., 1978). Others have observed alteration of conditioned unit responses in various cortical areas following iontophoretic administration of cholinergic agonists or antagonists (Aou et al., 1983; Inoue et al., 1983; Lénárd et al., 1989). However, very few studies have directly addressed the effects of altering basal forebrain function on conditioning-related electrophysiological responses in the

The Basal Forebrain, Edited by T.C. Napier *et al.*
Plenum Press, New York, 1991

cortex or other brain areas (Harrison et al., 1988; Pirch et al., 1986; Rigdon and Pirch, 1984, 1986).

Work in our laboratory has been directed toward the study of the role of various neurotransmitters, including acetylcholine, in modulation of cortical responses to conditioned stimuli. As part of those studies, we examined the effect of reversible suppression of basal forebrain neurons on cortical unit responses and the effect of nucleus basalis lesions on the percentage of cortical units that demonstrated conditioned responses. Additionally, we investigated the potential involvement of acetylcholine in cortical unit responses by giving local microinjection of atropine or microiontophoretic application of cholinergic antagonists (Rigdon and Pirch, 1984, 1986; Pirch et al., 1989; Turco et al., 1988).

METHODS

A simple associative conditioning procedure was used in these studies, in which a tone was paired with medial forebrain bundle (MFB) stimulation. A long-duration event-related potential recorded from the surface of the frontal cortex provided an index of conditioning. Several earlier studies demonstrated that these negative slow potential (SP) responses reflect "learning" of the association between the sensory cue and delivery of the reinforcer. These SP responses are elicited by tone, light or brain stimulation cues that precede food, foot shock, or rewarding MFB stimulation (Pirch, 1977, 1980; Pirch and Barnes, 1972; Pirch et al., 1981a, 1981b, 1986; Rowland et al., 1985).

The conditioned stimulus (CS) was a 2-second tone. A 2900 Hz tone served as the CS for single tone studies, while two different frequency tones (1 or 8 kHz) were used for the discrimination procedure employed in microiontophoretic experiments. MFB stimulation was applied immediately following the single tone cue; in the discrimination procedure, one of the two tones (CS+) was followed by MFB stimulation while no reinforcement was given after the other tone (CS-) (Rucker et al., 1986). Parameters for MFB stimulation via an implanted monopolar electrode were determined for each rat by testing for self-stimulation (generally a 500 msec train of 100 Hz, 0.5 msec square-wave pulses at 100-500 μA).

Intertrial intervals were variable and ranged between 6 and 40 seconds; the specific intervals differed among experiments. For discrimination, CS+ and CS- trials were presented in a pseudorandom sequence, with no more than two consecutive trials of the same type. Twenty-five trials were usually analyzed together as a set. For discrimination studies, a set consisted of approximately 25 trials of each stimulus. Details of paradigms and training procedures can be found elsewhere (Rigdon and Pirch, 1986, Rucker et al., 1986).

Some rats were acclimated to a restraint box for recording unit activity while the animals were awake but sedated with diazepam (10 mg/kg, s.c.). Recordings were obtained from other animals that were anesthetized with urethane, 1.2-1.5 g/kg, i.p. Conditioned slow potential and single unit responses which resemble those obtained in unanesthetized animals can be recorded from urethane anesthetized rats using MFB stimulation as reinforcement (Ebenezer, 1983; Pirch et al., 1985a, 1985b; Rucker et al., 1986). Other investigators have also reported conditioning-related changes in responses of neurons in various brain areas of urethane anesthetized rats when reinforcing MFB stimulation was paired with auditory (West and Michael, 1989) or olfactory stimuli (West et al., 1990; Wilson and Sullivan, 1990).

The center of the 2.3 mm diameter burr hole for the frontal cortex recording site was 2 mm anterior to bregma and 1.5 mm lateral to midline. Basal forebrain units were recorded approximately 1 mm posterior to bregma, 2.6-3.0 mm lateral and 6-8 mm below dura. Extracellular action potentials were recorded with tungsten microelectrodes except for microiontophoresis studies where the center barrel of a multibarrel glass pipette was filled with 4 M NaCl for recording unit activity. Responses of units to the two-second CS were divided into two one-second periods for statistical analysis. Time bins for acquisition of unit data were 100 msec.

RESULTS

Conditioned Unit Responses in Frontal Cortex and Nucleus Basalis Area

Some neurons in the frontal cortex responded to the tone CS with an increase in firing rate while others showed a decreased rate. The magnitude of response to the paired tone was significantly greater than the response to the same tone when it was not paired with MFB stimulation, and responses extinguished when MFB stimulation was withheld (Pirch and Peterson, 1981; Pirch et al., 1983, 1985b; Rigdon and Pirch, 1984, 1986).

If basal forebrain neurons are proposed to modulate the activity of cortical units, then basal forebrain neurons must demonstrate conditioning-related responses in the same paradigm. Seventy-four percent of recorded nucleus basalis units (28/38) responded to the CS (Rigdon and Pirch, 1986). Of the responding units, 19 were excited by the CS, 8 were inhibited and one had a biphasic response of inhibition followed by excitation. Extinction of the response occurred in all five units that were tested.

Effect of Inhibition of Basal Forebrain Neurons on Cortical Unit Responses

Depression of basal forebrain activity by microinjection of GABA or procaine into the nucleus basalis area significantly suppressed cortical unit responses to the CS (Table 1); both excitatory and inhibitory responses were affected (Rigdon and Pirch, 1984). Average changes in firing rate of excited units before and after treatment are illustrated in Fig. 1.

Effect of Nucleus Basalis Lesions on Cortical Unit Responses

Kainic acid (0.5 μg in 0.5 μl) was infused unilaterally into the nucleus basalis area (1.0 mm posterior to bregma, 2.6 mm lateral, 7.2 mm below dura) to lesion cell bodies of neurons that project to the cortex.

Table 1
Cortical Unit Responses After Microinjections into the Nucleus Basalis

GABA MICROINJECTION (5-20 μg)
  15/19 excited unit responses reduced
  _2/2_ inhibited unit responses reduced
  17/21 responses reduced (81%)

PROCAINE MICROINJECTION (6-20 μg)
  9/10 excited unit responses reduced
  10/12 inhibited unit responses reduced
  19/22 responses reduced (86%)

Table 2

Cortical Unit Responses After Nucleus Basalis Lesions

	Excited	Inhibited	No Response	% of Units Responding
Lesioned				
Ipsilateral	3	4	21	25% (7/28)
Contralateral	6	3	8	53% (9/17)
Non-Lesioned				
This study	11	8	8	70% (19/27)
Previous study	17	23	15	73% (40/55)

Such lesions selectively reduced choline acetyltransferase (ChAT) activity in the frontal cortex without decreasing ChAT in the hippocampus (Pirch et al., 1986). Furthermore, norepinephrine and serotonin were not reduced in the cortex, hippocampus, striatum or nucleus accumbens, nor was dopamine reduced in the striatum or nucleus accumbens. In animals used for these single unit studies, frontal cortex but not hippocampal ChAT activity was significantly reduced (Rigdon and Pirch, 1986).

The proportion of cortical units exhibiting a response to the CS was significantly altered by the nucleus basalis lesion (Table 2; Rigdon and Pirch, 1986). Only 25% of isolated units ipsilateral to the lesion responded to the CS, whereas 53% of contralateral units in the same animals responded. Because of the possibility of some contralateral influence of the lesion (Pirch et al., 1986), recordings were also made in untreated animals. Seventy percent of the cortical units in these animals responded to the CS, a percentage similar to that observed in other comparable experiments in our laboratory.

Effect of Local Microinjection of Atropine on Cortical Unit Responses

Microinjection of atropine into the recording area significantly suppressed responses of cortical units to the CS (Table 3); the magnitudes of both excitatory and inhibitory responses were reduced (Rigdon and Pirch, 1986). Average changes in firing rate before and after treatment are shown in Fig. 2.

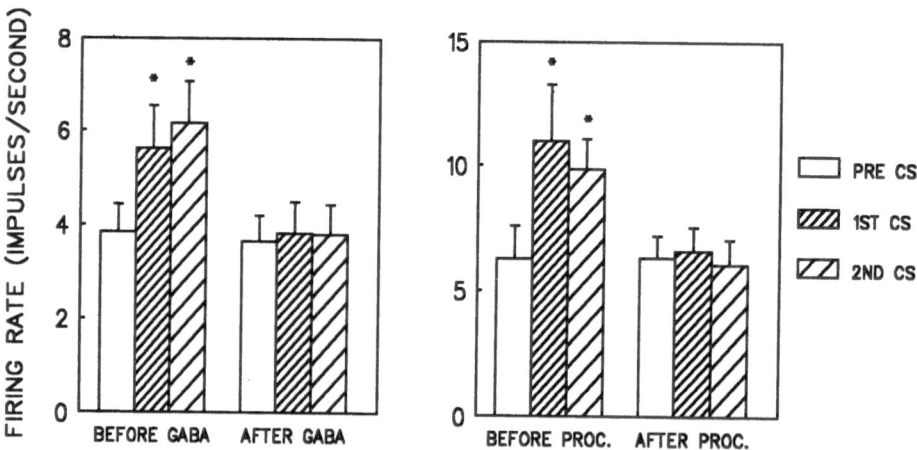

Fig. 1. Effect of GABA or procaine microinjection into nucleus basalis on average firing rate of cortical units excited by CS. Shown are rates ($\pm$ S.E.) during the pretone period, and 1st and 2nd half of the CS period. Asterisks indicate significant response to CS (p < 0.05, paired-t test).

Table 3

Cortical Unit Responses After Local Atropine Microinjection (4 μg)

9/10 excited unit responses reduced
13/15 inhibited unit responses reduced
22/25 responses reduced (88%)

Microiontophoretic Studies on Conditioned Cortical Units

In order to examine the role of cholinergic mechanisms in the cortex more closely, it was necessary to use iontophoretic techniques where the drug action would be more limited than with microinjection. We altered the conditioning procedure to one involving discrimination, thereby giving more assurance of the selectivity of responses of the cells that were studied. The iontophoretic studies were all performed in animals anesthetized with urethane.

After conditioning, differential SP responses and cortical single unit responses developed to the CS+ (reinforced tone) as compared to CS- (nonreinforced tone), with the magnitude of response to CS+ being significantly greater than the response to CS- (Rucker et al., 1986). Since the frequency of the tone used as CS+ varied between animals, the unit response depended upon association with the reinforcer rather than tone frequency. As with single tone cues, some units were excited by CS+ (41 of 56 units that had a differential response) while others were inhibited (15 units). The direction of response was independent of the frequency of the tone that served as CS+. The development and maintenance of a differential SP response indicated successful conditioning of the association between CS+ and MFB stimulation, and assured that the discrimination was maintained during the course of obtaining single unit data. Differential unit responses were not seen when SP responses indicated no discrimination between CS+ and CS-.

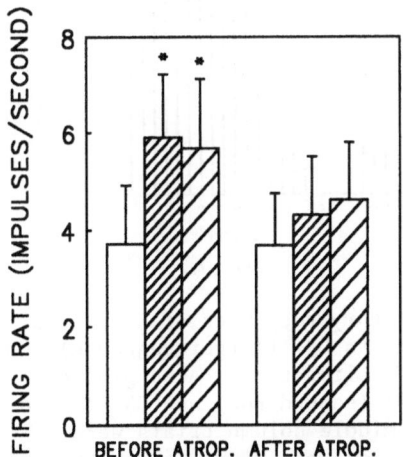 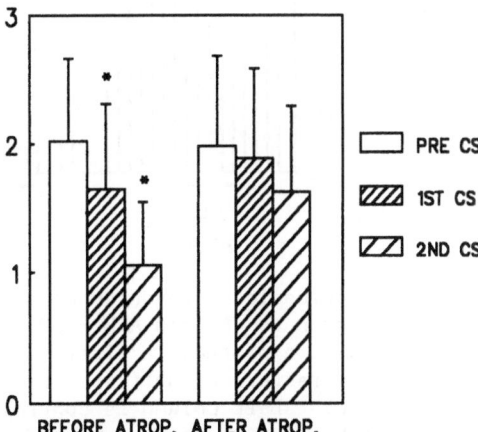

Fig. 2. Effect of local microinjection of atropine on average firing rate of cortical units excited or inhibited by CS (units from Table 3). Shown are rates (± S.E.) during the pretone period and during the 1st and 2nd half of the CS period. Asterisks indicate significant response to CS (p < 0.05, paired-t test).

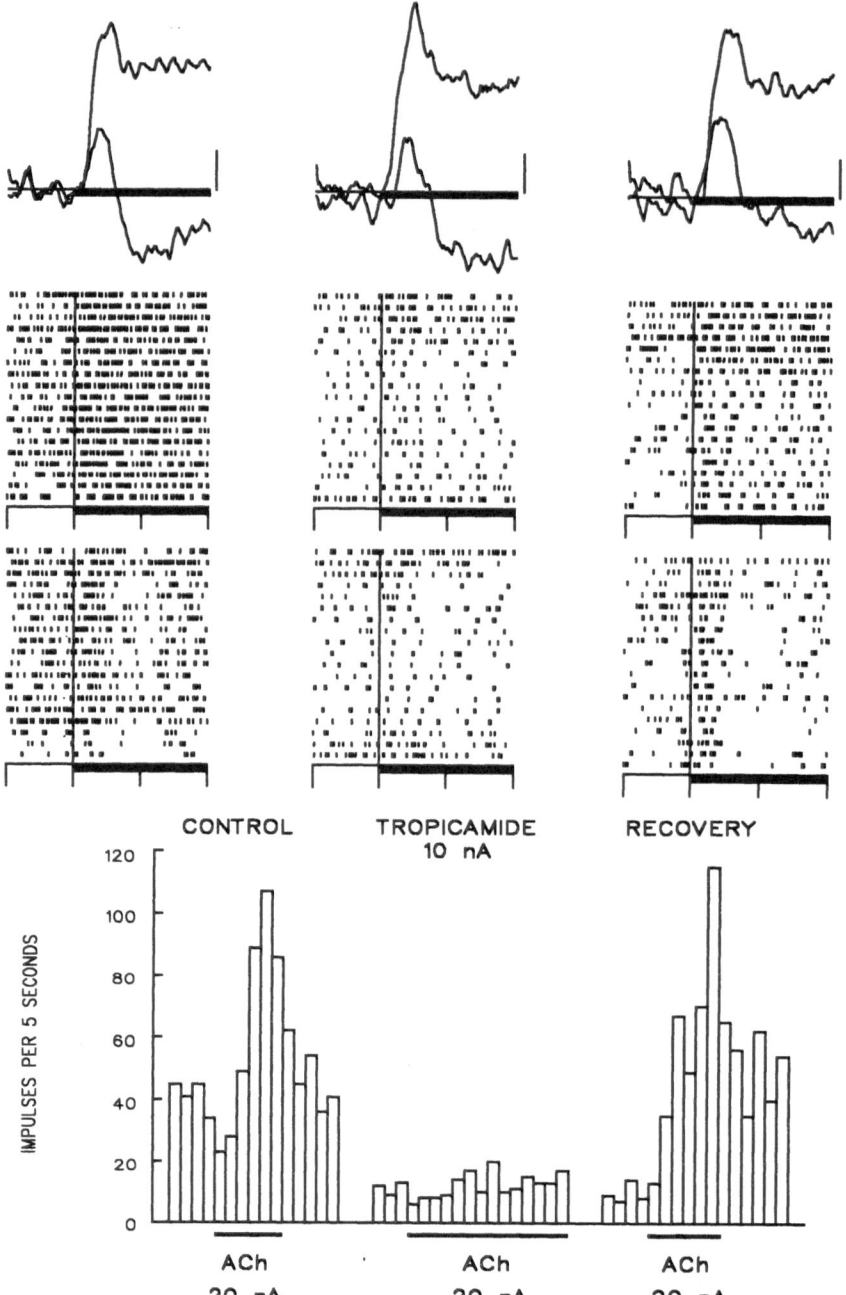

CONTROL

TROPICAMIDE
10 nA

RECOVERY

IMPULSES PER 5 SECONDS

120

100

80

60

40

20

0

ACh
20 nA

ACh
20 nA

ACh
20 nA

Fig. 3. Tropicamide antagonism of the excitatory responses of a cortical neuron to CS+ and ACh. SP responses to CS+ (larger response) and CS- are plotted in first row (10 μV calibration). Also plotted are rasters of unit responses to CS+ (second row) and CS- (third row). Left column is control set, middle column trials were taken during application of tropicamide, and right column trials were obtained following termination of drug application. Recovery for this set was only partial. Period of the tone is indicated by the wide horizontal bar on the SP response baseline or raster x-axis. Lower graph shows the effect of iontophoretic ACh on spontaneous activity before, during and after tropicamide.

After obtaining a set of control trials and testing the effect of iontophoretically administered acetylcholine on spontaneous activity of the unit, a cholinergic muscarinic antagonist (either atropine or tropicamide) was applied continuously by microiontophoresis during acquisition of another set of trials. Prior to termination of the antagonist, acetylcholine was again applied to test for adequacy of muscarinic receptor blockade. When possible, a third set of trials was obtained to ascertain recovery of the response of the unit to CS+ and acetylcholine. Recovery after atropine administration was prolonged, usually requiring more than 30 minutes before significant diminution of the blockade could be observed. However, recovery after treatment with tropicamide was often complete after 15 minutes. Acetylcholine produced excitation of 89% of all conditioned units, regardless of the direction of response to CS+; acetylcholine stimulated 37 of 41 cells that had an excitatory response to CS+ and 13 of 15 units inhibited by CS+. Only 4 of 56 units were inhibited by acetylcholine.

Fig. 3 illustrates tropicamide antagonism of the excitatory CS+ response of a frontal cortex unit that demonstrated discrimination between CS+ and CS-. Tropicamide also antagonized the acetylcholine-induced increase in spontaneous firing rate. At a time when recovery of the response to CS+ was observed following termination of tropicamide application, the response to acetylcholine also recovered. Tropicamide or atropine also antagonized inhibitory CS+ responses in some units, even though the response to acetylcholine in those units was excitation.

The graph in Fig. 4 depicts the different proportions of 56 conditioned frontal cortex units with respect to their responses to CS+ and acetylcholine, and the effect of the cholinergic antagonists on those responses. The major proportion (46%) showed excitatory responses to CS+ and acetylcholine, with both responses being suppressed by a muscarinic receptor antagonist. Units in a second group (16%) were excited by both CS+ and acetylcholine, but only the effect of acetylcholine was antagonized. Thirteen percent of the units had inhibitory responses to CS+ and excitatory responses to acetylcholine, with both responses antagonized by atropine or tropicamide. Cells in the fourth group (9%)

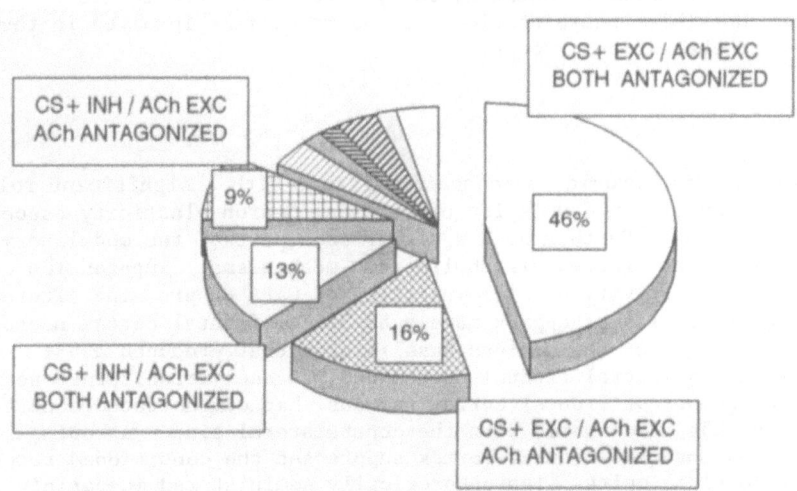

Fig. 4. Groups of frontal cortex neurons based on responses to CS+ and acetylcholine, and the effects of cholinergic muscarinic antagonists.

Table 4

Possible Interpretations of Microiontophoresis Results

Group 1 (46%): CS+ excited, ACh excited, both antagonized.

Acetylcholine, acting on muscarinic receptors, is directly involved in the conditioned response. Source of ACh may be either extrinsic or intrinsic cholinergic neurons.

Group 2 (16%): CS+ excited, ACh excited, only ACh antagonized.

a) A different transmitter is involved in the CS+ response.
b) A nicotinic receptor is involved in the CS+ response.
c) ACh is involved in the CS+ response, but the tip of the microiontophoretic pipette was not located in a position to allow the antagonist to reach the site of contact of cholinergic terminals.

Group 3 (13%): CS+ inhibited, ACh excited, both antagonized.

a) An inhibitory interneuron (GABAergic?) is involved in the CS+ response, and the antagonist spreads to block cholinergic excitation of the inhibitory interneuron.
b) The inhibitory response to CS+ is due to withdrawal of cholinergic tone, in which case the antagonist would diminish baseline tone and reduce the response to CS+.

Group 4 (9%): CS+ inhibited, ACh excited, only ACh antagonized.

a) A different transmitter is involved in the CS+ response.
b) A nicotinic receptor is involved in the CS+ response.

---

were inhibited by CS+ and excited by acetylcholine, but only the acetylcholine effect was antagonized. The remaining units fell into various other categories, with no more than two units per group. Including all groups, regardless of the effect of acetylcholine, 68% of the responses to CS+ were antagonized by atropine or tropicamide. Table 4 lists some possible interpretations of the mechanisms involved in the responses of the various groups.

DISCUSSION

These studies provide evidence consistent with a significant role of the basal forebrain in modulation of cortical neuron plasticity associated with conditioning. Furthermore, a significant part of the modulatory action seems to be exerted via cholinergic mechanisms. Suppression of basal forebrain activity by microinjection of GABA or procaine altered the conditioned responses of approximately 80-85% of frontal cortex neurons. Unilateral lesions of the nucleus basalis area that produced a 53% reduction in ipsilateral frontal cortex choline acetyltransferase activity reduced the number of frontal cortex neurons that demonstrated conditioned responses by 53% as compared with the contralateral side. Atropine microinjected into the frontal cortex suppressed the conditioned responses of 88% of cortical units. Iontophoretically administered muscarinic antagonists suppressed the conditioned responses of 68% of frontal cortex neurons. Nucleus basalis neurons demonstrated conditioned responses similar to those observed in the frontal cortex (Rigdon and Pirch, 1986; Pirch, 1990).

The importance of cholinergic mechanisms in the conditioned responses of frontal cortex neurons in these studies is indicated by the large proportion of units in the microiontophoretic study whose response to CS+ appeared to involve acetylcholine. However, the iontophoretic studies also indicate that noncholinergic mechanisms may be involved. Olton et al. (this book) discussed problems associated with determining the specificity of cholinergic mechanisms in functional changes induced by basal forebrain lesions or other experimental approaches, and also pointed out that intrinsic cholinergic neurons (Levey et al., 1984) complicate interpretations regarding the origin of changes in cholinergic function in the neocortex. It has been suggested that alterations in neurotensin rather than acetylcholine may account for certain memory deficits produced by basal forebrain lesions (Wenk et al., 1989). Local circuits utilizing GABA, acetylcholine, somatostatin or other transmitters may become secondarily involved in the overall cortical response to conditioned stimuli as a result of extrinsic input from cholinergic or noncholinergic neurons, either directly from the basal forebrain or indirectly from other areas that receive basal forebrain projections such as the mediodorsal nucleus or reticular nucleus of the thalamus (Cornwall et al., 1990; Jourdain et al., 1989; Mogenson et al., 1987; Vives and Mogenson, 1985). The results of the microinjection studies suggest that 80-85% of frontal cortex neurons that respond to a conditioned stimulus may be directly or indirectly influenced by the basal forebrain, whether or not the final input to these neurons is cholinergic.

Although other investigators have recorded both increases and decreases in basal forebrain neuron activity in response to reinforced stimuli, discussions concerning the role of acetylcholine in modulation of target neuron responsiveness generally focus on the facilitory effect. Our results suggest that inhibitory responses of some of the cortical units may involve withdrawal of cholinergic tone. Perhaps some cortical cells are innervated by cholinergic neurons in the basal forebrain which respond to the CS+ by decreasing activity. However, we cannot say whether nucleus basalis neurons that were inhibited by the CS+ were cholinergic, and it has been demonstrated that noncholinergic neurons in the basal forebrain project to the cerebral cortex (Rye et al., 1984; Woolf et al., 1986). Another interpretation of the CS+ inhibition in the third group of units described above is that a GABAergic interneuron was involved (McCormick, 1989).

The focus of this chapter has been on basal forebrain modulation of frontal cortex neuronal activity. However, it has also been suggested that "reinforcement-related" basal forebrain neurons receive information from various cortical and subcortical areas involved in motivated behavior (Wilson and Rolls, 1990). Thus, it would be important to determine how alteration of cortical activity influences responses of basal forebrain neurons to conditioned stimuli. Among the subcortical areas which innervate the nucleus basalis and substantia innominata are the nucleus accumbens, pedunculopontine and laterodorsal tegmental nuclei, dorsal and median raphe nucleus, ventral tegmental area, locus coeruleus, substantia nigra pars compacta, raphe magnus, and lateral hypothalamus (Haring and Wang, 1986; Jones and Cuello, 1989; Mogenson et al., 1983; Semba et al., 1988; Woolf and Butcher, 1986a; Zaborszky and Cullinan, 1989). There is also a need to understand how these subcortical areas which utilize a variety of neurotransmitters influence the response of nucleus basalis and substantia innominata neurons to conditioned stimuli.

ACKNOWLEDGEMENTS

Supported by NINDS **NS22408** and NIMH **MH29653**. A portion of this work was conducted at the Kirksville College of Osteopathic Medicine, Kirksville, Missouri. The authors wish to acknowledge the excellent technical assistance of Mary Jo Corbus and Dale Yadon.

REFERENCES

Aou, S., Oomura, Y., and Nishino, H., 1983, Influence of acetylcholine on neuronal activity in monkey orbitofrontal cortex during bar press feeding task, Brain Res., 275:178.

Ashe, J.H., McKenna, T.M., and Weinberger, N.M., 1989, Cholinergic modulation of frequency receptive fields in auditory cortex: II. Frequency-specific effects of anticholinesterases provide evidence for a modulatory action of endogenous ACh, Synapse, 4:44.

Boyd, E.H., Boyd, E.S., and Brown, L.E., 1982, Precentral cortex unit activity during the M-wave and contingent negative variation in behaving squirrel monkeys, Exp. Neurol., 75:535.

Cornwall, J., Cooper, J.D., and Phillipson, O.T., 1990, Projections to the rostral reticular thalamic nucleus in the rat, Exp. Brain Res., 80:157.

Delong, M.R., 1971, Activity of pallidal neurons during movement, J. Neurophysiol., 34:414.

Donoghue, J.P., and Carroll, K.L., 1987, Cholinergic modulation of sensory responses in rat primary somatic sensory cortex, Brain Res., 408:367.

Ebenezer, I.S., 1983, Cortical slow potentials recorded in rats anesthetized with urethane are affected by nicotine, British J. Pharmacol. Proceed. Suppl., 79:270P.

Fuster, J.M., Bauer, R.H., and Jervey, J.P., 1982, Cellular discharge in the dorsolateral prefrontal cortex of the monkey in cognitive tasks, Exp. Neurol., 77:679.

Greuel, J.M., Luhmann, H.J., and Singer, W., 1988, Pharmacological induction of use-dependent receptive field modifications in the visual cortex, Science, 242:74.

Haring, J.H., and Wang, R.Y., 1986, The identification of some sources of afferent input to the rat nucleus basalis magnocellularis by retrograde transport of horseradish peroxidase, Brain Res., 366:152.

Harrison, J.B., Buchwald, J.S., Kaga, K., Woolf, N.J., and Butcher, L.L., 1988, Cat 'P300' disappears after septal lesions, Electroencephalogr. Clin. Neurophysiol., 69:55.

Inoue, M., Oomura, Y., Nishino, H., Aou, S., Sikdar, S.K., Hynes, M., Mizuno, Y., and Katabuchi, T., 1983, Cholinergic role in monkey dorsolateral prefrontal cortex during bar-press feeding behavior, Brain Res., 278:185.

Jones, B.E., and Cuello, A.C., 1989, Afferents to the basal forebrain cholinergic cell area from pontomesencephalic catecholamine, serotonin, and acetylcholine neurons, Neuroscience, 31:37.

Jourdain, A., Semba, K., and Fibiger, H.C., 1989, Basal forebrain and mesopontine tegmental projections to the reticular thalamic nucleus: An axonal collateralization and immunohistochemical study in the rat, Brain Res., 505:55.

Kojima, S., and Goldman-Rakic, P.S., 1982, Delay-related activity of prefrontal neurons in Rhesus monkeys performing delayed response, Brain Res., 248:43.

Levey, A.I., Wainer, B.H., Rye, D.B., Mufson, E.J., and Mesulam, M.-M., 1984, Choline acetyltransferase-immunoreactive neurons intrinsic to rodent cortex and distinction from acetylcholinesterase-positive neurons, Neuroscience, 13:341.

Lénárd, L., Oomura, Y., Nakano, Y., Aou, S., and Nishino, H., 1989, Influence of acetylcholine on neuronal activity of monkey amygdala during bar press feeding behavior, Brain Res., 500:359.

Mauritz, K.-H., and Wise, S.P., 1986, Premotor cortex of the rhesus monkey: neuronal activity in anticipation of predictable environmental events, Exp. Brain Res., 61:229.

McCormick, D.A., 1989, Cholinergic and noradrenergic modulation of thalamocortical processing, TINS, 12:215.

McKenna, T.M, Ashe, J.H., Hui, G.K., and Weinberger, N.M., 1988, Muscarinic agonists modulate spontaneous and evoked unit discharge in auditory cortex of cat, Synapse, 2:54.

McKenna, T.M., Ashe, J.H., and Weinberger, N.M., 1989, Cholinergic modulation of frequency receptive fields in auditory cortex: I. Frequency-specific effects of muscarinic agonists, Synapse, 4:30.

Metherate, R., Tremblay, N., and Dykes, R.W., 1988a, The effects of acetylcholine on response properties of cat somatosensory cortical neurons, J. Neurophysiol., 59:1231.

Metherate, R., Tremblay, N., and Dykes, R.W., 1988b, Transient and prolonged effects of acetylcholine on responsiveness of cat somatosensory cortical neurons, J. Neurophysiol., 59:1253.

Metherate, R., and Weinberger, N.M., 1989, Acetylcholine produces stimulus-specific receptive field alterations in cat auditory cortex, Brain Res., 480:372.

Mitchell, S.J., Richardson, R.T., Baker, F.H., and Delong, M.R., 1987, The primate nucleus basalis of Meynert: neuronal activity related to a visuomotor tracking task, Exp. Brain Res., 68:506.

Mogenson, G.J., Ciriello, J., Garland, J., and Wu, M., 1987, Ventral pallidum projections to mediodorsal nucleus of the thalamus: an anatomical and electrophysiological investigation in the rat, Brain Res., 404:221.

Mogenson, G.J., Swanson, L.W., and Wu, M., 1983, Neural projections from nucleus accumbens to globus pallidus, substantia innominata, and lateral preoptic-lateral hypothalamic area: an anatomical and electrophysiological investigation in the rat, J. Neurosci., 3:189.

Peterson, S.L., 1986, Prefrontal cortex neuron activity during a discriminative conditioning paradigm in unanesthetized rats, Intern. J. Neuroscience, 29:245.

Pirch, J.H., 1977, Amphetamine effects on brain slow potentials associated with discrimination in the rat, Pharmacol. Biochem. Behav., 6:697.

Pirch, J.H., 1980, Event-related slow potentials in rat cortex during a reaction time task: Cortical area differences, Brain Res. Bull., 5:199.

Pirch, J.H., 1990, Basal forebrain neuron responses during right eye vs. left eye light discrimination, Soc. Neurosci. Abstr., In Press.

Pirch, J.H., and Barnes, P.R., 1972, Steady potential responses from the rat cortex during conditioning, Experientia, 28:164.

Pirch, J.H., Corbus, M.J., and Ebenezer, I., 1985a, Conditioned cortical slow potential responses in urethane anesthetized rats, Intern. J. Neuroscience, 25:207.

Pirch, J.H., Corbus, M.J., and Napier, T.C., 1981a, Auditory cue preceding intracranial stimulation induces event-related potential in rat frontal cortex: Alteration by amphetamine, Brain Res. Bull., 7:399.

Pirch, J.H., Corbus, M.J., and Rigdon, G.C., 1983, Single unit and slow potential responses from rat frontal cortex during associative conditioning, Exp. Neurol., 82:118.

Pirch, J.H., Corbus, M.J., and Rigdon, G.C., 1985b, Conditioning-related single unit activity in the frontal cortex of urethane anesthetized rats, Intern. J. Neuroscience, 25:263.

Pirch, J.H., Corbus, M.J., Rigdon, G.C., and Lyness, W.H., 1986, Generation of cortical event-related slow potentials in the rat involves nucleus basalis cholinergic innervation, Electroencephalogr. Clin. Neurophysiol., 63:464.

Pirch, J.H., Napier, T.C., and Corbus, M.J., 1981b, Brain stimulation as a cue for event-related potentials in rat cortex: amphetamine effects, Intern. J. Neuroscience, 15:217.

Pirch, J.H., and Peterson, S.L., 1981, Event-related slow potentials and activity of single neurons in rat frontal cortex, Intern. J. Neuroscience, 15:141.

Pirch, J.H., Yadon, D., and Turco, K., 1989, Acetylcholine (ACh) role in conditioning-related responses of cortical neurons: Microiontophoretic studies, FASEB Journal, 3:A394.

Richardson, R.T., and Delong, M.R., 1986, Nucleus basalis of Meynert neuronal activity during a delayed response task in monkey, Brain Res., 399:364.

Richardson, R.T., and Delong, M.R., 1988, A reappraisal of the functions of the nucleus basalis of Meynert, TINS, 11:264.

Rigdon, G.C., and Pirch, J.H., 1984, Microinjection of procaine or GABA into nucleus basalis magnocellularis affects cue-elicited unit responses in the rat frontal cortex, Exp. Neurol., 85:283.

Rigdon, G.C., and Pirch, J.H., 1986, Nucleus basalis involvement in conditioned neuronal responses in the rat frontal cortex, J. Neurosci., 6:2535.

Rolls, E.T., Sanghera, M.K., and Roper-Hall, A., 1979, The latency of activation of neurones in the lateral hypothalamus and substantia innominata during feeding in the monkey, Brain Res., 164:121.

Rowland, V., Gluck, H., Sumergrad, S., and Dines, G., 1985, Slow and multiple unit potentials in trace and temporal conditioning controlled by electrical reward in the rat, Electroencephalogr. Clin. Neurophysiol., 61:559.

Rucker, H.K., Corbus, M.J., and Pirch, J.H., 1986, Discriminative conditioning-related slow potential and single-unit responses in the frontal cortex of urethane-anesthetized rat, Brain Res., 376:368.

Rye, D.B., Wainer, B.H., Mesulam, M.-M., Mufson, E.J., and Saper, C.B., 1984, Cortical projections arising from the basal forebrain: a study of cholinergic and noncholinergic components employing combined retrograde tracing and immunohistochemical localization of choline acetyltransferase, Neuroscience, 13:627.

Semba, K., Reiner, P.B., McGeer, E.G., and Fibiger, H.C., 1988, Brainstem afferents to the magnocellular basal forebrain studied by axonal transport, immunohistochemistry, and electrophysiology in the rat, J. Comparative Neurol., 267:433.

Sillito, A.M., and Kemp, J.A., 1983, Cholinergic modulation of the functional organization of the cat visual cortex, Brain Res., 289:143.

Swartz, B.E., and Woody, C.D., 1979, Correlated effects of acetylcholine and cyclic guanosine monophosphate on membrane properties of mammalian neocortical neurons, J. Neurophysiol., 10:465.

Swartz, B.E., and Woody, C.D., 1984, Effects of intracellular antibodies to cGMP on responses of cortical neurons of awake cats to extracellular application of muscarinic agonists, Exp. Neurol., 86:388.

Travis, R.P.Jr., and Sparks, D.L., 1968, Unitary responses and discrimination learning in the squirrel monkey: the globus pallidus, Physiol. Behav., 3:187.

Turco, K., Yadon, D., and Pirch, J., 1988, Influence of microiontophoretic cholinergic drugs on conditioned cortical units, Soc. Neurosci. Abstr., 14:862.

Vives, F., and Mogenson, G.J., 1985, Electrophysiological evidence that the mediodorsal nucleus of the thalamus is a relay between the ventral pallidum and the medial prefrontal cortex in the rat, Brain Res., 344:329.

Watanabe, M., 1990, Prefrontal unit activity during associative learning in the monkey, Exp. Brain Res., 80:296.

Wenk, G.L., Markowska, A.L., and Olton, D.S., 1989, Basal forebrain lesions and memory: Alterations in neurotensin, not acetylcholine, may cause amnesia, Behav. Neurosci., 103:765.

West, C.H.K., and Michael, R.P., 1989, Responses of mesolimbic units to sensory input in rats conditioned by reinforcing brain stimulation, Soc. Neurosci. Abstr., 15:1012.

West, C.H.K., Schaefer, G.J., and Michael, R.P., 1990, Responses of single units in dopamine terminal areas to odors conditioned by reinforcing brain stimulation, FASEB Journal, 4:A978.

Wilson, D.A., and Sullivan, R.M., 1990, Olfactory associative conditioning in infant rats with brain stimulation as reward. I. Neurobehavioral consequences, Develop. Brain Res., 53:215.

Wilson, F.A.W., and Rolls, E.T., 1990a, Neuronal responses related to reinforcement in the primate basal forebrain, Brain Res., 509:213.

Wilson, F.A.W., and Rolls, E.T., 1990b, Learning and memory is reflected in the responses of reinforcement-related neurons in the primate basal forebrain, J. Neurosci., 10:1254.

Woody, C.D., 1982, Acquisition of conditioned facial reflexes in the cat: cortical control of different facial movements, Federation Proceedings, 41:2160.

Woody, C.D., Swartz, B.E., and Gruen, E., 1978, Effects of acetylcholine and cyclic GMP on input resistance of cortical neurons in awake cats, Brain Res., 158:373.

Woolf, N.J., and Butcher, L.L., 1986, Cholinergic systems in the rat brain: III. Projections from the pontomesencephalic tegmentum to the thalamus, tectum, basal ganglia, and basal forebrain, Brain Res. Bull., 16:603.

Woolf, N.J., Hernit, M.C., and Butcher, L.L., 1986, Cholinergic and non-cholinergic projections from the rat basal forebrain revealed by combined choline acetyltransferase and Phaseolus vulgaris leucoagglutinin immunohistochemistry, Neurosci. Lett., 66:281.

Zaborszky, L., and Cullinan, W.E., 1989, Hypothalamic axons terminate on forebrain cholinergic neurons: an ultrastructural double-labeling study using PHA-L tracing and ChAT immunocytochemistry, Brain Res., 479:177.

ELECTROPHYSIOLOGICAL STUDIES OF THE FUNCTIONS

OF THE NUCLEUS BASALIS IN PRIMATES

Russell T. Richardson and Mahlon R. DeLong

Department of Neurology
Johns Hopkins University
Baltimore, MD

INTRODUCTION

For nearly two decades, studies of the activity of nucleus basalis neurons in primates have provided insights to the possible functions of this brain structure. In 1971, DeLong noted that many basalis neurons respond vigorously when an animal receives a water reward in an operant conditioning task. Since the mid-1970's, Rolls and associates have studied basal forebrain neurons that respond selectively to the sight of food or to stimuli associated with food (Mora et al., 1976; Rolls et al., 1979; Wilson and Rolls, 1990). They have suggested that these neurons may be related to feeding and, more recently, to mnemonic processes. Other basalis neurons have response patterns that are somewhat similar to those of neighboring movement-related pallidal neurons, suggesting that the nucleus basalis may be related to basal ganglia function (Mitchell et al., 1987a,b). Subsequent studies from our laboratory have focused broadly on the entire nucleus basalis in an attempt to identify the processes in which it may be directly involved.

In brief, the findings from the studies reviewed here indicate that the nucleus basalis is probably not directly involved in trial specific memory, movement control, or sensory perception. Rather, the nucleus basalis may be related to the reinforcing properties of appetitive stimuli since a large proportion of basalis neurons have clear responses to rewards and stimuli that immediately precede rewards. In addition, many basalis neurons are responsive to an aversive air puff. For many basalis neurons, the responses to appetitive or aversive stimuli may be due to the fact that these stimuli are inherently arousing. Thus, the most salient feature of stimuli to which basalis neurons are particularly responsive (e.g. air puffs, water rewards, and stimuli associated with rewards) may be that they are particularly arousing to the animal.

The Basal Forebrain, Edited by T.C. Napier *et al.*
Plenum Press, New York, 1991

Our studies have attempted to evaluate the activity of
neurons throughout the entire nucleus basalis rather than
focusing on basalis neurons that have specific response patterns.
Neurons included in our studies are those that: (1) are located
in the region of the nucleus basalis; and (2) have spontaneous
firing patterns characteristic of basalis neurons. Neurons have
been recorded in all subdivisions of the nucleus basalis from
beneath the anterior commissure to the posterior extent of the
nucleus. The sample includes "border" (DeLong, 1971) or
"interstitial" neurons (Mesulam et al., 1983) that are located
along the anterior commissure and within the external and
internal medullary laminae of the globus pallidus. Basalis
neurons discharge spontaneously at a rate of about 20 spikes/sec
in a regular, steady firing pattern, as shown in Fig. 1. These
firing patterns are seen in neurons in the main body of the
basalis as well as along the medullary laminae. A few basalis
neurons have been activated antidromically from cortical
stimulation sites, and they have all had the moderate, steady
type of firing pattern, as shown in Fig. 2.

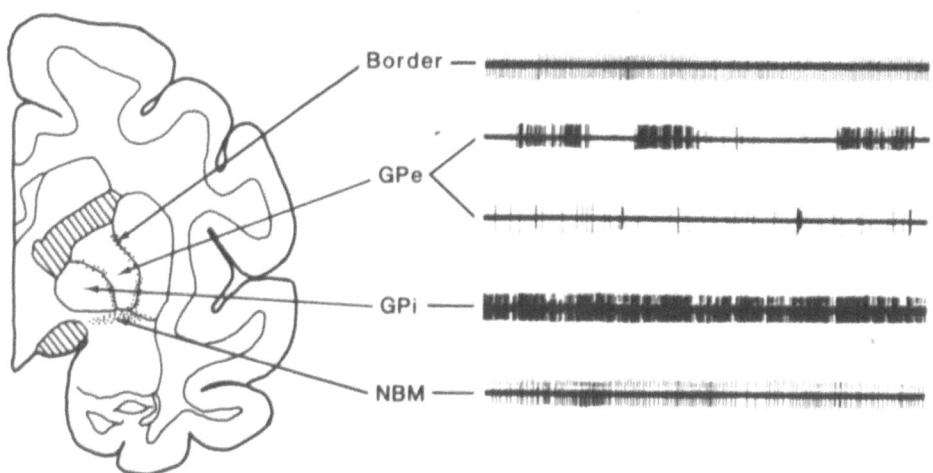

Fig. 1. Examples of characteristic neuronal discharge
         patterns in the nucleus basalis and globus
         pallidus. In these 3.5 sec traces, neurons within
         the medullary laminae (border) and in the main body
         of the nucleus basalis of Meynert (NBM) have fairly
         steady, regular discharge patterns in comparison to
         neighboring pallidal neurons. Two patterns are
         present in the external globus pallidus (GPe),
         fast, sustained firing punctuated by distinct
         pauses, and slow, sporadic firing punctuated by
         brief bursts of firing. In the internal pallidal
         segment (GPi), neurons have sustained, rapid firing
         (adapted from DeLong, 1971).

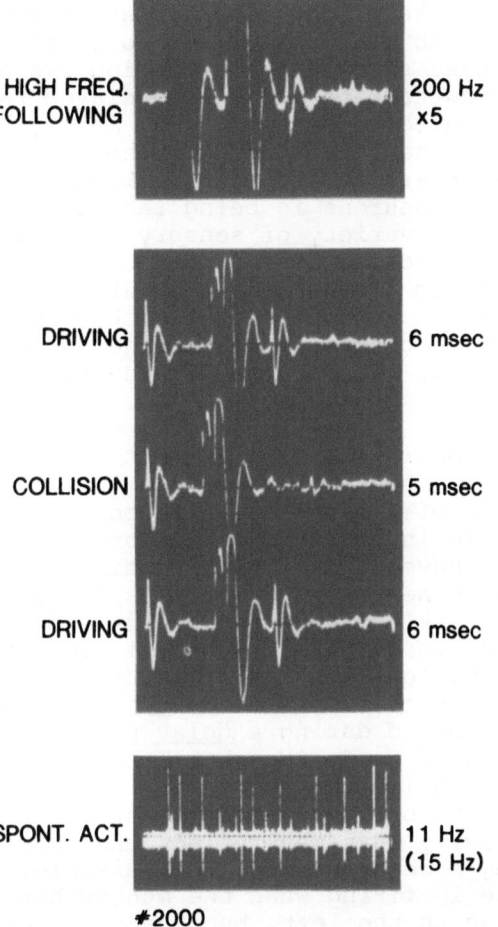

HIGH FREQ.
FOLLOWING   200 Hz
      x5

DRIVING   6 msec

COLLISION   5 msec

DRIVING   6 msec

SPONT. ACT.   11 Hz
      (15 Hz)

#2000

Fig. 2.   Example of a neuron in the anteromedial nucleus
basalis that is antidromically activated from
stimulating electrodes in the cingulum bundle. The
top four traces are 20 msec long and the bottom
trace is 1 sec long. The driven neuron followed
high frequency stimulation of 200 Hz at a constant
latency of 4.8 msec on five consecutive
stimulations. Collision of the orthodromic and
antidromic spikes occurred when the stimulation was
at 5 msec but not 6 msec after a spontaneous spike.
The neuron's firing rate was 11 spikes/sec in this
example, and an average of 15 spikes/sec over
several minutes. The steady, regular firing pattern
of this neuron is typical for basalis neurons.

The activity of each sampled neuron was recorded as the monkey performed a specific behavioral task. Each animal was trained to perform one of several behavioral tasks, all of which have been described in detail elsewhere (Richardson et al., 1988). The first task was a visuomotor tracking task in which the monkey had to make rapid, accurate arm movements to a visual target in order to obtain a water reward (Mitchell et al., 1987a). In the next study, a delay period was added to the visuomotor tracking task to produce a delayed response task, which is commonly used to study spatial memory (Richardson and DeLong, 1986). The next task was a go/no-go task which also had a delay period (Richardson and DeLong, 1990b). Currently, the activity of basalis neurons is being recorded in naive animals as they are presented a variety of sensory stimuli. Then some of the stimuli are paired together to create a classical conditioning paradigm (Richardson and DeLong, 1990a). The conclusions from this series of studies have provided some clear indications of the processes in which the nucleus basalis may (or may not) be involved, as discussed in the following sections.

TRIAL SPECIFIC MEMORY

Several lines of evidence suggest that the nucleus basalis may play a key role in learning and memory (Collerton, 1986; Smith, 1988). We have studied this possibility by recording the activity of basalis neurons in two types of memory tasks: a delayed response task and a go/no-go task. Both of these tasks use a delay period in which the monkey is required to remember some bit of information (i.e., the location of a visual stimulus) presented early in each trial in the cue phase. That information must then be remembered during a delay period lasting several seconds and then used to decide what behavioral response to make in the subsequent choice phase. In various parts of the brain that are believed to be involved in memory, neurons exhibit differential activity that may reflect the information being remembered during a delay period. For example, a neuron may have a robust increase in firing when the monkey has to remember that a cue has appeared on the left, but the same neuron may have no change in firing when the monkey remembers a cue on the right. Thus, this differential neuronal activity could be involved in the memory process that enables the monkey to remember the cue on the left. Such neurons can be found in the hippocampus (Watanabe and Niki, 1990), and they have long been found along the principal sulcus in the dorsolateral prefrontal cortex (Niki and Watanabe, 1976; Kojima and Goldman-Rakic, 1982).

Because the nucleus basalis projects to the dorsolateral prefrontal cortex (Mesulam et al., 1983) and may be involved in learning and memory, we expected to find numerous basalis neurons with differential delay period responses, but they are apparently quite rare (Richardson and DeLong, 1986; 1990b). In 608 basalis neurons sampled in two delay type memory tasks, only 2 (0.3%) had statistically significant differences in the delay period responses depending on the remembered cue location. Examples of delay period responses in basalis neurons are shown in Fig. 3. A number of the task-related neurons had significantly increased or decreased firing rates in the delay period of both the delayed

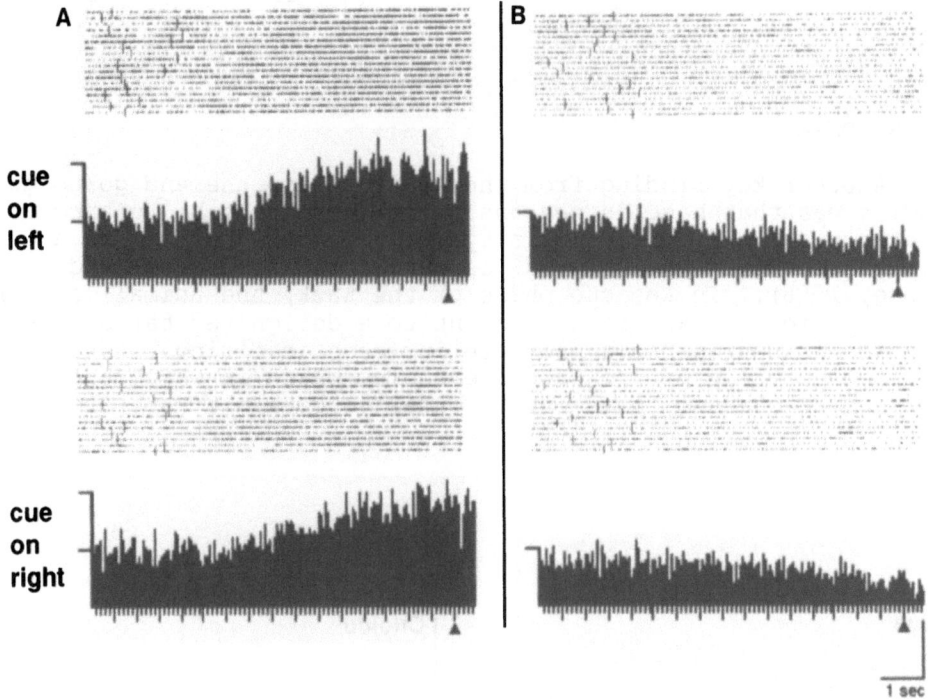

Fig. 3. Two examples of basalis neurons responding in the
delay period of a go/no-go task. In this and
subsequent figures, each dot in the raster
represents a single discharge of the neuron, and
each row of dots is from a single trial of the task.
The large vertical bars on each row indicate when
the visual cue appeared for 1.5 to 2 sec. The
histograms are formed by collapsing the rasters into
50 msec bins, and are centered on the onset of the
choice visual stimulus designated by the solid
triangle. Each mark on the horizontal axis
represents 100 msec, and each mark on the vertical
axis represents a mean bin firing rate of 20
spikes/sec. Trials on which the cue appeared on the
left were pseudorandomly mixed with trials with the
cue on the right. The basalis neuron in A had a
strong increase in firing in
the delay period that was the same magnitude
regardless of the position of the preceding cue.
The neuron in B had similar decreases in firing in
the delay period on both types of trials (from
Richardson and DeLong, 1990b).

response and the go/no-go tasks (14% in the delayed response task, 33% in the go/no-go task), but the same change in firing occurred whether the cue to be remembered was on the left or the right. Hence, there is little evidence that basalis neurons encode trial specific information in these types of memory tasks.

MOTOR CONTROL

Another key finding from the delayed response and go/no-go studies was that basalis neurons do not appear to be related to motor control. Fig. 4 shows a typical response pattern of a basalis neuron in the delayed response task (Richardson and DeLong, 1986). In the cue phase of the task, the animal made an elbow flexion or extension movement to a designated target, and the basalis neuron had a moderate increase in firing. After a 4 to 6 second delay period came the choice phase in which the monkey made the same movement in order to obtain a water reward.

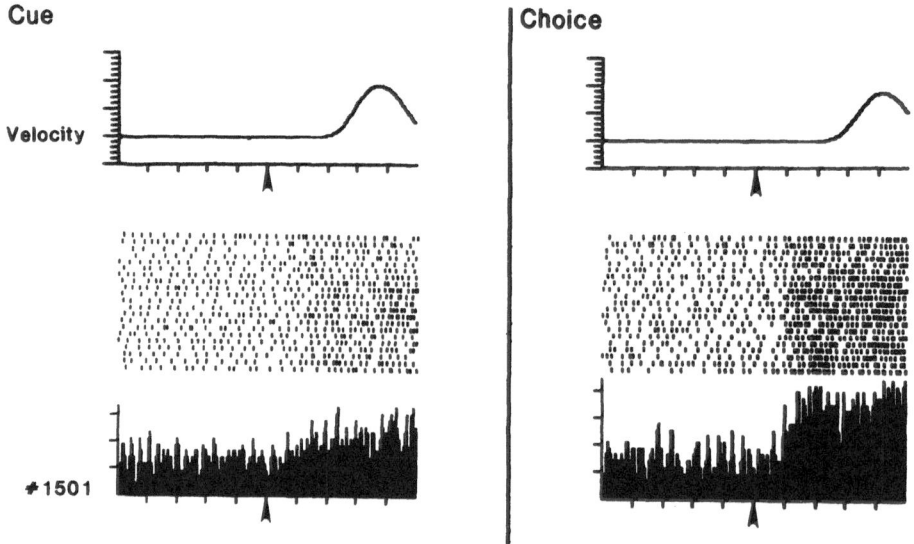

Fig. 4. Response pattern of a basalis neuron in the delayed response task. Rasters are aligned on the onset of the visual stimulus beginning the cue or choice phase of the task, and the histogram bins are 5 msec long. The average velocity traces show that the monkey's arm movements were similar in both phases, but the neuron's increase in firing in the choice phase is much larger than the response in the cue phase (from Richardson and DeLong, 1988).

In the choice phase, the neuron in Fig. 4 had an increase in firing that was much stronger than the response in the cue phase, although the monkey's movements were essentially the same in both conditions. Of the 135 task-related neurons in the delayed response task, 31% had significant changes in firing in the cue phase compared to 64% in the choice phase. Of the 89 basalis neurons that responded in either the cue or choice phases, two thirds had larger changes in discharge rates in the choice phase than in the cue phase. Although the arm movements were essentially constant in the two phases of the task, the neuronal responses varied; therefore, the basalis neuronal activity was dissociated from the movement.

A different type of dissociation occurred in the go/no-go task (Richardson and DeLong, 1990b). In the choice phase of this task, the animal had to decide whether to make an arm movement (similar to the movement in the delayed response task) or remain stationary in order to obtain the reward. Many basalis neurons responded like the one in Fig. 5. When the monkey made an arm movement (the go condition), there was a clear change in the neuron's firing. However, when the monkey chose to keep his arm motionless (no-go condition), the neuron had essentially the same change in firing. Of 253 basalis neurons that responded in the choice phase of the go/no-go task, 79% showed no significant difference between the changes in firing in the go condition when the monkey moved, and the no-go condition when he did not move. Hence, although the movements varied in the two conditions, the neuronal response remained constant; therefore, the basalis neuronal activity was again dissociated from the movement.

SENSORY PERCEPTION

The go/no-go task also provided evidence that basalis neurons are not related to simple sensory perception. In the cue phase, one of two visual stimuli was presented to the monkey, and he was only required to remember its location. In the subsequent choice phase, the same visual stimuli were used to signal the go and no-go conditions. Neuronal activity in the no-go condition (rather than the go condition) of the choice phase was used for comparisons with the cue phase since the monkey did not move his arm in either case. Many basalis neurons responded like the one represented in Fig. 6 in which there was a strong increase in firing in the no-go condition, but no significant change in firing in the cue phase. The monkey perceived the visual stimulus in both the cue and no-go conditions, but because the neuron responded only in the no-go condition, it was probably not involved in the perception of the stimulus. In the go/no-go task, of 251 basalis neurons that responded in the no-go condition, 58% had significantly larger responses than in the cue phase. Only 5% of the 326 task-related basalis neurons in the go/no-go task had significant changes in firing in the cue phase, compared to 70% that responded in the choice phase. Hence, although the same visual stimulus was perceived in the two conditions, the neuronal response varied; therefore, the basalis neuronal activity was dissociated from the sensory perception.

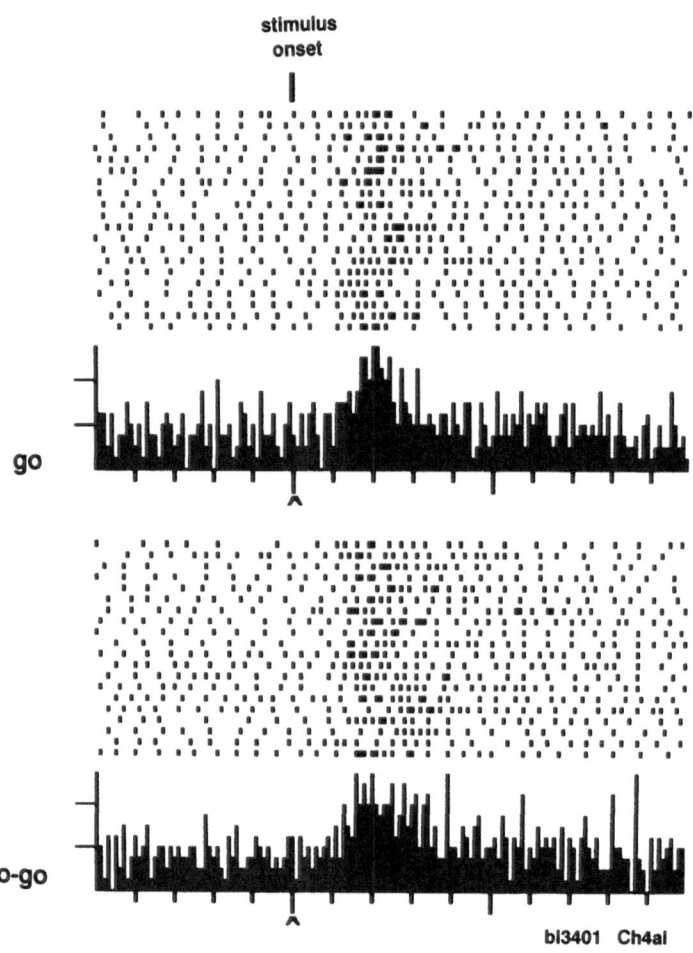

Fig. 5. Example of a basalis neuronal response that is not
related to movement in a go/no-go task. Conventions
are the same as in Figure 3, and the arrowheads
designate the onset of the visual stimulus in the
choice phase. On the go trials, the monkey moved
his arm to the left or right, and the neuron shown
here had a brisk increase in firing. On no-go
trials, although there was no movement, the neuron
had a similar increase in firing similar to the
response on go trials.

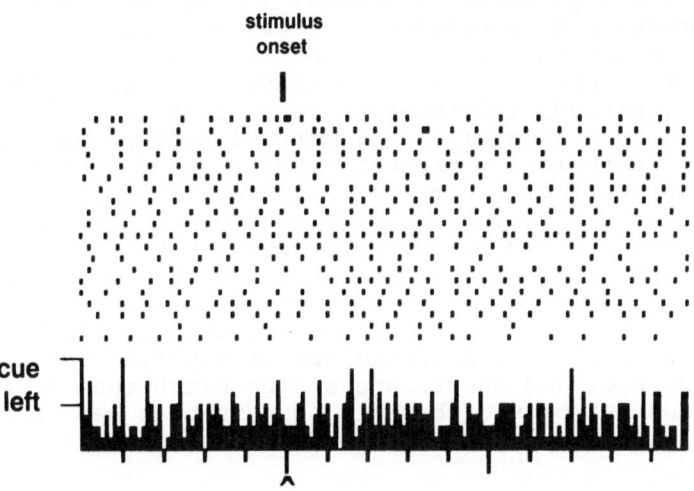

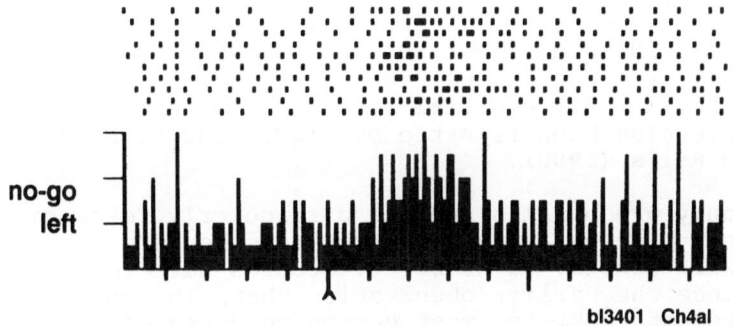

bl3401 Ch4al

Fig. 6. Example of a basalis neuronal response that is not
related to sensory perception in a go/no-go task.
Conventions are the same as in Figure 5 except that
the top graph is from the cue phase of the task. On
every trial, the identical stimulus appeared in the
animal's left visual field, and the animal made no
overt behavioral responses to the stimulus. The
neuron had no change in firing in the cue phase, but
it had a clear increase in firing in the no-go
condition of the choice phase.

In the delayed response and the go/no-go tasks, the majority of responses of basalis neurons occurred in the choice phase when the monkeys had to perform correctly in order to obtain a water reward. In other tasks as well, the majority of the responses occurred in the phase of the task that immediately preceded the water reward, regardless of whether the monkeys were required to make a choice, make a movement, or do anything at all (as in classical conditioning). It therefore appears that basalis neurons may be particularly responsive to stimuli that immediately precede the reward. This point is perhaps best exemplified in simple classical conditioning paradigms in which the monkeys were merely presented various auditory or somatosensory stimuli, some of which consistently preceded a water reward. For example, a brief torque that moved the arm to the left was followed by a water reward, but a similar torque to the right was not. Fig. 7 shows the activity of two basalis neurons that had robust changes in firing in response to the torque that preceded water, but no response to the torque not preceding water. In a separate set of experiments, a low pitch tone always preceded water, and a high pitch tone never did. Fig. 8 shows examples of basalis neurons that had changes in firing only in response to the auditory stimulus that preceded water.

In naive, untrained monkeys, there appear to be few basalis neurons that respond to simple sensory stimuli. In 141 basalis neurons recorded in naive monkeys, only 21% responded to brief perturbations of the arm or to pure auditory tones. However, after conditioning procedures, 66% of 161 basalis neurons responded to these stimuli. Of these responsive neurons, approximately 50% had larger responses to the stimulus that consistently preceded a water reward than to the stimulus that did not. Similar differential responses in basal forebrain neurons have also been reported by Rigdon and Pirch (1986) and Wilson and Rolls (1990).

Not only are basalis neurons particularly responsive to stimuli that precede an upcoming reward, they are also responsive to the reward itself. A number of studies have reported this finding since the earlier observation that, in monkeys performing an arm movement task, the most common response of basalis neurons appears to be to the water reward (DeLong, 1971). Basalis neurons have repeatedly been found to respond to a water reward, but such findings have usually been in the context of a behavioral task. Recently, basalis neurons in naive, thirsty monkeys have been found to respond to the delivery of water, as shown in Fig. 9 (Richardson and DeLong, 1990d). In 249 basalis neurons in naive monkeys, 57% responded to the water delivery. Hence, there is considerable evidence that many basalis neurons are responsive to this simple appetitive stimulus.

APPETITIVE AND AVERSIVE STIMULI

An important question is whether basalis neurons are <u>exclusively</u> responsive to appetitive stimuli, which would

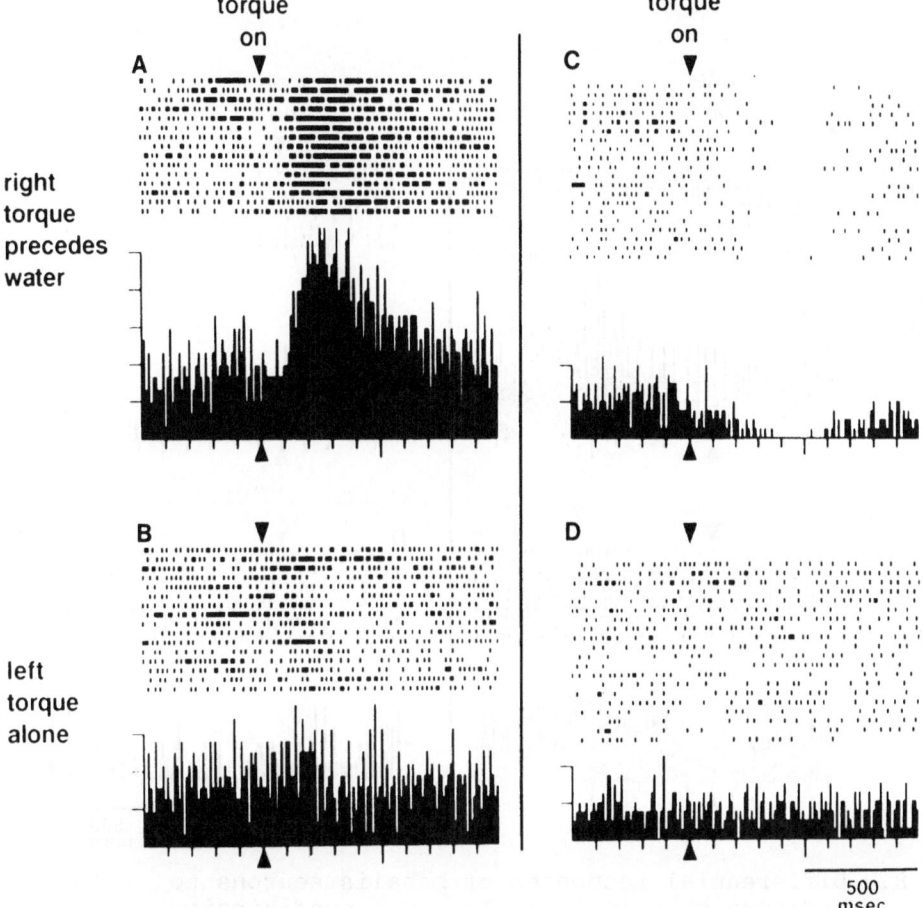

torque
on

torque
on

A

C

right
torque
precedes
water

B

D

left
torque
alone

500
msec

Fig. 7. Differential basalis neuronal responses to
somatosensory stimuli in a classical conditioning
paradigm. Rasters are aligned on the onset of a 50
msec torque that moved the monkey's arm to the left
or right. Only torques to the right were followed
1.5 - 2 sec later by a water reward. The neuron in
A and B had a strong response to the right torque
(A) and no response to the left torque (B).
Likewise, the neuron in C and D had a clear decrease
in firing to the torque preceding water (C) and no
response to the other torque (D) (from Richardson
and DeLong, 1990c).

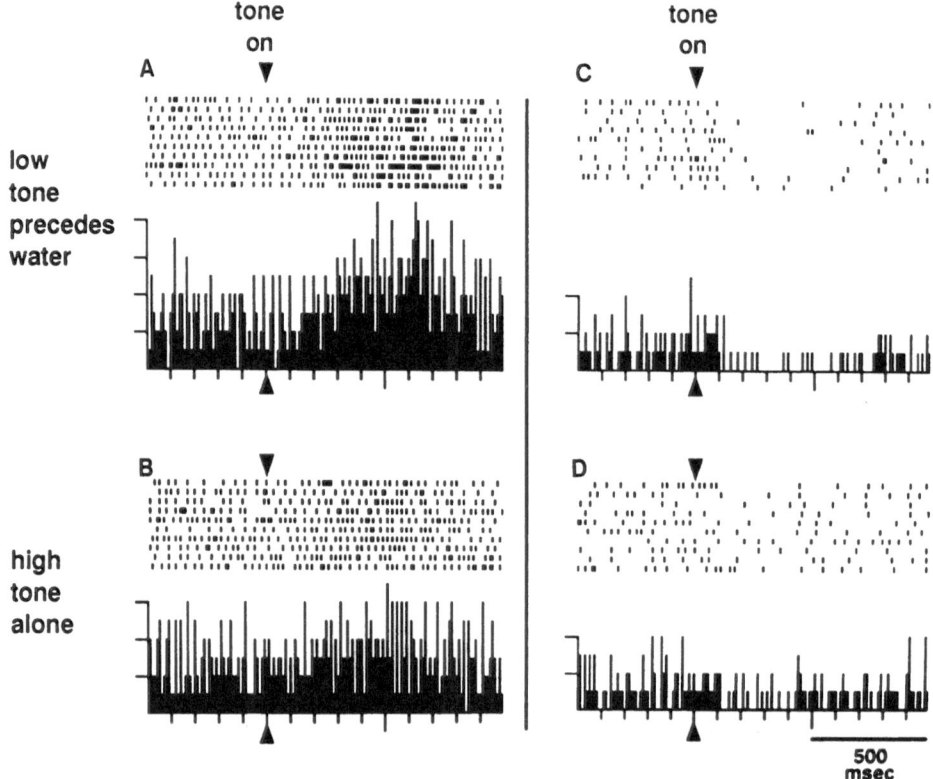

tone on ▼

A

low tone precedes water

B

high tone alone

tone on ▼

C

D

500 msec

Fig. 8. Differential responses of basalis neurons to
auditory stimuli in a classical conditioning
paradigm. Rasters are aligned on 50 msec tones of
500 Hz (low tone) or 1000 Hz (high tone). The low
tone preceded a water reward by 1.5 to 2 seconds,
but the high tone did not. The neuron in A and B
had an increase in firing following the low tone (A)
but no significant change in firing after the high
tone (B). In C and D, the basalis neuron responded
with a decrease in firing to the tone preceding
water (C) and did not respond to the tone not
preceding water (D).

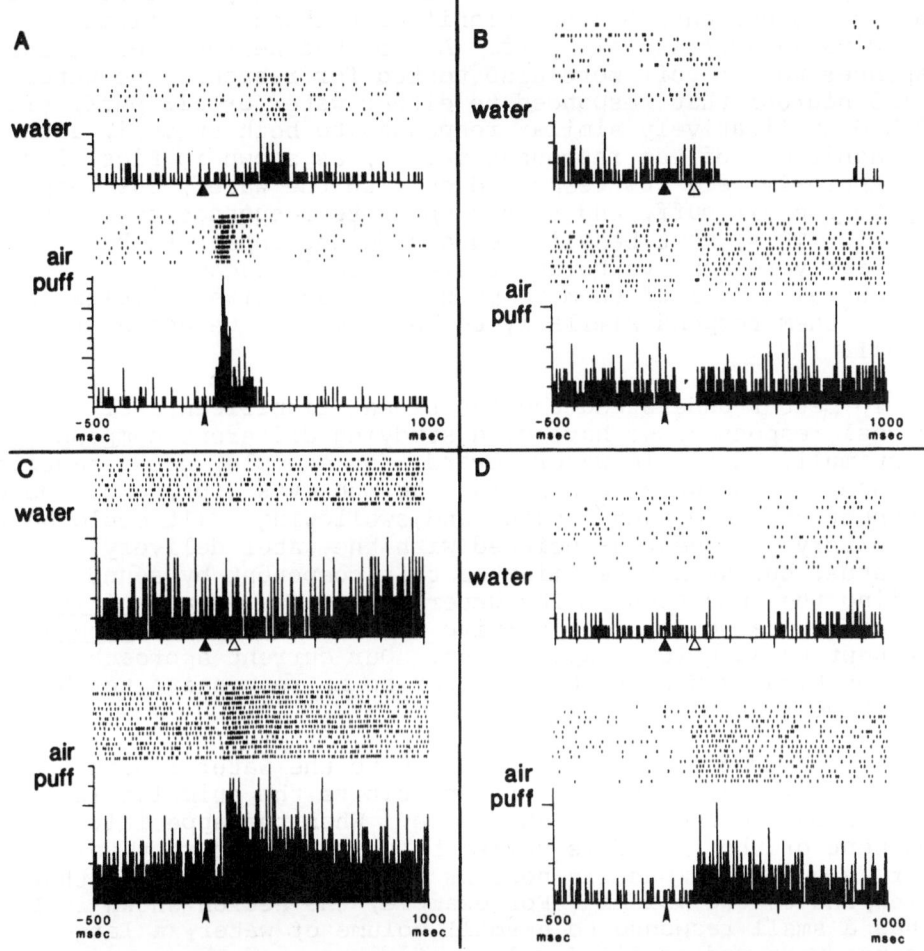

Fig. 9. Four examples of basalis neurons responding to water and/or air puff. The solid triangles indicate when the water delivery apparatus turned on, and the open triangles indicate approximately when the water arrived at the monkey's mouth. The arrow head indicates the onset of the air puff. The neuron in A had a larger increase in firing in response to the air puff than the water, whereas the neuron in B had a larger decrease in firing in response to the water than to the air puff. In C, the basalis neuron responded with an increase in firing only to the air puff. The neuron in D was one of the few that had opposite polarity responses to the two stimuli. It had a small decrease in firing in response to the water and a clear increase to the air puff (from Richardson and DeLong, 1990d).

indicate that the nucleus basalis is related primarily to reinforcement. If the nucleus basalis were related strictly to reinforcement, few of its neurons would respond to non-appetitive stimuli, such as an aversive air puff. However, in a sample of 190 basalis neurons, 53% had significant changes in firing following an aversive air puff. All of the neurons tested for responses to air puff were also tested for responses to water. Of 128 neurons that responded to either water or air puff, around 55% had qualitatively similar responses to both stimuli, although the magnitudes of the responses varied, as shown in Figs. 9 and 10. Approximately 20% responded only to the water, 20% responded only to the air puff, and 5% had opposite responses to the two stimuli. Thus, most basalis neurons do not appear to be selectively responsive to appetitive stimuli because a substantial number of them respond to an aversive stimulus, and many of them respond similarly to both appetitive and aversive stimuli.

To better understand the functional significance of these neuronal responses, we have been studying different components of the stimuli. A simple water reward, for example, has a tactile component of the water contacting the monkey's lips, and a motor component of the monkey licking and swallowing. (It could have an auditory component associated with the water delivery apparatus, but we have eliminated this component by sound proofing the apparatus.) The water reward has an appetitive component because it is a positive reinforcer, and an arousal component because it is appetitive. Our current approach is to focus on basalis neurons that appear to be responsive to the appetitive features of the reward.

If a neuron is indeed responsive to the water reward because it is appetitive, then the more appetitive the stimulus is, the more responsive the neuron should be. When the appetitive component of the reward is varied by using 3 different volumes of water, a number of basalis neurons have graded responses that correspond to the volume. For example, the neuron shown in Fig. 10 has a small response to a small volume of water, a larger response to a medium sized volume of water, and the largest response to a large volume of water. Thus, this neuron may be responsive to the appetitive component of the water. However, this neuron also has a clear increase in firing in response to the aversive air puff. Therefore, the neuron could not be selectively responsive to appetitive stimuli because it also responds to an aversive stimulus. The most parsimonious explanation of this phenomenon would be that the neuron is responsive to the arousal components of both stimuli.

This response pattern was fairly common in basalis neurons. In a group of 97 neurons tested under these conditions, 51% had increasing magnitudes of responses with increasing volumes of water. Of these 49 neurons, 80% had qualitatively similar responses to the air puff. Hence, the vast majority of basalis neurons that appeared to be related to the appetitive component of a water reward were also responsive to an aversive stimulus and were therefore probably responsive to the arousal component of these stimuli. This result implies that many of the responses of basalis neurons may actually be related to arousal mechanisms.

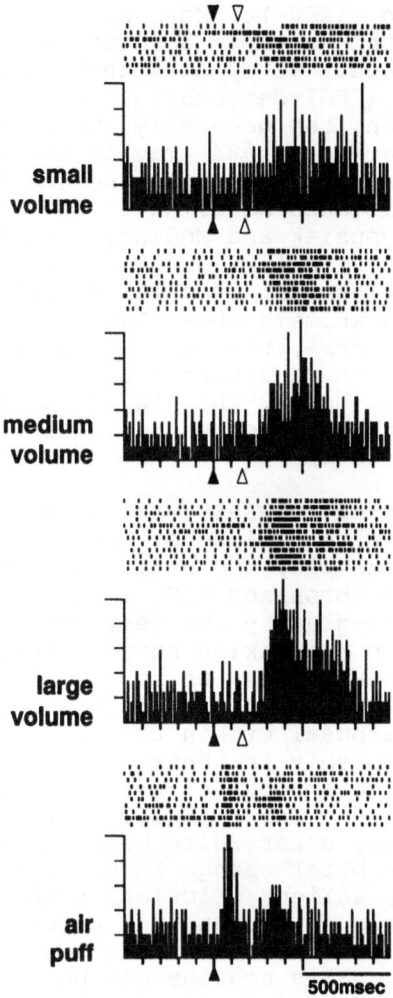

Fig. 10. Graded neuronal responses corresponding to the
          volume of water. Solid triangles indicate the
          onset of the water or air puff, and open triangles
          indicate the approximate time of water availability
          to the monkey. As the volume of water increased,
          the magnitude of the neuron's response also
          increased. This basalis neuron also responded to
          the aversive air puff with an increase in firing.

The nucleus basalis has previously been implicated in arousal processes because it is well suited both anatomically and physiologically to serve such a function (Buzsaki and Gage, 1990; Richardson and DeLong, 1990c). More specifically, the nucleus basalis may mediate the cortical activation component of the tonically aroused state. It appears that during electroencephalogram (EEG) desynchronization, which is indicative of cortical activation, the generally enhanced responsiveness of cortical neurons (Steriade, 1984) may be due in part to the increased tonic release of acetylcholine (ACh) from basalis neurons which have elevated baseline firing rates (Detari and Vanderwolf, 1987; Szymusiak and McGinty, 1989). In contrast to these sustained changes in firing, the responses of basalis neurons reviewed here consist of brief deviations from baseline firing rates in awake animals whose EEG is already desynchronized. Therefore, the phasic neuronal responses are probably not involved in mediating the cortical activation component of arousal. Hence, the kind of arousal process that could be related to the phasic responses of basalis neurons needs to be clarified.

The major states of arousal (waking, slow wave sleep, and paradoxical sleep) can be defined by specific criteria (Steriade et al., 1980). For example, the normal waking state is characterized by desynchronized EEG, increased electromyogram (EMG) activity in anti-gravity muscles, and directed eye movements, but, within the waking state, arousal levels may range from quiet, relaxed wakefulness to strong excitement or agitation. Although no consensus exists for how to categorize different levels of arousal within the waking state, they can obviously vary considerably in their intensity and duration. Furthermore, certain components of arousal may change independently of others. For example, a transient increase in cortical activation may occur while heart rate remains constant. It is possible that a brief change in cortical activation might follow a particularly salient stimulus, such as a water reward, and the mechanism for such a transient effect could be related to the phasic responses of nucleus basalis neurons. Hence, the phasic responses of basalis neurons may be related to brief changes in cortical activation, just as the tonic firing rates of basalis neurons may be related to sustained changes in cortical activation.

The concept of arousal is relevant to the studies reviewed here because it provides a parsimonious interpretation of the most frequent responses found in basalis neurons. A common characteristic of all the stimuli to which basalis neurons have been most responsive is that they are all potentially arousing to the animal. These responses of basalis neurons, which are so prevalent and consistent across several different testing conditions, are presumably related to some specific neural process, and current findings suggest that this process may be a transient increase in the cortical activation component of arousal.

Another process in which the phasic responses of basalis neurons may be involved is <u>attention</u>. The nucleus basalis appears to be well suited anatomically to mediate attentional mechanisms because, as a whole, it projects topographically to the entire cerebral cortex, but single basalis neurons appear to innervate small cortical areas of approximately 1-2 mm^2 (Price and Carnes, 1990). A possible mechanism for attention is that increased background discharge rates in a selected group of basalis neurons could increase ACh release in specific cortical subregions. The increased ACh release could enhance the response properties of cortical neurons in the affected area, thereby facilitating the functional properties of that cortical area. Selective facilitation of specific functional areas could possibly contribute to the neuronal substrate of attention, as suggested by recent blood flow studies in humans (Corbetta et al., 1990).

Despite the attractiveness of this attentional hypothesis, it is hindered by the lack of evidence that neurons in one region of the nucleus basalis respond differently from any other region. For every task in which we have recorded more than 100 basalis neurons, the same basic response patterns are found throughout the nucleus with no apparent clustering of responses in any subregion. A possible exception to this trend is that basalis neurons within the medullary laminae appear to have response properties that are more closely related to certain parameters of movement than do other basalis neurons (Mitchell et al., 1987a). These border neurons also have higher background discharge rates and more variable spontaneous discharge patterns than neurons in the main body of the nucleus basalis (Richardson and DeLong, 1990b). Nevertheless, the bulk of the electrophysiological findings thus far suggest that there is little regional variability in the responses of basalis neurons. Therefore, all cortical regions appear to receive similar influences from the nucleus basalis, which would preclude the possibility of a role in attentional mechanisms as described above. Hence, until there is better evidence for involvement in a selective process like attention, the nucleus basalis appears more likely to be involved in a non-selective process such as arousal.

SUMMARY

In summary, the studies reviewed here have indicated which neural functions might be directly influenced by the nucleus basalis. Basalis neurons do not appear to be directly involved in trial-specific memory because, in memory tasks, they have non-differential responses that do not correspond to the information being remembered by the monkey. Similarly, basalis neurons do not appear to be related to movements because, in a go/no-go task, similar neuronal responses occur whether the animal moves or does not move, and, in a delayed response task, different neuronal responses occur during the same arm movement made under different conditions. Basalis neurons also respond differently to the same sensory stimuli presented under different conditions, which indicates that the nucleus basalis is not involved in basic sensory perception. The responses of basalis neurons therefore appear to be strongly influenced by the context or behavioral significance of stimuli.

Many basalis neurons respond to appetitive stimuli. In trained animals, the most frequently observed responses have been to a water reward or to stimuli that consistently precede the reward. In naive, thirsty animals, a large proportion of basalis neurons respond to the delivery of water. However, a large number of neurons also respond to an aversive air puff, which indicates that the nucleus basalis cannot be exclusively related to appetitive stimuli. Although some basalis neurons apparently respond only to the appetitive stimulus and others respond only to the aversive stimulus, the majority appear to respond similarly to both stimuli. In particular, almost all of the neurons whose response magnitudes covary with the volume of the water respond similarly to the air puff. Hence, the neurons that appear most likely to be related to the appetitive component of the water are also responsive to an aversive stimulus. Basalis neurons may therefore be related to some common characteristic of aversive and appetitive stimuli, such as the arousing quality of these stimuli. The hypothesis that most basalis neurons are particularly responsive to arousing stimuli could account for the abundance of responses to rewards and stimuli associated with rewards. These phasic responses of basalis neurons are hypothesized to be related to a transient increase in the cortical activation component of arousal, just as the tonic activity of basalis neurons appears to be related to sustained cortical activation.

REFERENCES

Buzsaki, G., and Gage, F.H., 1990, Role of the basal forebrain cholinergic system in cortical activation and arousal, in: "Activation to Acquisition: Functional Aspects of the Basal Forebrain Cholinergic System," R.T. Richardson, ed., Birkhauser, Boston, MA.

Collerton, D., 1986, Cholinergic function and intellectual decline in Alzheimer's disease, Neurosci., 19:1.

Corbetta, M., Miezin, F.M., Dobmeyer, S., Shulman, G. and Petersen, S.E., 1990, Attentional modulation of neural processing of shape, color, and velocity in humans, Science, 248:1556.

DeLong, M.R., 1971, Activity of pallidal neurons during movement, J. Neurophysiol., 34:414.

Detari, L. and Vanderwolf, C.H., 1987, Activity of identified cortically projecting and other basal forebrain neurones during large slow waves and cortical activation in anaesthatized rats, Brain Res., 437:1.

Kojima, S. and Goldman-Rakic, P.S., 1982, Delay-related activity of prefrontal neurons in rhesus monkeys performing delayed response, Brain Res., 248:43.

Mesulam, M., Mufson, E.J., Levey, A.I. and Wainer, B.H., 1983, Cholinergic innervation of cortex by the basal forebrain: Cytochemistry and cortical connections of the septal area, diagonal band nuclei, nucleus basalis (substantia immominata), and hypothalamus in the rhesus monkey, J. Comp. Neurol., 214:170.

Mitchell, S.J., Richardson, R.T., Baker, F.H. and DeLong, M.R.,
    1987a, The primate nucleus basalis of Meynert: Neuronal
    activity related to a visuomotor tracking task, Exp. Brain
    Res., 68:506.

Mitchell, S.J., Richardson, R.T., Baker, F.H. and DeLong, M.R.,
    1987b, The primate globus pallidus: Neuronal activity
    related to direction of movement, Exp. Brain Res., 68:491.

Mora, F., Rolls, E.T. and Burton, M.J., 1976, Modulation during
    learning of the responses of neurons in the lateral
    hypothalamus to the sight of food, Exp. Neurol., 53:508.

Niki, H. and Watanabe, M., 1976, Prefrontal unit activity and
    delayed response. Relation to cue location versus direction
    of response, Brain Res., 105:79.

Price, J.L., and Carnes, K.M., 1990, Input/output relations of
    the magnocellular nuclei of the basal forebrain, in:
    "Activation to Acquisition: Functional Aspects of the Basal
    Forebrain Cholinergic System," R.T. Richardson, ed.,
    Birkhauser, Boston, MA.

Richardson, R.T. and DeLong, M.R., 1986, Nucleus basalis of
    Meynert neuronal activity during a delayed response task in
    monkey, Brain Res., 399:364.

Richardson, R.T., Mitchell, S.J., Baker, F.H., and DeLong, M.R.,
    1988, Responses of nucleus basalis of Meynert neurons in
    behaving monkeys, in: "Cellular Mechanisms of Conditioning
    and Behavioral Plasticity," C.D. Woody, D.L. Alkon, and J.L.
    McGaugh, eds., Plenum Publishing Corp, New York. p. 161.

Richardson, R.T. and DeLong, M.R., 1990a, Responses of primate
    nucleus basalis neurons related to appetitiveness,
    aversiveness, and arousal, Soc. Neurosci. Abstr., 16:

Richardson, R.T. and DeLong, M.R., 1990b, Context dependent
    responses of primate nucleus basalis neurons in a go/no-go
    task, J. Neurosci., 10:2528.

Richardson, R.T., and DeLong, M.R., 1990c, Functional
    implications of tonic and phasic activity changes in nucleus
    basalis neurons, in: "Activation to Acquisition: Functional
    Aspects of the Basal Forebrain Cholinergic System," R.T.
    Richardson, ed., Birkhauser, Boston, MA.

Richardson, R.T., and DeLong, M.R., 1990d, Responses of primate
    nucleus basalis neurons to water rewards and related
    stimuli, in: "Brain Cholinergic Systems," M. Steriade, and
    D. Biesold, eds., Oxford University Press, Oxford.

Rigdon, G.C. and Pirch, J.H., 1986, Nucleus basalis involvement
    in conditioned neuronal responses in the rat frontal cortex,
    J. Neurosci., 6:2535.

Rolls, E.T., Sanghera, M.K. and Roper-Hall, A., 1979, The latency
    of activation of neurones in the lateral hypothalamus and
    substantia innominata during feeding in the monkey, Brain
    Res., 164:121.

Smith, G., 1988, Animal models of Alzheimer's disease:
    experimental cholinergic denervation, Brain Res. Rev.,
    13:103.

Steriade, M., Ropert, N., Kitsikis, A., and Oakson, G., 1980,
    Ascending activating neuronal networks in midbrain reticular
    core and related rostral systems, in: "The Reticular
    Formation Revisted," J.A. Hobson, and M.A.B. Brazier, eds.,
    Raven Press, New York. p. 125.

Steriade, M., 1984, The excitatory-inhibitory response sequence in thalamic and neocortical cells: State-related changes and regulatory systems, in: "Dynamic Aspects of Neocortical Function," G.M. Edelman, W.E. Gall, and W.M. Cowan, eds., John Wiley & Sons, New York. p. 107.

Szymusiak, R. and McGinty, D., 1989, Sleep-waking discharge of basal forebrain projection neurons in cats, Brain Res. Bull., 22:423.

Watanabe, T. and Niki, H., 1990, Hippocampal unit activity and delayed response in monkey, Brain Res., 325:241.

Wilson, F.A.W. and Rolls, E.T., 1990, Neuronal responses related to reinforcement in the primate basal forebrain, Brain Res., 509:213.

THE RELATIONSHIP BETWEEN LEARNING, MEMORY AND NEURONAL

RESPONSES IN THE PRIMATE BASAL FOREBRAIN

Fraser A.W. Wilson

Section of Neuroanatomy, Yale University School of Medicine
New Haven, Connecticut 06510

INTRODUCTION

The purpose of this chapter is twofold: to describe the response properties of a class of *reinforcement-related neurons* (RRNs) in the basal forebrain of the monkey, and to provide a perspective on how processes of learning and memory influence these particular neurons. The basal forebrain is a complex region whose anatomical definition is being refined (see articles by Alheid, and Heimer in this book). In the context of this chapter, the term basal forebrain is restricted to the substantia innominata, the diagonal band of Broca, and a periventricular region adjacent to the walls of the third ventricle and anterior to the thalamus. These three regions share several common features: they contain cells that project to the cerebral cortex (Kievet and Kuypers, 1975; Mesulam et al., 1983), they receive afferent inputs from paralimbic regions of the cerebral cortex (Mesulam and Mufson, 1984; Russchen et al., 1985; Wilson and Rolls, 1990 a), and neurons with reinforcement-related activity are found throughout these three regions.

There are several lines of argument for the proposition that the basal forebrain has a role in learning and memory, but other arguments contradict or qualify this proposition. Firstly, there are severe losses of the magnocellular, mainly cholinergic neurons in the substantia innominata and diagonal band of Broca in the brains of patients with Alzheimer's disease (Whitehouse et al., 1982). As these patients have a marked memory deficit, it is possible that the loss of the cholinergic basal forebrain may be, at least in part, responsible for the deficit. However, such patients typically have massive pathology in cortical structures (e.g. Pearson et al., 1985) believed to be important for memory function. Furthermore, some pathology can occur in the magnocellular basal forebrain without the expression of frank dementia (Whitehouse, 1990). Secondly, many studies have shown that 'blockade' of acetylcholine (ACh) deleteriously affects the performance of memory tasks, but these deficits appear to differ from those in patients with Alzheimer's disease (Kopelman and Corn, 1988), indicating that the cognitive impairments of this disease are not attributable entirely to the cholinergic system. Thirdly, although lesions of the monkey basal forebrain produce deficits in a variety of tasks measuring recognition and associative memory (Aigner et al., 1984; Ridley et al., 1985; Roberts et al., 1990), these deficits require the complete destruction of the substantia innominata and diagonal band of Broca before a deficit of small magnitude appears in Macaques (Aigner et al., 1984), and recent data indicate that the results of ablation studies in Macaques may not be robust (Voytko et al., 1990), although this conclusion awaits histological confirmation of the locus of brain damage. Furthermore, other studies indicate that the loss of the cholinergic basal forebrain is not completely responsible for these behavioral impairments. Robbins et al. (1989a) have shown that a neurotoxic lesion of cholinergic basal forebrain neurons resulting in a reduction (44%) in cortical ChAT did not impair a conditional visual discrimination task in the rat. In contrast, a less selective lesion that additionally involved non-cholinergic basal forebrain neurons produced a behavioral impairment that was not ameliorated by the administration of physostigmine. Thus, it has not yet been established that either the cholinergic or the non-cholinergic groups of basal forebrain neurons play a role in learning and memory.

There do not appear to be documented cases describing the effects of brain damage restricted to the basal forebrain in man, although lesions involving the basal forebrain and the ventromedial temporal and prefrontal cortices produce a particularly severe amnesia (Damasio et al., 1985; Friedman and Allen, 1969; Gascon and Gilles, 1973). However, diencephalic amnesia following brain damage to structures adjacent to the walls of the anterior third ventricle is well documented (e.g. Wilson, 1989). In a study designed to examine possible memory functions of this periventricular region, monkeys performed recognition memory tasks while recordings of neuronal activity were made. A major finding was that neurons in the periventricular region of the basal forebrain encode the *familiarity* of visual stimuli (Rolls et al., 1982). These neurons are typically unresponsive to presentations of stimuli that are novel to monkeys, but respond very well to presentations of the same stimuli when they are familiar. How is it that these neurons, located in the center of the forebrain, have this memory-like property? In a thoughtful analysis of the neural systems involved in memory function, Horel (1978) suggested that pathways from the temporal lobe to the medial thalamus might have a function in memory, and cited the anatomical work of Whitlock and Nauta (1956) showing that lesions of the ventromedial temporal lobe resulted in the degeneration of fibers and terminals in the substantia innominata, the periventricular region and the medial thalamus. The temporal lobe is believed to be involved in visual memory (Mishkin, 1982), and this corticofugal pathway could provide information from memory to the periventricular region. If this hypothesis is correct, it follows that the substantia innominata should also be the recipient of information from the memory system. Consequently, recording studies were carried out in the substantia innominata and the diagonal band of Broca (Wilson and Rolls, 1990 a,b). These studies resulted in the finding of two additional groups of neurons whose activity reflected information from memory. One group responded on the basis of the *novelty* of the stimuli; the other group responded on the basis of the *reinforcement* value of the stimuli. In this chapter we focus on these reinforcement-related neurons because preliminary but firm evidence shows that these neurons are part of the basal nucleus of Meynert. This evidence is not yet available for the neurons that respond on the basis of novelty or familiarity, and therefore they are not discussed in detail.

The present study was designed to examine how learning and memory influence basal forebrain neurons that respond to reinforcing events. Data from two previous studies suggested that learning might influence such neurons. In a discrimination paradigm, Travis and Sparks (1968) showed that neurons in the border region between the globus pallidus and substantia innominata responded differentially to two sensory cues that signalled reward and punishment respectively. As the cues in the task were initially neutral, the differential neuronal responses may have reflected what the monkey had learned about the cue. In another study, Mora et al., (1976) showed that neurons in the hypothalamus and substantia innominata responded to the sight of a syringe used to deliver juice to the monkey, and ceased to respond to it when it no longer delivered juice. These results indicated that the monkey learned that the syringe tasted good (or had ceased to taste good). The present experiments were designed to analyze further any learning-related neuronal activity by looking at the time course of learning and the influence of memory on the neuronal activity; to ensure that 'learning-related' effects were not attributable to the sensory properties of syringes or other cues; and to ensure that changes in neuronal firing during learning (when the significance of a stimulus is changed) were not due simply to lapses of attention or lack of fixation.

The experiments had a number of other objectives that are worth stating. The first concerns the definition of a reinforcement-related neuronal response. Typically, neurophysiological studies require a thirsty monkey to perform a behavioral task for a water reward. The delivery of water at the successful completion of a trial is a potent reinforcer, and neurons in the substantia innominata respond to the delivery of water (DeLong, 1971; Richardson and DeLong, 1986). Other studies have reported that basal forebrain neurons respond to the sight and *before* the delivery of a primary reinforcer such as a piece of food (Rolls et al., 1976). However, a neuronal response to the sight or taste of an appetitive reinforcer does not show that its reinforcement value is encoded; it is necessary to show that such neurons respond *differentially* to appetitive and aversive reinforcers. Thus, to analyze further the relationship between the neuronal activity and reinforcement, the present study used appetitive and aversive reinforcers: fruit juice and hypertonic saline. As the concept of reinforcement implies a dimension along which a particular reinforcer may be rated, that is, its pleasurableness or its aversiveness, we hypothesized that basal forebrain neurons encoding information about the reinforcement value of a stimulus should respond differently *to the sight* of appetitive and aversive stimuli. It is these neurons that are termed *reinforcement-related* (RRN) in this paper.

A second feature of the study was that the monkeys were working for the palatable taste of the reinforcers. It is assumed that they were not hungry or thirsty at the start of an experiment as they had *ad libitum* food and water in their home cages. This feature was introduced because it is desirable to separate neural mechanisms (active in the hungry, thirsty monkey) that restore homeostasis through the initiation of feeding and drinking from those mechanisms related to the encoding of the reinforcement value of sensory stimuli. Consequently, in the present study neuronal activity that distinguishes between appetitive and aversive reinforcers may be related to the reinforcement value of the stimuli, but is not necessarily related to the ingestion of fluid and the restoration of homeostasis.

Thirdly, we wanted to know if the development of RRN activity required the monkey to taste as well as see a reinforcing stimulus, or whether it was sufficient for the stimuli to signal the availability of reinforcement contingent upon a behavioral response. To this end, monkeys were trained on tasks in which the stimuli were patterns on a video monitor, and which had a specific relationship with reinforcement, but which were never directly tasted or touched. Thus, these stimuli became cues which the monkey used to guide his behavior.

METHODS - the behavioral tasks

The monkeys were trained to perform two go/no-go *visual discrimination* tasks in which the delivery of fruit juice or aversive hypertonic saline was contingent on lick responses to a tube placed in front of the mouth. In one version of the task, an electromagnetic shutter was used to present two highly familiar syringes mounted in square plaques of different colors, one per trial. Lick responses at the presentation of a black syringe (the S-) resulted in the delivery of saline, while responses to the white syringe (S+) resulted in the delivery of juice. In the second discrimination task, two visual images equated for size, color and brightness, but differing in shape were displayed on a video monitor. Lick responses to the yellow circle (S+) produced juice, while responses to the yellow square (S-) produced saline. The monkeys learned to respond differentially to the S+ and S- stimuli in the two tasks.

In two *serial visual recognition memory* tasks, each stimulus was shown twice per day. Lick responses during the first, novel presentation of the stimuli elicited aversive saline, while lick responses to familiar stimuli resulted in the delivery of appetitive fruit juice. Thus, the monkeys had to determine the novelty or familiarity of the stimuli in order to obtain juice. The first (novel) presentation of a stimulus was followed by a second (familiar) presentation of the stimulus after 0 to 16 other trials, selected in pseudorandom order. A typical stimulus sequence was as follows: N1-->N2-->F2-->N3-->N4-->F1-->N5-->F4. The novel stimulus (N1) shown on trial 1 was shown again after four intervening trials as familiar (F1) on trial 6, while the novel stimulus (N2) shown on trial 2 was repeated without intervening trials on trial 3. Stimuli in the recognition and discrimination tasks were either 3-dimensional objects presented using the electromagnetic shutter, or were two-dimensional images presented on the video monitor. The recognition and discrimination tasks were performed concurrently: the S+, S-, a novel or familiar stimulus could be presented on any given trial. An analysis of variance was used to determine differences in responses to the various classes of stimuli. The differential responses of RRNs to the S+ and S-, and to novel and familiar stimuli, are significantly different.

In 'clinical' tests, the neuronal responses were examined under standardized conditions in which visual, auditory, gustatory and tactile stimuli were applied. For example, a protocol was used in which the monkey watched the experimenter while he presented his arm, reached for a piece of food, a syringe or object, brought it back, presented the stimulus, advanced it towards the monkey, and delivered the reinforcer to the monkey. Measurements of neuronal firing rate were taken during these different steps of the protocol.

RESULTS - responses of reinforcement-related neurons

We have described in detail the response properties of 120 neurons in the basal forebrain whose activity is most simply described as being related to the learned reinforcement value of sensory stimuli (Wilson and Rolls, 1990 b,c). These neurons constitute approximately 6% of all (n = 2119) recorded neurons, and appear to be a specific functional class; they respond differentially to stimuli that, through learning, signal the availability of appetitive or aversive reinforcement, irrespective of their physical appearance and sensory modality. Such neurons respond to the sight of foods, to syringes used to deliver fruit juice, to shapes presented on a video monitor signalling availability of juice, to familiar stimuli in the recognition memory task, and to auditory cues signalling juice. The same neurons respond differently to stimuli that are aversive: syringes containing saline, shapes signalling the availability of saline, and novel stimuli in the recognition task. The RRNs are distributed within the substantia innominata, the vertical and horizontal limbs of the diagonal band of Broca, and a region close to the walls of the anterior third ventricle in the region of the dorsal hypothalamus. The location of many of these neurons corresponds to the magnocellular basal nucleus of Meynert and there is direct evidence that RRNs project to the cerebral cortex and thus are part of the basal nucleus of Meynert (see below). In addition to these RRNs, a further 45% of the sample were active in the tasks and appear to be similar to 'choice-related' neurons described by Richardson and DeLong (1990), as noted in the Discussion.

Fig. 1 shows an example of the responses of an RRN. In most respects the pattern of responses of this neuron in the different behavioral tests is absolutely typical of RRNs. This neuron differed from the majority of other such neurons in that the spontaneous firing rate was high (but was not due to a difficulty in triggering on the action potential, which was 10 times the background noise); the responses to the sight of foods were weaker than usual; and it was located in the border region of the globus pallidus and substantia innominata in which

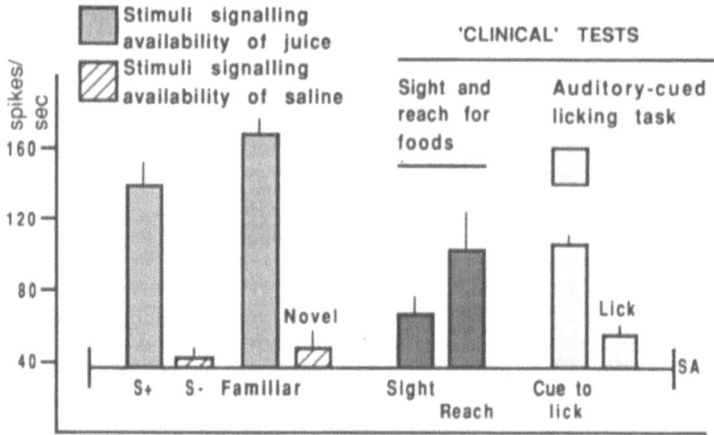

Fig.1. Responses of a single reinforcement-related neuron under different behavioral conditions. Each bar of the histogram represents the mean firing rate (and s.e.m.) of the neuron in a particular condition. Columns 1,2: the visual discrimination task; columns 3,4: the recognition memory task; columns 5,6: sight and reach for foods; columns 7,8: the auditory cue and licking in the absence of the cue. The neuron is responsive in those conditions in which the probability of receiving reinforcement is high. SA = spontaneous activity of the neuron recorded in the intertrial interval.

other studies (DeLong, 1971; Richardson and DeLong, 1986; Mitchell et al., 1987) have taken place. This neuron is used to illustrate the activity of RRNs because it was tested in many of the standard tests, and can be used to facilitate comparison with other studies.

When the monkey performs a *visual discrimination* task, the neuron responds *differentially*: a large increase in firing rate to the sight of the S+, and a significantly smaller response to the S-. The neuronal responses precede the lick movements and the delivery of the juice (not shown in the histograms). Some of these RRNs were tested in two different discrimination tasks; in most cases they responded differentially to the S+ and S- in both tasks. Measurements of the differential response latency (the time taken to respond differentially to the S+ and S-) yield average values of 180 ms, preceding lick responses that generally occur between 250 to 400 ms after the onset of a visual cue.

The neuron also responds differentially in the *recognition memory* task: a large increase in firing rate to the sight of familiar stimuli, and a significantly smaller response to the same stimuli when shown first as novel (Fig. 1). The neuron responds in the same way to the S+ and familiar stimuli, and in the same way to the S- and novel stimuli, and thus encodes the reinforcement value of the different stimuli. It is important to note that the monkey has (in practice) not seen the novel stimuli before, so on first sighting of these stimuli, the monkey has to judge that they are novel and use this information to make a decision about their reinforcement value. The differential neuronal response latency to novel and familiar stimuli is 250 ms on average. Thus, it takes the brain longer to determine whether a stimulus is novel or familiar and therefore signals a certain type of reinforcement than when it is a highly familiar appetitive or aversive stimulus with a fixed reinforcement value as are the S+ and S- in the discrimination task.

Interestingly, when the monkey recognizes a stimulus as familiar, RRNs respond to it *even before* he has obtained juice by making a lick response. Clearly, the knowledge that familiarity is the cue for obtaining reinforcement is sufficient for the occurrence of the neuronal responses, and prior delivery of reinforcement is not necessary. The most revealing aspect of these neuronal responses is that a familiar stimulus will elicit a response even when it has not been seen for many trials. A conservative estimate of the memory spans of these neurons indicates an average value of 30 trials. Thus, these neurons have access to a durable storage for information about the stimuli shown in the tasks, and therefore are the recipient of information from the neural substrate of recognition memory and the mechanisms that determine the novelty and familiarity of a stimulus.

In these behavioral tasks, the monkey has to make a decision on each trial as to whether these diverse stimuli signal the availability of juice or saline, and then make the appropriate behavioral response. To lick or not to lick - that is his question. However, there are several reasons why the differential responses of the RRNs are not related to lick movements, as shown in the 'clinical' tests (Fig. 1). Firstly, the neuron responds with an increase in firing at the sight of food, although the monkey is not required and does not make a behavioral response for its delivery. Thus, neuronal activity occurs when lick responses are *not* made. This test also shows that both food and fruit juice are effective in eliciting these neuronal responses. Secondly, the neuron is active when the monkey is cued to initiate an *arm movement* for a piece of food. Thirdly, the neuron responds to an *auditory cue* that signals the availability of juice contingent on a *lick movement*, but is significantly less active when the flow of juice is witheld and the monkey simply licks the tube. In additional tests, the firing rate increased to 79 spikes/sec at the sight and approach (but *before* its delivery) of a syringe known to the monkey to contain juice. In contrast, when the syringe was withdrawn from the monkey while drinking, the firing rate dropped to 21 spikes/sec, below the spontaneous firing rate of the neuron, even though the syringe was still in view. In summary, RRNs are active when visual and auditory cues signal the availability of reinforcement: in the *absence* of movements (sight of foods in 'clinical' tests), during *lick* movements (visual discrimination and recognition tasks; auditory cued task) and during *arm* movements ('clinical' tests). The actual form of the (mouth or arm) movement is not important: it is the likelihood of obtaining the reinforcer (juice or food) that is crucial for the neuronal responses.

The anatomical connections of the reinforcement-related neurons

The cerebral cortex is the recipient of ascending projections from the basal nucleus of Meynert within the basal forebrain (Mesulam et al., 1983). The location of histologically-confirmed RRNs corresponds to the basal nucleus of Meynert, so to determine if the axons of RRNs projected to cortex, we looked for the driving of basal forebrain neurons following electrical stimulation of discrete regions of cortex. In 6 of 7 cases (one neuron was not active in the tasks), RRNs recorded in the substantia innominata were driven at invariant latencies following stimulation pulses applied to area 46 of prefrontal cortex, or supplementary motor cortex, or area 4 of motor cortex. The collision test showed that the driving was antidromic (mean antidromic latency: 3 ms) and therefore that the RRN projected to the cortex, as opposed to receiving inputs from the cortex. The responses of one of these neurons is shown in Fig. 8 of Wilson and Rolls (1990b); the response profile of this neuron is identical to that shown in Fig. 1 in this chapter. These data show that the axons of RRNs ascend to the cerebral cortex, and that these functionally characterized neurons are part of the basal nucleus of Meynert.

Like the basal nucleus of Meynert, cells in the periventricular region are known to project to the cerebral cortex (Kievet and Kuypers 1975; Mesulam et al., 1983). To identify the brain structures that project to the periventricular region, an injection of Horseradish Peroxidase was made into this region bordered by the substantia innominata, the internal capsule, the walls of the third ventricle, in the region of the dorsal hypothalamus, anterior to the thalamus and posterior to the anterior commissure. This region was first mapped with microelectrodes and was chosen for the injection site because it is possible to record both RRNs and another class of neuron whose activity is a pure expression of information in memory: these latter neurons respond at the sight of visual stimuli when they are *familiar* to a monkey.

Retrogradely labeled cells were located in *ventromedial* regions of the brain. Many labeled cells were found in areas 13, 14, 25, and 32 of prefrontal cortex, lighter label was found in areas 24 and 38 of anterior cingulate and temporal polar cortex, and a few labeled cells were found in area 35, the perirhinal cortex. In subcortical structures, large numbers of labeled cells were found in the midline thalamic nuclei, the magnocellular mediodorsal nuclei of the thalamus, and a region dorsal and lateral to the mammillary bodies. Moderate numbers of labeled cells were found in the amygdala, substantia innominata, hypothalamus, nucleus accumbens and ventral and medial regions of the midbrain.

DISCUSSION

A major property of RRNs is that the neuronal responses occur at the sight of stimuli that signal the availability of reinforcement, *before* the delivery of the reinforcer, and not necessarily to the delivery of the reinforcer itself, although some may do so. This type of neuronal activity is distinct from neural mechanisms encoding the hedonic attributes of reinforcement delivery, such as their taste and palatability, and sensations resulting from addictive, stimulant drugs which involve dopaminergic systems (see Kalivas and Austin; Koob et. al., Napier et al., Porrino et al. and Smith et al., this book) and which may interact with the RRNs described above, although this remains to be shown. In contrast, the properties of RRNs indicate the operation of an analytical rather than a hedonic mechanism. The neuronal responses reflect learning because they occur to stimuli that, although initially neutral, have become signals for the availability of reinforcement. The neurons respond differentially to novel and familiar stimuli, reflecting access to information from recognition memory that is used to determine the reinforcement value of the stimuli. The definition of a reinforcement-related neuron requires that the neuronal responses are independent of the properties of the sensory stimuli, the sensory modality of the stimuli, the type of behavioral response, and the type of reinforcer to which the neuron responds. The observations on which these conclusions are based are outlined below.

Firstly, it is evident that the reinforcement-related neuronal responses do not encode information about sensory properties such as shape, color and size of visual stimuli to which RRNs respond. Secondly, the responses to auditory cues signalling reinforcement also support this contention, and show that information in several modalities is able to influence these neurons. It was possible, in many cases, to record the responses of the neurons in two visual discrimination tasks, two recognition memory tasks, a visual discrimination reversal task and to the sight of foods. In each case, the neurons responded on the basis of the reinforcement value of the stimuli, not the sensory appearance of the stimuli. This is particularly clear in the case of tests using the recognition memory task where each stimulus is different. These responses occur *at the sight* of the reinforcer, before its delivery and before the behavioral response required to obtain it. It is these 'visual' responses that suggest that the neuronal activity reflects what the monkey has learned about the stimuli, rather than a sensory response to the delivery of the reinforcer. Furthermore, it is evident that a stimulus that the monkey has seen, but never touched or tasted, can through learning, come to signal the availability of reinforcement which is then reflected in the neuronal responses.

Thirdly, the neuronal activity can be dissociated from motor responses, as the neuronal activity occurs when no movement is made, and when a lick response is required, and when an arm movement is required. Although neurons in other brain structures may respond to visual stimuli that signal the availability of reinforcement, it remains to be shown that this neuronal activity does not depend upon the type of stimulus or behavioral response being employed, or the type of reinforcer used. Neurons in premotor cortex and the caudate nucleus, for example, may show a response to a signal for reinforcement, but the neuronal responses may be dependent upon the elicitation of a particular response (mouth, arm or eye movement) or some parameter of movement such as direction, force or velocity.

Fourthly, the neuronal activity does not simply reflect mechanisms to restore caloric, salt, or fluid imbalances. The monkeys had water and food available *ad libitum* in their home cages, and therefore worked for the palatability of the juices and fruits used in the experiments, rather than to redress homeostatic imbalances. It follows that the neurons encode the reinforcement value of the stimuli, rather than provide a signal to initiate feeding or drinking. This is supported by the fact that these neurons responded both to signals for juice availability and to the sight of food, and thus are not a corrective mechanism *per se*, as feeding and drinking are clearly controlled by very different mechanisms. However, the activity of RRNs is modified by the willingness of the monkey to work for a reinforcer, as measured by the satiation of the monkey for a particular reinforcer (Burton et al., 1976), and by the use of water as reinforcement, which results in poor responses of the RRNs in monkeys with *ad libitum* water, as well as a clear lack of motivation on the part of the monkey to work for such reinforcers.

Fifthly, the neuronal responses are not related to behavioral arousal as all the stimuli used in the tasks are arousing for the monkeys, yet the neurons respond *differentially* to appetitive and aversive stimuli. Several observations show that the activity of RRNs is dissociable from arousal: (1) monkeys will reach to obtain the S+ syringe, but will turn their head and bodies away from the S- syringe, the stimuli eliciting approach and avoidance behavior, respectively. Although the S+ and S- arouse the monkeys, the neurons respond differentially to them; (2) novel stimuli are likely to be more arousing than familiar stimuli, yet the neurons respond to novel stimuli in the same way that they respond to the S- (Fig.1), while the same stimuli shown as familiar elicit responses similar to that of the S+; (3) RRNs do not respond to the delivery of the aversive saline when the monkeys make errors, nor are they particularly responsive to the delivery of juice or chewing food, which one would have thought is very arousing; (4) RRNs are not notably responsive during the manipulation of the trunk and limbs of the monkey by the experimenter, nor at the sight of his face. These tests are very arousing for monkeys, who may vocalize and lip smack; and (5) the neurons do not respond to loud noises which are not correlated with the delivery of reinforcement.

The neuronal responses reflect knowledge of the experimental contingencies

The differential responses of RRNs to any stimulus with a specific reinforcement value show that the basis of the neuronal responses is the learned reinforcement value and therefore the neuronal activity is not a pure expression of memory. The neuronal responses are better described as reflecting the rules or even the knowledge that the monkey has acquired in learning the contingencies of various behavioral tasks. Three examples of the knowledge reflected in the neuronal responses are discussed below.

Firstly, in the recognition memory task the responses to novel stimuli resemble those to the S-, although in practice the monkey is looking at these stimuli for the first time. Yet the neuronal activity encodes their reinforcement value within 250 ms. In contrast, the responses to familiar stimuli resemble those of the S+, yet the monkey is looking at these stimuli for the second time but has never been reinforced directly for responding to the stimulus. Clearly, the differential responses reflect the monkeys' recognition of the stimulus and the use of the abstract quality of familiarity to infer the reinforcement value of the stimuli. Thus, these neurons can be said to reflect a rule-based form of learning, as opposed to (but not excluding) a form of learning based on the development of associations that accrue between stimuli, responses and the learning-enhancing effect of the delivery of reinforcement.

A second example is the response of RRNs to the approach and withdrawal of the S+ syringe that contains juice. When the syringe approaches the monkey, RRNs respond with increases in firing rate, but withdrawal of the syringe results in a large decrease in firing, similar to the response elicited by the sight of an S- that contains saline, even though the S+ syringe is still in view. A simple explanation of this effect is that the monkeys have learned that the probability of juice decreases to zero when the syringe is withdrawn and thus the approach and withdrawal of the syringes are signals for the differential availability of the reinforcer. These tests also show that the neuronal responses are not locked to the stimulus itself, but presumably reflect the monkeys' cognizance of the contingencies in effect during the behavioral testing.

A third example is found in the responses to the S+ and the S- which are differentially reinforcing, but are also highly familiar. The neuronal and behavioral responses show that the monkeys ignore the familiarity of these stimuli as a cue for obtaining reinforcement, and instead use information about the fixed relationship that these two stimuli have with reinforcement; the converse is true when a novel or familiar stimulus is presented. Evidently these neuronal responses reflect a selection between the differing attributes and the relevant dimensions of the stimuli, such as their prior reinforcement history and their familiarity, which are stored in memory. Most importantly, the same stimulus can be interpreted to mean something different dependent upon the context (see also Richardson and DeLong, 1990). Thus, a stimulus that is judged to be novel is negatively reinforcing, whereas the same stimulus presented as familiar is positively reinforcing.

The functional significance of afferents to the Basal Forebrain

The retrograde tracing study indicated that paralimbic cortical structures located in the ventromedial temporal and frontal lobes project to the periventricular region of the basal forebrain. These same regions also project to the substantia innominata (Mesulam and Mufson, 1984; Russchen et al., 1985), although it remains to be shown that these corticofugal fibers synapse on basal forebrain neurons and their dendrites. The significance of these findings is threefold. Firstly, damage to the paralimbic cortices produces severe memory dysfunction in man (Damasio et al., 1985; Friedman and Allen, 1969; Gascon and Gilles, 1973). Lesion studies in monkeys have shown that these cortices - the perirhinal and entorhinal cortex - surrounding and extending anterior to the hippocampus play a role in recognition memory (Horel et al., 1987; Murray and Mishkin, 1986; Zola-Morgan et al., 1989). Neurophysiological studies also support this view, for neurons responsive to novel stimuli are found in the ventromedial temporal cortex (Brown et al., 1987; Riches et al., 1990; Wilson et al., 1988). The direct projections from the ventromedial temporal cortex to the substantia innominata could provide information from recognition memory to the basal forebrain, resulting in the differential neuronal responses to novel and familiar stimuli. It is also likely that the ventromedial prefrontal cortex is involved in such functions, as damage to this region also impairs the performance of object recognition tasks (Bachevalier and Mishkin, 1986; Voytko, 1985), and neurons in the orbitofrontal cortex respond differentially in visual discrimination tasks (Thorpe et al., 1983). In man, memory impairments due to frontal lobe damage take the form of a failure to utilize information that an individual has learned in order to organize goal-directed behavior (e.g. Stuss and Benson, 1986, chapters 11 and 12). Secondly, these paralimbic cortices are severely affected early in the course of Alzheimer's disease (Pearson et al., 1985). This cortical damage may be the basis of the prominent memory impairment observed in the disease, rather than the basal forebrain lesion.

Thirdly, these ventromedial regions of the prefrontal and temporal cortices receive the most intense cholinergic innervation of all cortical regions (Mesulam et al., 1986), providing an anatomical substrate for the basal forebrain influence on cortical mechanisms underlying memory function and the control of motivated behavior.

There were no retrogradely labeled cells in the hippocampus in the present study and as shown by Russchen et al. (1985) for the substantia innominata. This suggests that the functions of the hippocampus are not directly and solely responsible for the memory-related activity in the basal forebrain. This is consistent with the repeated demonstrations that ablations largely restricted to the hippocampus do not severely impair the ability of monkeys to perform recognition memory tasks (Mishkin, 1982; Murray and Mishkin 1984), and that the ventromedial temporal cortex plays a role in recognition memory (Horel et al., 1987; Murray and Mishkin, 1986; Zola-Morgan and Squire, 1989). Studies to date have searched and failed to find evidence for the neuronal encoding of familiarity in the hippocampus, although such neurons are found in the ventromedial paralimbic and association cortices surrounding the hippocampus (see above).

In addition, there are substantial inputs from diverse diencephalic, mesencephalic and rhombencephalic regions to the basal forebrain (Russchen et al., 1985; Mesulam and Mufson, 1984; present study). These regions are implicated in the control of feeding behavior and in the regulation of caloric and fluid intake. It is possible that pathways from these structures to the basal forebrain carry information about taste and current homeostatic states. Such information is necessary to establish the homeostatic need to obtain food, fluid, the monkey's state of satiation and therefore the desirability of particular reinforcers.

Comparisons of Basal Forebrain neuronal activity with other brain areas

Neurons with reinforcement-related activity have not been found in the amygdala, hippocampus, medial temporal cortex, inferior temporal cortex, areas 9 and 46 of prefrontal cortex, and the substantia nigra. These other structures contribute to generation of behavior to obtain reinforcement, but those contributions are specific to the individual structure, and differ from the basal forebrain. Thus, neurons in the amygdala respond to stimuli such as faces that monkeys will innately approach or avoid, particularly those stimuli that elicit emotional reactions (Leonard et al., 1985; Wilson and Rolls, 1990d); the hippocampus and dorsolateral prefrontal cortex are important for the short term representation of the spatial location of a stimulus or spatially-directed responses to these stimuli (Funahashi et al., 1989; Olton et al., 1979; Wilson and Goldman-Rakic, 1989; Wilson et al., 1990); and the perirhinal and medial inferotemporal cortices are important for recognition memory (Brown et al., 1987; Horel et al., 1987; Murray and Mishkin, 1986; Riches et al., 1990; Wilson et al., 1988; Zola-Morgan et al., 1989). In a study of the substantia nigra, primarily the pars compacta (A9), Schultz and Romo (1990) reported that putative dopaminergic neurons respond *equally* to sensory stimuli signalling the availability (go trial) of food or the absence (no-go trial) of food in a go/no-go task that resembles in many respects the visual discrimination task used in the present study. This contrasts with RRNs which respond *differentially* on go and no-go trials. A firm conclusion on the role of dopaminergic neurons and reinforcement will require an analysis of the A10 dopaminergic system, however.

Comparisons with other neurophysiological studies

In a series of experiments, Richardson and DeLong (1986, 1990) trained monkeys to perform a variety of tasks for water reinforcement. They have consistently found that most basal forebrain neurons are responsive in the tasks at the point at which a monkey has received all the information necessary for making a choice about a behavioral response that will result in the delivery of water. The elegant design of their experiments enabled them to conclude that the neuronal responses are not related to the sensory properties of the stimuli, nor to the motor responses, but occur when monkey makes a choice about a behavioral response. They have recently reported that 70% of all basal forebrain neurons respond during the choice, and 25% respond after the delivery of water (Richardson and DeLong, 1990). Studies in the rat show that tone cues can elicit responses in basal forebrain neurons when the cue precedes the delivery of reinforcement. It is possible that this neuronal activity reflects the learned reinforcement value of the tone, and clear that the neuronal activity is not related to movement, as reinforcement delivery did not require an operant response (Rigdon and Pirch, 1986).

How do these studies compare with the present experiments? There is reason to believe that our various studies agree well, despite differences in the experimental contingencies. Whereas Richardson and DeLong did not use stimuli signalling aversive reinforcement, we established contingencies in which appetitive and aversive reinforcers were delivered contingent upon specific behavioral responses. Under these conditions, in which the dimension of reinforcement is explicitly examined, we find that only 6% of all neurons are *differentially* responsive to the appetitive and aversive stimuli. However, 51% of all recorded neurons in our study respond at or preceding the point in the task when the monkey makes a choice resulting in the delivery of reinforcement, a proportion approaching that reported by Richardson and DeLong. It is probable that RRNs are a functional subset of all choice-related basal forebrain neurons recorded by ourselves, and by Richardson and DeLong. It remains to be shown what information is carried by the activity of the choice-related basal forebrain neurons, which in our experiments occurred when the monkey made a choice (go and no-go), but evidently did not encode the reinforcement value of the stimuli *per se*.

Richardson and DeLong conclude that their data does not support the possibility that the activity of basal forebrain neurons directly reflects memory. They reasoned that memory-related neuronal activity should occur in the delay period of their go/no-go task when information about the spatial location of a cue must be remembered by the monkey. Although 33% of basal forebrain neurons were responsive during the delay, their activity did not reflect the spatial information necessary to perform the task. Clearly, the performance of this task is not completely dependent upon the basal forebrain as the neuronal activity does not carry information about the spatial location of cues or the direction of spatial responses. However, the large proportion of 'delay-related' neurons is indicative of a contribution to task performance in the period in which the monkey must remember the location of the cue.

Our experiments concentrated on the analysis of the RRNs, in which information stored in memory - the novelty or familiarity of the stimuli - was linked to the dimension of aversive or appetitive reinforcement. Under these conditions it is clear that the resources of recognition memory influence basal forebrain neurons. However, the responses of RRNs reflect the learned reinforcement value of the stimuli and are based on, but are not a pure expression of memory, in agreement with Richardson and DeLong.

What is the function of task-related basal forebrain neuronal activity?

A satisfactory answer to this question requires information about the neurotransmitter released by basal forebrain neurons and the structures to which they project. There is limited evidence that RRNs and choice-related neurons project to the cerebral cortex and thus may be cholinergic (Richardson and DeLong, 1987; present study), a reasonable proposition as more than 70% of basal forebrain neurons are cholinergic (Mesulam et al., 1983). Grant and Aston-Jones (1986) have also recorded from cortically-projecting basal forebrain neurons, but behavioral correlates of these neurons were not apparent; the animal in their study was untrained, which raises the interesting possibility that basal forebrain neuronal activity occurs only when an animal is engaged in purposive, goal-directed behavior. An account of the functions of RRNs and choice-related neurons does not require the assumption that their activity is an index of cholinergic release in the cerebral cortex, as their behavioral correlates need interpretation irrespective of their neurochemical nature. However, the probable consequences of this activity are consistent with the known effects of ACh, as discussed below. Nevertheless, it should be stressed that the relationship between basal forebrain neuronal activity and the release of ACh is currently unknown.

It is important to note that RRNs and choice-related neurons are found throughout the basal forebrain, and thus the same signal may be broadcast to sensory, motor and association cortices simultaneously. The signal does not carry sensory, motor or information about the type (food or fluid) of reinforcer, but appears to reflect a state reached when monkeys are making or have made decisions about appropriate behavioral responses. Presumably this state has an enabling effect on the efferent target neurons, providing an optimizing function for diverse processes carried out in sensory, motor and association cortices during brief periods of time when rapid analysis of the significance of a stimulus and the appropriate response will be of maximum benefit. This hypothesis about function does not require the neuron to project to cortex or to release ACh. However, this facilitatory role is consistent with the known effects of ACh on cortical neurons. The action of ACh operates through several biophysical mechanisms to make neurons more responsive to their inputs over a time period of several hundreds of milliseconds (Brown, 1983), and this facilitatory role is apparent for both sensory and motor systems (Sillito and Kemp, 1983; Lamour et al., 1988). There are several lines of evidence indicating that cholinergic mechanisms facilitate the operation of motor systems: there are substantial basal forebrain projections to 'motor' regions of the cerebral cortex (Kievet and Kuypers, 1975) and brainstem (Semba et al., 1989); basal forebrain stimulation potentiates the occurrence of movement (Murphy and Gellhorn, 1945); and there is a high incidence of cholinoceptive neurons in layer 5 of somatomotor cortex projecting through the pyramidal tract (Lamour et al., 1988). Although the activity of RRNs is clearly dissociable from the occurrence of movement (see also Richardson and DeLong, 1990), it is evident that they are most active in situations in which movements are often required or would be useful in obtaining reinforcement. Similar arguments can be made for the functional facilitation of the sensory and associational cortices, although the data are less complete.

RRNs are phasically active at the time when decisions are being made about appropriate behavioral responses, and the latencies of these differential neuronal responses are short, occurring even before monkeys have been able to initiate eye movements in behavioral tasks (Wilson and Goldman-Rakic, 1990), indicating that pattern recognition can precede scanning eye movements. Thus, it seems reasonable to conclude that the output of these neurons may optimize the functions of efferent structures. On the assumption that these neurons are cholinergic, it may be that acetylcholine simply increases the efficiency of processing when it is most needed - when monkeys have to make behavioral decisions about a course of action. A corollary is that the basal forebrain is an auxiliary system that facilitates, but is not a prerequisite for goal-directed behavior as such, only for optimal *performance* of behavioral tasks, the learning of which may be facilitated by but is not dependent upon the basal forebrain. It follows that lesions of cholinergic neurons should not produce severe behavioral impairments, that animals will recover from such lesions, and that one might expect to see changes in the latencies in which lesioned animals make decisions, rather than an inability to make the decisions at all. This is consistent with the reduction in speed of information processing associated with patients with Alzheimer's disease (Flicker et al., 1985), with the effects of cholinergic 'blockade' (Kopelman and Corn, 1988), and with the effects of neurotoxic lesions of rat basal forebrain (Robbins et al., 1989b). A second corollary is that the phasic responses of RRNs and choice-related neurons suggest that the timing of the neuronal activity is an important feature of this system. If these neurons are cholinergic, it follows that the timing of the release of ACh is important, and therefore a therapy based on chronic administration of cholinergic agonists is unlikely to result in complete restoration of function.

CONCLUSION

Neurons in the basal forebrain respond to visual stimuli that, through learning, have become signals for the availability of reinforcement. The responses occur when monkeys are intelligently employing information stored in memory in accordance with current goals and rules used to perform behavioral tasks. These rules may vary, as in the recognition task, the discrimination task, or the delayed response task (Richardson and DeLong, 1986). On the basis of anatomical and lesion studies it appears that the basal forebrain is the recipient of information from neural mechanisms that allow monkeys to learn complex, rule-based relations about their behavior and sensory stimuli, and reflect the utilization of recognition memory when this is important for obtaining reinforcement. It is conceivable that damage to this region results in the disconnection of learning and memory mechanisms from one of their functional outputs. Basal forebrain damage may reduce performance dependent upon memory mechanisms, although as these mechanisms are likely to be diverse and distributed across many structures and with several outputs, a major deficit may not occur. In this conception, basal forebrain neurons are the recipients of learning and memory, and not of major importance for it. The cortical projections of basal forebrain neurons may provide a mechanism for the optimization of diverse functions in the sensory, motor and association cortices through the decoding of the reinforcement value of information from recognition and spatial memory mechanisms when it is relevant to the rules for performing a behavioral task. This architecture obviates the need for extensive connections between each and every functionally distinct region in the brain; for example, there is no need for the neural substrates of recognition memory to project directly to motor structures. Finally, there are other groups of basal forebrain neurons whose activity encodes the novelty or familiarity of visual stimuli (Rolls et al., 1982; Wilson and Rolls, 1990a), and these neurons may play a role in memory function or in the use of information from memory.

Acknowledgements

The present study has benefitted by the contributions of Colleen Phillips who assisted with the data collection; the laboratory facilities of Edmund Rolls, Oxford University; the financial support of the Medical Research Council (U.K.); and by the optimizing comments of Amy Arnsten, Yale University, and Steven Grant, University of Delaware.

REFERENCES

Aigner T., Mitchell S., Aggleton J.P., DeLong M., Struble R., Wenk G., Price D., and Mishkin M., 1984, Recognition deficit in monkeys following neurotoxic lesions of the basal forebrain, Soc. Neurosci. Abstr., 10: 116.11.

Bachevalier J., and Mishkin M., 1986, Visual recognition impairment follows ventromedial but not dorsolateral prefrontal lesions in monkeys. Behav. Brain Res., 20: 249.

Brown M.W., Wilson F.A.W., and Riches I.P., 1987, Neuronal evidence that inferomedial temporal cortex is more important than hippocampus in certain processes underlying recognition memory. Brain Res., 409: 158.

Brown D.A., 1983, Slow cholinergic excitation - a mechanism for increasing neuronal excitability, Trends in Neuroscience, 6:302.

Burton M.J., Rolls E.T., and Mora F., 1976, Effects of hunger on the responses of neurones in the lateral hypothalamus to the sight and taste of food, Exp. Neurol., 5: 668.

Damasio A.R., Graff-Radford N.R., Eslinger P.J., Damasio H., and Kassal N., 1985, Amnesia following basal forebrain lesions. Arch. Neurol., 42: 263.

DeLong M.R., 1971, Activity of pallidal neurons during movement, J. Neurophysiol. 34: 414.

Flicker C., Ferris S.H., Crook T., Bartus R.T., and Reisberg B., 1985, Cognitive function in normal aging and early dementia, in: "Senile Dementia of the Alzheimer type", Traber J., Gispen W.H., eds, Springer-Verlag, Berlin, pp. 269.

Friedman H.M., and Allen N., 1969, Chronic effects of complete limbic lobe destruction in man. Neurology, 19: 679.

Funahashi S., Bruce C.J., and Goldman-Rakic P.S., 1989, Mnemonic coding of visual space in the monkey's dorsolateral prefrontal cortex. J. Neurophysiol., 61: 331.

Gascon G.G., and Gilles F. , 1973, Limbic dementia. J. Neurol. Neurosurg. Psychiat., 36: 421.

Grant S.J., and Aston-Jones G., 1986, Discharge properties of cortically projecting nucleus basalis neurons in behaving animals. Soc Neurosci. Abstr., 12:158.7

Horel J.A., 1978, The neuroanatomy of amnesia: a critique of the hippocampal memory hypothesis. Brain, 101: 403.

Horel J.A., Pytko-Joiner D.E., Voytko M., and Salsbury K., 1987, The performance of visual tasks while segments of the inferotemporal cortex are suppressed by cold. Behav. Brain Res., 23: 29.

Kievet J., and Kuypers H.G.J.M., 1975, Basal forebrain and hypothalamic connections to prefrontal and parietal cortex in the rhesus monkey, Science, 187: 660.

Kopelman M.D., and Corn T.H., 1988, Cholinergic 'blockade' as a model for cholinergic depletion. Brain 111: 1079.

Lamour Y., Dutar P., Jobert A., and Dykes R.W., 1988, An iontophoretic study of single somatosensory neurons in rat granular cortex serving the limbs: a laminar analysis of glutamate and acetylcholine effects on receptive-field properties, J. Neurophysiol., 60: 725.

Leonard C.M., Rolls E.T., Wilson F.A.W. and Baylis G.C., 1985, Neurons in the amygdala of the monkey with responses selective for faces. Behav. Brain Res., 15: 159.

Mesulam M-M., Mufson E.J., Levey A.I., and Wainer B.H., 1983, Cholinergic innervation of cortex by the basal forebrain: cytochemistry and cortical connections of the septal area, diagonal band nuclei, nucleus basalis (substantia innominata) and hypothalamus in the rhesus monkey, J. Comp. Neurol., 214: 170.

Mesulam M-M., and Mufson E.J., 1984, Neural inputs into the nucleus basalis of the substantia innominata (CH4) in the rhesus monkey, Brain, 107: 257.

Mesulam M-M., Volicer L., Marquis J.K., Mufson E.J., and Green R.C., 1986, Systematic regional differences in the cholinergic innervation of the primate cerebral cortex: distribution of enzyme activities and some behavioral implications. Ann. Neurol., 19: 144.

Mishkin M., A memory system in the monkey, 1982, Phil. Trans. Roy. Soc. Lond., B298: 89.

Mitchell S.J., Richardson R.T, Baker F.H., and Delong M.R., 1987, The primate nucleus basalis of Meynert: neuronal activity related to a visuomotor tracking task, Exp. Brain Res., 68: 506.

Mora F., Rolls E.T., and Burton M.J., 1976, Modulation during learning of the responses of neurones in the lateral hypothalamus to the sight of food. Exp. Neurol., 53: 508.

Murphy J.P., and Gellhorn E., 1945, The influence of hypothalamic stimulation on cortically-induced movements and on action potentials of the cortex, J. Neurophysiol., 8: 339.

Murray E.A., and Mishkin M., 1984, Severe tactual as well as visual memory deficits follow combined removals of the amygdala and hippocampus in monkeys. J. Neurosci., 4: 2580

Murray E.A., and Mishkin M., 1986, Visual recognition in monkeys following rhinal cortical ablations combined with either amygdalectomy or hippocampectomy. J. Neurosci., 6: 1991.

Olton D.S., Becker J.T., and Handelman G.E., 1979, Hippocampus, space and memory. Behav. Brain Sciences 2: 313.

Pearson R.C.A., Esiri M.M., Hiorns R.W., Wilcock G.K., and Powell T.P.S., 1985, Anatomical correlates of the distribution of the pathological changes in the neocortex in Alzheimer's disease, Proc. Nat. Acad. Sci. U.S.A., 82: 4531.

Richardson R. T., and DeLong M.R., 1986, Nucleus basalis of Meynert neuronal activity during a delayed response task in monkey, Brain Res., 399: 364.

Richardson R. T., and DeLong M.R., 1987, Tonically active nucleus basalis neurons in the awake monkey project to cerebral cortex, Soc. Neurosci. Abstr., 13: 1027.

Richardson R. T., and DeLong M.R., 1990, Context dependent responses of primate nucleus basalis neurons during a go/no go task, J. Neurosci. 10: 2528.

Riches I.P., Wilson F.A.W. and Brown M.W., 1990, The effects of visual stimulation and memory on neurones of the primate hippocampal formation and the neighbouring parahippocampal gyrus and inferior temporal cortex of the primate (submitted for publication).

Ridley R.M., Baker H.F., Drewett B., and Johnson J.A., 1985, Effects of ibotenic acid lesions of the basal forebrain on serial reversal learning in marmosets, Psychopharmacol., 86: 438.

Rigdon G.C., and Pirch J.H., 1986, Nucleus basalis involvement in conditioned neuronal responses in the rat frontal cortex. J. Neurosci., 6: 2535.

Robbins T.W., Everitt B.J., Ryan C.N., Marston H.M., Jones G.H., and Page K.J., 1989a, Comparative effects of quisqualic and ibotenic acid-inducing lesions of the substantia innominata and globus pallidus on the acquisition of a conditional visual discrimination: differential effects on cholinergic mechanisms. Neuroscience, 28: 337.

Robbins T.W., Everitt B.J., Marston H.M., Wilkinson J., Jones G.H., and Page K.J., 1989b, Comparative effects of ibotenic acid and quisqualic acid-inducing lesions of the substantia innominata on attentional function in the rat: futher implications for the role of cholinergic neurons of the nucleus basalis in cognitive processes. Beh. Brain Res., 35: 221.

Roberts A.C., Robbins T.W., Everitt B.J., Jones G.H., Sirkia T.E., Wilkinson J., and Page K., 1990, The effects of excitotoxic lesions of the basal forebrain on the acquistion, retention and serial reversal of visual discrimination in marmosets. Neuroscience, 34: 311.

Rolls E.T., Burton M.J., and Mora F., 1976, Hypothalamic neuronal responses associated with the sight of food, Brain Res., 11: 53.

Rolls E.T., Perrett D.I., Caan A.W., and Wilson F.A.W., 1982, Neuronal responses related to visual recognition, Brain, 105: 611.

Russchen F.T., Amaral D.G., and Price J.L., 1985, The afferent connections of the substantia innominata in the monkey, macaca fascicularis, J. Comp. Neurol,. 242: 1.

Semba K., Reiner P.B., McGeer E.G., and Fibiger H.C., 1989, Brainstem projecting neurons in the rat basal forebrain: neurochemical, topographical and physiological distinctions from cortically projecting cholinergic neurons. Brain Res. Bull. 22: 501.

Schultz W., and Romo R., 1990, Dopamine neurons of the monkey midbrain: contingencies of responses to stimuli eliciting immediate behavioral reactions. J. Neurophysiol., 63; 607.

Sillito A.M., and Kemp J.A., 1983, Cholinergic modulation of the functional organisation of the cat visual cortex, Brain Res., 289: 143.

Stuss D.T., and Benson D.F., 1986, "The Frontal Lobes", Raven Press, New York.

Thorpe S.J., Rolls E.T., and Maddison S., 1983, The orbitofrontal cortex: neuronal activity in the behaving monkey, Exp. Brain Res., 49: 93.

Travis R.P., and Sparks D.L., 1969, Unitary responses and discrimination learning in the squirrel monkey: the globus pallidus, Physiol. Behav., 3: 187.

Voytko M.L., 1985, Cooling orbitofrontal cortex disrupts matching-to-sample and visual discrimination learning in monkeys. Physiol. Psychol. 13: 219.

Voytko M.L., Olton D.S., Richardson R.T., Wenk D.L. and Price J.L., 1990, Lack of memory impairment following basal forebrain lesions in monkeys. Soc. Neurosci. Abstr., 16: 258.9.

Whitehouse P.J., 1990, Pathology in basal forebrain in dementia: implication for treatment. The Neurotransmitter, 3: 3, Loyola University Chicago Medical Center.

Whitehouse P.J., Price A.W., Struble R.G., Clark A.W., Coyle J.T., and DeLong M.R., 1982, Alzheimer's disease and senile dementia: loss of neurons in the basal forebrain, Science, 215: 1237.

Whitlock D.G., and Nauta W.J.H. , 1956, Subcortical projections from the temporal neocortex in macaca mulatta. J. Comp. Neurol., 106: 184.

Wilson F.A.W., 1989, Cortical and subcortical structures involved in recognition memory: physiological and anatomical studies. Int. J. Neurol. 21-22.

Wilson F.A.W., M.W. Brown and I.P. Riches, 1988, Neuronal activity in the inferomedial temporal cortex compared to that in the hippocampal formation: implications for amnesia of medial temporal lobe origin. in: "Cellular Mechanisms of Conditioning and Behavioural Plasticity", C.D. Woody, D.L. Alkon and J.L. McGaugh, eds, Plenum Press, New York, pp. 313.

Wilson F.A.W., and Goldman-Rakic P.S., 1989, Effect of spatial and color cues on delay-related neuronal responses in prefrontal cortex. Soc. Neurosci. Abstr., 20: 33.5

Wilson F.A.W., and Goldman-Rakic P.S., 1990, Viewing preferences of rhesus monkeys related to memory for complex pictures, colours and faces (in preparation).

Wilson F.A.W., I.P. Riches, and M.W. Brown, 1990, Medial temporal neuronal activity related to behavioural responses during the performance of memory tasks by primates. Behav. Brain Res. (in press).

Wilson F.A.W., and Rolls E.T., 1990a, Neuronal responses related to novelty and familiarity of visual stimuli in the substantia innominata, diagonal band of Broca and periventricular region of the primate basal forebrain. Exp. Brain Res., 80: 104.

Wilson F.A.W., and Rolls E.T., 1990b, Neuronal responses related to reinforcement in the primate basal forebrain. Brain Res., 509: 213.

Wilson F.A.W., and Rolls E.T., 1990c, Learning and memory is reflected in the responses of reinforcement-related neurons in the primate basal forebrain. J. Neurosci., 10: 1254.

Wilson F.A.W., and Rolls E.T., 1990d, The primate amygdala and reinforcement: a dissociation between rule-based and associatively-mediated memory reflected in neuronal activity (submitted for publication).

Zola-Morgan S., Squire L.R., and Amaral D.G. (1989) Lesions of the amygdala that spare adjacent cortical regions do not impair memory or exacerbate the impairment following lesions of the hippocampal formation. J. Neurosci., 9: 1922.

THE CONTRIBUTION OF BASAL FOREBRAIN TO LIMBIC-MOTOR INTEGRATION

AND THE MEDIATION OF MOTIVATION TO ACTION

Gordon J. Mogenson and Charles R. Yang*

Department of Physiology
University of Western Ontario
London, Ontario
CANADA, N6A 5C1

* Neuroscience Unit
Loeb Research Institute
Ottawa Civic Hospital
1053 Carling Avenue
Ottawa, Ontario
CANADA, K1Y 4E9

INTRODUCTION

A major approach of our laboratory has been the use of electrophysiological recording techniques to investigate the effects on the electrical activity of neurons of the basal forebrain of electrical stimulation of the amygdala and hippocampus. As shown in Fig. 1, highly reliable electrophysiological responses to inputs from these two prominent limbic structures are excitation of accumbens neurons and inhibition of subpallidal neurons. Since the ventral striatum receives strong mesolimbic dopamine projections we have also investigated the effects on these electrophysiological responses of dopamine, either applied exogenously to accumbens neurons by micro-iontophoresis, or released endogenously from electrical stimulation of the ventral tegmental area of the midbrain. Dopamine has been shown to modulate the excitatory responses of accumbens neurons to stimulation of the amygdala and hippocampus and, in turn, to influence the electrophysiological responses of subpallidal neurons. The functional implications for limbic-motor integration of the interaction of dopamine inputs to the accumbens with inputs from amygdala and hippocampus have been investigated in complementary behavioral experiments. Before considering the results of our research in more detail some background is needed.

HISTORICAL PERSPECTIVE

The basal forebrain has gained prominence during the last ten or fifteen years, especially as increasing experimental and clinical evidence has suggested a key for this region in debilitating neurological and psychiatric disorders, e.g. Alzheimer's Disease or schizophrenia. The basal forebrain has become one of the most active and important areas of neuroscience research. Although a good deal was known twenty years ago

The Basal Forebrain, Edited by T.C. Napier *et al.*
Plenum Press, New York, 1991

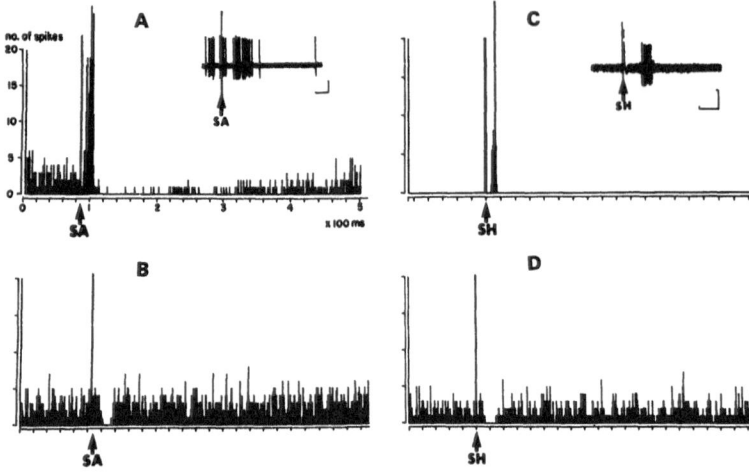

Fig. 1. Peristimulus time histograms showings the effects of single-
pulse stimulation of limbic sites (basolateral amygdala-SA,
ventral subiculum of hippocampus-SH) on the activity of neurons
in the basal forebrain. Stimulation of the basolateral amygdala
activated a neuron in the nucleus accumbens shown in A and
stimulation of the ventral subiculum activated a neuron in the
nucleus accumbens shown in C. Neurons in the subpallidal region
were inhibited by stimulation of the basolateral amygdala as
shown in B and by stimulation of the ventral subiculum of the
hippocampus as shown in D. Insets to A and C are oscilloscope
recordings. Scale in A is 200 μV and 5 ms and in C is 100 μV and
10 ms. (Based on Yim and Mogenson, 1982 and Yang and Mogenson,
1984).

about limbic influences on the hypothalamus and brainstem in the mediation
of autonomic and endocrine responses, little was known about the neural
mechanisms by which the amygdala and other limbic structures influence the
skeletomotor system and adaptive behaviours, such as food procurement or
predatory escape.

The classical experiments of Hess (1954) utilizing chronic electrical
stimulation techniques in cats provided evidence that the amygdala
contributes to "fight and flight" reactions and to the integration of
autonomic, endocrine and skeletomotor responses (Mogenson, 1987). The
anatomical demonstrations of neural projections from the amygdala to the
basal forebrain and, in particular, to the nucleus accumbens (Kelley, et
al., 1982; Krettek and Price, 1978) have provided an important lead for
us to undertake investigation of the neural mechanisms by which the
amygdala contributes to the motor components of "fight and flight"
reactions and of other adaptive behaviours (Fig. 2).

It was more than a decade ago that we began to investigate the functional role of the nucleus accumbens in integrating limbic inputs. In one
of our earlier series of experiments, it was demonstrated that the release
of endogenous dopamine via pharmacological blockade of an inhibitory
GABAergic input to the mesolimbic dopamine neurons in the ventral tegmental area (VTA)(Dahlstrom and Füxe, 1964) increased locomotor activity
in rats (Mogenson et al., 1979). This hypermotility from dopaminergic
activation of accumbens neurons was similar to that resulting from the
direct injection of exogenous dopamine into the nucleus accumbens (Jones
and Mogenson, 1980b; Jones et al., 1981; Pijnenberg and Van Rossum, 1973).

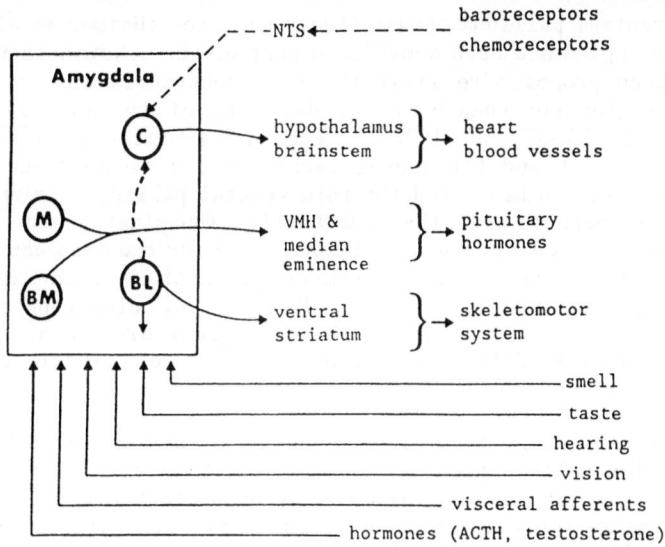

Fig. 2. The amygdala sends neural projections to the brain stem and ventral striatum as well as to the hypothalamus. These projections enable the amygdala to influence endocrine responses via hypothalamic projections, autonomic responses via hypothalamic and brainstem efferents, and skeletomotor responses via striatal and basal ganglia projections. Inputs to the amygdala include internal hormones. So the amygdala is in a position to contribute to higher order integrative activity for emotional expressionas illustrated by the classical fight and flight reaction elicited by electrical stimulation of the amygdala. Abbreviations: BL, basolateral nucleus; BM, basomedial nuclei; C, central nucleus; M, medial nucleus; NTS, nucleus tractus solitarius. (Based on Mogenson, 1987).

Recent studies utilizing anatomical, electrophysiological and behavioral techniques suggest that amygdala → accumbens → subpallidal projections contribute to limbic-motor integration (Mogenson, 1987; Mogenson et al., 1988; Cador et al., 1989; Yim and Mogenson, 1989). In a later section of this article, an overview of experimental evidence will show that another major limbic structure, the hippocampus, also contributes to behavioral response initiation via the hippocampal → accumbens → subpallidal pathways. Before considering this evidence it is necessary to provide more background about the basal forebrain.

THE BASAL FOREBRAIN FROM A FUNCTIONAL PERSPECTIVE

1. Chemical Neuroanatomy Led to Functional Analysis

Neuroanatomy serves as a prerequisite for neurophysiological investigation. So before undertaking electrophysiological studies of the basal forebrain it was necessary for us to review what was known from neuroanatomical studies. Research in my laboratory has been strongly influenced by the findings of Heimer and his colleagues.

A number of years ago, Heimer and Wilson (1975) drew attention to a rostral extension of the globus pallidus to a region ventral to the anterior commissure that receives strong neural projections from the accumbens (Nauta et al., 1978). In the years that followed the designation ventral pallidum became widely accepted (Heimer et al., 1982). Earlier, this region had been considered part of the substantia innominata but it has been proposed to limit the term substantia innominata to "a homogeneous region underneath the caudal part of the globus pallidus" (Switzer et al., 1982, p. 1902). For our functional studies using electrophysiological and behavioral techniques, in accordance with the proposal of Heimer, we have used the term ventral pallidum for the rostral subcommissural portion of the substantia innominata and the term subpallidal/sublenticular substantia innominata for the more caudal region (see Fig. 3) that also receives neural projections from the nucleus accumbens (Mogenson et al., 1983). Examples of electrophysiological recordings from these two basal forebrain regions are shown in Fig. 4. Note the distinctly different firing patterns of action potentials recorded from the two regions.

Recently this subpallidal region has been considered to be part of a complex region called "the extended amygdala" in view of the emerging anatomical data which demonstrated the cytoarchitectonic coextension of the subpallidal region with the centromedial amygdala (Alheid and Heimer, 1988; Grove, 1988a). In the following sections, however, we continue to address this region by the more familiar 'subpallidal/ sublenticular substantia innominata' as used in earlier publications.

The use of immunohistochemical techniques in recent years provided chemical neuroanatomical maps of the basal forebrain. It was observed that both the ventral pallidum and the sublenticular substantia innominata contain high concentrations of iron which serve as a reliable marker of these regions (Hill and Switzer, 1984). The subpallidal regions also contain high concentrations of GABA and glutamic acid decarboxylase (GAD), reflecting the dense innervation by GABA projections (Walaas and Fonnum, 1979). It has been suggested that "the presence of iron...is related, at least in part, to the utilization of GABA" (Hill, 1985, p.20). In addition, there are GABAergic cell bodies in the subpallidal region as well as GABA axon terminals (Brashear et al., 1986). The axons of these GABA neurons are components of subpallidal projections to the mediodorsal thalamus, the habenular nucleus, and possibly to the region of the pedun-

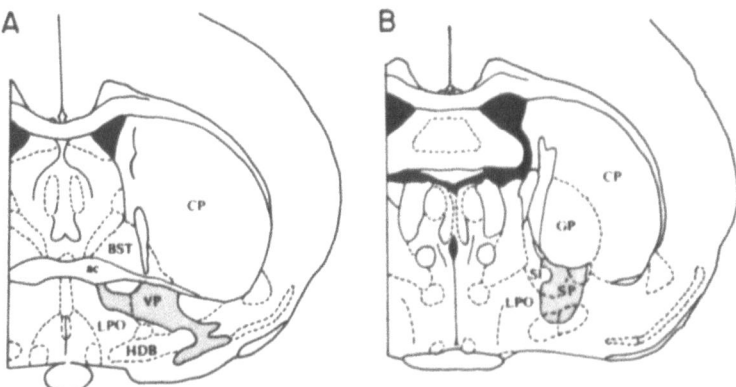

Fig. 3. A. The ventral pallidum (VP), as proposed by Heimer and coworkers in recent years (Heimer et al., 1982), is shown as the stippled region below the anterior commissure. B. A second subpallidal area (SP), which Heimer designated the sublenticular substantia in nominata, is shown as the stippled area ventral to the globus pallidus; A and B are modified from section 18 and 22 of the Paxinos and Watson (1986) atlas of the rat brain.

culopontine nucleus (Swanson et al., 1984; Zahm et al., 1985; Vives and Mogenson, 1986). Thus, a significant portion of the outputs from the basal forebrain are GABAergic. Besides GABA, compartmentalized ventral and subpallidal regions are richly innervated by a population of accumbens → subpallidal output neurons which are immunoreactive to substance P, enkephalins, neurotensin and their co-existence within the same neuron has also been described (Haber and Nauta, 1983; Zahm et al., 1985; Sugimoto and Mizuno, 1987; Zahm and Heimer, 1988). The functional significance of some subpallidal efferent projections are considered by Koob (1990, this book).

In recent years investigation of the role of the mesolimbic dopaminergic pathway from the VTA to the nucleus accumbens (Dahlstrom and Füxe, 1964) has extended our understanding of the accumbens outputs that convey dopamine-mediated signals. Direct interaction of dopamine synapses with accumbens GABAergic output neurons (Onteniente et al., 1987; Pickel et al., 1988; Yang and Mogenson, 1989) suggests the possible actions of dopamine on GABAergic accumbens outputs to the ventral pallidum and subpallidal region (Yang and Mogenson, 1989). The presence of a major cholinergic cell group (CH4) in the nucleus basalis which is intermingled with the subpallidal region (Divac, 1975; Lehmann et al., 1980; Mesulam et al., 1983; Zaborszky et al., 1984; Semba and Fibiger, 1988) has raised the possibility that dopamine-mediated signals are transmitted from the accumbens to cholinergic neurons of the subpallidal area (Zaborszky et al., 1986; Haber, 1987; Onteniente et al., 1987)(see Fig. 5). The major target of these ascending basal forebrain cholinergic neurons is the cortex, including the prefrontal cortex (Yang and Mogenson, 1990). These cholinergic projections of the basal forebrain are concerned with cognitive and other higher brain functions and their pathology is linked

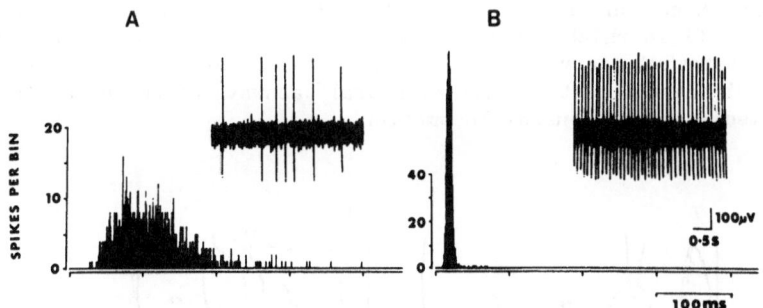

Fig. 4.   Inter-spike interval histograms recorded from ventral and subpallidal neurons. A: inter-spike interval histogram showing a skewed distribution of inter-spike intervals from a typical ventral pallidum neuron. The mean interval was 113.4 ms and the histogram was complied from 822 spikes. The inset shows a photographic record taken from the oscilloscope showing the typical slow and irregular firing pattern of the VP neuron B: inter-spike interval histogram showing action potentials recorded from the subpallidal/sublenticular substantia innominata as a comparison. The fast-firing neurons in this region have short but regular inter-spike intervals and thus the histogram has a tight and narrow distribution of the intervals. The mean interspike intervals was 20.1 ms and the histogram was complied from 797 spikes. The inset shows the photographic record of the same fast-firing neuron from the oscilloscope. The binwidth of the interspike histograms are 1 ms. (Modified from Yang and Mogenson, 1989).

to Alzheimer's disease (Coyle et al., 1983). It is possible that some of our electrophysiological recordings shown in Fig. 4 were from these cholinergic neurons of the subpallidal regions. However, this does not seem to be the case because a substantial number of neurons in the sublenticular substantia innominata and ventral pallidum were activated antidromically from the pedunculopontine nucleus (Swanson et al., 1984; Mogenson and Wu, 1986; Yang and Mogenson, 1987)(See section 2c).

2. Critical Evaluation of the Functional GABAergic Accumbens → Subpallidal Link In Limbic-Motor Integration

Our initial studies of the accumbens focused largely on endogenous release of dopamine from mesolimbic dopamine projections on locomotor activity (Mogenson et al., 1979), and on inputs from the amygdala (Mogenson and Yim, 1981; Yim and Mogenson, 1982; Yim, 1990, this book). The findings from these earlier studies have been extended to investigations of hippocampal inputs to the accumbens and the accumbens outputs to the subpallidal area which mediate locomotor activity.

The hippocampus contributes to the initiation of locomotor activity (Vanderwolf, 1971; Schacter et al., 1989), to exploration (Isaacson, 1982), and to the processing of memory of spatial information (O'Keefe and Nadel, 1978). The accumbens is located strategically to receive much of the excitatory glutamatergic hippocampal output (Kelley and Domesick, 1982; Christie et al., 1987; Groenewegen et al., 1987) and contributes to the integration of these limbic signals into motor behavioral acts, e.g. the hippocampus-dependent radial arm maze memory task (Schacter et al., 1989). We have conducted extensive electrophysiological and behavioral investigations of the mechanisms by which dopamine and hippocampal inputs interact to modify subpallidal functions. Locomotion seemed appropriate for the behavioral studies because it is a fundamental component for food and water procurement, exploration, predatory escape and other adaptive behaviors (Mogenson, 1987). Ambulatory locomotor activity was recorded in an open field which did not detect different components of movement. This measurement serves as a simple "behavioral assay" to assess quantitatively the contribution of neural pathways of the basal forebrain to the mediation of general locomotion.

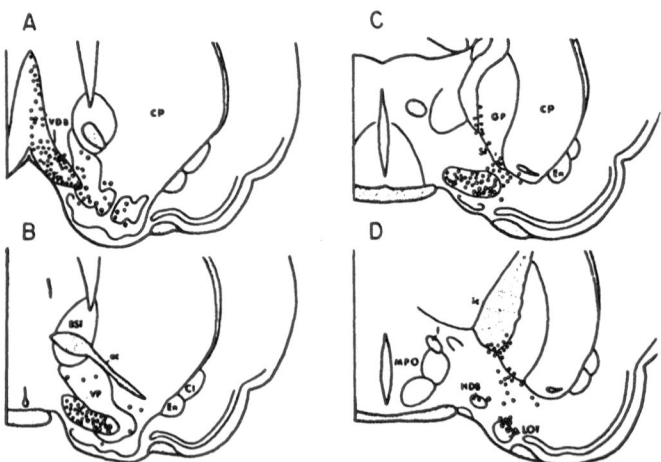

Fig. 5. Sections of the basal forebrain showing cholinergic and GABAergic neurons in the ventral pallidum (A and B) and in the sublenticular substantia innominata (C and D). (After Brashear et al., 1986).

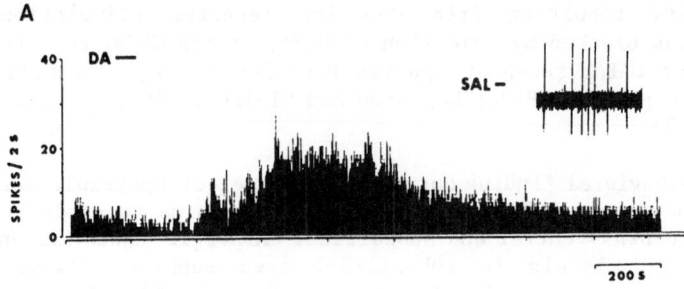

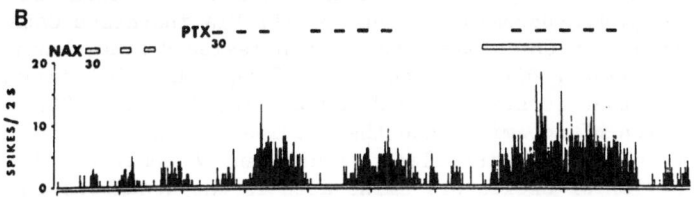

Fig. 6. Frequency-time histograms showing responses of neurons in the
ventral pallidum to A) dopamine injection into the accumbens; B)
direct iontophoretic application of naloxone (NAX) and picrotoxin
(PTX). When dopamine (DA) was microinjected into the accumbens
the discharge rate of this ventral pallidal neuron increased
substantially after a delay of 100 s. Inset shows the oscil-
lographic record of this neuron. B) In another ventral pallidal
neuron which previously responded to accumbens dopamine
injection, blockade of GABA receptor by direct iontophoretic
application of PTX produced a greater increase in the firing rate
than blockade of opiate receptor by direct iontophoretic
application of naloxone (NAX). When both GABA and opiate
receptors were blocked, there was a further increase in firing
rate beyond that produced by PTX or NAX alone. Solid lines are
drug applications (After Yang and Mogenson, 1989).

Influence of dopaminergic input to the accumbens on ventral pallidal function

The converging mesolimbic dopaminergic neurons from the VTA to the
accumbens (Totterdell and Smith, 1989; Yang and Mogenson, 1984) appear to
contribute to limbic-motor integrative mechanisms (Mogenson, 1987;
Mogenson et al., 1988). Electrophysiological recording obtained in the
accumbens have demonstrated several effects of dopamine: (i) inhibition
of the excitatory transmission of hippocampal-accumbens neurons presynaptically (Mogenson et al., 1988; Yang and Mogenson, 1984; 1986); (ii)
suppression of the excitatory response of accumbens neurons to iontophoretic application of glutamate post-synaptically (White and Wang
1985); and (iii) enhancement of the firing rate of the slow, randomly
discharging, ventral pallidal neurons via action on post-synaptic dopamine
receptors located in the accumbens (Fig. 6). It appears that dopamine
in the accumbens also inhibits the GABAergic accumbens output neurons to
the ventral pallidum, thus removing the inhibitory influence that GABA
exerts on these neurons and triggers locomotor activity via ventral
pallidal output pathways (Yang and Mogenson, 1989). Hence, dopamine can
either limit excitatory hippocampal-accumbens transmission, or initiate
activity in presumed motor-related accumbens-ventral pallidal neurons.

Direct stimulation of dopamine receptors in the accumbens triggers
a prolonged episode of increased ambulatory locomotor activity (Pijnenberg
and van Rossum, 1973; Jones et al., 1981; Jones and Mogenson, 1980b). The

hypermotility resulting from dopamine receptor stimulation in the accumbens was blocked by injection of GABA, or its GABA$_A$ receptor agonist muscimol, or GABA$_B$ receptor agonist baclofen into the subpallidal site (Jones and Mogenson, 1980a; Mogenson and Nielsen, 1983; Patel and Slater, 1988)(Fig. 7).

This behavioral finding corroborates our electrophysiological results (Yang and Mogenson, 1989), suggesting that dopamine receptor stimulation reduces accumbens-ventral and subpallidal GABAergic transmission and that increased GABA levels in subpallidal area suppress transmission of accumbens dopaminergic signals through the subpallidal region. This interpretation is supported by neurochemical evidence showing that unilateral 6-hydroxydopamine lesions of the VTA increased GABA levels in the subpallidal region compared to the non-lesioned side, but neurochemical confirmation of a direct association of dopamine inhibiting accumbens-subpallidal output is still needed (Austin and Kalivas, 1987). Furthermore, injection of dopamine into the accumbens in a dose subthreshold to that which induced hypermotility significantly enhanced the dopamine-dependent action of excitatory amino acid in the accumbens in triggering an increase in locomotor activity (Boldry and Uretsky, 1988). This provides an example of an additional modulatory interaction of dopamine and glutamate in the accumbens.

```
DA → - ACCUMBENS → (↓ GABA IN VP & SP?) ---------→ ↑ LOCOMOTION.
                    [for VP and SP see Fig.3]
```

Influence of glutamatergic hippocampal input to the accumbens on subpallidal functions

The glutamatergic hippocampal-accumbens pathway has been characterized anatomically and electrophysiologically (Groenewegen et al., 1987; Yang and Mogenson, 1984; 1985). More recently, this pathway was shown to be involved in the hippocampal-initiated locomotor activity that requires spatial orientation. Bilateral injection of kynurenic acid, a glutamate antagonist, into the accumbens in food-deprived rats significantly increased the latency of the hippocampal-initiation of movement from the central platform toward the baited arms of an eight-arm radial maze. In addition, rats with glutamate receptor blockade in the accumbens also made more 'reference memory' errors, i.e. by visiting unbaited arms in the partially baited eight-arm radial maze (Schacter et al.,1989)(Fig. 8).

This glutamatergic hippocampal-accumbens pathway may thus be the key link for conveying processed hippocampal signals which code for the spatial environment and relay them to the motor effector sites in the ventral and subpallidal areas.

In contrast to ventral pallidal neurons recorded in the subcommissural area, the more caudal subpallidal neurons have a typically fast discharge rate (>75 Hz) (Yang and Mogenson, 1985; 1987; 1989), as shown earlier in Fig. 4. The major extrinsic source(s) of afferent inputs which entrain the fast firing rate of the subpallidal neurons has not been identified, and the possibility of intrinsic mechanisms which generate this fast frequency must also be considered. Electrical stimulation of accumbens (Jones and Mogenson, 1980a), or an excitatory hippocampal input to the accumbens (Yang and Mogenson, 1985), result in an abrupt cessation of firing of subpallidal neurons (see Fig. 9) and this inhibition is (i) antagonized by iontophoretic application of GABA antagonist picrotoxin on subpallidal neurons (Jones and Mogenson, 1980a); and (ii) attenuated by microinjection of glutamate acid diethyl ester (GDEE), a non-selective

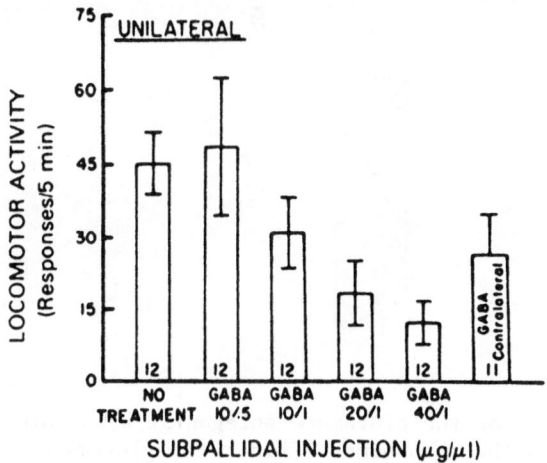

Fig. 7. Open-field locomotor hyperactivity elicited by dopamine injection into the accumbens is mediated via the subpallidal area since the dopamine-induced hypermotility was attenuated dose-dependently by GABA injections into the subpallidal area. GABA was injected into the subpallidal region ipsilateral to the dopamine (20 μg/0.5 μl) injection into the accumbens. As a control, GABA (40 μg/μl) was also injected into the subpallidal region contralateral to the site of the injection of dopamine into the accumbens (Modified from Mogenson and Nielsen, 1983).

glutamate antagonist, into the nucleus accumbens (Yang and Mogenson, 1985). Since ventral pallidal neurons fire very slowly (see Fig. 4A), it is difficult to demonstrate electrophysiologically that stimulation of the glutamatergic hippocampal input to the accumbens actually activates the GABAergic accumbens output to ventral pallidal neurons, i.e. inhibition of an already slow firing neuron. Moreover, it is not known whether this population of accumbens-ventral pallidal neurons mediates the hippocampal-induced behaviour. Nevertheless, it is reasonable to suggest that hippocampal input to the accumbens is glutamatergic and accumbens output to the ventral or subpallidal areas is GABAergic.

The accumbens output neurons that receive hippocampal inputs project disproportionately to the ventral pallidum and subpallidal area. Over 4 times as many accumbens neurons, activated by hippocampal stimulation, were activated antidromically by stimulation of the ventral pallidum than to stimulation of the subpallidal area. Thus hippocampal inputs have a preferential input onto accumbens output neurons which project monosynaptically to the ventral pallidum (Yang and Mogenson, 1985).

Findings from behavioral studies have shown that this GABAergic accumbens-subpallidal projection links limbic inputs and expresses them in movements. Increased locomotion induced by chemical stimulation of the glutamatergic hippocampal outputs to the accumbens (Yang and Mogenson, 1987), or novelty-induced exploration in an open field (Mogenson and Nielsen, 1984a) was attenuated by injection of nipecotic acid, a GABA uptake inhibitor, or GABA, into the subpallidal area (Yang and Mogenson, 1987, Mogenson and Nielsen, 1984b).

Based on the additional findings that procaine injection into the subpallidal area significantly attenuated the hypermotility induced by

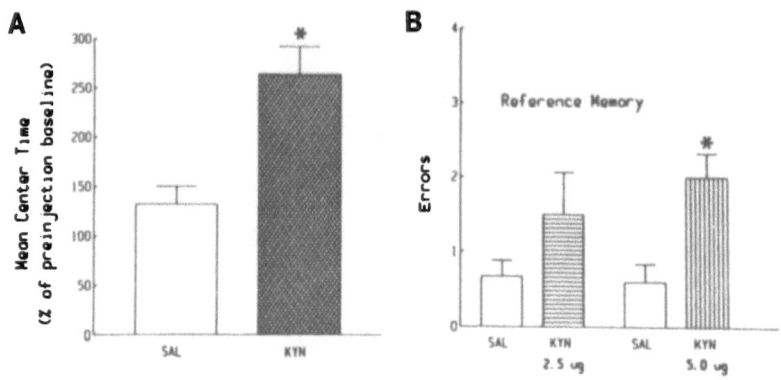

Fig. 8. Injection of the glutamate antagonist kynurenic acid (5 μg/0.2 μl) into the accumbens to block the glutamatergic hippocampal-accumbens pathway resulted in: A) a delayed initiation of the hippocampus-mediated locomotor activity from the central platform to the baited arms of an eight-arm radial maze. B) Rats with kynurenic acid injection into the accumbens also made significantly more 'reference memory' errors i.e. errors made due to entering the unbaited arms of a four-arm baited eight-arm radial maze. *p<0.005 (Modified from Schacter et al., 1989).

stimulation of the glutamatergic hippocampal-accumbens pathway (Yang and Mogenson, 1987), it appears likely that the subpallidal area is required for transmission of hippocampal-mediated locomotion. Nevertheless, neurochemical data showing release of GABA from accumbens-subpallidal pathway during the hippocampus-induced locomotion is still not available. From the existing evidence, however, one is tempted to generalize by stating that activation of accumbens output by excitatory glutamatergic hippocampal inputs turns subpallidal neurons 'OFF' while inhibition of the same pathway by dopamine turns subpallidal neurons 'ON' electrophysiologically. Nonetheless, in both conditions, there is an overall increase of locomotor activity in an open field.

HIPPOCAMPUS → + ACCUMBENS → (↑GABA IN SP?)-------→↑ LOCOMOTION.

Taken together, these studies show that, although neurons in the ventral pallidal and subpallidal areas are both involved in the accumbens-initiated locomotor activity, the cellular mechanisms which mediate the locomotion are functionally heterogenous. For instance, if hippocampal-initiated increase in locomotor activity was associated with an <u>activation</u> of accumbens → subpallidal GABAergic neurons, one would expect that a potentiation of GABAergic transmission in the SP should potentiate the increased locomotor activity. Instead, an increase in GABA availability by injecting nipecotic acid, a GABA uptake inhibitor, into the SP actually <u>suppressed</u> the hippocampal-initiated locomotion (Fig. 10A). What is not known, however, is whether hippocampal-initiated locomotion through the accumbens only activated GABAergic accumbens output neurons alone. As mentioned above, accumbens → subpallidal neurons also contain neurotensin, substance P and enkephalins. There are several questions which still need to be answered:

(i) Are any of these peptidergic neurons activated independently or together with the GABAergic output neurons by the glutamatergic hippocampal input to the accumbens?

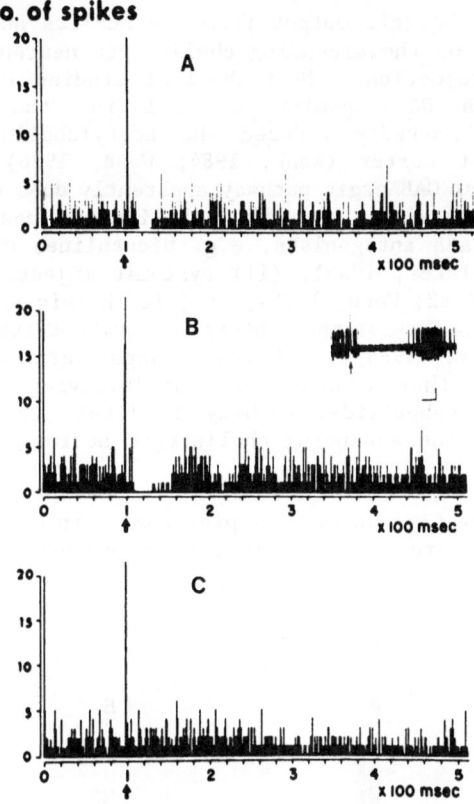

no. of spikes

Fig. 9. Peristimulus time histograms showing inhibitory responses of
neurons in the subpallidal area to single-pulse stimulation of:
A) ventral subiculum of the hippocampus, B) medial aspect of the
nucleus accumbens. Inset in B is an oscilloscope record of the
inhibitory response. The peristimulus time histogram in C is a
comparison in which a subpallidal neuron did not respond to
single-pulse stimulation of the dorsolateral caudate nucleus
(After Yang and Mogenson, 1985 in A and after Mogenson et al.,
1983 in B and C). Calibrations in B are 1 mV and 20 ms

(ii) Do the neuropeptides interact with GABAergic neurons presynatically
and/or postsynaptically in the subpallidal area to trigger locomotion (see
Austin and Kalivas, 1990)?

Undoubtedly, future studies that evaluate separate populations of ventral
pallidal and subpallidal outputs which mediate locomotion initiated by
hippocampal stimulation, or dopamine receptor activation in the accumbens,
may clarify these discrepancies.

Do accumbens-subpallidal GABAergic neurons require ascending cholinergic neurons from the adjacent nucleus basalis to express its functions?

The subpallidal region is intermingled with the nucleus basalis of
which the cortical-projecting CH4 cholinergic cell group is the principal
site of lesion in Alzheimer Disease (Coyle et al., 1983; Lehman et al.,
1980; Mesulam et al., 1983). Chronic extracellular single unit recordings
in the nucleus basalis of behaving monkeys showed that the spontaneous
firing rate is regular, and changes in the firing rate are associated with
water drinking reward or motor movement that is linked to a reward
(Richardson and DeLong, 1988).

In rats, the GABAergic output from the nucleus accumbens exerts a powerful inhibition on the ascending cholinergic neurons of the nucleus basalis-cortical projection. Neurochemical studies demonstrated that microinjection of the GABA agonist, muscimol, into the nucleus basalis/subpallidal region markedly reduced the acetylcholine turnover rate measured in the rat cortex (Wenk, 1984; Wood, 1986). However, the accumbens-subpallidal GABAergic pathway apparently does not exert a tonic inhibitory effect on these ascending cholinergic neurons since: (i) microinjection of GABA antagonists, e.g. bicuculine, directly into the subpallidal area (Blaker, 1985), (ii) systemic injection of picrotoxin (Wood and Richard, 1982; Wood, 1986), or (iii) kainic acid lesion of the accumbens where the source of the GABAergic accumbens-subpallidal neurons are located (Blaker, 1985), did not change cortical acetylcholine turnover. It appears that locomotor activity involving a tonically active GABAergic accumbens-subpallidal pathway is likely to be mediated via pathways other than the ascending cholinergic neurons from the nucleus basalis in rats.

Injection of the GABA antagonist picrotoxin, in doses as low as 50ng into the subpallidal area, on the other hand, produced a dose-dependent

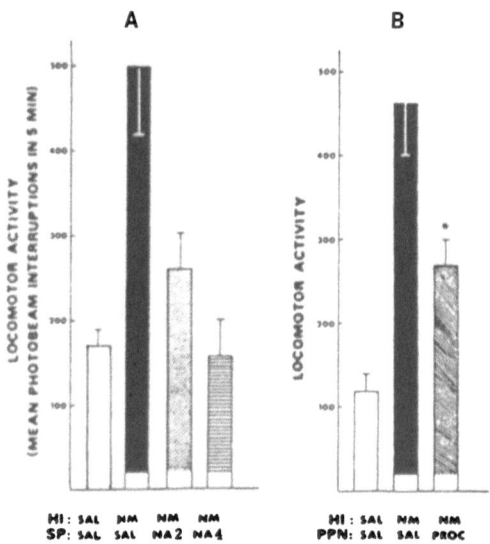

Fig. 10. Locomotor activity was increased by more than three-fold when N-Methyl-D-Aspartic Acid (NM, 0.5 μg) was injected bilaterally into the ventral subiculum of the hippocampus of rats. In A the NMDA-elicited locomotion was reduced significantly when the subpallidal area was treated bilaterally with nipecotic acid (NA, 2 and 4 μg). In B the NMDA-elicited locomotion was reduced significantly when procaine (PROC, 20% W/V, 0.2 μl), a reversible neural blocker, was injected bilaterally into the pedunculopontine nucleus. These observations complement the results of a number of studies which suggest that neural projections from hippocampus, to nucleus accumbens, to subpallidal region to pedunculopontine nucleus contribute to locomotor activity and, thereby, to behaviors such as exploration and food procurement in which locomotion is an important component. Abbreviations: HI, hippocampus; SAL, saline; SP, subpallidal region; PPN, pedunculopontine nucleus. (Modified from Yang and Mogenson, 1987).

278

increase of rat spontaneous ambulatory locomotor activity which was sometimes associated with galloping in the open field (Austin and Kalivas, 1990; Mogenson and Nielsen, 1983; Scheel-Kruger, 1986). Scheel-Kruger (1986) has also shown that a variety of behavioral stimulations, in addition to hypermotility, was induced by injection of higher doses of picrotoxin (250 ng) into the ventral pallidum. There is some topographical differentiation of the picrotoxin injection sites which induce behavioral changes. Thus, injection of picrotoxin into the ventral pallidum induced hypermotility, rearing, sniffing, episodic licking/biting activity. Picrotoxin injection to the lateral portion of subcommissure ventral pallidum induced behavioural responses resembling those following an apomorphine challenge and consisted of continuous stereotyped locomotion, accompanied by rearing, sniffing and licking/biting activities. We have found that the hyperactivity induced by subpallidal injections of picrotoxin was markedly attenuated by microinjection of procaine into an area called pedunculopotine nucleus within the "mesencephalic locomotor region" in the brainstem (Mogenson and Wu, 1988a; also see next section). These results indicate the possibility that suppression of a tonic GABAergic inhibitory accumbens-subpallidal pathway which synapses on a descending pathway initiates hypermotility, e.g. from dopamine receptor activation in the accumbens.

3. Descending Limbic Signals To The Mesencephalic Locomotor Region By Way Of The Subpallidal Area

Animals deprived of cerebral cortex can perform adaptive behaviours such as eating, drinking, mating and nursing of young (Bjursten et al., 1976; Grillner, 1985) although some movements might be contextually inappropriate. This interesting finding has prompted us to consider the possibility of limbic signals integrated into behavioral acts through basal forebrain circuits that descend to the brainstem motor effector sites without cortical execution. In 1966, Shik and co-workers demonstrated that in cats with precollicular-postmammilary transections, electrical stimulation of a brainstem area called 'mesencephalic locomotor region' (MLR) using low amplitude current pulses (20-60 μA) at high frequency (20-60 Hz), triggered rhythmic limbic movements on a treadmill (Grillner, 1985; Shik et al., 1966). In the similar brainstem transected cats, rhythmically firing single units were located in the pedunculopontine nucleus. Using spike-triggered averaging recording, the firing of 23% of these units during spontaneous treadmill locomotion was shown to correlate with one or more limb electromyographic recordings (post-spike mean onset of EMG $\approx$7ms) (Garcia-Rill et al., 1983). It is likely that PPN may serve as the 'final common pathway' towards which signals from the basal forebrain converge before descending to the spinal motor circuits.

In the rat, the MLR has been located in a loosely defined region of the brainstem called pedunculopontine nucleus (Skinner and Garcia-Rill, 1984; Garcia-Rill, 1986). The PPN has reciprocal connections with major outputs sites of the basal ganglia, e.g. substantia nigra pars reticulata, endopendunculous nucleus and subthalamic nucleus, as well as the cortex. In turn, efferents from PPN descend to the mesencephalic trigeminal nucleus and reticularis gigantocellularis nucleus that mediate stepping and locomotion via their relays to the spinal cord (Garcia-Rill, 1986; Moon-Edley and Graybiel, 1983; Shefchyk et al., 1984). In the PPN the distribution of a compact group of cholinergic neurons was found to overlap with the stimulation sites where locomotion was elicited (Garcia-Rill, 1986). Although limbic and cortical projections to the PPN probably determine the final locomotor response by initiating, further increasing locomotion, or decreasing and stopping it altogether, activity of the PPN of the MLR probably determines the pattern of locomotion itself (Brudzynski and Mogenson, 1986). Thus, the MLR must not be treated merely as a relay site, for it is also an important integrative structure for locomotion.

By combining axonal tracing and electrophysiological mapping techniques, it has been shown that the PPN also receives a substantial direct input from the subpallidal areas (Grove, 1988b; Swanson et al., 1984). With intact brainstem in anaesthetized rat, the PPN is generally a 'quiet' area electrophysiologically, as shown by its slow irregular spontaneous discharge rate (8.3 ± 0.7 Hz) (Brudzynski and Mogenson, 1986). However, when the ventral pallidum is stimulated electrically, short-latency (10 ms) bursts of excitatory responses were recorded in 80% of the PPN neurons (Swanson et al., 1984). On the other hand, stimulation of the subpallidal areas evoked both short-latency excitatory and inhibitory responses in the PPN (Swanson et al.,1984). In both cases, the transmitters in these pallidal-PPN pathways which mediate these responses are currently unknown.

The subpallidal-PPN pathway forms a serial link that conveys output signals from the nucleus accumbens to the spinal motor mechanisms. Two major inputs to the nucleus accumbens have been investigated:

Transmission of accumbens dopaminergic signals to the PPN by way of the accumbens-subpallidal pathway

Stimulation of dopamine receptors in the accumbens via microinjection of amphetamine into the accumbens (to release endogenous dopamine from the nerve terminals of the VTA dopaminergic projection) significantly increased the firing rate of the PPN neurons (Brudzynski and Mogenson, 1986). This was correlated with a 2.5 fold increase in locomotor activity in an open-field (Fig. 11A). In addition, this dopamine-induced hypermotility can be attenuated markedly by procaine injection into the PPN (Brudzynski and Mogenson, 1985), substantiating its functional importance in conveying dopaminergic signals from the accumbens (Fig. 11B).

Transmission of glutamatergic hippocampal signals from the accumbens to the PPN by way of the accumbens-subpallidal pathway

Increase in open-field locomotor activity was also induced by NMDA injection into the ventral subiculum of the hippocampus, the origin of the hippocampal-accumbens pathway (Yang and Mogenson, 1987). The hypermotility was shown to be elicited via the accumbens → subpallidal neurons to the pedunculopontine nucleus, since injection of procaine in the supallidal area or the PPN significantly attenuated the hypermotility triggered by hippocampal stimulation (Fig. 10B). This experimental evidence thus strongly indicate that converging hippocampal and dopaminergic signals to the accumbens are relayed via the subpallidal area and descend to the PPN.

4. Parallel projections for accumbens outputs ascending through the subpallidal-mediodorsal thalamus projection

Apart from conveying accumbens output through serial linkages to the brainstem motor effector sites, accumbens output signals to the ventral pallidum may also ascend in parallel to the mediodorsal thalamus (MD)(Groenewegen, 1988; Mogenson et al., 1987; Young et al., 1984). Experimental evidence has strongly implicated that MD, as a major relay site for limbic signals, is necessary for the maintenance of the flow of memory-related processes, perhaps via its reciprocal connection to the prefrontal cortex which is also known to handle memory in sequential order (Beckstead et al., 1979; Fuster, 1989; Krettek and Price, 1978).

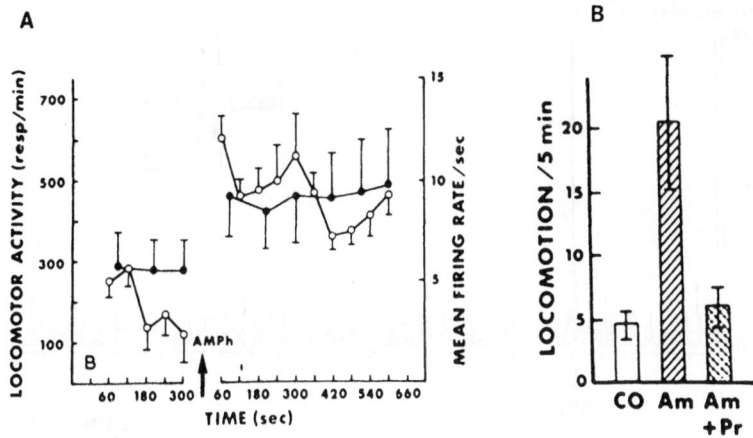

Fig. 11. Hypermotility initiated by amphetamine injection into the accumbens was mediated via PPN in the brainstem. A). Amphetamine injection (20 μg/0.2 μl) into the nucleus accumbens increased the firing rate of PPN neurons (filled circles), parallel to the increase in locomotion following the same injection into the accumbens (open circles). Both locomotion and firing rate were measured 5 min before and 10 min after the injection (arrow= 0 s). B). The increase in locomotor activity from amphetamine injection in the accumbens was markedly reduced to control level following procaine (Pr) injection (30 μg/0.2 μl) into the PPN. CO (open bar)= saline control. Am (hedged bar)= amphetamine injection into the accumbens and procaine injection into the PPN (Modified from Brudzynski and Mogenson, 1985; 1986).

Neurological symptoms such as memory loss, retrograde amnesia, temporal-spatial disorientation, confabulation and false recognition are hallmarks of Korsakoff Syndrome patients who have MD lesions (Victor, et al., 1971). This has reinforced the notion of MD's involvement in memory processes (Markowitsch, 1982). Heimer and associates have added that the ventral striatal → ventral pallidal → MD projections may have a role "...in initiating movements in response to emotionally or motivationally powerful stimuli" (Heimer et al., 1982; page 87). These speculations, based on neuroanatomical observations, have prompted us to examine functional evidence from neurophysiological and behavioral studies of the subpallidal → MD projections.

The majority of MD neurons (45%) fire in single spikes, but alternate with rapid bursting periods. Single-pulse electrical stimulation of the ventral pallidum and subpallidal region inhibited about 40% of MD neurons (see Fig. 12). Iontophoretic application of picrotoxin enhanced the basal firing rate, but the inhibitory response of the MD neuron to ventral- or sub-pallidal stimulation remained unchanged. It appears that the pallidal-MD pathway is not entirely GABAergic. The lack of immunohistochemical evidence of a GABAergic ventral pallidal-MD pathway supports the electrophysiological data (Vives and Mogenson, 1985; Young et al., 1984).

The number of MD neurons inhibited by stimulation of the subpallidal area exceeded slightly the number inhibited by stimulation of the subcommissural ventral pallidum (95% versus 85%). This is noteworthy since substantially fewer subpallidal neurons were activated antidromically by MD stimulation than VP neurons activated antidromically by MD stimulation (Fig. 13). Also the retrograde transport of horseradish per-

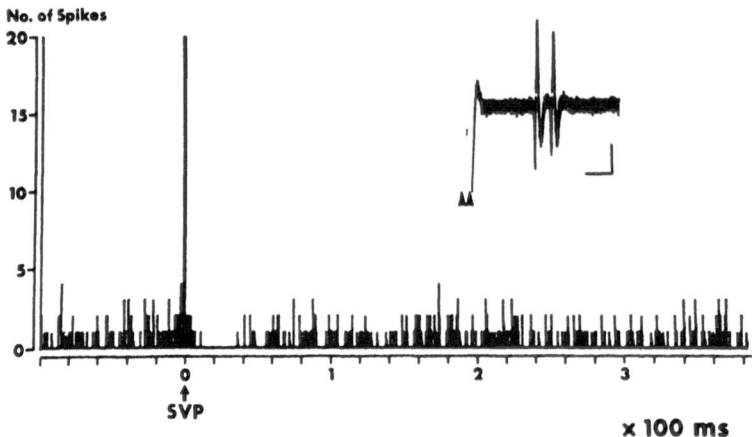

Fig. 12. This peristimulus time histogram shows the action potentials recorded from a neuron in the mediodorsal thalamus (MD) of a rat anaesthetized with urethane. Single pulse stimulation of ventral pallidum (SVP) inhibited this MD neuron with an onset latency of 8 ms. In a series of over 200 MD neurons 26 were activated antidromically by stimulation of the medial prefrontal cortex (MPC). The criteria for antidromic activation were cancellation by collision with spontaneous action potentials and the following of twin stimulus pulses delivered at 200 Hz, as shown in inset (Scale=100 µV, 5 ms). Twenty-four of these 26 antidromically activated MD neurons were inhibited by single-pulse stimulation of the SVP (Modified from Vives and Mogenson, 1985).

oxidase from the MD labelled more neurons in the VP than in the sub-lenticular subpallidal area (as shown in Fig. 4 of Mogenson et al., 1987). These results suggest that some of the inhibitory responses of MD neurons to stimulation of the sublenticular subpallidal area may not be monosynaptic.

MD output neurons project to the medial prefrontal cortex (MPC), which in turn, send reciprocal connections back to the MD (Beckstead et al., 1979; Krettek and Price, 1978). Output neurons from the MD to the MPC were identified electrophysiologically by their antidromic responses to stimulation of the MPC (see inset to Fig. 12). Ninety-two percent of these MD-MPC neurons were inhibited by stimulation of the VP, suggesting that MD output neurons to the MPC are influenced by an inhibitory input from the VP. Since MPC project to the nucleus accumbens (Christie et al., 1985), this completes a circuit linking accumbens output, ventral and subpallidal areas, to the MD and MPC and back to the accumbens. This may be part of a circuit which served as a "...differentiated 'return loop', enabling the limbic system to monitor the effects of its outputs to the striatum and thus possibly acting as an aid in adjusting the organism's motivational set to the central somatic-motor system" (Haber et al., 1985).

```
|                             MPC  ←- - - -→  MD               |
|                              ↓               ↑               |
| LIMBIC (HIPP, VTA, AMY) → ACCUMBENS  →  VP/SP → PPN |
```

The contribution of the subpallidal → MD projections to behaviour was studied in two series of experiments. Locomotor activity was enhanced by injecting picrotoxin, a GABA antagonist, into the subpallidal region. As

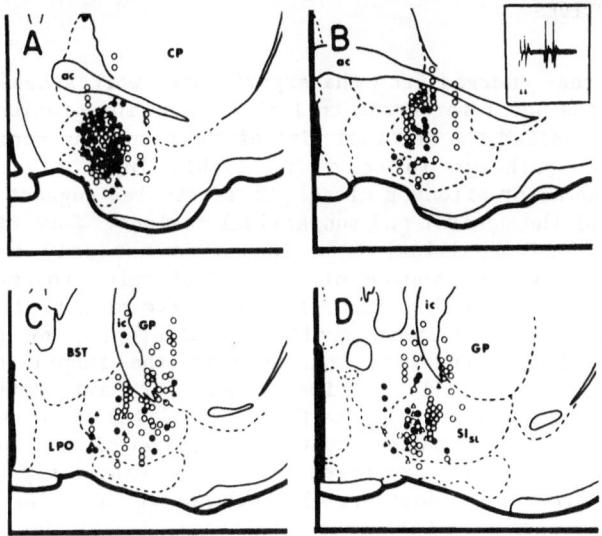

Fig. 13. Hemisections of the basal forebrain of the rat show the recording
sites of neurons tested for antidromic activation to single pulse
stimulation of the mediodorsal thalamus (MD). A and B are at the
level of the ventral pallidum (subcommissural substantia innominata) and C and D are at the level of the subpallidal area.
The criteria for antidromic activation were cancellation by
collision with spontaneous action potentials pulses delivered at
high frequency, as shown in the inset of panel B. The latencies
of the antidromic responses are designated: ●<5 ms, ▲6-10 ms,
△ >15 ms. The sites of neurons not antidromically activated by
MD stimulation are shown by O. (After Mogenson et al., 1987).

indicated earlier, the locomotion is attributed to the removal of the
inhibitory GABAergic ventral striatal-ventral pallidal projections (Jones
and Mogenson, 1980b; Swerdlow et al., 1984). This picrotoxin-elicited
locomotion was not reduced when procaine, a reversible neuronal blocker,
was administered to the MD, although the elicited locomotion was reduced
significantly by the administration of procaine to the pedunculopontine
nucleus, the region of mesencephalic locomotor region (Mogenson and Wu,
1986). In contrast, hypermotility induced via stimulation of 'denervation
supersensitive' dopamine receptors in the nucleus accumbens by systemic
injection of apomorphine, was attenuated by ibotenate lesions of the MD,
but not MPC or PPN (Swerdlow and Koob, 1987b).

Hoarding of food pellets from an open field back to its home cage is
a well organized sequence of goal-oriented responses in rats. The
quantity of food pellets hoarded was significantly reduced following: (a)
bilateral microinjection of the dopamine antagonist haloperidol into the
nucleus accumbens (Mogenson and Wu, 1988b); (b) 6-hydroxydopamine lesions
of the accumbens dopaminergic neurons that project from the VTA to the
nucleus accumbens (Kelley and Stinus, 1985); (c) GABA injection into the
subpallidal area; and (d) procaine injection into the MD (Mogenson and Wu,
1988b)(Fig. 14). Since MD is directly connected with the medial
prefrontal cortex which has also been implicated in hoarding (Kolb, 1974)
and temporally organized behaviours (Fuster, 1989), the available
experimental data suggest the possibility that the subpallidal-MD-MPC
circuits contribute to the temporal sequencing of response components
necessary for food hoarding (Mogenson and Wu, 1988b).

Findings that emerged from the experimental work assessing the contributions of the basal forebrain to limbic-motor integration in the last
decade have emphasized the critical link of the accumbens-ventral pallidal
and subpallidal pathways in conveying limbic input signals from the
accumbens to the motor effector sites. These studies suggest a functional
heterogeneity of the ventral and subpallidal regions. Some of our results
also suggest that subpallidal GABA inputs do not only come from the
accumbens. What is the source of other GABAergic projections to the
subpallidal area? The notably fast discharge rate of subpallidal neurons
can be entrained by extrinsic excitatory afferents, or generated by
electrophysiological mechanisms intrinsic to the subpallidal neurons.
Where are the sources of the extrinsic excitatory afferents and how do
intrinsic membrane currents regulate the excitability of these subpallidal
neurons? The answers to these questions may contribute to the understanding of the integrative processes by which accumbens signals reach the
motor effector sites. In addition, although one of the major transmitters
of the accumbens output is GABA, the functional role of the neuropeptides
which either co-exist in this accumbens-pallidal pathway or emerge in
parallel from the accumbens to the subpallidal area are not well-
characterized. To be able to measure GABA release in the subpallidal area
in behaving animals is thus a priority in future studies. Furthermore,
most of our electrophysiological studies, based on neuroanatomical
findings were conducted in anaesthetized preparations. There is risk in
drawing the conclusions about motor function from results obtained from
such 'static' experiments! Future electrophysiological studies should
employ chronic recording techniques in behaving animals to investigate the
neural mechanisms that contribute to the translation of "motivation to
action". (Mogenson et al., 1980; Nauta et al., 1978). Elucidation of the
mechanisms of multiple transmitter actions in the accumbens, ventral and
subpallidal areas, PPN and MD is a daunting task. Some beginning using
in vitro intracellular recording techniques in brain slice preparations
has been made in the accumbens (Uchimura and North, 1990). It is only
through these multi-disciplinary approaches that the mechanisms can be
elucidated that underlie the processes that determine "why we want to do
the things that we do". More importantly, these findings may enable us
to understand the mysteries of the basal forebrain and to remove the
miseries of Parkinson's Disease, Schizophrenia, Alzheimer's Disease or
Korsakoff Syndromes (Swerdlow and Koob, 1987a).

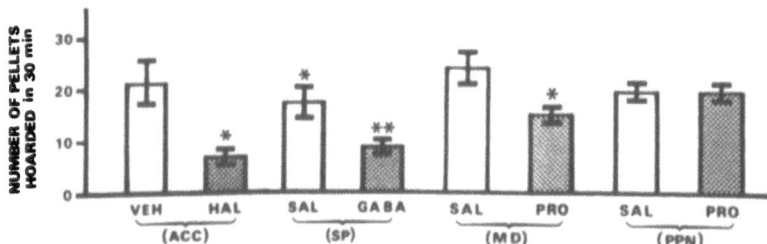

Fig. 14. Involvement of accumbens, subpallidal area, MD but not PPN in
hoarding behaviour. Hoarding of food pellets from an open space
back to the home cage was significantly reduced following:
haloperidol (0.2 μg in 0.2 μl 1% lactic acid) injection into the
accumbens (ACC) and GABA (10 μg/0.2 μl) injection into the
subpallidal area. Procaine (20%, w/v, 0.2 μl) injection into the
MD, but not into the PPN, also reduced the hoarding of the food
pellets. *p<0.02, **p<0.001 (Modified from Mogenson and Wu,
1988b).

SUMMARY

The contribution of hippocampal glutamatergic and VTA dopaminergic inputs to the nucleus accumbens and the role of accumbens → ventral and subpallidal GABAergic pathway in integrating the limbic signals into motor responses via pedunculopontine nucleus were examined with electrophysiological and behavioural techniques. Stimulation of hippocampal input to the accumbens activates GABAergic output to the subpallidal area which leads to suppresion of spontaneous firing of subpallidal neurons, while activation of dopamine receptors in the accumbens suppresses GABAergic output to subpallidal area and thus increases the firing of picrotoxin-sensitive ventral pallidal neurons. However, both treatments induced hypermotility suggesting the functional heterogeneity of the ventral and subpallidal areas in "limbic-motor integration". Furthermore, both hippocampal output signals and dopaminergic input to the accumbens descend via ventral and subpallidal areas serially to the pedunculopontine nucleus, the region of the mesencephalic locomotor region. In addition, a parallel ascending pathway from the subpallidal area to the mediodorsal nucleus, and subsequently to the medial prefrontal cortex, probably mediates behaviour, e.g. food hoarding, that requires higher cognitive processing.

ACKNOWLEDGEMENT

We thank Michael Wu for preparing the figures and assisting with the manuscript. The research was supported by grants from the MRC of Canada and the NSERC of Canada.

REFERENCES

Alheid, G.F. and Heimer, L., 1988, New perspectives in basal forebrain organization of special relevance for neuropsychiatric disorders: the striatopallidal, amygdaloid and corticopetal components of substantia innominata. Neurosci., 27: 1-39.

Austin, M.C. and Kalivas, P.W., 1987, Modulation of GABAergic function on the substantia innominata by the mesolimbic dopamine system. Proc. Soc. Neurosci., 13: 958, Abstr. 264.16.

Austin, M.C. and Kalivas, P.W., 1990, Enkephalinergic and GABAergic modulation of motor activity in the ventral pallidum. J. Pharmacol. Exp. Therap. 252: 1370-1377.

Beckstead, R.M., Domesick, V. B. and Nauta, W.J.H., 1979, Efferent connections of the substantia nigra and ventral tegmental area in the rat. Brain Res., 175: 191-217.

Bjursten, L.-M., Norrsell, K., and Norrsell, U., 1976, Behavioral repertory of cats without cerebral cortex from infancy. Exp. Brain Res. 25: 115-130.

Blaker, W.D., 1985, GABAergic control of the cholinergic projections to the frontal cortex is not tonic. Brain Res., 325: 389-390.

Boldry, R.C. and Uretsky, N.J., 1988, The importance of dopaminergic neurotransmission in the hypermotility response produced by the administration of N-methyl-D-aspartic acid into the nucleus accumbens. Neuropharmacol., 27: 569-577.

Brashear, H.R., Zaborszky, L. and Heimer, L., 1986, Distribution of GABAergic and cholinergic neurons in the rat diagonal band. Neurosci., 17: 439-451.

Brudzynski, S.M. and Mogenson, G.J. 1985, Assocation of the mesencephalic locomotor region with locomotor region with locomotor activity induced by injections of amphetamine into the nucleus accumbens. Brain Res., 334: 77-84.

Brudzynski, S.M. and Mogenson, G.J. 1986, Inhibition of amphetamine-induced locomotor activity by injection of carbachol into the anterior hypothalamic/preoptic area: pharmacological and electrophysiological studies in the rat. Brain Res. 376: 47-56.

Cador, M., Robbins, T.W. and Everitt, B.J. 1989, Involvement of the amygdala in stimulus-reward associations: interaction with the ventral striatum. Neurosci., 30: 77-86.

Christie, M.J., James, L.B. and Beart, P.M., 1985, An excitatory amino acid projection from the medial prefrontal cortex to the anterior part of nucleus accumbens in the rat. J. Neurochem., 45: 477-482.

Christie, M.J., Summers, R.J., Stephenson, J.A., Cook, C.J. and Beart, P.M., 1987, Excitatory amino acid projections to the nucleus accumbens septi in the rat: a retrograde transport study utilizing D-[^{3}H]aspartate and [^{3}H]GABA. Neurosci., 22: 425-439.

Coyle, J.T., Price, D.L. and DeLong, M.R., 1983, Alzheimer's disease: a disorder of cortical cholinergic innervation. Science, 219: 1184-1190.

Dahlstrom, A. and Fuxe, K., 1964, Evidence for the existence of monoamines in the central nervous system. I. Determination of monoamines in the cell bodies of brainstem neurones. Acta Physiol. Scand. Suppl., 232: 1-25.

Divac, I., 1975, Magnocellular nuclei of the basal forebrain project to neocortex, brain stem, and olfactory bulb. Review of some functional correlates. Brain Res., 93:385-398.

Fuster, J.M., 1989, "The Frontal Cortex", 2nd Edition, Raven Press, New York.

Garcia-Rill, E., Skinner, R.D. and Fitzgerald, J.A., 1983, Activity in the mesencephalic locomotor region during locomotion. Exp. Neurol. 82: 609-622.

Garcia-Rill, E., 1986, The basal ganglia and the locomotor regions. Brain Res. Rev. 11:47-63.

Grillner, S., 1985, Neurobiological bases of rhythmic acts in vertebrates. Science, 228: 143-149.

Groenewegen, H.J., 1988, Organization of the afferent connections of the mediodorsal thalamic nucleus in the rat, related to the mediodorsal-prefrontal topography. Neurosci., 24: 379-431.

Groenewegen, H.J., Vermeulen-Van der Zee, E., te Kortschot, A. and Witter, M.P., 1987, Organization of the projections from the subiculum to the ventral striatum in the rat. A study using anterograde transport of Phaseolus vulgaris leucoglutinin. Neurosci., 23: 103-120.

Grove, E.A., 1988a, Neural associations of the substantia innominata in the rat: afferent connections. J. Comp. Neurol. 277: 315-346.

Grove, E.A., 1988b, Efferent connections of the substantia innominata in the rat. J. Comp. Neurol. 277: 347-164.

Haber, S..N., 1987, Anatomical relationship between the basal ganglia and the basal nucleus of Meynert in human and monkey forebrain. Proc. Natl. Acad. Sci. USA, 84: 1408-1412.

Haber, S.N. and Nauta, W.J.H., 1983, Ramifications of the globus pallidus in the rat as indicated by patterns of immunohistochemistry. Neurosci., 9: 245-260.

Haber, S.N., Groenewegen, H.J., Grove, E.A. and Nauta, W.J.H., 1985, Efferent connections of the ventral pallidum: evidence of a dual striato pallidofugal pathway. J. Comp. Neurol., 235: 322-335.

Heimer, L. and Wilson, R.D., 1975, The subsortical projections of the allocortex: similarities in the neural associations of the hippocampus, the piriform cortex, and the neocortex. In: "Golgi Centennial Symposium", M. Santini, ed., Raven Press, N. Y., pp.177-193.

Heimer, L., Switzer, R.D. and Van Hoesen, G.W., 1982, Ventral striatum and ventral pallidum: Components of the motor system? Trends in Neurosci., 5: 83-87.

Hess, W.R., 1954, "Das Zwischenhirn" 2nd Ed. Schwabe, Basel.

Hill, J.M., 1985, Iron concentration reduced in ventral pallidum, globus pallidus, and substantia nigra by GABA-transaminase inhibitor, Gamma-vinyl GABA. Brain Res., 342: 18-25.

Hill, J.M. and Switzer, R.C., 1984, The regional distribution and cellular localization of iron in the rat brain. Neurosci., 11: 595-603.

Isaacson, R.L., 1982, "Limbic System" 2nd Ed., Plenum, New York.

Jones, D.L. and Mogenson, G.J., 1980a, Nucleus accumbens to globus GABA projection: electrophysiological and iontophoretic investigations. Brain Res., 188: 93-105.

Jones, D.L. and Mogenson, G.J., 1980b Nucleus accumbens to globus pallidus GABA projection subserving ambulatory activity. Am. J. Physiol., 238: 65-69.

Jones, D.L. Mogenson, G.J. and Wu, M., 1981, Injections of dopaminergic, cholinergic, serotoninergic and GABAergic drugs into the nucleus accumbens: effects on locomotor activity in the rat. Neuropharmacol., 20: 29-37.

Kalivas, P.W., 1990, This book.

Kelley, A.E. and Domesick, V.B., 1982, The distribution of the projection from the hippocampal formation to the nucleus accumbens in the rat: an anterograde and retrograde horseradish peroxidase study. Neurosci., 7: 2321-2335.

Kelley, A.E., Domesick, V.B. and Nauta, W.J.H., 1982, The amygdalostriatal projection in the rat--- an anatomical study by anterograde and retrograde tracing methods. Neurosci., 7: 615-630.

Kelley, A.E., and Stinus, L., 1985, Disappearance of hoarding behaviour after 6-hydroxydopamine lesions of the mesolimbic dopamine neurones and its reinstatement with L-dopa. Behav. Neurosci. 99: 531-545.

Kolb, B., 1974, Prefrontal lesions alter eating and hoarding behaviour in rats. Physiol. Behav. 12: 507-511.

Koob, G.F., 1990, This book.

Krettek, J.E. and Price, J.L., 1978, Amygdaloid projections to subcortical structures within the basal forebrain and brainstem in the rat and cat. J. Comp. Neurol., 178: 225-254.

Lehmann, J., Nagy, J.I., Atmadja, S. and Fibiger, H.C., 1980, The nucleus basalis magnocellularis: the origin of a cholinergic projection to the neocortex of the rat. Neurosci., 5: 1161-1174.

Markowitsch, J.J., 1982, Thalamic mediodorsal nucleus and memory: a critical evaluation of studies in animals and man. Neurosci. Biobehav. Rev., 6: 351-381.

Mesulam, M.M., Mufson, E.J., Wainer, B.H. and Levey, A.I. 1983, Central cholinergic pathways in the rat, an overview based on an alternative nomenclature (Ch1-Ch6). Neuroscience, 10: 1185-1201.

Mogenson, G.J., 1987, Limbic-motor integration. Prog. Psychobiol. Physiol. Psychol., 12: 117-170.

Mogenson, G.J. and Nielsen, M., 1983, Evidence that an accumbens to subpallidal GABAergic projection contributes to locomotor activity. Brain Res. Bull., 11: 309-314.

Mogenson, G.J. and Nielsen, M., 1984a, A study of the contribution of hippocampal-accumbens-subpallidal projections to locomotor activity. Behav. Neural. Biol. 42: 38-51.

Mogenson, G.J. and Nielsen, M., 1984b, Neurochemical evidence to suggest that the nucleus accumbens and subpallidal regions contribute to exploratory locomotion. Behav. Neural Biol., 42: 52-60.

Mogenson, G.J. and Wu, M., 1986, Subpallidal projections to the mesencephalic locomotor region investigated with a combination of behavioral and electrophysiological recording techniques. Brain Res. Bull., 16: 383-390.

Mogenson, G.J. and Wu, M., 1988a, Differential effects on locomotor activity of injections of procaine into mediodorsal thalamus and pedunculopontine nucleus. Brain Res. Bull. 20: 241-246.

Mogenson, G.J. and Wu, M., 1988b,   Disruption of food hoarding by injections of procaine into mediodorsal thalamus, GABA into subpallidal region and haloperidol into the accumbens. Brain Res. Bull. 20: 247-251.

Mogenson, G.J. and Yim, C.Y., 1981, Electrophysiological and neuropharmacological-behavioral studies of the nucleus accumbens: implications for its role as a limbic-motor interface. In "The Neurobiology of the Nucleus Accumbens", R. B. Chronister and J. F. DeFrance, eds., Haer Institute, New Brunswick, pp. 210-229.

Mogenson, G.J., Jones, D.L. and Yim , C.Y., 1980,   From motivation to action: Functional interface between the limbic system and the motor system. Prog. Neurobiol., 14: 69-97.

Mogenson, G.J., Swanson, L.W. and Wu, M., 1983, Neural projections from nucleus accumbens to globus pallidus, substantia innominata, and lateral preoptic-lateral hypothalamic area: an anatomical and electrophysiological investigation in the rat. J. Neurosci., 3: 189-202.

Mogenson, G.J., Wu, M. and Manchanda, S.K., 1979,   Locomotor activity initiated by microinfusions of picrotoxin into the ventral tegmental area. Brain Res., 161: 311-319.

Mogenson, G.J., Yang, C.R. and Yim, C.Y., 1988,   Influence of dopamine on limbic inputs to the nucleus accumbens. Ann. N. Y. Acad. Sci., 537: 86-100.

Mogenson, G.J., Ciriello, J., Garland, J. and Wu, M., 1987,   Ventral pallidum projections to mediodorsal nucleus of the thalamus: an anatomical and electrophysiological investigation in the rat. Brain Res., 404: 221-230.

Moon-Edley, S. and Graybiel, A.M. 1983,   The afferent and efferent connections of the feline nucleus tegmenti pedunculopontinus pars compacta. J. Comp. Neurol. 217: 187-215.

Nauta, W.J.H., Smith, G.P., Faull, R.L.M. and Domesick, V.B., 1978, Efferent connections and nigral afferents of the nucleus accumbens septi in the rat. Neurosci., 3: 385-401.

O'Keefe, J. and Nadel, L., 1978,   "The Hippocampus as a Cognitive Map" Clarendon Press, Oxford.

Onteniente, B., Simon, H., Taghzouti, K., Geffard, M., Le Moal, M. and Calas, A., 1987,   Dopamine-GABA interactions in the nucleus accumbens and lateral septum of the rat. Brain Res. 421: 391-396.

Patel, S. and Slater, P., 1988,   Effects of GABA compounds injected into the subpallidal regions of rat brain on nucleus accumbens evoked hyperactivity. Behav. Neurosci., 102: 596-600.

Paxinos, G. and Watson, C., 1986,   "The Rat Brain in Stereotaxic Co-ordinates", 2nd edition, Academic Press, N. Y.

Pickel, V.M., Towle, A.C., Joh, T.H., and Chan, J., 1988,   GABA in the medial rat nucleus accumbens: ultrastructural localization in neurones receiving monosynaptic input from catecholaminergic afferents.   J. Comp. Neurol., 272: 1-14.

Pijnenberg, A.J.J. and van Rossum, J., 1973,   Stimulation of locomotor activity following injection of dopamine into the nucleus accumbens. J. Pharm. Pharmacol. 25: 1003-1005.

Richardson, R.T. and DeLong, M.R., 1988,   A reappraisal of the functions of the nucleus basalis of Meynert. Trends in Neurosci. 11: 265-267.

Schacter, G.B., Yang, C.R., Innis, N.K. and Mogenson, G.J., 1989,   The role of the hippocampal-nucleus accumbens pathway in radial-arm maze performance. Brain Res. 494: 339-349.

Scheel-Kruger, J., 1986,   Dopamine-GABA interactions: evidence that GABA transmits, modulates and mediates dopaminergic functions in the basal ganglia and the limbic system. Acta Neurol. Scanda Suppl., 107: 1-54.

Semba, K. and Fibiger, H.C., 1988,   Time of origin of cholinergic neurons in the rat basal forebrain. J. Comp. Neurol., 269: 87-95.

Shefchyk, D.J., Jell, R.M. and Jordan, L.M., 1984, Reversible cooling of the brainstem reveals areas required for mesencephalic locomotor region evoked treadmill locomotion. Exp. Brain Res. 56, 257-262.

Shik, M.L., Severin, F.V. and Orlovsky, G.N., 1966, Control of walking and running by means of electrical stimulation of the mid-brain. Biophysics 11: 756-765.

Skinner, R.D. and Garcia-Rill E., 1984, The mesencephalic locomotor region (MLR) in the rat. Brain Res. 323: 385-389.

Sugimoto, T. and Mizuno, N., 1987, Neurotensin in projection neurons of the striatum and nucleus accumbens, with reference to co-existence with enkephalin and GABA: immunohistochemical study in the cat. J. Comp. Neurol., 257: 383-395.

Swanson, L.W., Mogenson, G.J., Gerfen, C.R. and Robinson, P., 1984, Evidence for a projection from the lateral preoptic area and substantia innominata to the "mesencephalic locomotor region" in the rat. Brain Res., 295: 161-178.

Swerdlow, N.R. and Koob, G.F., 1987a, Dopamine, schizophrenia, mania, and depression: toward a unified hypothesis of cortical-striato-pallido-thalamic function. Brain Behav. Sci., 10: 197-245.

Swerdlow, N.R. and Koob, G.F., 1987b, Lesions of the dorsomedial nucleus of the thalamus, medial prefrontal cortex and pedunculopontine nucleus: effects on locomotor activity mediated by nucleus accumbens-ventral pallidal circuitry. Brain Res., 412: 233-243.

Swerdlow, N.R., Swanson, L.W. and Koob, G.F., 1984, Electrolytic lesions of the substantia innominata and lateral preoptic area attenuate the 'supersensitive' locomotor response to apomorphine resulting from denervation of the nucleus accumbens. Brain Res. 306: 141-148.

Switzer, R.C., Hill, J. and Heimer, L., 1982, The globus pallidus and its rostroventral extension into the olfactory tubercle of the rat: a cyto- and chemoarchitectural study. Neurosci., 7: 1891-1904.

Totterdell, S. and Smith, A.D., 1989, Convergence of hippocampal and dopaminergic input onto identified neurons in the nucleus accumbens of the rat. J. Chem. Neuroanat., 2: 285-298.

Uchimura, N. and North, R.A., 1990, Muscarine reduces inwardly rectifying potassium conductance in rat nucleus accumbens neurones. J. Physiol. (London) 422: 369-380.

Vanderwolf, C.H., 1971, Limbic-diencephalic mechanisms of voluntary movement. Psychol. Rev., 78: 83-113.

Victor, M., Adams, R.D. and Collins, G.H., 1971, The Wernicke-Korsakoff Syndrome. Oxford, Blackwell, 1971.

Vives, F. and Mogenson, G.J., 1985, Electrophysiological evidence that mediodorsal nucleus of the thalamus is a relay of the pathway between the ventral pallidum and the medial prefrontal cortex in the rat. Brain Res., 344: 329-337.

Vives, F. and Mogenson, G.J., 1986, Electrophysiological study of the effects of D_1 and D_2 dopamine antagonists on the interaction of converging inputs from the sensory-motor cortex and substantia nigra neurons in the rat. Neurosci., 17: 349-359.

Walaas, I. and Fonnum, F., 1979, The distribution and origin of glutamate decarboxylase and choline acetyltransferase in ventral pallidum and other basal forebrain regions. Brain Res., 177: 325-336.

Wenk, G.J., 1984, Pharmacological manipulations of the substantia innominata - cortico cholinergic pathway. Neurosci. Lett., 51: 99-103.

White, F.J. and Wang, R.X., 1985, Electrophysiological evidence for the existence of both D_1 and D_2 dopamine receptors in the rat nucleus accumbens. J. Neurosci., 6: 274-280.

Wood, P.L., 1986, Pharmacological evaluation of GABAergic and glutamatergic inputs to the nucleus basalis - cortico and the septal - hippocampal cholinergic projections. Canadn. J. Physiol. Pharmacol., 64: 325-328.

Wood, P.L. and Richard, J., 1982, GABAergic regulation of the substantia innominata - cortico cholinergic pathway. Neuropharmacol., 21: 969-972.

Yang, C.R. and Mogenson, G.J., 1984, Electrophysiological responses of neurones in the nucleus accumbens to hippocampal stimulation and the attenuation of the excitatory responses by mesolimbic dopaminergic system. Brain Res., 324: 69-84.

Yang and Mogenson, G.J., 1985, An electrophysiological study of the neural projections from the hippocampus to the ventral pallidum and the subpallidal areas by way of the nucleus accumbens. Neurosci., 15: 1015-1024.

Yang, C.R. and Mogenson, G.J., 1986, Dopamine enhances terminal excitability of hippocampal-accumbens neurones via D2 receptor: role of dopamine in presynaptic inhibition. J. Neurosci., 6: 2470-2478.

Yang, C.R. and Mogenson, G.J., 1987, Hippocampal signal transmission to the pedunculopontine nucleus and its regulation by dopamine D2 receptors in the nucleus accumbens: an electrophysiological and behavioral study. Neurosci., 23: 1041-1055.

Yang, C.R. and Mogenson, G.J., 1989, Ventral pallidal responses to dopamine receptor stimulation in the nucleus accumbens. Brain Res., 489: 237-246.

Yang, C.R. and Mogenson, G.J., 1990, Dopaminergic modulation of cholinergic responses in rat medial prefrontal cortex. Brain Res., (in press).

Yim, C.Y., 1990, This book.

Yim, C.Y. and Mogenson, G.J., 1982, Responses of nucleus accumbens neurones to amygdala stimulation and its modification by dopamine. Brain Res., 239: 401-415.

Yim, C.Y. and Mogenson, G.J., 1989, Low doses of accumbens dopamine modulate amygdala suppression of spontaneous exploratory activity in rats. Brain Res., 477: 202-210.

Young, W.S., Alheid, G.F. and Heimer, L., 1984, The ventral pallidal projection to the mediodorsal thalamus: a study with fluorescent retrograde tracers and immunohistofluorescence. J. Neurosci. 4:1626-1638.

Zaborszky, L., Leranth, C. and Heimer, L. 1984, Ultrastructural evidenceof amygdalofugal axons terminating on cholinergic cells of the rostral forebrain. Neurosci. Lett., 52: 219-225.

Zaborszky, L., Heimer, L., Eckenstein, F. and Leranth, C., 1986, GABAergic input to cholinergic basal forebrain neurones: an untrastructural study using retrograde tracing of HRP and double immunolabelling. J. Comp. Neurol., 250: 282-295.

Zahm, D.S. and Heimer, L., 1988, The ventral striatopallidal parts of the basal ganglia in the rat: I. Neurochemical compartmentation as reflected by the distributions of neurotensin and substance P immunoreactivity. J. Comp. Neurol., 272: 516-535.

Zahm, D.S., Zaborsky, L., Alones, V.E. and Heimer, L., 1985, Evidence for the coexistence of glutamate decarboxylase and met-enkephalin immunoreactivities in axon terminals of rat ventral pallidum. Brain Res., 325: 317-321.

FUNCTIONAL OUTPUT OF THE BASAL FOREBRAIN

George F. Koob
Neal R. Swerdlow
Franco Vaccarino
Carol Hubner
Luigi Pulvirenti
Friedbert Weiss

Department of Neuropharmacology
Research Institute of Scripps Clinic
La Jolla, CA   92037

INTRODUCTION

The basal forebrain is composed of many important components, one of which is the ventral striatum including the nucleus accumbens and olfactory tubercle. Both neuroanatomical and behavioral studies have provided important evidence implicating the ventral striatum as an interface between the limbic system and the extrapyramidal motor system (Kelley and Stinus, 1984; Heimer and Wilson, 1975; Mogenson and Nielson, 1984a). The ventral striatum receives allocortical projections from the hippocampus and amygdala (Kelley and Domesick, 1982), and a major dopaminergic projection from the ventral midbrain, especially the region of the ventral tegmental area. The availability of reliable behavioral measures and powerful neuropharmacological probes has allowed substantial progress to be made in the understanding of the functional significance of the ventral striatum and its circuitry.

The dopamine projection to the nucleus accumbens appears to be an important substrate for the activating and reinforcing actions of psychomotor stimulants (Koob and Swerdlow, 1988; Koob and Goeders, 1988). Denervation of the nucleus accumbens with the neurotoxin 6-hydroxydopamine blocks the locomotor activation produced by amphetamine and cocaine (Kelly et al., 1975; Roberts et al., 1975; Kelly and Iversen, 1976). In addition, the reinforcing actions of amphetamine and cocaine, as measured by intravenous self-administration, are blocked by 6-hydroxydopamine lesions of the nucleus accumbens (Roberts et al., 1977; Roberts et al., 1980; Lyness et al., 1979).

Non drug-induced activation is also attenuated by denervation of the dopamine projection to the region of the nucleus accumbens. Food deprived rats show decreases in locomotor activity associated with feeding following 6-hydroxydopamine lesions of the nucleus accumbens (Koob et al., 1978). Similar 6-hydroxydopamine lesions to the region of the nucleus accumbens produce decreases in locomotor activity in an open field test (Joyce et al., 1983; Taghzouti et al., 1985), and decreases in acquisition of schedule-induced polydipsia (Robbins and Koob, 1980).

The Basal Forebrain, Edited by T.C. Napier *et al.*
Plenum Press, New York, 1991

Denervation of the dopamine projection to the region of the nucleus accumbens also produces a dramatic increased behavioral responsiveness to direct dopamine agonists. This increased responsiveness is attributed to a postsynaptic receptor supersensitivity (Staunton et al., 1982) and is reflected in a greatly potentiated locomotor response to systemic injections of apomorphine. Animals with nucleus accumbens 6-hydroxydopamine lesions show a ten-fold increase in sensitivity to apomorphine when compared with sham lesioned animals with significant increases in locomotor activity at a dose of 0.1 mg/kg subcutaneously, a dose that actually decreases activity in sham lesioned animals (Van der Kooy et al., 1983). Since only "supersensitive" receptors are activated by such low doses of systemically administered apomorphine, this increased locomotor response provides a powerful dependent variable for reflecting activation of the dopamine receptors in a select brain area, i.e. in the region of the nucleus accumbens.

EFFERENT PROJECTIONS FROM NUCLEUS ACCUMBENS-LOCOMOTOR ACTIVITY

A series of studies have characterized the functional efferent output of the nucleus accumbens using locomotor activity as the dependent variable (Jones and Mogenson, 1980; Mogenson et al., 1980; Mogenson and Nielson, 1983). Dopamine injected into the nucleus accumbens produced a hyperactivity and this hyperactivity was reversed by injecting gamma-aminobutyric acid (GABA) into the region of the ventral pallidum (Jones and Mogenson, 1980). Injection of the GABA antagonist picrotoxin into the region of the ventral tegmental area increased locomotor activity and this hyperactivity was also attenuated by injections of GABA into the pallidum (Mogenson et al., 1980). Similar blockade of dopamine stimulation of the nucleus accumbens was observed with injection of the GABA transaminase inhibitor, ethanolamine O-sulfate into the ventral pallidum (Pycock and Horton, 1976).

Substantia Inominata-Ventral Pallidum

Similar results were obtained using a different model of dopamine receptor activation, i.e. the augmented locomotor response to systemic injections of dopamine agonists in nucleus accumbens denervated animals. This model has the advantage of an exaggerated locomotor response produced by activation of a select group of dopamine receptors, presumably only those having been denervated, see above.

Animals receiving 6-hydroxydopamine lesions of the nucleus accumbens showed a significantly potentiated locomotor response to apomorphine, as is shown in Fig. 1. Electrolytic lesions of the substantia inominata/lateral preoptic area [SI/LPO] significantly depressed the locomotor response to apomorphine in dopamine denervated rats (Swerdlow et al., 1984a). A similar, but an even larger, reversal of the supersensitive apomorphine response was observed in rats receiving ibotenic acid lesions of the SI/LPO (Swerdlow et al., 1984b), see Fig. 1.

These results confirm that the first-order efferent projection from the nucleus accumbens onto cells within the SI/LPO forms an important output for the behavioral expression of nucleus accumbens dopamine receptor stimulation. Locomotor activation produced by stimulation of "supersensitive" dopamine receptors within the nucleus accumbens was significantly decreased by destruction or either cells and fibers within the SI/LPO region--using electrolytic lesions--or by destruction of only cells in this area--using the cell body specific neurotoxin ibotenic acid.

Projections from the nucleus accumbens that innervate the substantia inominata-ventral pallidum contain GABA, enkephalin, and substance P (Zaborsky et al., 1982). High doses of naloxone do not alter the super-sensitive locomotor response, suggesting that enkephalin-containing fibers do not contribute to the "supersensitive" locomotor response to apomorphine (Swerdlow et al., 1987). However, locomotor activation produced by direct application of dopamine (DA) into the nucleus accumbens is attenuated by infusion of GABA into the region of the ventral pallidum (Mogenson and Nielson, 1983). While this DA-stimulated locomotion is distinct from the "supersensitive"response that follows 6-hydroxydopamine-induced denervation, the findings nevertheless suggest that the locomotor-activating properties of DA stimulation within the nucleus accumbens are mediated by the inhibition of the release of GABA from terminals within the substantia inominata-ventral pallidum.

To test this hypothesis, GABA receptors in the SI/LPO were stimulated with a GABA agonist in rats showing a supersensitive locomotor response to apomorphine following DA denervation of the nucleus accumbens. Injection of low doses of the GABA-agonist muscimol into the SI/LPO dose-dependently decreased the locomotor response to apomorphine in 6-hydroxydopamine-injected animals, but had no reliable effect on the locomotor response to apomorphine in vehicle-injected animals, see Fig. 2. Higher doses of mus-cimol (>10 ng) produced an initial blockade of apomorphine-stimulated

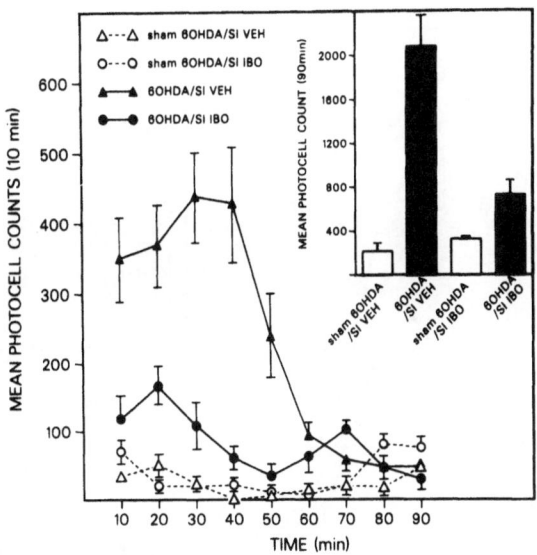

Fig. 1.   Locomotor activity following subcutaneous (s.c.) injection of 0.1 mg/kg apomorphine. Ordinate refers to mean photocell counts for each 10 min period for the groups represented as follows: Δ---Δ, sham 6OHDA/SI vehicle (n=6); o---o, sham 6OHDA/SI ibotenic acid (n=6); Δ---Δ, 6OHDA/SI vehicle (n=10), •---•, 6PHDA/SI ibotenic acid (n=9). Statistical analysis was accomplished with a two-way analysis of variance with repeated measures on time, with significance taken at p < 0.05. (SI - substantia inominata; IBO - ibotenic acid; for lesion histology, taken with permission from Swerdlow, Swanson and Koob, 1984b).

locomotion, however these higher doses eventually produced a prolonged increase in locomotor activity (Swerdlow and Koob, 1984). The locomotor response to amphetamine and heroin is also significantly decreased by muscimol injections into the SI/LPO, however the locomotor activation produced by caffeine and corticotropin releasing factor is not blocked by SI/LPO injections of muscimol (Swerdlow and Koob, 1985).

The specific anatomical nomenclature for the functional output of the nucleus accumbens has varied somewhat. Recent anatomical data suggests that a major output from the nucleus accumbens projects to the sub-commissural part of the substantia inominata, and is now called the ventral pallidum (Alheid and Heimer, 1988). However, lesions and injections in the above mentioned studies fall more in the sublenticular part of the substantia inominata which is caudal and just ventral to the globus

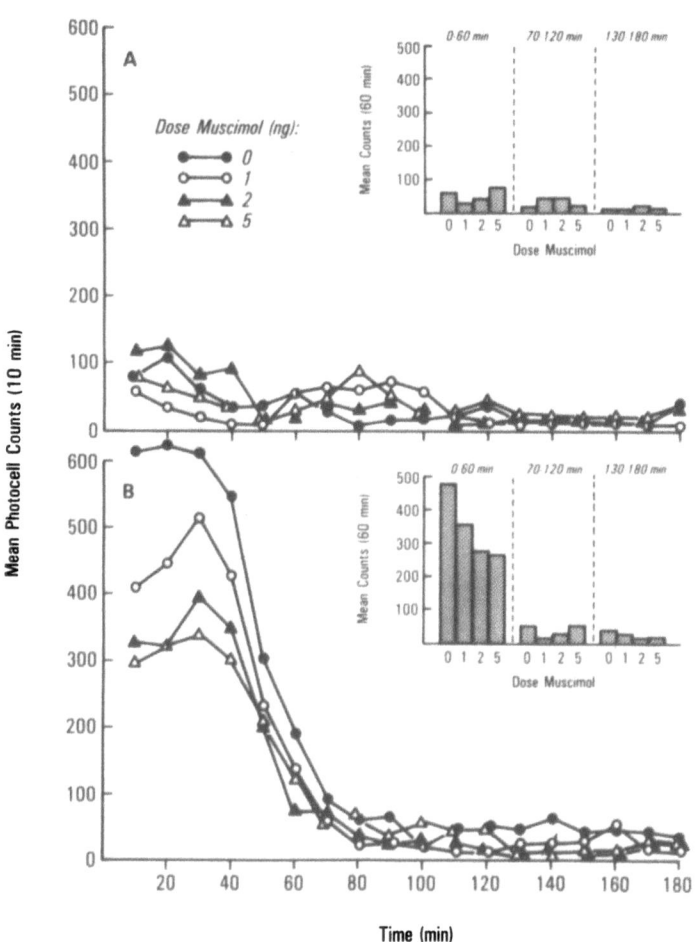

Fig. 2.   Locomotor response to 0.1 mg/kg apomorphine s.c. in vehicle-(A) and 6OHDA-injected (B) rats following intracerebral injection of 0, 1, 2, or 5 μg muscimol into the SI/LPO regions. Insert histogram indicates total locomotor activity collapsed over 60 min intervals. *Significantly different from 0 dose muscimol (p < 0.05, Newman-Keuls test following a significant dose x time interaction). Taken with permission from Swerdlow and Koob, 1984.

pallidus. This region is continuous with the bed nucleus of the stria terminalis and the centromedial part of the amygdala and thus forms part of the "extended amygdala" (Alheid and Heimer, 1988). Future studies will be necessary to delineate exactly what part of the basal forebrain forms the output of the nucleus accumbens. Indeed, some of the components of the medial nucleus accumbens may also represent a rostral part of the extended amygdala (Alheid and Heimer, 1988).

Efferent Projections From the Ventral Pallidum

The question of the circuitry involved in further processing of the locomotor stimulation associated with activation of DA receptors in the nucleus accumbens remains a current area of research. Efferent projections from the ventral pallidum include a major cholinergic projection which traverses through the medial prefrontal cortex (MPC) and then spreads caudally to innervate most of the neocortex (Divac et al., 1978), a projection to the pedunculopontine nucleus (PPN) (Swanson et al., 1984) which is considered a homolog in the rat to the mesencephalic locomotor region (Skinner and Garcia-Rill, 1984; Grillner and Shik, 1973), and a projection to the dorsomedial thalamus (DMT) (Young et al., 1984).

To examine which of these projections forms the next important link from the ventral pallidum to lower motor circuitry responsible for conveying the effects of nucleus accumbens DA receptor stimulation into locomotor activation, the effects of lesions of MPC, DMT and PPN on the "supersensitive" locomotor response to apomorphine were studied. Of the three regions tested, only destruction of the DMT produced a reliable decrease in the "supersensitive" apomorphine stimulated locomotion, (see Fig. 3). Large electrolytic lesions of the MPC and large ibotenic acid lesions of the PPN failed to alter the apomorphine stimulated locomotion (Swerdlow and Koob, 1987a). In addition picrotoxin, when injected into the ventral pallidum, produced a dose-dependent increase in locomotor activity that was also blocked by lesions of the DMT (Swerdlow and Koob, 1987a). Thus, it appears that the locomotor stimulation of DA receptors in the region of the nucleus accumbens is dependent on the integrity of cells within the region of the ventral pallidum. This accumbens-ventral pallidum connection appears to include GABAergic projections. While further work is necessary, preliminary results with the "supersensitive" model suggest that the third link in this "output" circuitry is the region of the DMT.

AFFERENT PROJECTIONS TO NUCLEUS ACCUMBENS-LOCOMOTOR ACTIVITY

Glutamate

The nucleus accumbens receives afferent projections from a number of allocortical structures, including the hippocampus and amygdala. There is some evidence for a role for a glutamate projection from the hippocampus in locomotor activation. Locomotor activation produced by infusion of carbachol directly into the dentate gyrus of the hippocampus is blocked by infusion of the glutamate antagonist glutamic acid diethyl ester (GDEE) into the nucleus accumbens (Mogenson and Nielson, 1984a). Also, rat exploratory activation within a novel open-field is blocked by injection of the glutamate antagonist into the nucleus accumbens (Mogenson and Nielson, 1984b). In a recent study, the glutamate antagonist (GDEE) blocked cocaine-induced locomotor activity, but not caffeine-induced locomotor activity (Pulvirenti et al., 1989).

Other possible neurotransmitters involved in modulating the function of the nucleus accumbens, particularly as regards locomotor activation associated with enhanced dopaminergic activity, include neuropeptides such as the enkephalins, neurotensin and cholecystokinin. An opiate antagonist, methylnaloxonium, injected into the nucleus accumbens, blocks opiate-induced locomotor activity (Amalric and Koob, 1985), but this effect appears to be independent of the dopamine projection to the nucleus accumbens (Vaccarino et al., 1986). Neurotensin injected into the nucleus accumbens blocks the locomotor activation associated with amphetamine, cocaine, and dopamine (Kalivas et al., 1984). However, the origin of the neurons containing these two neurotransmitters is not clear.

Cholecystokinin

Cholecystokinin octapeptide (CCK) has been shown to coexist with DA in the mesocorticolimbic dopamine system. Electrophysiological, biochemical and neuropharmacological studies suggest that CCK antagonizes functionally the postsynaptic effects of dopamine in the region of the nucleus accumbens (Agnati and Fuxe, 1983; White and Wang, 1984; Wang and Hu, 1986). Using the "supersensitive" model where low doses of apomorphine induce a significant hyperactivity in rats with DA denervation of the nucleus accumbens, CCK injected into the nucleus accumbens attenuates this

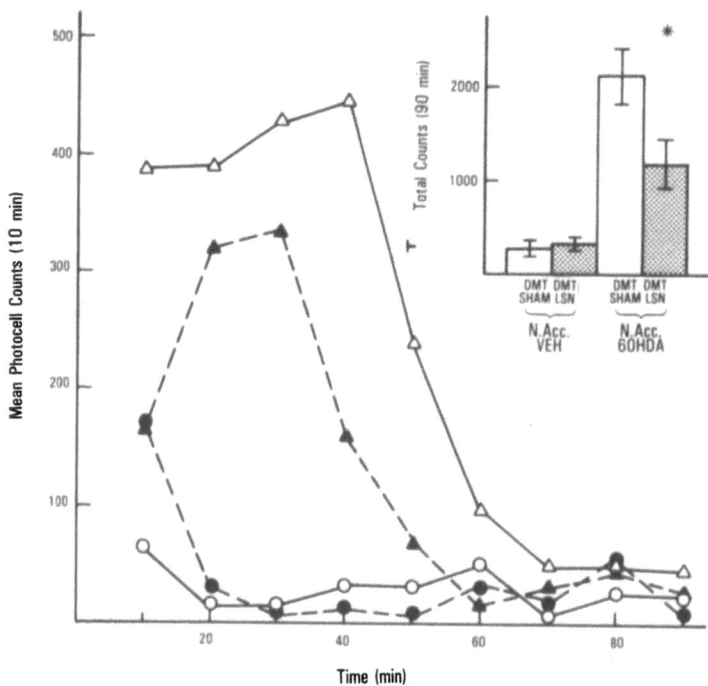

Fig. 3.   Locomotor response after s.c. injection of 0.1 mg/kg apomorphine. Animals had received either vehicle (circles) or 6-OHDA (triangles) injections into the nucleus accumbens (N.Acc.) and either sham (open figures) or electrolytic lesions (solid figures) of the dorsomedial thalamus (DMT). *Significantly different from N.Acc. 6-hydroxydopamine/DMT sham-lesioned animals, $P < 0.05$, two-way ANOVA with repeated measures on time. (Taken with permission from Swerdlow and Koob, 1987a). LSN refers to lesion.

hyperactivity at doses of 1, 10, 100 ng and 1 ug (Weiss et al., 1989). These results suggest that CCK is functionally opposed to dopamine in the nucleus accumbens, but while it is clear that CCK can attenuate the effects of postsynaptic dopamine receptor activation, it is unclear at exactly what neuropharmacological site CCK exerts its antagonistic action.

EFFERENT PROJECTIONS FROM NUCLEUS ACCUMBENS-DRUG REINFORCEMENT

Significant evidence exists which also implicates a role for the nucleus accumbens in drug reinforcement. Denervation of the DA projection to the nucleus accumbens blocks established cocaine and amphetamine self-administration (Roberts et al., 1977; Roberts et al., 1980; Lyness et al., 1979). Using a progressive ratio schedule of reinforcement for cocaine, destruction of presynaptic terminals in the region of the nucleus accumbens produced significant decreases in the highest ratio for which the rats would respond for cocaine, suggesting that the reinforcing value of cocaine is decreased following disruption of nucleus accumbens DA activity (Koob et al., 1987).

The nucleus accumbens has also been implicated in opiate reinforcement. Central injections of a quaternary derivative of naloxone, methylnaloxonium, increases intravenous heroin self-administration similar to systemic naloxone (Vaccarino et al., 1985). These increases in drug intake in single lever situations are thought to represent an attempt by the animal to compensate for a competitive receptor blockade and as such are thought to reflect a decrease in the reinforcing properties of the drug. While injection at other sites produces such increases in opiate self-administration, the nucleus accumbens appeared to be particularly sensitive (Vaccarino et al., 1985). Rats will also self-administer opioids into the nucleus accumbens (Goeders et al., 1984) and the ventral tegmental area (Bozarth and Wise, 1981). This, and other evidence, has led to the hypothesis that opioid reinforcement is also dependent on the release of DA in the nucleus accumbens (Bozarth and Wise, 1981). However, some portion of the reinforcing properties of heroin appear to be independent of the release of DA since heroin self-administration persists following 6-hydroxydopamine lesions of the nucleus accumbens that block cocaine self-administration (Pettit et al., 1984). Kainic acid lesions of the nucleus accumbens disrupt heroin as well as cocaine self-administration (Zito et al., 1985), again suggesting a role for the nucleus accumbens in opiate reinforcement.

Substantia Inominata-Ventral Pallidum

Relatively little information has been obtained regarding the efferent anatomical substrates through which the nucleus accumbens may process drug reinforcement. Since previous work has established the substantia inominata-ventral pallidum as an important connection in the expression of behavioral stimulation produced by activation of the nucleus accumbens, and since there are established efferent connections between the nucleus accumbens and ventral pallidum, a logical hypothesis was that the region of the ventral pallidum may also be involved in the processing of the reinforcing properties of cocaine and heroin.

To test this hypothesis rats were trained to intravenously self-administer heroin or cocaine. Following establishment of a stable baseline (fixed ratio-5 schedule), the rats received bilateral ibotenic acid lesions of the region of the substantia inominata-ventral pallidum (Hubner and Koob, 1990). The substantia inominata-ventral pallidum lesions significantly decreased baseline cocaine and heroin self-administration, and when the rats were subjected to a progressive ratio

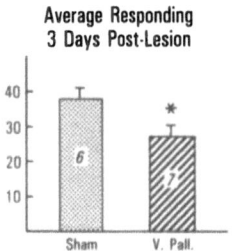

Average Responding
3 Days Post-Lesion

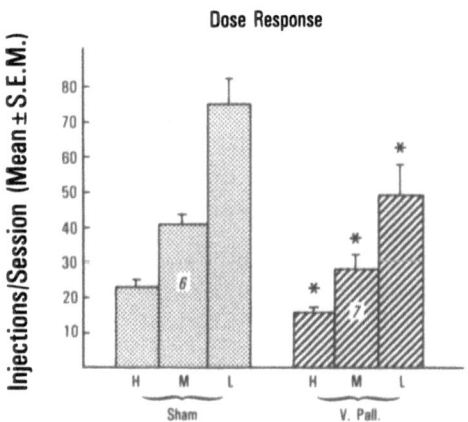

Dose Response

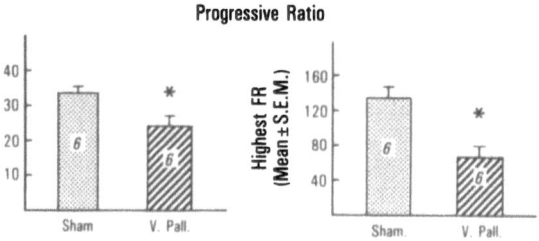

Progressive Ratio

Fig. 4. Effects of bilateral ibotenic acid lesions of the ventral pal-
        lidum on responding in rats self-administering cocaine. Sham,
        vehicle (pH 7.4 phosphate buffer solution) injected controls.
        V. Pall., 5 μg/0.5 μl (expressed as salt) ibotenic acid injected
        into the ventral pallidum. Top panel shows the mean number of
        injections maintained on an FR 5 reinforcement schedule averaged
        over the first 3 days postlesion. Middle panel shows the dose-
        effect function, expressed as mean number of injections, where
        the normal dose of cocaine (0.75 mg/kg/inj.) was doubled to 1.5
        mg/kg/inj. (H), returned to 0.75 mg/kg/inj. (M), and halved to
        0.375 lmg/kg/inj. (L) on successive days, all on an FR 5
        schedule. Bottom panel shows the mean number of injections and
        the mean highest ratio completed on the progressive-ratio
        schedule. Error bars reflect mean ± S.E.M. *Significantly dif-
        ferent from sham group, $P < 0.05$ t-test following significant
        ANOVA main effect. Taken with permission from Hubner and Koob,
        1990.

procedure, lesions of the substantia inominata-ventral pallidum produced a significant decrease in the highest ratio obtained for both cocaine and heroin (Hubner and Koob, 1990) (see Fig. 4). These results suggest that the substantia inominata-ventral pallidum may be an important site in the processing of the reinforcing effects of drugs and that the nucleus accumbens-ventral pallidum connection may be a common pathway for both stimulant and opiate reinforcement. Again, the precise part of the substantia inominata ventral pallidum regions critical for this functional output of the nucleus accumbens will need to be delineated in future studies.

AFFERENT PROJECTIONS TO NUCLEUS ACCUMBENS-DRUG REINFORCEMENT

As discussed above, the major inputs to the nucleus accumbens originate in limbic areas such as the hippocampus, frontal cortex, thalamus, and amygdala (Kelley and Stinus, 1984). Particularly striking is the correspondence between the thalamic and hippocampal afferents which project to the medial nucleus accumbens. There is a high density of opiate receptors in the medial nucleus accumbens (Herkenham and Pert, 1981) and also the medial nucleus accumbens contains a significant amount of cholecystokinin which appears to be co-localized with dopamine in the mesocorticolimbic dopaminergic neurons (Hokfelt et al., 1980). High concentrations of glutamate receptors, particularly the NMDA subtype, have been localized to the nucleus accumbens (Cotman et al., 1987), and at least one projection to the nucleus accumbens, the hippocampal projection, is thought to be glutamatergic (Walaas and Fonnum, 1979).

Glutamate

Glutamate antagonists, when injected into the nucleus accumbens, block the locomotor activation produced by carbachol injections into the dentate gyrus of the hippocampus (Mogenson and Nielson, 1984a) and block the locomotor activation produced by systemic injection of psychomotor stimulants (Pulvirenti et al., 1989). Also, administration of the selective NMDA antagonist, APV, into the nucleus accumbens in rats trained to self administer cocaine intravenously produced an increase in cocaine self-administration (Pulvirenti, Rassnick, and Koob, unpublished results). This increase occurred at doses similar to those used to block cocaine-induced activation. Again, this increase was interpreted as a decrease in the reinforcing effects of cocaine analogous to the increases in self-administration observed with dopamine receptor blockade (Ettenberg et al., 1982). These results suggest that glutamate may modulate dopamine release or the effects of dopamine release in the nucleus accumbens, and thus may modulate the reinforcing actions of cocaine.

Cholecystokinin

Immunocytochemical studies have shown that a subset of neurons in the ventral tegmental area contain both dopamine and CCK (Hokfelt et al., 1980), and significant evidence has been generated to show that CCK and dopamine may be functionally opposed to each other in the nucleus accumbens (Schneider et al., 1983; Van Ree et al., 1983). Not only does CCK injected into the nucleus accumbens attenuate the locomotor activating effects of psychomotor stimulants (Weiss et al., 1989), but CCK also attenuates the rewarding effects of intracranial self-stimulation from the ventral tegmental area. In preliminary results from our laboratory CCK, when injected into the nucleus accumbens, increases cocaine self-administration, again suggesting a decrease in the reinforcing effects of cocaine (Weiss, Vaccarino, and Koob, unpublished results).

DISCUSSION

The "supersensitive" locomotor response preparation allows a means by which to trace the neural substates mediating an activation of a specific population of dopamine receptors through the extrapyramidal motor system. The stimulant effects resulting from activation of dopamine receptors in the region of the nucleus accumbens depend on the integrity of cells within the substantia inominata-ventral pallidum. This nucleus accumbens-ventral pallidum connection appears to be GABAergic since low doses of the GABA agonist muscimol, when injected into the region of the ventral pallidum, will reverse the supersensitive locomotor response. While the third link in this output remains under study, the supersensitive locomotor response appears to depend, at least in part, on the integrity of the DMT.

These results are consistent with earlier work by Mogenson and colleagues showing that various forms of locomotor activity can be reversed by infusion of GABA agonists into the ventral pallidum. For example, DA injected into the nucleus accumbens produces hyperactivity and this hyperactivity is reversed by administration of GABA into the ventral pallidum (Jones and Mogenson, 1980). Locomotor activity elicited by injections of the GABA antagonist picrotoxin into the ventral tegmental area also is inhibited by injection of GABA into the ventral pallidum (Mogenson et al., 1980), and exploratory behavior produced by exposure to a novel open field is inhibited by a ventral pallidal infusion of GABA (Mogenson and Nielson, 1984b).

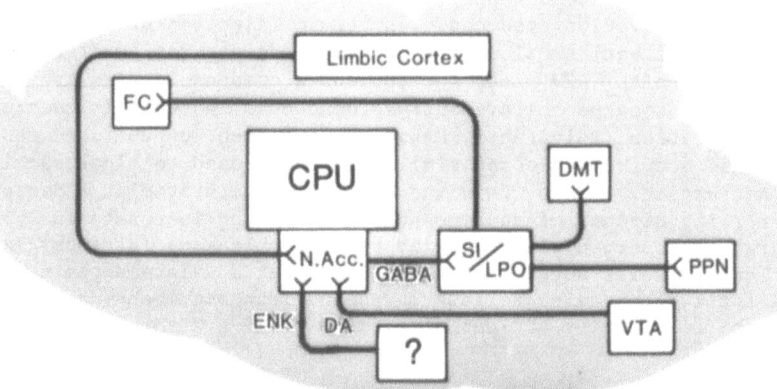

Fig. 5.   A model showing the efferent connections of the nucleus accumbens. Results from Mogenson and colleagues and our laboratory concur in describing a functionally significant GABAergic projection from the nucleus accumbens to the ventral pallidum (SI/LPO). The projection from the ventral pallidum to the mesencephalic motor region (PPN in this model) is proposed to be functionally important by Mogenson and colleagues (Mogenson, 1987) and a projection from ventral pallidum to the dorsal medial thalamus (DMT) has been proposed to be functionally important by Swerdlow and Koob (1987a). Taken with permission from Koob and Swerdlow, 1988.

The ultimate processing of locomotor activation beyond the region of the ventral pallidum is not well understood. The data presented here suggest that the DMT may at least be part of the output system, and thus the ultimate final common pathway may be the corticospinal pathway. Activity from pallidothalamic projections and subsequently thalamocortical projections may thus direct limbic and cortical information to appropriate motor groups ultimately through the corticospinal system (see Fig. 5). Others have provided evidence that the PPN may be involved in the output of ventral pallidal activity. Locomotor activation stimulated by infusion of amphetamine into the nucleus accumbens is blocked by injection of procaine into the region of the PPN (Brudynski and Mogenson, 1985). Kainic acid lesions of the PPN block the effects of amphetamine injected into the nucleus accumbens but also produce a significant hyperactivity in control animals. In contrast, ibotenic acid lesions of this region in DA denervated rats failed to block the supersensitive locomotor response to apomorphine (Swerdlow and Koob, 1987a).

Of particular interest is the recent data showing that the reinforcing properties of both indirect psychomotor stimulants (cocaine) and opiates (heroin) may also be processed through the ventral pallidum (Hubner and Koob, 1990). This parallel circuitry for locomotor activation and reinforcement at the level of the pallidum is consistent with earlier observations at the level of the nucleus accumbens. Denervation of DA in the nucleus accumbens blocks cocaine locomotor activation (Kelly and Iversen, 1976) and cocaine reinforcement (Roberts et al., 1980), but fails to block opiate locomotion (Vaccarino et al., 1986) or opiate reinforcement (Pettit et al., 1984). However, kainic acid lesions of the nucleus accumbens block both cocaine and heroin self-administration (Zito et al., 1985).

The processing of drug reinforcing stimuli beyond the ventral pallidum is still largely unexplored. Recent data show that lesions of the PPN block the conditioned place preferences produced by morphine and amphetamine (Bechara and Van der Kooy, 1989). These data again suggest a possible role for the PPN in mediating the behavioral effects of limbic activation. What role the DMT has in processing drug reinforcing stimuli remains to be determined.

As discussed above, the afferent projections to the nucleus accumbens originate in limbic areas such as the hippocampus and amygdala. However, largely unknown at this time is the neurochemical content of these afferents and their role, if any, in modulating the psychostimulant and reinforcing properties of indirect sympathomimetics and opiates. Preliminary results suggest that both CCK and glutamate may selectively decrease DA release and thus modulate the locomotor and reinforcing actions of indirect sympathomimetics. There is some evidence to suggest that the hippocampal input to the nucleus accumbens is glutaminergic, and that CCK is co-localized with DA in the mesocorticolimbic system.

The circuitry described here that uses the nucleus accumbens as a focal point may be involved not only in drug activation but also in the activation associated with motivated behavior, particularly that arising from the presentation of positively reinforcing stimuli. This close correspondence of the neural substrates for locomotor activation and for the reinforcing properties of drugs is intriguing, and whether a similar correspondence extends to primates and man remains to be determined. In addition, such a neural circuitry important for motivated behavior may have significance for understanding the pathophysiology of drug dependence and mental disorders (Koob and Bloom, 1988; Swerdlow and Koob, 1987a; Swerdlow and Koob, 1987b).

ACKNOWLEDGEMENTS

This work was supported in part by NIDA grants DA-04398 and DA-04043. We thank the MEM Word Processing Center for manuscript preparation.

REFERENCES

Agnati, L.F., and Fuxe, K., 1983, Subcortical limbic ^{3}H-N-propylnorapomorphine binding sites are markedly modulated by cholecystokinin-8 in vitro, Biosci. Rep., 3:1101.

Alheid, G.F., and Heimer, L., 1988, New perspectives in basal forebrain organization of special relevance for neuropsychiatric disorders: the striato pallidal, amygdaloid, and corticopetal components of substantia inominata, Neuroscience, 27:1.

Amalric, M., and Koob, G., 1985, Low doses of methylnaloxonium in the nucleus accumbens antagonize hyperactivity induced by heroin in the rat, Pharmacol. Biochem. Behav., 23:411.

Bechara, A., and Van der Kooy, D., 1989, The tegmental pedunculopontine nucleus: A brain-stem output of the limbic system critical for the conditioned place preferences produced by morphine and amphetamine, J. Neurosci., 9:3400.

Bozarth, M.A., and Wise, R.A., 1981, Intracranial self-administration of morphine into the ventral tegmental area in rats, Life Sci., 28:551.

Brudynski, S.M., and Mogenson, G.J., 1985, Association of the mesencephalic locomotor region with locomotor activity induced by injections of amphetamine into the nucleus accumbens, Brain Res., 334:77.

Cotman, C.W., Monaghan, D.T., O'Hersen, O.P., and Storm-Mathisen, J., 1987, Anatomical organization of excitatory amino acid receptors and their pathways, Trends Neurosci., 10:273.

Divac, I., Kosmal, A., Bjorklund, A., and Lindvall, D., 1978, Subcortical projections to the prefrontal cortex in the rat as revealed by the horseradish peroxidase technique, Neuroscience, 3:785.

Ettenberg, A., Pettit, H.O., Bloom, F.E., and Koob, G.F., 1982, Heroin and cocaine self-administration in rats: Mediation by separate neural systems, Psychopharmacology, 78:204.

Goeders, N.E., Lane, J.D., and Smith, J.E., 1984, Intracranial self-administration of methionine enkephalin into the nucleus accumbens, Pharmacol. Biochem. Behav., 20:451.

Grillner, S., and Shik, M.I., 1973, On the descending control of the lumbrosacral spinal cord from the mesencephalic locomotor region, Acta Physiol. Scand., 87:320.

Heimer, L., and Wilson, R.D., 1975, The subcortical projections of the allocortex: Similarities in the neural associations of the hippocampus, the piriform cortex, and the neocortex. In: "Golgi Centennial Symposium Proceedings," M. Santini, ed., Raven Press, New York. p. 177.

Herkenham, M., and Pert, C.B., 1981, Mosaic distribution of opiate receptors, parafascicular projections and acetylcholinesterase in rat striatum, Nature, 291:415.

Hokfelt, T., Skirboll, R., Rehfeld, J.F., Goldstein, M., Markey, K., and Dann, O., 1980, A subpopulation of mesencephalic dopamine neurons projecting to limbic areas contains a cholecystokinin-like peptide: Evidence from immunohistochemistry combined with retrograde tracing, Neuroscience, 5:2093.

Hubner, C.B., and Koob, G.F., 1990, The ventral pallidum plays a role in mediating cocaine and heroin self-administration in the rat, Brain Res., 508:20.

Jones, D.L., and Mogenson, G.J., 1980, Nucleus accumbens to globus pallidus GABA projection subserving ambulatory activity, Am. J. Physiol., 238:R63.

Joyce, E.M., Stinus, L., and Iversen, S.D., 1983, Effect of injections of 6-OHDA into either nucleus accumbens septi or frontal cortex on spontaneous and drug-induced activity, Neuropharmacology, 22:1141.

Kalivas, P.W., Nemeroff, C.B., and Prange, A.J.Jr., 1984, Neurotensin microinjection into the nucleus accumbens antagonizes dopamine-induced increase in locomotion and rearing, Neuroscience, 4:919.

Kelley, A.E., and Domesick, V.B., 1982, The distribution of the projection from the hippocampal formation to the nucleus accumbens in the rat: An anterograde and retrograde horseradish peroxidase study, Neuroscience, 7:2321.

Kelley, A.E., and Stinus, L., 1984, The distribution of the projection from the parataenial nucleus of the thalamus to the nucleus accumbens in the rat: An autoradiographic study, Exp. Brain Res., 54:499.

Kelly, P.H., and Iversen, S.D., 1976, Selective 6-OHDA-induced destruction of mesolimbic dopamine neurons: Abolition of psychostimulant-induced locomotor activity in rats, Eur. J. Pharmacol., 40:45.

Kelly, P.H., Seviour, P.W., and Iversen, S.D., 1975, Amphetamine and apomorphine responses in the rat following 6-OHDA lesions of the nucleus accumbens septi and corpus striatum, Brain Res., 94:507.

Koob, G.F., and Bloom, F.E., 1988, Cellular and molecular mechanisms of drug dependence, Science, 242:715.

Koob, G.F., and Goeders, N., 1988, Neuroanatomical substrates of drug self-administration. In: "Neuropharmacological Basis of Reward," J.M. Liebman, and S.J. Cooper, eds., Oxford University Press, Oxford. p. 214.

Koob, G.F., Riley, S.J., Smith, S.C., and Robbins, T.W., 1978, Effects of 6-hydroxydopamine lesions of the nucleus accumbens septi and olfactory tubercle on feeding, locomotor activity, and amphetamine anorexia in the rat, J. Comp. Physiol. Psychol., 92:917.

Koob, G.F., and Swerdlow, N.R., 1988, Functional output of the mesolimbic dopamine system, Ann. NY Acad. Sci., 537:216.

Koob, G.F., Vaccarino, F.J., Amalric, M., and Bloom, F.E., 1987, Positive reinforcement properties of drugs: Search for neural substrates. In: "Brain Reward Systems and Abuse," J. Engel, and L. Oreland, eds., Raven Press, New York. p. 35.

Lyness, W.H., Friedle, N.M., and Moore, K.E., 1979, Destruction of dopaminergic nerve terminals in nucleus accumbens: Effect on d-amphetamine self-administration, Pharmacol. Biochem. Behav., 11:553.

Mogenson, G.J., and Nielson, M., 1984a, A study of the contribution of hippocampal-accumbens-subpallidal projections to locomotor activity, Behav. Neural Biol., 42:38.

Mogenson, G.J., and Nielson, M., 1984b, Neuropharmacological evidence to suggest that the nucleus accumbens and subpallidal regions contribute to exploratory locomotion, Behav. Neural Biol., 42:52.

Mogenson, G.J., and Nielson, M.A., 1983, Evidence that an accumbens to subpallidal GABAergic projection contributes to locomotor activity, Brain Res. Bull., 11:309.

Mogenson, G.J., Wu, M., and Jones, D.L., 1980, Locomotor activity elicited by injections of picrotoxin into the ventral tegmental area is attenuated by injections of GABA into the globus pallidus, Brain Res., 191:569.

Pettit, H.O., Ettenberg, A., Bloom, F.E., and Koob, G.F., 1984, Destruction of dopamine in the nucleus accumbens selectively attenuates cocaine but not heroin self-administration in rats, Psychopharmacology, 84:167.

Pulvirenti, L., Swerdlow, N.R., and Koob, G.F., 1989, Microinjection of a glutamate antagonist into the nucleus accumbens reduces psychostimulant locomotion in rats, Neurosci. Lett., 103:197.

Pycock, C., and Horton, R., 1976, Evidence for an accumbens-pallidal pathway in the rat and its possible gabaminergic control, Brain Res., 110:629.

Robbins, T.W., and Koob, G.F., 1980, Selective disruption of displacement behaviour by lesions of the mesolimbic dopamine system, Nature, 285:409.

Roberts, D.C.S., Corcoran, M.E., and Fibiger, H.C., 1977, On the role of ascending catecholaminergic systems in intravenous self-administration of cocaine, Pharmacol. Biochem. Behav., 6:615.

Roberts, D.C.S., Koob, G.F., Klonoff, P., and Fibiger, H.C., 1980, Extinction and recovery of cocaine self-administration following 6-hydroxydopamine lesions of the nucleus accumbens, Pharmacol. Biochem. Behav., 12:781.

Roberts, D.C.S., Zis, A.P., and Fibiger, H.C., 1975, Ascending catecholamine pathways and amphetamine-induced locomotor activity: Importance of dopamine and apparent non-involvement of norepinephrine, Brain Res., 93:441.

Schneider, L.H., Alpert, J.E., and Iversen, S.D., 1983, CCK-8 modulation of mesolimbic dopamine: Antagonism of amphetamine-stimulated behaviors, Peptides, 4:749.

Skinner, R.D., and Garcia-Rill, E., 1984, The mesencephalic locomotor region (MLR) in the rat, Brain Res., 323:385.

Staunton, D.A., Magistretti, P.J., Koob, G.F., Shoemaker, W.J., and Bloom, F.E., 1982, Dopaminergic supersensitivity induced by denervation and chronic receptor blockade is additive, Nature, 299:72.

Swanson, L.W., Mogenson, G.J., Gerfen, C.R., and Robinson, P., 1984, Evidence for a projection from the lateral preoptic area and substantia innominata to the "mesencephalic locomotor region" in the rat, Brain Res., 295:161.

Swerdlow, N.R., Amalric, M., and Koob, G.F., 1987, Nucleus accumbens opiate-dopamine interactions and locomotor activation in the rat: Evidence for a pre-synaptic locus, Pharmacol. Biochem. Behav., 26:765.

Swerdlow, N.R., and Koob, G.F., 1984, Neural substrates of apomorphine-stimulated locomotor activity following denervation of the nucleus accumbens, Life Sci., 35:2537.

Swerdlow, N.R., and Koob, G.F., 1985, Separate neural substrates of the locomotor-activating properties of amphetamine, caffeine and corticotropin releasing factor (CRF) in the rat, Pharmacol. Biochem. Behav., 23:303.

Swerdlow, N.R., and Koob, G.F., 1987a, Lesions of the dorsomedial nucleus of the thalamus, medial prefrontal cortex and pedunculopontine nucleus: Effects on locomotor activity mediated by nucleus accumbens-ventral pallidal circuitry, Brain Res., 412:233.

Swerdlow, N.R., and Koob, G.F., 1987b, Dopamine schizophrenia, mania and depression: Towards a unified hypothesis of cortico-striato-pallido-thalamic function, Behav. Brain Sci., 10:197.

Swerdlow, N.R., Swanson, L.W., and Koob, G.F., 1984a, Electrolytic lesions of the substantia inominata and lateral preoptic area attenuate the "supersensitive" locomotor response to apomorphine resulting from denervation of the nucleus accumbens, Brain Res., 306:141.

Swerdlow, N.R., Swanson, L.W., and Koob, G.F., 1984b, Substantia innominata: Critical link in the behavioral expresssion of mesolimbic dopamine stimulation in the rat, Neurosci. Lett., 50:19.

Taghzouti, K., Simon, H., Louilot, A., Herman, J.P., and Le Moal, M., 1985, Behavioral study after local injection of 6-hydroxydopamine into the nucleus accumbens in the rat, Brain Res., 344:9.

Vaccarino, F.J., Amalric, M., Swerdlow, N.R., and Koob, G.F., 1986, Blockade of amphetamine- but not opiate-induced locomotion following antagonism of dopamine function in the rat, Pharmacol. Biochem. Behav., 24:61.

Vaccarino, F.J., Bloom, F.E., and Koob, G.F., 1985, Blockade of nucleus accumbens opiate receptors attenuates intravenous heroin reward in the rat, Psychopharmacology, 85:37.

Van der Kooy, D., Swerdlow, N.R., and Koob, G.F., 1983, Paradoxical rein-forcing properties of apomorphine: Effects of nucleus accumbens and area postrema lesions, Brain Res., 259:111.

Van Ree, J.M., Gaffori, O., and De Wied, D., 1983, In rats the behavioral profile of CCK-8 related peptides resembles that of antipsychotic agents, Eur. J. Pharmacol., 93:63.

Walaas, I., and Fonnum, F., 1979, The effects of surgical and chemical le-sions on neurotransmitter caudidates in the nucleus accumbens of the rat, Neuroscience, 4:209.

Wang, R.Y., and Hu, X-T, 1986, Does cholecystokinin potentiate dopamine action in the nucleus accumbens?, Brain Res., 380:363.

Weiss, F., Ettenberg, A., and Koob, G.F., 1989, CCK-8 injected into the nucleus accumbens attenuates the supersensitive locomotor response to apomorphine in 6-OHDA and chronic neuroleptic treated rats, Psychopharmacology, 99:409.

White, F.J., and Wang, R.Y., 1984, Interactions of cholecystokinin oc-tapeptide and dopamine on nucleus accumbens neurons, Brain Res., 300:161.

Young, W.S., Alheid, G.F., and Heimer, L., 1984, The ventral pallidal projection to the mediodorsal thalamus: A study with fluorescent retrograde tracers and immunohistofluorescence, J. Neurosci., 4:1626.

Zaborsky, L., Alheid, G.F., Alones, V.E., Oertel, W.H., Schmechel, D.E., and Heimer, L., 1982, Afferents of the ventral pallidum studied with a combined immunohistochemical-anterograde degeneration method, Soc. Neurosci. Abs., 8:218.

Zito, K.A., Vickers, G., and Roberts, D.C.S., 1985, Disruption of cocaine and heroin self-administration following kainic acid lesions of the nucleus accumbens, Pharmacol. Biochem. Behav., 23:1029.

EFFECT OF GABAERGIC AND GLUTAMATERGIC DRUGS INJECTED INTO THE VENTRAL

PALLIDUM ON LOCOMOTOR ACTIVITY

Lane J. Wallace and Norman J. Uretsky

Ohio State University
College of Pharmacy
Columbus, OH

There has been a great deal of recent interest in elucidating the
behavioral role of the basal forebrain in cognitive function (Alheid and
Heimer, 1988). The ventral pallidum, a part of the basal forebrain, has
received a great deal of attention because of the relationship between the
degeneration of neurons in this region and the symptoms of cognitive dysfunction associated with Alzheimer's disease (Whitehouse et al., 1983; Coyle et
al., 1983; Ezrin-Waters and Resch, 1986; Smith, 1988). However the ventral
pallidum has also been implicated in other functions. Thus, the results of
several studies have suggested that the ventral pallidum plays an important
role in the regulation of movement or locomotion. This hypothesis is
supported by the observations that specific neurotransmitters injected into
the ventral pallidum markedly affect normal locomotion (Scheel-Kruger, 1983,
1986; Baud et al., 1988) and the locomotor stimulation produced by drugs that
act on neurons in the nucleus accumbens (Mogenson et al., 1980; Heimer et
al., 1982; Mogenson and Nielsen, 1983; Swerdlow et al., 1986; Mogenson, 1987;
Swerdlow et al., 1986; Austin and Kalivas, 1988). In this chapter we will
review these studies and report on our work indicating that quisqualic acid
receptors in the ventral pallidum are involved in the regulation of locomotion induced by psychostimulant drugs.

EFFECT OF GABAERGIC AGONISTS ADMINISTERED INTO THE VENTRAL PALLIDUM ON
LOCOMOTOR ACTIVITY

The activation of GABA receptors in the ventral pallidum can produce
both stimulatory and inhibitory effects on locomotor activity, depending on
the site of drug injection. Thus, it has been reported that the local
administration of either high doses (25-50 μg) of muscimol, a GABA receptor
agonist, or ethanolamine-o-sulfate, an inhibitor of GABA metabolism, into
specific sites in the ventral pallidum produces locomotor stimulation in rats
(Scheel-Kruger, 1983, 1986; Baud et al., 1988). A similar hypermotility
response is reported to occur after ibotenic acid lesions of this region
(Dubois et al., 1985; Hepler and Lerer, 1986; Baud et al., 1988). According
to the histological examination of the needle tracks, the sites of these
injections were ventral to the globus pallidus, below the anterior commiss-
ures in the region of the cholinergic cells of the nucleus basalis magnocellularis. This region is thought to be the counterpart of the cholinergic
cells of the nucleus basalis of Meynert in humans; the cells that degenerate
in Alzheimer's disease (Alheid and Heimer, 1988). The local administration

of muscimol into this site of the ventral pallidum has been reported to inhibit cortical acetylcholine turnover (Wood and Richard, 1982), which is consistent with an inhibitory effect of muscimol on nucleus basalis choliner-gic neurons, which project to the cortex. Therefore, the stimulation of locomotion by muscimol may be related to its ability to inhibit the firing rate of cholinergic neurons of the nucleus basalis, resulting in an inhibi-tion of the release of acetylcholine from nerve terminals in the cerebral cortex (Baud et al., 1988).

In contrast, the administration of either GABA or muscimol into more medial sites that fall within the substantia innominata - lateral preoptic region (SI/LPO) of the ventral pallidum produces an inhibitory effect on locomotor stimulation. For example, the systemic administration of amphet-amine, heroin, or apomorphine (in animals with 6-hydroxydopamine-induced lesions in the nucleus accumbens) produces a stimulatory effect on locomotion (Swerdlow et al., 1986). This stimulation, which is thought to be due to an action of these drugs on neurons in the nucleus accumbens, was antagonized by the direct administration of low doses of muscimol (2-25 µg) into the SI/LPO (Swerdlow and Koob, 1984, 1985). Similarly, the injection of muscimol or GABA into the SI/LPO inhibited the stimulation of locomotion produced by the direct administration of dopamine, amphetamine, picrotoxin, carbachol or excitatory amino acids into the nucleus accumbens (Mogensen and Nielsen, 1983; Shreve and Uretsky, 1988; Austin and Kalivas, 1988; see also Table 1). The finding that the activation of GABAergic mechanisms in the SI/LPO inhibits locomotion is in agreement with the finding that the administration into the SI/LPO of picrotoxin, a GABAergic antagonist, stimulates locomotion (Mogenson and Nielsen, 1983; Shreve and Uretsky, 1988).

The inhibitory effect of GABA agonists in the SI/LPO has led to the proposal that the locomotor stimulation produced by drugs which act in the nucleus accumbens is mediated by a decrease in the activity of a nucleus

Table 1. Effect of muscimol injected into the SI/LPO on locomotor activity elicited by injection of various drugs into the nucleus accumbens.

SI/LPO	DRUG IN NUCLEUS ACCUMBENS				
	Amphetamine 25 µg	Picrotoxin 0.5 µg	AMPA 0.5 µg	Kainic Acid 15 ng	NMDA 2.5 µg
Saline	7530±1020	7050±1600	9420±1190	3890±1190	2950±410
Muscimol 25µg	1320±779	1000±369	579±246	684±266	211±205

Rats anesthetized with halothane were injected with either saline or muscimol in the SI/LPO and indicated compound in the nucleus accumbens. Following recovery (5-10 min. after injection), locomo-tor activity was assessed as number of interruptions of photocell beams in an activity cage. Numbers shown are means ± sem for one hour of observation with 4 or 5 observations per group. Abbrevia-tions are: AMPA, α-amino-3-hydroxy-5-methylisoxazole-4-propionic acid; NMDA, N-methyl-D-aspartic acid; SI/LPO, substantia innominata/ lateral preoptic region.

accumbens-ventral pallidal GABAergic pathway, resulting in a decreased release of GABA and a decreased activation of GABAergic receptors in the ventral pallidum (Swerdlow et al., 1986; Mogenson, 1987). This would eliminate an inhibitory action of GABA in the SI/LPO, resulting in locomotor stimulation. The inhibitory effect of GABA in the SI/LPO is consistent with biochemical and physiological evidence supporting the existence of a large GABAergic neuronal projection, originating in the nucleus accumbens and terminating in the ventral pallidum (Zahm et al.,1985; Churchill et al., 1990).

EFFECT OF EXCITATORY AMINO ACIDS INJECTED INTO THE SI/LPO ON LOCOMOTOR ACTIVITY

Excitatory neurotransmitters are present in the ventral pallidum and may be involved in the expression of locomotor behavior. The results of autoradiographic studies indicate that the ventral pallidum contains a high concentration of excitatory amino acid (glutamic acid) receptors (Monaghan and Cotman, 1985). In addition, glutamic acid, an endogenous excitatory amino acid, has been shown to be released in a calcium dependent manner in response to depolarization from synaptosomes prepared from the ventral pallidum (Davies et al., 1984). Finally, the direct administration of the excitatory amino acid agonists, (R,S)-α-amino-3-hydroxy-5-methylisoxazole-4-propionic acid (AMPA), kainic acid, or N-methyl-D-aspartic acid (NMDA) into the SI/LPO region of the ventral pallidum produces an intense stimulation of locomotor activity (Shreve and Uretsky, 1989; see also Fig. 1). Each of these compounds is a specific agonist for one of the three subtypes of glutamate receptors. Thus, all three subtypes appear to be capable of regulating locomotor activity, but activation of quiqualate receptors (by AMPA) produces the largest locomotor stimulation.

The locomotor stimulation produced by the excitatory amino acids in the SI/LPO can be characterized as intense, but coordinated and controlled. Thus, although the animals were very active, they were able to avoid contact with the walls of the activity cages and obstacles placed in their path. Higher doses produced seizure activity that interfered with locomotion. The dose of NMDA that stimulated locomotion was closer than that of AMPA and kainic acid to the dose that produced seizures.

The hypermotility responses elicited by agonists for each of the three subclasses of excitatory amino acid receptor were attenuated by antagonists used at doses that show appropriate selectivity. Thus, gamma-glutamylamino methanesulfonic acid (GAMS), an antagonist of quisqualic acid receptors, blocked the hypermotility response to AMPA (Shreve and Uretsky, 1989; see also Fig. 2). 6,7-Dinitroquinoxaline-2,3-dione (DNQX), a compound that blocks both quisqualic acid and kainic acid receptors, attenuated the hypermotility response to AMPA and kainic acid. Finally, D-α-aminoadipic acid (DAA), an antagonist of the NMDA receptor, inhibited the response to NMDA. These results suggest that the locomotor stimulation produced by the administration of excitatory amino acids into the ventral pallidum is mediated by activation of specific receptors.

The SI/LPO was the site where injection of excitatory amino acids or picrotoxin produced locomotor stimulation and where injection of GABA agonists produced locomotor inhibition. These observations are consistent with the concept that GABAergic and glutamatergic mechanisms in the SI/LPO may interact to regulate locomotor activity. This concept is supported by the observation that the stimulation of locomotor activity produced by the injection of AMPA into the SI/LPO was antagonized by the coinjection of muscimol (Table 2). Conversely, the stimulation of locomotor activity produced by the GABAergic antagonist, picrotoxin, was inhibited by the

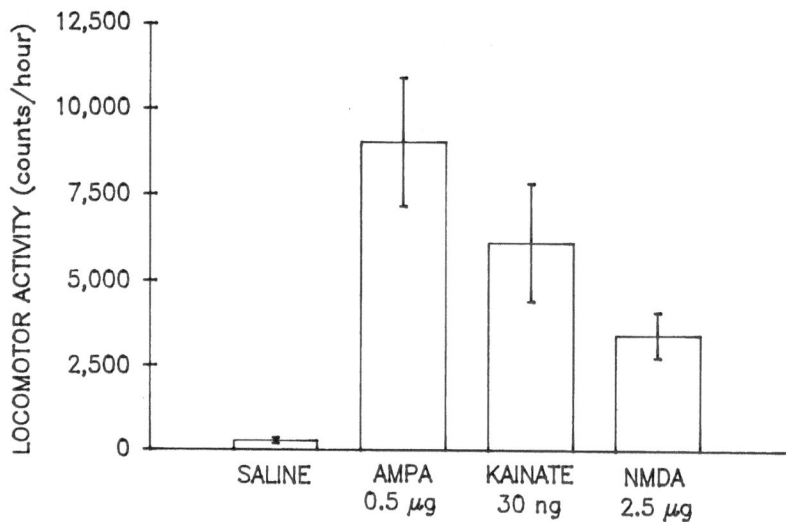

Fig. 1. Effect of excitatory amino acid agonists injected into the SI/LPO on locomotor activity of rats. Rats were anesthetized with a halothane-oxygen mixture and injected into the SI/LPO with various excitatory amino acid agonists or vehicle. Locomotor activity was monitored for one hour after administration of drugs. Data shown are means ± sem of numbers of interruptions of photocell beams for 5 observations per group.

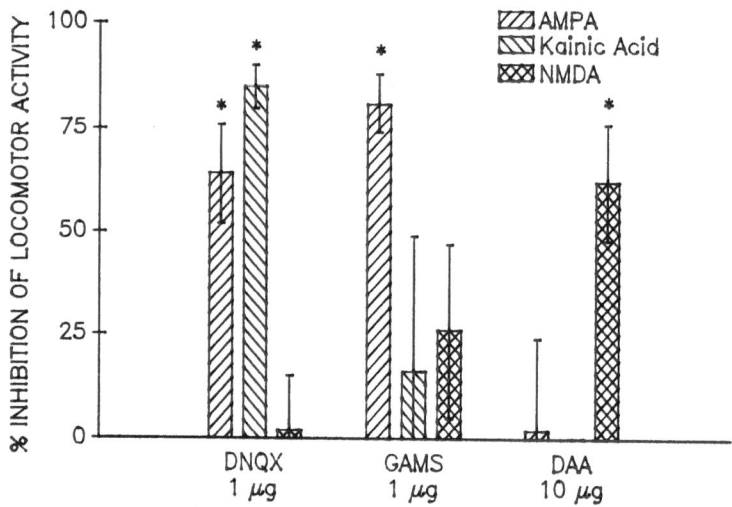

Fig. 2. Effect on locomotor activity of excitatory amino acid antagonists coinjected with excitatory amino acid agonists into the SI/LPO. Rats were anesthetized with halothane-oxygen, and saline, DNQX, GAMS, or DAA was coinjected with either 0.5 μg AMPA, 30 ng kainic acid, or 2.5 μg NMDA. Locomotor activity was monitored for one hour. Data show the mean percent reduction in locomotor activity in the presence of the antagonists as compared to the saline injections, means ± sem for 4-8 observations per group.

Table 2. Interaction between glutamatergic and GABAergic systems
in the SI/LPO on regulation of locomotor activity.

Locomotor Stimulant	Locomotor Inhibitor	% Inhibition
AMPA (0.5 µg)	Muscimol (25 ng)	98 ± 29
Picrotoxin (0.5 µg)	DNQX (1 µg)	66 ± 15

Either saline, muscimol, or DNQX was co-injected into the SI/LPO
with AMPA or picrotoxin. Motility was monitored for 1 hour
after administration of drugs. Data show the mean percent reduction
in locomotor activity produced by muscimol or DNQX when compared to
saline controls, means ± sem for 4 or 5 observations per group.

coinjection of DNQX (Table 2). Thus, locomotor stimulation may require both
a decrease of GABAergic and an increase of glutamatergic transmission in the
SI/LPO.

        Since hypermotility elicited by drugs that act in the nucleus accumbens
is thought to be mediated by a decrease in GABAergic transmission to the
SI/LPO. The data presented above suggest that this hypermotility should be
accompanied by an increase in glutamatergic transmission in the SI/LPO. If
correct, then excitatory amino acid antagonists in the SI/LPO might inhibit
the response. To test this hypothesis, we determined the effect of the
administration of various excitatory amino acid antagonists into the SI/LPO
on the locomotor stimulation produced by the direct administration of
amphetamine and AMPA into the nucleus accumbens. We found that the direct
administration into the SI/LPO of either GAMS or DNQX, inhibitors of the
quisqualic and kainic acid receptors, attenuated the locomotor stimulation
produced by the injection of amphetamine or AMPA into the nucleus accumbens
(Fig. 3). However, DAA, an inhibitor of the NMDA receptor, did not significantly alter the hypermotility response to the intra-accumbens administration
of either amphetamine or AMPA. These results suggest that the locomotor
stimulation produced by drugs administered into the nucleus accumbens may
require the activation of quisqualic acid or kainic acid receptors in the
SI/LPO.

        The locomotion produced by amphetamine administered systemically is
thought to be mediated by an action of amphetamine in the nucleus accumbens
(Pijnenburg et al., 1976). Therefore, DNQX and GAMS, which inhibited the
hypermotility response to amphetamine after intra-acumbens administration,
should also inhibit the hypermotility response produced by amphetamine
administered systemically. To test this hypothesis, DNQX or GAMS was
administered into the SI/LPO just before the systemic administration of
amphetamine. As expected, the locomotor stimulation produced by amphetamine
was inhibited by both DNQX and GAMS injected into the SI/LPO at doses that
produced no change in control locomotor activity (Willins et al., 1990; see
also Fig. 4).

        To test whether the inhibitory effects of DNQX were due to a blockade
of receptors for AMPA (quisqualic acid receptors), we compared DNQX and two
of its structural analogs for their ability to inhibit the locomotor response
to amphetamine and to inhibit the binding of tritiated AMPA to brain membrane

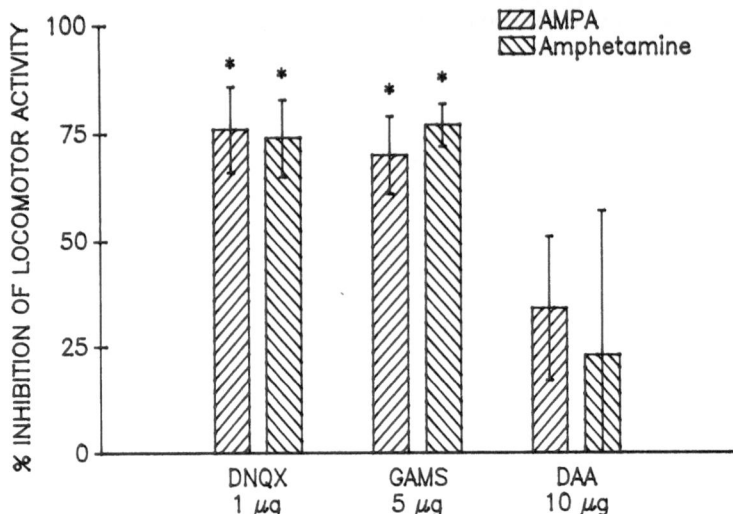

Fig. 3.   Effect of excitatory amino acid antagonists injected into the
SI/LPO on locomotor activity elicited by injection of AMPA or
amphetamine into the nucleus accumbens.   Rats were anesthetized
with a halothane-oxygen mixture.   Saline, DNQX, GAMS, or DAA was
injected into the ventral pallidum immediately prior to the
injection of either 0.25 μg AMPA or 10 μg amphetamine into the
nucleus accumbens.   Locomotor activity was monitored for one hour.
Data show the mean percent reduction in locomotor activity in the
presence of the antagonists as compared to saline controls, means ±
sem for 4-7 observations per group.

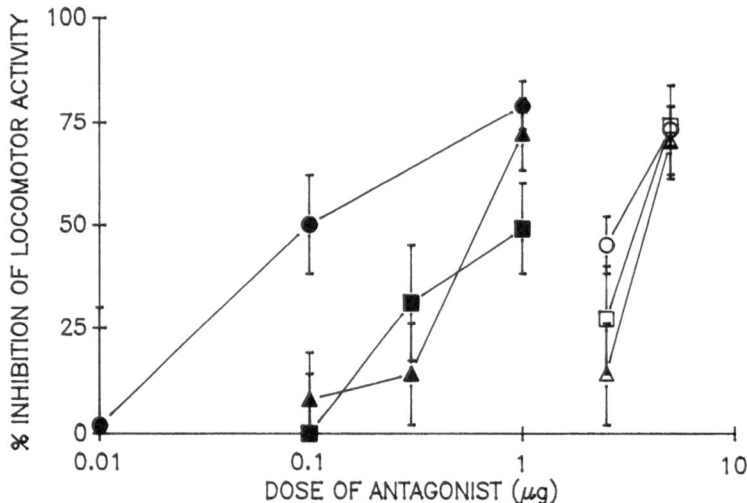

Fig. 4.   Effect of quisqualic acid receptor antagonists injected into the
SI/LPO on locomotor activity elicited by systemic injection of
amphetamine, scopolamine, and caffeine.   Either saline, DNQX
(solid symbols), or GAMS (open symbols) was injected just prior to
the administration of 0.5 mg/kg d-amphetamine (circles), 0.5 mg/kg
scopolamine (squares), or 20 mg/kg caffeine (triangles).   Locomotor activity was monitored for one hour.   Data show the mean
percent reduction in locomotor activity in the presence of the
antagonists, means ± sem for at least 5 observations per group.

fractions (Supko et al., 1990; Willins et al., 1990). 6-Amino-7-fluoroquin-oxaline-2,3-dione (AFQX) is an analog of DNQX that was found to weakly inhibit [^{3}H]-AMPA binding with an IC$_{50}$ of 100 uM, which is 200 times less potent than DNQX. 2-hydroxy-3,5-dinitrophenylalanine was found to inhibit [^{3}H]-AMPA binding with an IC$_{50}$ of 9 uM, which is 18 times less potent than DNQX. The potency of these compounds for AMPA binding was found to parallel their ability to inhibit the locomotor stimulation produced by amphetamine. Thus, while 2-hydroxy-3,5-dinitrophenylalanine injected into the SI/LPO at a dose of 3.25 nmoles was able to inhibit the locomotor stimulation produced by amphetamine, AFQX at a dose of 5 nmoles was ineffective. These result support the concept that AMPA (quisqualic acid) receptors in the SI/LPO are important for the locomotor stimulation produced by amphetamine.

Glutamatergic mechanisms in the ventral pallidum may also be involved in regulation of locomotion elicited by drugs that are thought to produce their effects by acting in brain areas other than the nucleus accumbens. Thus, systemic administration of caffeine or scopolamine produces a stimu-lation of locomotor activity that apparently does not involve the nucleus accumbens. The locomotor stimulation produced by the systemic administration of both drugs was found to be inhibited by the injection of either DNQX or GAMS into the SI/LPO (Willins et al., 1990; see also Fig. 4). These observa-tions are consistent with the hypothesis that projections regulating locomo-tion that originate from the nucleus accumbens and from other areas of the brain converge in the ventral pallidum and that activation of receptors for AMPA (quisqualate) receptors in the ventral pallidum is a necessary condition for stimulation of locomotion.

REFERENCES

Alheid, G.F. and Heimer, L., 1988, New perspectives in basal forebrain organization of special relevance for neuropsychiatric disorders: the striatopallidal, amygdaloid, amd corticopetal components of the substantia innominata, Neuroscience, 27:1.
Austin, M. and Kalivas, P., 1988, The effect of cholinergic stimulation in the nucleus accumbens on locomotor behavior, Brain Res., 441:209.
Baud, P., Mayo, W., Le Moal, M. and Simon, H., 1988, Locomotor hyperactivi ty in the rat after infusion of muscimol and [D-Ala^2]Met-enkephalin into the nucleus basalis magnocellularis. Possible interaction with central cholinergic projections, Brain Res., 452:203.
Churchill, L., Dilts, R.P. and Kalivas, P., 1990, Changes in γ-aminobutyric acid, u-opioid and neurotensin receptors in the accumbens-pallidal projection after discrete quinolinic acid lesions in the nucleus accumbens, Brain Res., 511:41.
Coyle, J., Price, D., and Delong, M., 1983, Alzheimer's disease: a disorder of cortical cholinergic innervation, Science, 219:1184.
Davies, S., McBean, G. and Roberts,P., 1984, A glutamatergic innervation of the nucleus basalis/substantia innominata, Neurosci. Lett., 45:105.
Dubois, B., Mayo, W., Agid,Y., Lemoal,M., and Simon, H., 1985, Profound disturbances of spontaneous and learned behaviors following lesions of the nucleus basalis magnocellularis in the rat, Brain Res., 338:249.
Ezrin-Waters, C. and Resch, L., 1986, The nucleus basalis of Meynert, Cand. J. Nerosci., 13:8.
Heimer,L., Switzer, R.D. and Hoesen, G.,1982, Ventral striatum and ventral pallidum. Components of the motor system, Trends Neurosci., 5:83.
Hepler, D. and Lerer, B., 1986, The effect of magnocellular basal forebrain lesions on circadian locomotor activity in the rat, Pharmacol. Biochem. Behav., 25:293.

Mogenson, G., Jones, D. and Yim, C.Y., 1980, From motivation to action: Functional interface between the limbic system and the motor system, Prog. Neurobiol., 14:69.

Mogenson, G., 1987, Limbic-motor integration, Prog. Psychobiol. Physiol. Psychol., 12:117.

Mogenson, G. and Nielsen M., 1983, Evidence that an accumbens to subpal lidal GABAergic projection contributes to locomotor activity, Brain Res. Bull., 11:309.

Monaghan, D.T. and Cotman, C.W., 1985, Distribution of N-methyl-D-aspartate sensitive L-^{3}H-glutamate binding sites in rat brain. J. Neurosci., 5:2909.

Pijnenburg, A., Honig, W., Van der Heyden, J. and Van Rossum, J.M., 1976, Effects of chemical stimulation of the mesolimbic dopamine system upon locomotor activity. Eur. J. Pharmacol., 35:45.

Scheel-Kruger, J., 1983, The GABA receptor and animal behavior, in: "The GABA Receptors," S. Enna, ed., The Humana Press, Clifton, New Jersey, p. 215.

Scheel-Kruger, J., 1986, Dopamine-GABA interactions: evidence that GABA transmits, modulates, and mediates dopaminergic functions in the basal ganglia and the limbic system. Acta Neurol. Scand. suppl 107, 73:1.

Shreve, P. and Uretsky, N., 1988, Effect of GABAergic transmission in the subpallidal region on the hypermotility response to the administration of excitatory amino acids and picrotoxin into the nucleus accumbens, Neuropharmacol., 27:1271.

Shreve, P. and Uretsky, N., 1989, AMPA, kainic acid, and N-methyl-D-aspartic acid stimulate locomotor activity after injection into the substantia innominata/lateral preoptic area, Pharmacol. Biochem. Behav., 34:101.

Smith, G., 1988, Animal models of Alzheimer's disease: experimental cholinergic denervation, Brain Res. Rev., 13:103.

Supko, D., Hill, R.., Uretsky, N., Miller, D., and Wallace, L., 1990, Substituted O-tyrosines and quinoxaline-2,3-diones as potential quisqualate analogs. FASEB J., 4:A603.

Swerdlow, N. and Koob,G., 1984, The neural substrates of apomorphine-stimulated locomotor activity, Life Sci., 35:2537.

Swerdlow, N. and Koob, G., 1985, Seperate neural substrates of the locomo tor-activating properties of amphetamine, heroin, caffeine, and corticotropin releasing factor (CRF) in the rat, Pharmacol. Biochem. Behav., 23:303.

Swerdlow, N., Vaccarino, F., Amalric, M. and Koob,G., 1986, The neural substrates for the motor activating properties of psychostimulants: a review of recent findings, Pharmacol. Biochem. Behav., 25:233.

Whitehouse, P., Price, D., Struble, R. Clark,A., Coyle, J., and DeLong, M., 1982, Alzheimer's disease and senile dementia: loss of neurons in the basal forebrain, Science, 215:1237.

Willins, D., Wallace, L.J., and Uretsky, N., 1990, The role of excitatory amino acid receptors within the subpallidum in the hypermotility response induced by systemically administered stimulants of locomotor activity. FASEB J., 4:A995.

Wood, P.L. and Richard, J., 1982, GABAergic regulation of the substantia innominata-cortical cholinergic pathway, Neuropharmacol., 21:969.

Zahm, D.S., Zaborszky, L., Alones, V. and Heimer,L., 1985, Evidence for the coexistence of glutamate decarboxylase and met-enkephalin immunoreactivities in axon terminals of rat ventral pallidum, Brain Res., 325:317.

GABAERGIC AND ENKEPHALINERGIC REGULATION OF LOCOMOTION IN THE

VENTRAL PALLIDUM: INVOLVEMENT OF THE MESOLIMBIC DOPAMINE SYSTEM

Peter W. Kalivas, Mark A. Klitenick, Hanni Hagler

and Mark C. Austin

Department of Veterinary Comparative Anatomy,
Pharmacology and Physiology, Washington State
University, Pullman WA and Unit on Behavioral
Neuropharmacology, National Institute of Mental
Health, Bldg 10, Rm 4N214, Bethesda, MD

It is well established that the mesolimbic dopamine
projection from the A10 dopamine region in the ventromedial
mesencephalon to the nucleus accumbens is important in mediating
spontaneous and psychostimulant-induced locomotion (Fink and
Smith, 1980; Koob et al., 1981; Clarke et al., 1988). The
nucleus accumbens has a dense projection to the ventral pallidum
(Nauta et al., 1978; Mogenson et al., 1983; Heimer et al., 1987;
Churchill et al., 1990) which utilizes both GABA and enkephalin
as neurotransmitters (Walaas and Fonnum, 1979; Zahm et al.,
1985). Modulation of this projection to the ventral pallidum by
dopamine in the nucleus accumbens mediates dopamine-dependent
locomotion. Thus, locomotion elicited by stimulation of
dopamine receptors in the nucleus accumbens is abolished
following a lesion of the ventral pallidum (Swerdlow et al.,
1984a). A role for GABA transmission in the accumbens-ventral
pallidal pathway has been shown by the observation that
stimulation of $GABA_A$ receptors in the ventral pallidum blocks
motor activity produced by dopamine agonist injection into the
nucleus accumbens (Jones and Mogenson, 1980; Mogenson and
Nielsen, 1983; Swerdlow et al., 1984b; Austin and Kalivas,
1988). More recently, it was found that locomotion elicited by
microinjection of other neurotransmitter agonists or antagonists
into the nucleus accumbens, including a *mu* opioid agonist, $GABA_A$
antagonist, nicotinic agonist and glutamate agonist, is also
blocked by muscimol injection into the ventral pallidum (Austin
and Kalivas, 1989; Shreve and Uretsky, 1988). Therefore, it
appears that locomotion elicited pharmacologically from the
nucleus accumbens involves modulation of a GABAergic projection
to the ventral pallidum, regardless of the neurochemical
stimulus.

Further evidence of a role for the ventral pallidum in
modulating locomotion is that microinjection of opioid agonists
or $GABA_A$ antagonists into this structure will elicit motor
activation (Mogenson and Nielsen, 1983; Napier and Marx, 1987;

The Basal Forebrain, Edited by T.C. Napier *et al.*
Plenum Press, New York, 1991

Swerdlow and Koob, 1987; Napier et al., 1988; Baud et al., 1988; Patel and Slater, 1988; Austin and Kalivas, 1990). A variety of behavioral and electrophysiological data suggest that projections from the ventral pallidum to the pedunculopontine nucleus and/or mediodorsal thalamus further mediate locomotion (Brudzynski and Mogenson, 1985; Haber et al., 1985; Mogenson et al., 1985; Vives and Mogenson, 1985; Swerdlow and Koob, 1987). Thus, it is generally believed that the ventral pallidum and its descending projections constitute a neuronal substrate mediating locomotion that is distal to the release of dopamine in the nucleus accumbens. However, the ventral pallidum, pedunculopontine area and mediodorsal thalamus have projections to the A10 dopamine region and/or nucleus accumbens (Phillipson and Griffiths, 1985; Hallanger and Wainer, 1988; Niijima and Yoshida, 1988), posing the possibility that locomotion elicited from the ventral pallidum may involve feedback onto the mesoaccumbens dopamine pathway. Furthermore, contralateral rotation produced by the unilateral injection of *mu* agonist into the ventral pallidum was blocked by peripheral administration of haloperidol or the D_1 receptor antagonist, SCH-23390 (Napier et al., 1988). Thus, just as locomotion stimulated pharmacologically from the nucleus accumbens is dependent upon GABA transmission in the ventral pallidum, perhaps locomotion elicited from the ventral pallidum may be dependent upon dopamine transmission in the nucleus accumbens.

The present series of studies was designed to evaluate: 1) the role of GABA and enkephalin in the ventral pallidum in mediating locomotion; and 2) the importance of the mesoaccumbens dopamine system in locomotion elicited pharmacologically from the ventral pallidum.

METHODS

Surgery and Housing. Male Sprague-Dawley rats (Laboratory Animal Resource Center, Pullman, WA) were anesthetized, and implanted stereotaxically with injection cannulae over the ventral pallidum or nucleus accumbens. At least seven days were allowed to pass prior to beginning behavioral studies. In some rats, the dopamine system was lesioned by injection of 6-hydroxydopamine (6-OHDA) into the nucleus accumbens 10 days prior to initiating behavioral experiments. Control rats in the 6-OHDA study received a microinjection of ascorbate-saline vehicle. Microinjection of drugs into the ventral pallidum was made bilaterally with a stainless steel injector (33 gauge) connected to a one µl syringe that was mounted in an infusion pump. Injections were made in a total volume of 0.5 µl given over 60 sec. The details of the surgical and injection procedures are described elsewhere (Austin and Kalivas, 1990).

Behavioral Measurements. Eight photocell cages (Omnitech, Columbus, OH) were located in individual wooden boxes with separate light and air sources (Austin and Kalivas, 1990). The day prior to the first behavioral trial, the rats were placed in the photocell cage for 60 min and given a sham injection by inserting an injection needle into the ventral pallidum. The next day, the rats were adapted to the photocell cage for 60 min, removed for microinjection and returned to the same photocell cage for 120 min. The drugs administered included the mu opioid agonist, Tyr-D-Ala-Gly-NMe-Phe-Gly-ol (DAMGO), the indirect GABA$_A$ antagonist, picrotoxin, and the dopamine antagonists, fluphenazine and haloperidol.

Neurochemical Measurements. Rats were decapitated and various regions of the rat forebrain dissected (Austin and Kalivas, 1991). The caudal portion of the brain was placed in formalin for later histological examination. The dissected tissue was prepared for analysis of dopamine and its metabolites, 3,4-dihydroxyphenylacetic acid (DOPAC) and homovanillic acid (HVA), using HPLC with electrochemical detection, as described elsewhere (Austin and Kalivas, 1991).

Histology and Data Analysis. Brains were fixed with formalin and coronal slices (100 μm thick) made with a vibratome. Cannulae tracks in the ventral pallidum were identified as described elsewhere (Austin and Kalivas, 1990). The data were statistically evaluated using one-way or two-way analysis of variance (ANOVA). Where appropriate, repeated measures analysis was applied. Post hoc multiple comparisons tests were either Newman-Kuels or a Dunnett's test if comparison was made to a single control group.

RESULTS

Enkephalinergic and GABAergic modulation of motor activity in the ventral pallidum. In a previous report, we demonstrated a dose-dependent increase in locomotion following the microinjection of DAMGO, picrotoxin or bicuculline into the ventral pallidum (Austin and Kalivas, 1990). Furthermore, we found that pretreatment of the ventral pallidum with the GABA$_A$ agonist, muscimol abolished the motor stimulant effect of either DAMGO or picrotoxin. Although the opioid antagonist, naloxone completely antagonized the motor stimulant effect of DAMGO, it was only partially effective in blocking picrotoxin-induced motor activity (Austin and Kalivas, 1990).

The microinjection of opioids into the ventral tegmental area produces motor stimulation similar to that observed in the ventral pallidum, and this response is augmented with daily administration (Joyce and Iversen, 1979; Vezina and Stewart, 1984; Kalivas and Bronson, 1985). Figure 1 shows that daily administration of DAMGO into the ventral pallidum also results in an augmented motor stimulant response to a subsequent acute microinjection.

Role of the mesolimbic dopamine system. Figure 2 shows that microinjection of either DAMGO or picrotoxin into the ventral pallidum significantly elevates the postmortem tissue concentration of the dopamine metabolites, DOPAC and HVA in the nucleus accumbens 30 min after injection.

Peripheral administration of the dopamine antagonist, haloperidol (0.1 mg/kg, ip) antagonized the locomotor stimulant effect of either DAMGO or picrotoxin microinjection into the ventral pallidum (Austin and Kalivas, 1991). Also, microinjection of the dopamine antagonist, fluphenazine, directly into the nucleus accumbens, abolishes the motor stimulant effect of DAMGO or picrotoxin injection into the ventral pallidum (Austin and Kalivas, 1991).

Table 1. Effect of 6-OHDA lesions of the nucleus accumbens on locomotion induced by the injection of picrotoxin or DAMGO into the ventral pallidum and the corresponding concentrations of biogenic amines in the nucleus accumbens.

Lesion	N	Treatment and Dose	Horizontal Photocell Counts	Concentration (pmole/mg protein)				
				NE	DA	DOPAC	HVA	5-HT
Sham	5	Saline	4357± 784	8.2±0.8	548.4±33.9	131.5±4.4	30.7±0.7	35.0±2.5
		Picrotoxin 0.17[a]	15746±1260††					
		DAGO 0.03	14386±2998††					
6-OHDA	7	Saline	4826±1091	9.2±0.7*	15.9± 6.5*	14.3±5.9*	4.1±2.7*	25.6±4.0
		Picrotoxin 0.17	13550±1583††					
		DAGO 0.33	15818±2620††					

[a] Doses are given as nmole/side

* $p < 0.05$ comparing lesion with sham lesion.

†† $p < 0.005$ statistically compared the saline treatment group.

In contrast to this apparent dependence on mesoaccumbens dopamine transmission, table 1 shows that lesioning dopamine innervation to the nucleus accumbens with 6-OHDA did not alter the motor stimulant response to either DAMGO or picrotoxin microinjection in the ventral pallidum. Table 1 also shows that the depletion of dopamine in the nucleus accumbens was extensive, averaging 97%. Lesser depletions of norepinephrine and serotonin were also noted.

DISCUSSION

A GABAergic and enkephalinergic projection to the ventral pallidum has been demonstrated (Zahm et al., 1985), and this projection plays an important role in the regulation of locomotion (Mogenson et al., 1980; Swerdlow et al., 1984a). Stimulation of enkephalinergic transmission or inhibition of GABAergic transmission in the ventral pallidum facilitates locomotion. Furthermore, *mu* opioid and $GABA_A$ receptors appear to be most significant in this regard.

Injection of either DAMGO or picrotoxin into the ventral pallidum activates mesoaccumbens dopamine metabolism, and acute blockade of dopamine transmission in the nucleus accumbens by fluphenazine attenuates the locomotor stimulant effect of these treatments. This extends the previous study by Napier et al. (1988) who found that peripheral administration of D_1 or D_2

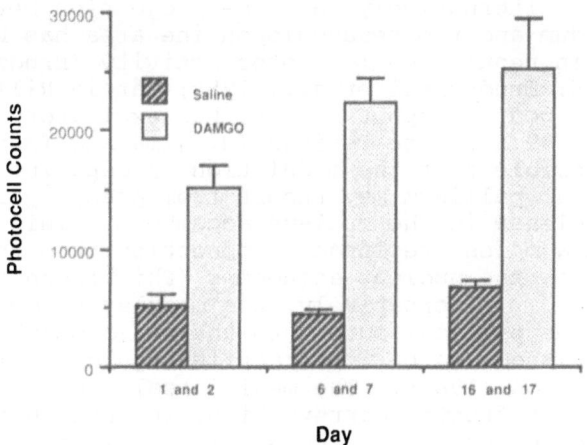

Figure 1. Behavioral sensitization following daily DAMGO administration into the VP. DAMGO (0.1 nmol) was administered daily for 5 days (days 2-6). Data are shown as mean ± sem photocell counts for 120 min. *$p < 0.05$ comparing the first DAMGO to DAMGO on days 6 and 17 using a one-way repeated measures ANOVA followed by a Dunnett's test.

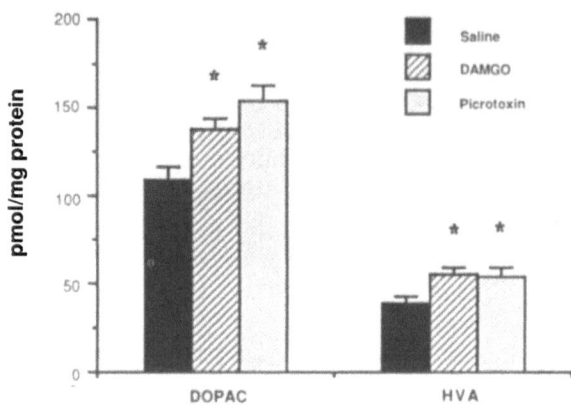

Figure 2. Increase in dopamine metabolites in the nucleus
accumbens following DAMGO (0.03 nmol) or picrotoxin
(0.17 nmol) microinjection into the VP. Data are
shown as the mean ± sem pmol/mg protein. *p < 0.05
compared to saline injection using a one-way ANOVA
followed by a Dunnett's test.

antagonists block DAMGO-induced contralateral rotation, and
argues that feedback from the ventral pallidum to the mesolimbic
dopamine system is important in regulating locomotion elicited
pharmacologically from the pallidum. Efferent projections from
the ventral pallidum offer a variety of anatomical substrates
whereby this interaction could occur. The first is a direct
connection between the ventral pallidum and A10 dopamine region
(Zahm, 1989). Alternatively, a dense projection between the
ventral pallidum and the pedunculopontine area has been found to
be important in regulating locomotor activity (Brudzynski and
Mogenson, 1985; Brudzynski et al., 1986; Garcia-Rill et al.,
1983), and the pedunculopontine area has excitatory projections
to the A10 and A9 dopamine regions (Niijima and Yoshida, 1988).
It is also possible that the modulation of dopamine transmission
from the ventral pallidum may result from presynaptic regulation
of dopamine release in the nucleus accumbens. This could occur
directly from a modest reciprocal connection from the ventral
pallidum back to the nucleus accumbens (Phillipson and
Griffiths, 1985). Alternatively, the modulation could be
indirect via the pallidal output to the mediodorsal thalamus,
which has been shown to be important in regulating locomotion
(Swerdlow and Koob, 1987). The mediodorsal thalamus projects
densely to the prefrontal cortex which, in turn, has descending
projections to the nucleus accumbens (Groenewegen, 1988; Jones
et al., 1987, 1988). Future studies will be oriented towards
evaluating which of these projections is responsible for the
regulation of mesolimbic dopamine transmission by the ventral
pallidum.

When DAMGO was injected daily into the ventral pallidum behavioral sensitization was produced. This is similar to what is observed following daily opioid injection into the A10 dopamine region (Joyce and Iversen, 1979; Vezina and Stewart, 1984; Kalivas, 1985). While sensitization in the A10 region has been linked to an augmentation in mesolimbic dopamine transmission (Kalivas and Duffy, 1990), it remains to be determined if this is also true for sensitization elicited from the ventral pallidum. However, the activation of mesoaccumbens dopamine transmission following acute injection of DAMGO into the ventral pallidum supports this possibility. Interestingly, daily microinjection of *mu* opioids into the nucleus accumbens does not augment the behavioral stimulation seen after a single acute injection (Vezina et al., 1987). This indicates that locomotion elicited by DAMGO in the nucleus accumbens may involve a different anatomical or neurochemical substrate than the motor stimulation produced by DAMGO in either the ventral pallidum or A10 region. This is supported by the fact that the stimulant action of opioids in the nucleus accumbens is not blocked by dopamine receptor antagonists, unlike their effect in either the ventral pallidum or A10 region (Pert and Sivit, 1977; Kalivas et al., 1983). However, the fact that muscimol injection into the ventral pallidum prevents the acute locomotor stimulant effect of DAMGO injection into the nucleus accumbens argues that some overlap in anatomical and/or neurochemical substrate exists (Austin and Kalivas, 1989).

Perhaps the most surprising finding in this report is that although acute interruption of mesoaccumbens dopamine transmission attenuates the locomotor response elicited from the ventral pallidum, when the dopamine neurons are destroyed, motor activity stimulated from the ventral pallidum proceeds normally. One possible explanation is that the dopamine remaining in the forebrain after an average 97% depletion of the nucleus accumbens is sufficient to maintain the motor response elicited in the ventral pallidum. Chronic dopaminergic denervation of the nucleus accumbens causes a supersensitive locomotor effect in response to peripheral administration of apomorphine (Koob et al., 1981). Thus, although we reported a 97% depletion of dopamine in the nucleus accumbens, the remaining dopamine released upon ventral pallidal stimulation may be sufficient to stimulate the supersensitive postsynaptic system and permit the picrotoxin- and DAMGO-induced locomotor response. This possible mechanism is not supported in a microdialysis study conducted by Robinson and Whishaw (1988) showing that unilateral 6-OHDA lesions of the substantia nigra resulting in greater than a 95% reduction in striatal tissue dopamine content decreased the basal extracellular levels of dopamine in the denervated striatum. However, it remains possible that dopamine terminals outside the nucleus accumbens which were not as severely lesioned may mediate the motor stimulant response to intra-ventral pallidal injection of DAMGO and picrotoxin.

Another possible explanation for the lack of effect by 6-OHDA lesions on the motor stimulant response elicited from the ventral pallidum is that a permanent reduction of dopamine neurotransmission in the nucleus accumbens and medial striatum may induce a compensatory change in other neurotransmitters or neural pathways to allow the expression of locomotion in the absence of dopamine. Such an alternative, nondopaminergic mechanism, could be by way of projections to the pedunculopontine area. The pedunculopontine area, via a synapse in the nucleus reticularis gigantocellularis, activates spinal locomotor generators (Garcia-Rill et al., 1983). Also, it is possible that ventral pallidal projections to the nucleus accumbens could influence nondopaminergic neurotransmission to elicit enhanced motor activity in lesioned rats. The injection of opioids and enkephalin analogues into the nucleus accumbens produces an increase in motor activity that is not mediated by enhanced dopamine transmission (Pert and Sivit, 1978; Kalivas et al., 1983). Furthermore, in 6-OHDA lesioned rats the motor stimulant response following intra-accumbens injection of opioids is augmented (Kalivas and Bronson, 1985; Stinus et al., 1985). This effect has recently been linked to the delta opioid receptor (Dauge et al., 1988), and unilateral 6-OHDA lesions upregulate *delta* receptor density in the rostral nucleus accumbens (Dilts and Kalivas, 1990). Also, chronic disruption of dopamine neurotransmission elevates the levels of both immunoreactive enkephalin and proenkephalin messenger RNA in the nucleus accumbens (Hong et al., 1978; Morris et al., 1988). Finally, we observed that the motor stimulant effect of picrotoxin injected into the ventral pallidum was partially antagonized by peripheral naloxone administration (Austin and Kalivas, 1990). Thus, it is conceivable that in the absence of dopamine, enhanced release of enkephalin in the nucleus accumbens could mediate the motor stimulation following intra-ventral pallidum stimulation. Another site of alteration may be in the ventral pallidum itself. Pan and Walters (1988) observed that unilateral destruction of dopamine innervation to the striatum results in an altered firing pattern of neurons in the globus pallidus. Similar lesions reduce $GABA_A$ and opioid receptor densities in the globus pallidus. Considering the similarities between the projections from the striatum to the globus pallidus and from the nucleus accumbens to the ventral pallidum (Haber et al., 1985), it is possible that analogous transsynaptic alterations could result from the dopamine lesions in the present study.

In summary, both enkephalin and GABA transmission in the ventral pallidum can modulate locomotor activity. This action appears to at least partially involve activation of dopamine transmission in the nucleus accumbens.

Acknowledgements: We thank Jenny Baylon for assistance in the preparation of this manuscript. This research was supported in part by USPHS grant #'s **DA-03906**, **MH-40817**, a Research Career Development Award, **DA-00153** (PWK) and a National Research Service Award, DA-05391 (MAK).

REFERENCES

Austin, M.C. and Kalivas, P.W., 1988, The effect of cholinergic stimulation in the nucleus accumbens on locomotor behavior, Brain Res., 441:209-214.
Austin, M.C. and Kalivas, P.W., 1989, Blockade of enkephalinergic and GABAergic mediated locomotion in the nucleus accumbens by muscimol in the ventral pallidum, Jpn. J. Pharmacol., 50:487-490.

Austin, M.C. and Kalivas, P.W., 1990, Enkephalinergic and GABAergic modulation of motor activity in the ventral pallidum, J. Pharmacol. Exp. Ther., 252:1370-1377.

Austin, M.C. and Kalivas, P.W., 1991, Possible dopaminergic involvement in locomotion elicited from the ventral pallidum, Brain Res., in press.

Baud, P., Mayo, W., Le Moal, M. and Simon, H., 1988, Locomotor hyperactivity in the rat after infusion of muscimol and [D-Ala^2]Met-enkephalin into the nucleus basalis magnocellularis. Possible interaction with cortical cholinergic projections, Brain Res. 452:203-211.

Brudzynski, S.M. and Mogenson, G.J., 1985, Association of the mesencephalic locomotor region with locomotor activity induced by injections of amphetamine into the nucleus accumbens, Brain Res., 334:77-84.

Brudzynski, S.M., Houghton, P.E., Brownlee, R.D. and Mogenson, G.J., 1986, Involvement of neuronal cell bodies of the mesencephalic locomotor region in the initiation of locomotor activity of freely behaving rats, Brain Res. Bull. 16:407-381.

Churchill, L., Dilts, R.P. and Kalivas, P.W., 1990, Changes in gamma-aminobutyric acid, mu-opioid and neurotensin receptors in the accumbens-pallidal projection after discrete quinolinic acid lesions in the nucleus accumbens, Brain Res. 511:41-54.

Clarke, P.B.S., Jakubovic, A. and Fibiger, H.C., 1988, Anatomical analysis of the involvement of mesolimbocortical dopamine in the locomotor stimulant actions of d-amphetamine and apomorphine, Psychopharmacology 96:511-520.

Dauge, V., Rossignol, P. and Roques, B.P., 1988, Comparison of the behavioral effects induced by administration in rat nucleus accumbens or nucleus caudatus of selective mu and delta opioid peptides or kelatorphan an inhibitor of enkephalin-degrading-enzymes, Psychopharmacology 96:343-352.

Dilts, R.P. and Kalivas, P.W., 1990, Autoradiographic localization of delta opioid receptors within the mesocorticolimbic dopamine system using ^{125}I - DPDPE, Synapse, 6:121-132.

Fink, J.S. and Smith, G.P., 1980, Mesolimbicocortical dopamine terminal fields are necessary for normal locomotor and investigatory exploration in rats, Brain Res., 199:359-384.

Garcia-Rill, E., Skinner, R.D., Gilmore, S.A. and Owings, R., 1983, Connections of the mesencephalic locomotor region (MLR) II. Afferents and efferents, Brain Res. Bull., 10:63-71.

Groenewegen, H.J., 1988, Organization of the afferent connections of the mediodorsal thalamic nucleus in the rat, related to the mediodorsal-prefrontal topography, Neuroscience, 24:379-431.

Haber, S.N., Groenewegen, H.J., Grove, E.A. and Nauta, W.J.H., 1985, Efferent connections of the ventral pallidum: evidence of a dual striatopallidofugal pathway, J. Comp. Neurol., 235:322-335.

Hallanger, A.E. and Wainer, B.H., 1988, Ascending projections from the pedunculopontine tegmental nucleus and the adjacent mesopontine tegmentum in the rat, J. Comp. Neurol., 274:483-515.

Heimer, L., Zaborszky, L., Zahm, D.S. and Alheid, G.F., 1987, The ventral striatopallidothalamic projection: I. The striatopallidal link originating in the striatal parts of the olfactory tubercle, J. Comp. Neurol., 255:571-591.

Hong, J.S., Yang, M.-Y., Fratta, W. and Costa, E., 1978, Rat striatal met-enkephalin content after chronic treatment with cataleptogenic and non-cataleptogenic antischizophrenia drugs, J. Pharmac. Exp. Ther., 205:141-147.

Jones, D.L. and Mogenson, G.J., 1980, Nucleus accumbens to globus pallidus GABA projection subserving ambulatory activity, Am. J. Physiol., 238:R63-R96.

Jones, M.W., Kilpatrick, I.C. and Phillipson, O.T., 1987, Regulation of dopamine function in the prefrontal cortex of the rat by the thalamic mediodorsal nucleus, Brain Res. Bull., 19:9-17.

Jones, M.W., Kilpatrick, I.C. and Phillipson, O.T., 1988, Dopamine function in the prefrontal cortex of the rat is sensitive to a reduction of tonic GABA-mediated inhibition in the thalamic mediodorsal nucleus, Exp. Brain Res., 69:623-634.

Joyce, E.M. and Iversen, S.D., 1979, The effect of morphine applied locally to mesencephalic dopamine cell bodies on spontaneous motor activity in the rat, Neurosci. Lett., 14:207-212.

Kalivas, P.W., Widerlov, E., Stanley, D., Breese, G. and Prange, A.J., 1983, Enkephalin action on the mesolimbic system: A dopamine-dependent and a dopamine-independent increase in locomotor activity, J. Pharmacol. Exp. Ther., 227:229-237.

Kalivas, P.W., 1985, Sensitization to repeated enkephalin administration into the ventral tegmental area of the rat. II. Involvement of the mesolimbic dopamine system, J. Pharmacol. Exp. Ther., 235:544-550.

Kalivas, P.W. and Bronson, M., 1985, Mesolimbic dopamine lesions produce an augmented behavioral response to enkephalin, Neuropharmacology, 24:931-936.

Kalivas, P.W. and Duffy, P., 1990, The effect of acute and daily neurotensin and enkephalin on extracellular dopamine in the nucleus accumbens, J. Neurosci., 10:2940-2949.

Koob, G.F., Stinus, L. and Le Moal, M., 1981, Hyperactivity and hypoactivity produced by lesions to the mesolimbic dopamine system, Behav. Brain Res., 3:341-359.

Mogenson, G.J., Jones, D.L. and Yim, C.Y., 1980, From motivation to action: functional interface between the limbic system and the motor system, Prog. Neurobiol., 14:69-97.

Mogenson, G.J. and Nielsen, M.A., 1983, Evidence that an accumbens to subpallidal GABAergic projection contributes to locomotor activity, Brain Res. Bull., 11:309-314.

Mogenson, G.J., Swanson, L.W. and Wu, M., 1983, Neural projections from the nucleus accumbens to globus pallidus, substantia innominata, and lateral preoptic-lateral hypothalamic area: an anatomical and electrophysiological investigation in the rat, J. Neurosci., 3:189-202.

Mogenson, G.J., Swanson, L.W. and Wu, M., 1985, Evidence that projections from the substantia innominata to zona incerta and mesencephalic locomotor region contribute to locomotor activity, Brain Res., 334:65-76.

Morris, B.J., Hollt, V. and Herz, A., 1988, Dopaminergic regulation of striatal proenkephalin mRNA and prodynrophin mRNA: contrasting effects of D1 and D2 antagonists, Neuroscience, 25:525-532.

Napier, T.C. and Marx, K., 1987, Enkephalin unilaterally microinjected into the ventral pallidum/nucleus basalis induces circling, Neuroscience Abstracts 13:445.

Napier, T.C., An, D., Austin, M.C. and Kalivas, P.W., 1988, Opiates microinjected into the ventral pallidum/substantia innominata (VP/SI) produce locomotor responses that involve dopaminergic systems, Neuroscience Abstracts 14:293.

Nauta, W.J.H., Smith, G.P., Faull, R.L.M. and Domesick, V.B., 1978, Efferent connections and nigral afferents of the nucleus accumbens septi in the rat, Neuroscience 3:385-401.

Niijima, K. and Yoshida, M., 1988, Activation of mesencephalic dopamine neurons by chemical stimulation of the nucleus tegmenti pedunculopontinus pars compacta, Brain Res., 345:163-171.

Pan, H.S. and Walters, J.R., 1988, Unilateral lesion of the nigrostriatal pathway decreases the firing rate and alters the firing pattern of the globus pallidus neurons in the rat, Synapse 2:650-656.

Patel, S. and Slater, P.L., 1988, Effects of GABA compounds injected into the subpallidal regions of rat brain on nucleus accumbens evoked hyperactivity, Behav. Neurosci. 102:596-600.

Pert, A. and Sivit, C., 1977, Neuroanatomical focus for morphine and enkephalin-induced hypermotility, Nature (Lond), 265:645-647.

Phillipson, O.T. and Griffiths, A.C., 1985, The topographic order of inputs to the nucleus accumbens in the rat, Neuroscience, 16, 275-296.

Robinson, T.E. and Whishaw, I.Q., 1988, Normalization of extracellular dopamine in striatum following recovery from a partial unilateral 6-OHDA lesion of the substantia nigra: a microdialysis study in freely moving rats, Brain Res. 450:209-224.

Shreve, P.E. and Uretsky, N.J., 1988, Effect of GABAergic transmission in the subpallidum on the hypermotility response to administration of excitatory amino acids and picrotoxin into the nucleus accumbens, Neuropharmacology, 27:1271-1277.

Stinus, L., Winnock, M. and Kelly, A.E., 1985, Chronic neuroleptic treatment and mesolimbic dopamine denervation induce behavioral supersensitivity to opiates, Psychopharmacology , 85:323-328.

Swerdlow, N.R., Swanson, L.W. and Koob, G.F., 1984a, Electrolytic lesions of the substantia innominata and lateral preoptic area attenuate the supersensitive locomotor response to apomorphine resulting from denervation of the nucleus accumbens, Brain Res., 306:141-148.

Swerdlow, N.R., Swanson, L.W. and Koob, G.F., 1984b, Substantia innominata: critical link in the behavioral expression of mesolimbic dopamine stimulation in the rat, Neurosci. Lett., 50:19-24.

Swerdlow, N.R. and Koob, G.F., 1987, Lesions of the dorsomedial nucleus of the thalamus, medial prefrontal cortex and pedunculopontine nucleus: effects on locomotor activity mediated by nucleus accumbens-ventral pallidal circuitry, Brain Res., 412:233-243.

Vezina, P. and Stewart, J., 1984, Conditioning and place-specific sensitization of increases in activity induced by morphine in the VTA, <u>Pharmacol. Biochem. Behav.</u>, 20:925-934.

Vezina, P., Kalivas, P.W. and Stewart, J., 1987, Sensitization occurs to the locomotor effects of morphine and the specific *mu* opioid receptor agonist, DAGO, administered repeatedly to the ventral tegmental area but not to the nucleus accumbens, <u>Brain Res.</u>, 417:51-58.

Vives, F. and Mogenson, G.J., 1985, Electrophysiological evidence that the mediodorsal nucleus of the thalamus is a relay between the ventral pallidum and the medial prefrontal cortex in the rat, <u>Brain Res.</u>, 344:329-337.

Walaas, I. and Fonnum, F., 1979, The distribution and origin of glutamate decarboxylase and choline acetyltransferase in ventral pallidum and other basal forebrain regions, <u>Brain Res.</u>, 177:325-336.

Zahm, D.S., Zaborsky, L., Alones, V.E. and Heimer, L., 1985, Evidence for the coexistence of glutamate decarboxylase and met-enkephalin immunoreactivities in axon terminals of rat ventral pallidum, <u>Brain Res.</u>, 325:317-321.

Zahm, D.S., 1989, The ventral striatopallidal pars of the basal ganglia in the rat. II. Compartmentation of ventral pallidal efferents, <u>Neurosci.</u>, 30:33-50.

PHARMACOLOGY OF BASAL FOREBRAIN INVOLVEMENT

IN REINFORCEMENT

Steven I. Dworkin, Linda J. Porrino, and James E. Smith

Wake Forest University Bowman Gray School of Medicine
Winston-Salem, NC, The Unit on Brain Imaging
NINDS, Bethesda, MD

INTRODUCTION

The search for the substrates of reinforcement within the organism began shortly after the initial demonstration that electrical stimulation of discrete brain regions could serve as a reinforcer (Olds and Milner, 1954). This intracranial electrical self-stimulation (ICSS) was assumed to result in the activation of the same neuronal pathways responsible for the reinforcing effects of natural environmental events. One of the major goals of the neurosciences is the elucidation of these basic neurobiological substrates of reinforcement. The identification of these basic processes includes the identification of both the neuronal pathways involved as well as the neurochemical characteristics of each neuron in the circuit. Knowledge of the location of the cell bodies and the projection fields for each neuron as well as the neurohumor(s) released would also be necessary to provide a complete account of the central processes of reinforcement. Information concerning the location of receptors critical to these processes including the types of presynaptic and postsynaptic binding sites and the nature of the neurons on which these were located would be required. The modulatory effects of other inputs to each component neuron in the circuit mediating effects like satiation, deprivation or conditioning influences would also be needed if these basic mechanisms were to be understood. Ultimately, the relationship between various patterns of activity of these inputs on the output of each component neuron would be necessary (the sum total of activity in the mediating neuronal networks). This information is not available for even the most simple mammalian behaviors. However, substantial progress has been made in investigations of the nature of these brain processes.

THEORIES OF THE BIOLOGICAL SUBSTRATES OF REINFORCEMENT

Several theories have been set forth suggesting that one or two neuronal systems are responsible for positive "hedonic" phenomena. Two major current theories include the noradrenergic-opioid hypothesis (Stein, 1975) and the dopaminergic hypothesis (Wise, 1978) both of which could be designated as "pathway hypotheses". The noradrenergic-opioid theory contends that since all forebrain sites that support ICSS contain norepinephrine and that pharmacological blockade of noradrenergic neuronal activity results in decreases in ICSS maintained responding, noradrenergic neurons are responsible for positive hedonic processes. The discovery of opioid binding sites and endogenous ligands have been incorporated into the theory on the basis of the reinforcing efficacy of opiates. However, later studies showed noradrenergic systems to be sufficient but not necessary (Wise, 1978), and thus the theory has been all but discarded. The dopamine hypothesis has received

The Basal Forebrain, Edited by T.C. Napier *et al.*
Plenum Press, New York, 1991

substantial attention, but it too has significant problems. For example, brain sites devoid of or depleted of dopamine still support ICSS (Ott et al., 1980; Phillips and Fibiger, 1978). Clearly, subsets of dopamine, norepinephrine and opioid releasing neurons are involved in the overall processes underlying reinforcement. Although it is likely that both theories may have some correct elements, neither qualify as universal comprehensive theories of reinforcement. The neuronal events involved in reinforcement are likely to be substantially more complex than can be accounted for by the activation of two or three neuronal systems.

A potential limitation of these theories is that usually only the "pleasing effects" of an event are considered without an analysis of the response strengthening properties (the increase in the probability of the response that preceded stimulation). Thus, other conditioned and unconditioned factors including satiation, deprivation and conditioned stimulus effects are not generally addressed. Reinforcement is a process through which response probabilities are modified while the term "reward" often imparts immutable properties to environmental events independent of interactions with the organism. The effects of the "rewarding" stimulus on the frequency of the response (strengthening) is often used as the measure of the "pleasing" effects (increased lever pressing, faster maze running, place conditioning). The response strengthening effects themselves and potential pleasing effects of the removal of response suppressing stimuli (negative reinforcement) are issues that are not dealt with by these "reward" hypothesis. The basic assumption that the neurons activated by intracranial electrical self-stimulation are uniquely involved in the pleasurable effect is not yet conclusive. In fact, rats will respond to terminate electrical stimulation for which they have previously responded if its presentation is no longer under their control (Steiner et al., 1968).

Another issue indicating the complexity of ICSS is the multitude of sites that support responding maintained by contingent electrical stimulation. Although many studies have addressed the sites and pathways responsible for ICSS, there is still controversy as to the necessary and sufficient status of particular innervations or projections. In spite of the controversy, discrete, and perhaps even dedicated, neuronal circuits may mediate at least some of the neuronal processes responsible for reinforcement. This is apparent from studies showing both positive and negative ICSS sites and from data showing lesions of some brain regions to modulate the response contingent presentation of a reinforcer without producing confounding deficits in the animal's ability to emit the required response.

An alternative to the pathway hypotheses are the circuit-network hypotheses. These more sophisticated notions have been developed to include recent findings that complex neuronal circuits and networks mediate the underlying events of reinforcement. This type of approach can account more adequately for the diverse neuroanatomical substrates and neurochemical mechanisms known to be involved in reinforcement and is more consistent with the multitude of variables that influence reinforcement processes. These topics will be developed and discussed in the concluding portions of this chapter.

BASAL FOREBRAIN SITES SUPPORTING INTRACRANIAL SELF-STIMULATION

The animal laboratory model used to investigate reinforcement has been predominantly ICSS. What is known of the neurobiological basis of reinforcement? The first wave of investigations attempted to identify the neuronal pathways and brain sites that supported intracranial self-stimulation. In the ventral forebrain these included the lateral hypothalamus (Olds, 1962), nucleus accumbens (Rolls, 1971), olfactory tubercle (Prado-Alcala and Wise, 1984), septal area (Olds and Milner, 1954), amygdala (Wurtz and Olds, 1963), caudate-putamen (Phillips et al., 1976b), mid-thalamus (Cooper and Taylor, 1967) ventral hypothalamus (Ball, 1972),and pyriform, entorhinal (Prado-Alcala et al., 1984) and sulcal prefrontal cortices (Clavier and Gerfen, 1979). Other studies attempted to identify the types of neurons involved by using neurotoxin lesions or pharmacologic blockade. These studies were followed by theories of the neurobiological substrates of reinforcement.

As stated above, investigators attempted to explain the complex events involved in these processes through limited data obtained from experiments often designed to corroborate rather than test the hypothesis at hand.

PHARMACOLOGICAL MANIPULATIONS OF ICSS

Initial experiments to identify the nature of the neuronal circuits mediating the reinforcing effects of electrical brain stimulation involved systemic pharmacological manipulations. One of the first studies evaluated the effects of chlorpromazine on ICSS (Olds and Travis, 1960), this drug decreases responding maintained by many drug and non-drug environmental events. More recent studies include evaluations of antagonists for acetylcholine, dopamine, GABA, glycine, serotonin and opiates on ICSS at several forebrain sites (Table 1). Antagonism of dopamine actions with either pimozide, haloperidol or spiroperidol were shown to antagonize rates of responding maintained by ICSS in the lateral hypothalamus (Fouriezos and Wise, 1976), medial prefrontal cortex (Ferrer et al., 1983), nucleus accumbens (Phillips et al., 1975; Roberts and Mogenson, 1978) and dorsal noradrenergic bundle (Roberts and Mogenson, 1978). ICSS thresholds were increased by blockade of dopaminergic actions with haloperidol when the stimulating electrode was located in the lateral hypothalamus (Esposito et al., 1979) and by pimozide and trifluoperazine (Lynch and Wise, 1985). D_1 antagonists decreased the rates of ICSS in the medial forebrain bundle, ventral tegmental area and dorsal raphe', while sulpiride, a D_2 antagonist, or metergoline, a serotonin receptor antagonist, had no effect (Nakijima and McKenzie, 1986). One could conclude from these data that dopamine has a primary role in the neuronal processes that underlie ICSS. However, there was a significant body of evidence showing antagonism of acetylcholine, GABA, opiates, serotonin and benzodiazepine receptors to modify ICSS.

Lateral hypothalamus ICSS rates were increased by cholinergic blockade (Stephens and Herberg, 1979) and decreased by GABA receptor antagonism while glycine antagonists had no effect (Porrino and Coons, 1980). An opioid antagonist and benzodiazepine agonist also decreased ICSS rates at this site (Ichitani et al., 1985). GABA receptor blockade decreased ICSS thresholds in the ventral tegmental area (Nazzaro and Gardner, 1980) while opioid antagonists increased thresholds in the nucleus accumbens and lateral hypothalamus to a greater degree than in the ventral tegmental area (West and Wise, 1988). Naloxone also antagonized ICSS rates in the medial entorhinal cortex (Reymann et al., 1986). Antagonism of serotonergic neurotransmission with para-chlorophenylalanine increased rates of dorsal raphe' ICSS (Phillips et al, 1976a) while administration of serotonin directly into the cerebral ventricles decreased rates (Simon et al., 1976). Injections of metergoline decreased rates of ICSS in the lateral habenula and medial raphe' nucleus with lesser effects seen with stimulating electrodes in the lateral hypothalamus (Nakijima, 1984).

TABLE 1

EFFECTS OF SYSTEMIC NEUROTRANSMITTER RECEPTOR ANTAGONISTS ON REINFORCING EFFICACY OF ICSS

	Antagonists						
	ACh	D_1	D_2	GABA	Gly	5-HT	Opioid
ICSS Site							
Dorsal NA Bundle		$Dec_{R\&T}$					
Dorsal Raphe'		Dec_T	NE_T				
Lateral Hypothalamus	Inc_R	Dec_R		Dec_R	NE	NE	Dec_R
Medial Prefrontal Cortex		Dec_R					
Nucleus Accumbens		Dec_R					Dec_T
VTA		Dec_T	NE_T	Inc_T		NE_T	NE_T

Abbreviations= Dec - decrease; Inc - increase; NE - no effect; $_T$ - threshold measure; $_R$ - rate measure; NA - noradrenergic; VTA - ventral tegmental area.

Table 2
EFFECTS OF CENTRAL ADMINISTRATION OF NEUROTRANSMITTER RECEPTOR ANTAGONISTS ON THE REINFORCING EFFICACY OF ICSS

	Antagonists			
	D_1	D_2	Opioid$_{delta}$	Opioid$_{mu}$
ICSS Site				
Lateral Hypothalamus			$_{mTh}Dec_T$	$_{mTh}Inc_T$
Medial Prefrontal Cortex	$_{mpc}Dec_R$	$_{mpc}NE_R$		
VTA		$_{cp}Dec_R$		
		$_{aa}Dec_R$		

Abbreviations = The same as Table 1 plus the site of intracranial antagonist administration which included: $_{mpc}$ - medial prefrontal cortex; $_{aa}$ - nucleus accumbens; $_{mTh}$ - mid thalamus.

Several studies attempted to characterize the sites in the brain that these antagonists were acting by injecting them directly into various brain regions and assessing effects upon ICSS (Table 2). Injections of the dopamine receptor antagonists pimozide or spiroperidol into the medial prefrontal cortex decreased rates of self-stimulation at that site while injections of sulpiride had no effect (Ferrer et al., 1983) further supporting an important role for D_1 receptors in these processes and not D_2 receptors. However, haloperidol injected into the striatum or nucleus accumbens decreased rates of ICSS in the VTA (Broekkamp and Van Rossum, 1975). Antagonism of mu opiate receptors in the medial thalamus increases thresholds for ICSS of the lateral hypothalamus while a kappa antagonist decreases such thresholds (Carr and Bak, 1988). A number of these investigation only reported the effects of drugs on rates of responding maintained by ICSS. Thus, these data are difficult to interpret since many pharmacological agents have both rate-dependent and unconditioned effects on performance that can result in decreased response rates independent of any effect on reinforcing efficacy. More recently developed ICSS procedures provide some control for rate effects by the use of threshold measures and signal detection procedures (for review see, Steller and Steller, 1985). In general, these data would suggest a role for several different neurotransmitter releasing neuronal systems in the biological substrates of reinforcement.

INTRACRANIAL INFUSIONS OF NEUROTRANSMITTER EFFECT ICSS

Several studies have attempted to identify the neuronal systems involved in ICSS at several different brain sites by injecting neurotransmitters directly into a discrete brain region (Table 3). Dopamine injections into the nucleus accumbens increased rates of ICSS in the ventral anterior striatum while injections of serotonin or GABA decreased rates (Neill et al., 1978). Dopamine injections into the nucleus accumbens or caudate-putamen also increased ICSS rates from electrodes implanted in the lateral hypothalamus or ventral tegmental area while serotonin injections into the nucleus accumbens decreased these rates (Redgrave, 1978). It is interesting that norepinephrine injections were seen sometimes to have similar effects as dopamine by these investigators. When stimulating electrodes were in the lateral hypothalamus, serotonin injections into the ventral tegmental area increased rates of ICSS while acetylcholine had an inverted "U" shaped dose effect (Redgrave and Horrell, 1976). Taken together, these data indicate that several neuronal systems in several brain regions are involved in the neurobiological substrates of ICSS.

Neurotoxin lesions of some brain regions have been used to characterize the neuronal systems that mediate the effects of ICSS (Table 4). Again, a number of these studies only assessed effects on rates. As previously stated, such data are difficult to interpret. 6-OHDA has been used to selectively remove dopaminergic innervations of specific brain regions while kainic acid or ibotenic acid have been used to remove local neurons or neurons of origin in discrete brain regions. Lesions of the lateral hypothalamus, substantia nigra, ventral tegmental area, medial prefrontal and sulcal prefrontal cortex have been evaluated on ICSS at that site or at a projection area for that region. Unilateral 6-OHDA induced lesions of the substantia nigra abolished ICSS in the ipsilateral caudate-putamen and decreased rates on the contralateral side (Phillips et al., 1976b). This would suggest an important role for the nigrostriatal dopamine pathway in the neuronal events maintaining ICSS at this site. Similar lesions of the substantia nigra pars compacta or ventral tegmental area had less conclusive effects on ICSS at a medial prefrontal cortical site (Carey, 1982). The unilateral ventral tegmental area lesion had no effect while three of six animals with unilateral lesions of the substantia nigra pars compacta showed a decrease in rates of ICSS bilaterally while the remaining three animals showed no effects. This would suggest that the neuronal events mediating the effects of ICSS of the medial prefrontal cortex are different than those mediating ICSS at caudate-putamen neurons. However, similar lesions of the ventral tegmental area did decrease rates of ICSS at lateral hypothalamic sites (Koob et al., 1978). Bilateral lateral hypothalamic 6-OHDA lesions also decreased rates of ICSS at ventral tegmental area sites with little effect on ICSS at nucleus accumbens or medial prefrontal cortex sites (Phillips and Fibiger, 1978; Simon et al., 1979). More recently, unilateral 6-OHDA induced lesions of the lateral hypothalamus was shown to flatten rate-intensity functions for ICSS at ipsilateral ventral tegmental area sites without affecting such for contralateral sites more strongly supporting the conclusion that the reinforcing efficacy of stimulation was reduced by the lesion (Fibiger et al., 1987). Removal of local neurons and projections from lateral hypothalamic sites of ICSS with ibotenate or kainate either had no effect (Sprick et al., 1985) or decreased rates on the ipsilateral side (Nassif et al., 1985). Similar lesions of the medial prefrontal cortex also decreased rates on the side ipsilateral to the lesion (Lestang et al., 1985; Nassif et al., 1985). The same effects were seen after kainate induced lesions of the medial prefrontal cortex (Ferrer et al., 1985) and the sulcal prefrontal cortex for ICSS at that site (Gerfen and Clavier, 1981). These data clearly indicate that local neurons and projections from an ICSS site are necessary for the neuronal events that mediate the effects of ICSS with brainstem pathways having selective roles in different forebrain regions.

Table 3

EFFECTS OF CENTRAL ADMINISTRATION OF NEUROTRANSMITTERS ON ICSS

	ACh	Neurotransmitter DA	NE	5-HT	GABA
ICSS Site					
Lateral Hypothalamus	$_{vta}$I-D$_R$	$_{na}$Inc$_R$	DA Like	$_{vta}$Inc$_R$	
		$_{cp}$Inc$_R$	DA Like		
Ventral Anterior Striatum		$_{na}$Inc$_R$		$_{na}$Dec$_R$	$_{na}$Dec$_R$
VTA		$_{na}$Inc$_R$	DA Like	$_{na}$Dec$_R$	
		$_{cp}$Inc$_R$	DA Like		

Abbreviations= Same as Tables 1 & 2 plus the following: $_{cp}$ - caudate nucleus-putamen; $_{vta}$ - ventral tegmental area; I-D - increase at lower doses and decreases at higher doses.

Table 4
EFFECTS OF CENTRAL NEUROTOXIN LESIONS ON ICSS

| | Neurotoxin | | |
	6-OHDA	Kainate	Ibotenate
ICSS Site			
Caudate-putamen	$_{sn}Dec_{Ri}$ $_{sn}NE_{Rc}$		
Lateral Hypothalamus	$_{vta}Dec_R$	$_{lh}NE_R$ $_{lh}Dec_{Ri}$ $_{lh}NE_{Rc}$	$_{lh}NE_R$
Medial Prefrontal Cortex	$_{sn}\tfrac{1}{2}Dec_{Ri}$ $_{sn}\tfrac{1}{2}Dec_{Rc}$	$_{mpc}Dec_{Ri}$ $_{mpc}NE_{Rc}$	
Nucleus Accumbens	$_{lh}NE_R$		
Sulcal Prefrontal Cortex		$_{spc}Dec_R$	
VTA	$_{lh}Dec_R$ $_{lh}Dec_{Ti}$ $_{lh}NE_{Tc}$		

Abbreviations= Same as Tables 1,2 & 3 plus the following: $_{Ri}$ or $_{Ti}$ effects on stimulation ipsilateral to a unilateral lesion; $_{Rc}$ or $_{Tc}$ effects on stimulation contralateral to a unilateral lesion; $_{lh}$ - lateral hypothalamic lesion; $_{sn}$ - substantia nigra lesion; $_{spc}$ -sulcal-pyriform cortex lesion.

GLUCOSE UTILIZATION CORRELATED WITH ICSS

Several studies were initiated to identify brain regions involved in ICSS using 2-deoxyglucose autoradiography for assessment of glucose utilization (Esposito et al., 1984; Porrino et al., 1984; 1987). Data from these studies has significantly advanced knowledge of the neuronal systems involved in these processes.

A. Substantia Nigra ICSS

The distribution of 2-deoxyglucose has been evaluated in brain regions of rats receiving either response dependent or response independent brain stimulation of the substantia nigra (Porrino, 1987). Rats receiving noncontingent stimulation of this region demonstrated significant increases in glucose utilization bilaterally in the entopeduncular nucleus, subthalamic nucleus, and dorsal raphe' and ipsilateral to the stimulating electrode in the caudate nucleus, lateral septum and globus pallidus compared to non-stimulated controls. Rats receiving contingent stimulation showed significant increases in glucose utilization bilaterally in the prefrontal and anterior cingulate cortices, nucleus accumbens, ventral pallidum, mediodorsal thalamus and lateral habenulae and significant increases contralateral to the stimulating electrode in the caudate nucleus, lateral septum and globus pallidus, when compared to controls receiving response independent stimulation.

B. Ventral Tegmental Area ICSS

This same laboratory has also investigated brain glucose utilization in rats receiving contingent and noncontingent stimulation of the ventral tegmental area (Porrino et al., 1984). Rats receiving noncontingent stimulation of these regions showed increased glucose utilization bilaterally in the locus coeruleus and dorsal raphe' nucleus and ipsilaterally to the stimulating electrode in the lateral septum, mediodorsal thalamus and hippocampus CA3 region compared to non-stimulated controls. In the rats receiving response contingent stimulation, increases in activity occurred bilaterally in the nucleus accumbens, bed nucleus of the stria terminalis and medial brachial nucleus, ipsilaterally to the stimulating electrode in the medial prefrontal cortex, central amygdala and basolateral amygdala, and contralateral to the stimulating electrode in the lateral septum, mediodorsal thalamus and CA3 area of the hippocampus. Increases in activity that occurred bilaterally in the locus coeruleus and dorsal raphe' and ipsilaterally to the stimulating electrode in the lateral septum, mediodorsal thalamus and CA3 area of the hippocampus were related to the electrical stimulation itself, since these were seen in the noncontingently stimulated rats as well.

C. Medial Forebrain Bundle ICSS

Similar investigations of cerebral glucose utilization have been recently completed in rats receiving ICSS of the medial forebrain bundle at the level of the lateral hypothalamus (Porrino et al., 1990). The animals receiving response contingent stimulation showed similar changes to those seen with ventral tegmental area stimulation, except that the increased utilization was greater in the olfactory tubercle.

Some regions showed increased glucose utilization that were unique to the site of stimulation (Porrino, 1987). For substantia nigra stimulation these areas included the caudate nucleus and the anterior cingulate cortex. For ventral tegmental area and medial forebrain bundle stimulation these included the central and medial nuclei of the amygdala, the hippocampus, bed nucleus of the stria terminalis and the olfactory tubercle. The terminal field of the dopaminergic innervations of the caudate-putamen from the substantia nigra appears to be necessary to the neuronal events mediating ICSS of the caudate-putamen since lesions of this system abolish ICSS of this region as previously stated (Phillips et al., 1976b) further supporting the accuracy of these changes in glucose utilization. There are nine brain regions that showed increased glucose utilization with ICSS of either brain site. These include the prefrontal cortex, nucleus accumbens, lateral septum, globus pallidus, mediodorsal thalamus, entopeduncular nucleus, subthalamic nucleus, ventrolateral thalamus and sensorimotor cortex. These data suggest that reinforcement produced by brain stimulation may involve some shared or common components as well as some unique elements. This would clearly argue for a complexity of systems involved in reinforcement processes.

NEUROTRANSMITTER / METABOLITE RATIOS AND ICSS

Several experiments have attempted to assess the neurochemical effects of ventral tegmental area ICSS using several procedures. Ventral tegmental area ICSS was shown to increase dopamine metabolism in the olfactory tubercle while ICSS at posterior hypothalamic sites did not (Mitchell et al., 1982). Ventral tegmental area ICSS was also shown to increase tyrosine hydroxylase activity *in vivo* in the nucleus accumbens, olfactory tubercle and striatum (Phillips et al., 1987). The effects of ICSS of the same region on the ratio of the major metabolites of dopamine (either dihydroxyphenylacetic acid or homovanilic acid) in the striatum, nucleus accumbens or olfactory tubercle were recently assessed (Fibiger et al., 1987). Increases in the ratio were seen in self-stimulating animals in all three of these areas, however, the same changes were also seen in yoked stimulated animals, suggesting these changes to be the result of the electrical stimulation alone. In general, these data indicate that the utilization of dopamine in some forebrain regions may be increased with ICSS of the ventral tegmental area, but that ICSS of other sites may not produce the same changes.

NEUROTRANSMITTER TURNOVER RATES WITH ICSS

The authors have recently completed an experiment in which both glucose utilization and biogenic amine turnover has been concurrently assessed in the brains of rats exposed to ICSS of the ventral tegmental area and in non-stimulated and noncontingently stimulated controls (Smith,Dworkin,Co and Porrino, 1990). Changes in glucose utilization in the nucleus accumbens and caudate-putamen seen in the self-stimulating animals were correlated with changes in the turnover of biogenic amine neurotransmitters. Increases in the turnover rates of dopamine were seen in the contralateral nucleus accumbens while decreases in the turnover of serotonin were observed. Increases in the turnover of dopamine and norepinephrine were seen in the ipsilateral caudate-putamen which were also correlated with the self-stimulation. These data suggest that neuronal systems other that dopamine are involved in the neuronal processes that mediate ICSS of the ventral tegmental area. Dopamine neurons in the substantia nigra may also be important to these processes, although the contribution of motor effects required by the operant response to these effects in the striatum cannot be discounted at this time.

INTRACRANIAL SELF-ADMINISTRATION OF NEUROTRANSMITTERS

In the last several years a new approach to characterizing brain reinforcement systems has been utilized. Intracranial self-administration procedures have been used to identify neurotransmitters involved in neuronal networks mediating reinforcement by assessing discrete brain regions to determine whether the response contingent presentation of agonists for receptors for neurotransmitters released in those regions will initiate reinforcing neuronal activity and maintain self-administration. Biogenic monoamine, amino acid and peptide neurohumor receptor agonists have been shown to be intracranially self-administered by rats into discrete brain regions (Table 5). Dopamine is self-administered into the nucleus accumbens at 750 pmol doses but not into the ventral tegmental area or the medial prefrontal cortex of naive animals (Dworkin and Smith, 1986). However, dopamine is self-administered into the latter area after 6-OHDA lesions by animals that have previously received response contingent infusions of cocaine into this region (Goeders and Smith, 1986). However, naive rats will not self-administer dopamine into the medial prefrontal cortex likely because the mechanisms for inactivation (reuptake) are still operational and because a normal density of D_2 receptors is likely present and not increased as may occur after the removal of dopamine input by the lesion. Methionine enkephalin (Goeders et al., 1984) and cholecystokinin (Hoebel and Aulisi, 1984) are also self-administered into the nucleus accumbens at doses of 500 pmol and 25 nmol respectively. Neurotensin is self-administered into the ventral tegmental area at 1-2 nmol doses (Glimcher et al., 1987) and glutamate into the lateral preoptic nucleus at 500 pmol doses (Goeders et al., 1987). Although these data are provocative, necessary controls such as pharmacological blockade studies to demonstrate receptor mediated processes and obviate nonspecific chemical stimulated activation of the neurons at the cannulae tip are generally missing. Direct chemical stimulation of these neurons would be similar to the nonspecific activation that occurs with ICSS. When appropriate controls are used and self-administration of a neurohumor demonstrated at a discrete brain site, then it suggests that the release of that substance from innervating neurons may be part of the neuronal processes occurring with reinforcement. These criteria have been satisfied only for methionine enkephalin self-administration in the nucleus accumbens (Goeders et al., 1984).

TABLE 5

NEUROHUMORS AND BRAIN REGIONS SUPPORTING INTRACRANIAL
SELF-ADMINISTRATION IN RATS

Brain Region	Neurohumor
Nucleus Accumbens	Dopamine
Nucleus Accumbens	Methionine Enkephalin
Nucleus Accumbens	Cholecystokinin
Ventral Tegmental Area	Neurotensin
Lateral Preoptic Nucleus	Glutamate

However, demonstration of direct release by microdialysis in the behaving animal would be necessary for evidence of release with endogenous reinforcement processes.

CONCLUSIONS

To describe and investigate the neurons that mediate reinforcement in terms of neuronal systems or networks has distinct advantages over attributing such complex processes to one or two pathways or neurotransmitter releasing systems. First of all, it is more parsimonious with the data that indicate a variety of pathways and neurotransmitters are involved in the reinforcing effects of ICSS. Secondly, it is more consistent with what is known about complex brain function and the reinforcing effects of other environmental events. There are few if any neuronal systems in the CNS that are independent of influences from other neurons. It is unlikely, perhaps impossible for intricate behavioral processes that can be altered by experiences to be conveyed by one or two neuronal pathways. It is likely that a single neuronal pathway has multiple effects depending on a number of environmental and historical variables. Reinforcement involves the detection of environmental events from the broad stream of stimuli occurring at any moment in addition to some type of analysis of temporal contiguity or contingencies. Environmental events which can serve as reinforcers do not have inherent reinforcing qualities. The state of the organism (deprivation, satiation), its history and the current context can all modulate the saliency of a putative reinforcer. Many reinforcers are not universal but develop reinforcing efficacy through conditioning. All of these environmental factors are important components of the processes that comprise reinforcement. It is doubtful that one or two neuronal pathways control and integrate these complex phenomena. Neuronal circuits or networks involving a number of pathways and neurotransmitters are necessary for such complex actions. A theorist may potentially attribute the pleasing or "hedonic" effects of the direct electrical stimulation of a brain region to the activation of a specific pathway or group of pathways since this is exactly what this procedure involves. However, theories based on such phenomena will not likely account for the complex processes of reinforcement. Limited theories may potentially explain intracranial electrical self-stimulation but may not generalize to overall processes. When both the pleasing and response strengthening properties of reinforcement are addressed irrespective of the hedonic value of the stimulus (positive versus negative stimuli), then a hypothesis with greater utility and applicability to a broader spectrum of behaviors will result.

Discrete and perhaps dedicated neuronal networks likely mediate reinforcement processes. These neuronal networks may be artificially activated by direct electrical stimulation or by chemical stimulation with drugs. Drugs may initiate activity or modulate activity in these circuits by interacting with receptors on component neurons or on neurons that modulate these circuits. The circuits identified in the morphine self-administering rats (see Porrino et al. this book) may be portions of neuronal networks that mediate reinforcement for positive stimuli since these brain regions generally show increased glucose utilization in animals electrically activating reinforcement systems as outlined above. These neuronal networks at minimum include acetylcholine, dopamine, norepinephrine, serotonin, aspartate, glutamate, methionine enkephalin and gamma aminobutyric acid releasing neurons. It is likely that other neurohumor releasing neurons will be identified as important components of reinforcement in experiments designed to do such. Like most theories, the current favored conviction that dopamine is the mediator of neuronal events related to hedonic processes has to some degree retarded the investigation of other functional neuronal networks in these processes. The existing literature includes many demonstrations of the involvement of neuronal systems releasing other neurotransmitters. Hopefully, a more comprehensive experimental approach to this complex area of research will likely be beneficial to understanding the basic mechanisms of behavioral processes.

ACKNOWLEDGEMENTS

Supported in part by USPHS Research Grants DA-01999, DA-03628, DA-03832, DA-03631 and DA KO5 00114.

REFERENCES

Ball, G.G., 1972, Self-stimulation in the ventromedial hypothalamus, Science 178:72-73.

Broekkamp, C.L.E., and Van Rossum, J.M., 1975, The effect of microinjections of morphine and haloperidol into the neostriatum and the nucleus accumbens on self-stimulation behavior, Arch Int. Pharmacodyn. Ther., 217:110-116.

Carey, R.J., 1982, Unilateral 6-hydroxydopamine lesions of dopamine neurons produce bilateral self-stimulation deficits, Behav. Brain Res., 6:101-114.

Carr, K.D., and Bak, T.H., 1988, Medial thalamic injection of opioid agonists: μ-agonist increases while kappa agonist decreases stimulus thresholds for pain and reward, Brain Res., 441:173-184.

Clavier, R.M., and Gerfen, C.R., 1979, Self-stimulation of the sulcal prefrontal cortex in the rat: direct evidence for ascending dopaminergic mediation, Neurosci. Lett., 12:183-187.

Cooper, R.M., and Taylor, L.M., 1967, Thalamic reticular system and central gray: Self-stimulation, Science, 156:102-103.

Dworkin, S.I., and Smith, J.E., 1986, The reinforcing, rate effects and stimulus properties of intracranial dopamine administration, NIDA Research Monograph 67, Problems of Drug Dependence-1985, 242-248.

Esposito, R.U., Faulkner, W., and Kornetsky, C., 1979, Specific modulation of brain stimulation reward by haloperidol, Pharmacol. Biochem. Behav., 10:937-940.

Esposito, R.U., Porrino, L.J., Seeger, T.F., Crane, A.M., Everist, H.D., and Pert, A., 1984, Changes in local cerebral glucose utilization during rewarding brain stimulation, Proc. Nat. Acad. Sci., USA, 81:635-639.

Ferrer, J.M.R., Myers, R.D., and Mora, F., 1985, Suppression of self-stimulation of the medial prefrontal cortex after local microinjection of kainic acid in the rat, Brain Res. Bull., 15:225-228.

Ferrer J.M.R., Sanguinetti, A.M., Vives, F., and Mora, F., 1983, Effects of agonists and antagonists of D_1 and D_2 dopamine receptors on self-stimulation of the medial prefrontal cortex in the rat, Pharmacol. Biochem. Behav., 19:211-217.

Fibiger, H.C., LePiane, F.G., Jakubovic, A., and Phillips, A.G., 1987, The role of dopamine in intracranial self-stimulation of the ventral tegmental area, J. Neurosci., 7:3888-3896.

Fouriezos, G., and Wise, R.A., 1976, Pimozide-induced extinction of intracranial self-stimulation: response patterns rule out motor or performance deficits, Brain Res., 103:377-380.

Gerfen, C.R., and Clavier, R.M., 1981, Intracranial self-stimulation from the sulcal prefrontal cortex in the rat: the effect of 6-OHDA or kainic acid lesions at the site of stimulation, Brain Res., 224:291-304.

Glimcher, P.W., Giovino, A.A., and Hoebel, B.G., 1987, Neurotensin self-injection in the ventral tegmental area, Brain Res., 403:147-150.

Goeders, N.E., Lane, J.D., and Smith, J.E., 1984, Intracranial self-administration of methionine enkephalin into the nucleus accumbens, Pharmacol. Biochem. Behav., 20:451-455.

Goeders, N.E., Van Osdell, C.A., McAllister, K.H., Dworkin, S.I., and Smith, J.E., 1987, Responding maintained by intracranial delivery of L-glutamate, Fed. Proc., 46:548.

Goeders, N.E., and Smith, J.E., 1986, Reinforcing properties of cocaine in the medial prefrontal cortex: Primary action on presynaptic dopaminergic terminals, Pharmacol. Biochem. Behav., 25:191-199.

Hoebel, B.G., and Aulisi, E., 1984, Cholecystokinin self-injection in the nucleus accumbens and block with proglumide, Neurosci. Abst., 10:694.

Ichitani, Y., Iwasaki, T., and Satoh, T., 1985, Effects of naloxone and chlordiazepoxide on lateral hypothalamic self-stimulation in rats, Physiol. Behav., 34:779-782.

Koob, G.F., Fray, P.J., and Iversen, S.D., 1978, Self-stimulation at the lateral hypothalamus and locus coeruleus after specific unilateral lesions of the dopamine system, Brain Res., 146:123-140.

Lestang, I., Cardo, B, Rox, M.T., and Vellex, L, 1985, Electrical self-stimulation deficits in the anterior and posterior parts of the MFB after ibotenic acid lesion of the middle lateral hypothalamus, Neurosci., 15:379-388.

Lynch, M.R., and Wise, R.A., 1985, Relative effectiveness of pimozide, haloperidol and trifluoperazine on self-stimulation rate-intensity functions, Pharmacol. Biochem. Behav. 23:777-780.

Mitchell, M.J., Nicolaou, N.M., Arbuthnott, G.W., and Yates, C.M., 1982, Increases in dopamine metabolism are not a general feature of intracranial self-stimulation, Life Sci., 30:1081-1085.

Nakajima, S., 1984, Serotonergic mediation of habenular self-stimulation in the rat, Pharmacol. Biochem. Behav., 20:859-862.

Nakajima, S., and McKenzie, G.M., 1986, Reduction of the rewarding effect of brain stimulation by a blockade of dopamine D_1 receptor with SCH 23390, Pharmacol. Biochem. Behav., 24:919-923.

Nassif, S., Cardo, B., Libersat, F., and Velley, L., 1985, Comparison of deficits in electrical self-stimulation after ibotenic acid lesions of the lateral hypothalamus and the medial prefrontal cortex, Brain Res., 332:247-257.

Nazzaro, J., and Gardner, E.L., 1980, GABA antagonism lowers self-stimulation thresholds in the ventral tegmental area, Brain Res., 189:279-283.

Neill, D.B., Peavy, L.A., and Gold, M.S., 1978, Identification of a subregion within rat neostriatum for the dopaminergic modulation of lateral hypothalamic self-stimulation, Brain Res., 153:515-528.

Olds, J., 1962, Hypothalamic substrates of reward, Physiol. Rev., 42:554-604.

Olds, J., and Milner, P., 1954, Positive reinforcement produced by electrical stimulation of septal area and other regions of rat brain, J. Comp. Physiol. Psychol., 47:419-427.

Olds, J., and Travis, R.P., 1960, Effects of chlorpromazine, meprobamate, pentobarbital and morphine on self-stimulation, J. Pharmacol. Exp. Ther., 128:397-404.

Ott, T., Destrade, C., and Rüthrich, H., 1980, Introduction of self-stimulation behavior derived from a brain region lacking in dopaminergic innervation, Behav. Neurol. Biol., 28:512-516.

Phillips, A.G., Brooke, S.M., and Fibiger, H.C., 1975, Effects of amphetamine isomers and neuroleptics on self-stimulation from the nucleus accumbens and dorsal noradrenergic bundle, Brain Res., 85:13-22.

Phillips, A.G., Carter, D.A., and Fibiger, H.C., 1976a, Differential effects of parachloro-phenylalanine on self-stimulation in caudate putamen and lateral hypothalamus, Psychopharmacology (Berlin), 49:23-27.

Phillips, A.G., Carter, D.A., and Fibiger, H.C., 1976b, Dopaminergic substrates of intracranial self-stimulation in caudate-putamen, Brain Res., 104:222-232.

Phillips, A.G., and Fibiger, H.C., 1978, The role of dopamine in maintaining intracranial self-stimulation in the ventral tegmentum, nucleus accumbens and medial prefrontal cortex, Canad. J. Psychol., 32:58-66.

Phillips, A.G., Jakubovic, A., and Fibiger, H.C., 1987, Increased in vivo tyrosine hydroxylase activity in rat telencephalon produced by self-stimulation of the ventral tegmental area, Brain Res., 402:109-116.

Porrino, L.J., 1987, Cerebral metabolic changes associated with activation of reward systems, In: Brain Reward Systems, J. Engel and L. Oreland (Eds.), Raven Press, New York, 51-60.

Porrino, L.J., and Coons, E.E., 1980, Effects of GABA receptor blockade on stimulation-induced feeding and self-stimulation, Pharmacol. Biochem. Behav., 12:125-130.

Porrino, L.J., Esposito, R.U., Seeger, T.F., Crane, A.M., Pert, A., and Sokoloff, L., 1984, Metabolic mapping of brain during rewarding self-stimulation, Sci., 224:306-309.

Porrino L.J., Huston-Lyons, D., Bain, G., Sokoloff, L., and Kornetsky, C., 1990, The distribution of changes in local cerebral energy metabolism associated with brain stimulation reward to the MFB in the rat, Brain Res., 511:1-6.

Prado-Alcala, R., Streather, A., and Wise, R.A., 1984, Brain stimulation reward and dopamine terminal fields, II. Septal and cortical projections, Brain Res., 301:209-219.

Prado-Alcala, R., and Wise, R.A., 1984, Brain stimulation reward and dopamine terminal fields, I. Caudate-putamen, nucleus accumbens and amygdala. Brain Res., 297:265-273.

Redgrave, P., 1978, Modulation of intracranial self-stimulation behavior by local perfusions of dopamine, noradrenaline and serotonin within the caudate nucleus and nucleus accumbens, Brain Res., 155:277-295.

Redgrave, P., and Horrell, R.I., 1976, Potentiation of central reward by localized perfusion of acetylcholine and 5-hydroxytryptamine, Nature (London), 262:305-306.

Reymann, K.G., Mulcko, S., Ott, T., and Matthies, H., 1986, Opioid-receptor blockade reduces nose-poke self-stimulation derived from medial entorhinal cortex, Pharmacol. Biochem. Behav., 24:439-443.

Roberts, A., and Mogenson, G.J., 1978, Evidence for a role for dopamine in self-stimulation of the nucleus accumbens of the rat, Can. J. Psychol., 32:67-76.

Rolls, E.T., 1971, Contrasting effects of hypothalamic and nucleus accumbens septi self-stimulation on brain stem single unit activity and cortical arousal, Brain Res., 31:275-285.

Simon, H., LeMoal, M., and Cardo, B., 1976, Intracranial self-stimulation for the dorsal raphe nucleus of the rat: effects of the injection of para-chlorophenylalanine and of α-methylparatyrosine, Behav. Biol., 16:353-364.

Simon, H., Stinus, L., Tassin, J.P., Lavielle, S., Blanc, G., Thierry, A.M., Glowinski, J., and LeMoal, M., 1979, Is the dopaminergic mesocorticolimbic system necessary for intracranial self-stimulation?, Behav. Neurol. Biol., 27:125-145.

Smith, J.E., Dworkin, S.I., Co, C., and Porrino, L.J., 1990, Concurrent brain glucose utilization and biogenic monoamine neurotransmitter turnover rates with VTA brain stimulation reinforcement, Neurosci. Abst., In Press.

Sprick, U., Munoz, C., and Huston, J.P., 1985, Lateral hypothalamic self-stimulation persists in rats after destruction of lateral hypothalamic neurons by kainic acid or ibotenic acid, Neurosci. Lett., 56:211-216.

Stein, L., 1975, Norepinephrine Reward Pathways, Role in self-stimulation, memory consolidation and schizophrenia, In: D.K. Cole and T.B. Sonderegger (Eds.), Nebraska Symposium on Motivation (Vol. 27), Lincoln: University of Nebraska Press.

Steiner, S.S., Beer, B., and Shaefer, M.M., 1968, Escape from self-produced rates of brain stimulation, Sci., 163:90-91.

Stellar, J.R., and Stellar, E., 1985, The Neurobiology of Motivation and Reward, Springer-Verlag, New York, pp. 255.

Stephens, D.N., and Herberg, L.J., 1979, Dopamine-acetylcholine "balance" in nucleus accumbens and corpus striatum and its effect on hypothalamic self-stimulation, Eur. J. Pharmacol., 54:331-339.

West, T.E.G., and Wise, R.A., 1988, Effects of naltrexone on nucleus accumbens, lateral hypothalamic and ventral tegmental self-stimulation rate-frequency functions, Brain Res., 462:126-133.

Wise, R.A., 1978, Catecholamine theories of reward: A critical review, Brain Res., 152:215-247.

Wurtz, R.H., and Olds, J., 1963, Amygdaloid stimulation and operant reinforcement in the rat, J. Comp. Physiol. Psychol., 56:941-949.

BASAL FOREBRAIN INVOLVEMENT IN SELF-ADMINISTRATION OF DRUGS OF ABUSE

Linda J. Porrino, Steven I. Dworkin, and James E. Smith

Unit on Brain Imaging, NINDS, Bethesda, MD
and the Department of Physiology and Pharmacology
Wake Forest University
Bowman-Gray School of Medicine, Winston-Salem, NC

INTRODUCTION

There are a wide variety of chemical substances that have been classified as abused drugs on the basis of their excessive or inappropriate use. Although they may have distinct neurochemical actions and come from different pharmacological classes, what these drugs have in common is their capacity to engender pleasurable feelings or sensations in those individuals that use them. It is commonly believed that it is this property, their euphorigenic action, that causes their abuse.

The administration, however, of any abused substance, whether self-administered or not, can result in a variety of pharmacological actions which, in addition to its reinforcing effects, may be either central or peripheral. The nature of these effects depends on many factors which include, among others, the dose of the drug, the route of administration, the physiological status of the organism, and the environmental setting or context in which the drug is administered. The multiplicity of effects and the number of potentially important determining variables make the identification of the neural circuits that mediate the rewarding properties of abused drugs a rather complicated question. Paradigms must be employed in which the reinforcing effects can be separated from other actions of a drug such as effects on responding, stimulus properties, or effects on the physiological status of the organism.

In considering the neuroanatomical substrates of reward another question that should also be addressed is that of unitary vs. multiple reward systems. Are the neuroanatomical circuits which subserve self-administration of drugs such as psychostimulants, opiates, ethanol, and benzodiazepines the same? If not, do they contain common elements or connections, or are independent systems responsible for self-administration of abused drugs from different pharmacological classes?

Drugs that are self-administered by humans have generally been shown to be self-administered by animals as well (Weeks, 1962). Intravenous self-administration of drugs by animals, therefore, provides an excellent model for the assessment of the reinforcing properties of a given

The Basal Forebrain, Edited by T.C. Napier *et al.*
Plenum Press, New York, 1991

substance, but it can also be used as a model for the study of the
neuroanatomical substrates that mediate the reinforcing properties of
abused drugs. We will briefly review several approaches to this question
with a particular focus on methods that make it possible to examine the
functioning of entire circuits or pathways in the intact animal.

ELECTROLYTIC AND NEUROTOXIN LESIONS

One of the most commonly used strategies for the identification of
the brain regions or pathways that participate in self-administration
behavior is to remove or lesion specific brain areas. Recent reviews have
described in some detail the effects of lesions of various brain regions
on opiate and psychostimulant self-administration (cf. Koob and Goeders,
1989; Koob, this book). We will, therefore, be brief and selective.

Early studies, many of them done by Glick and his associates, made
use of electrolytic lesions to identify criticial substrates of
intravenous self-administration (Table 1). In general, similar results
were found for opiate and psychostimulant self-administration. Lesions of
the frontal cortex, for example, shifted the dose response curve to the
right, so that higher doses of both morphine and amphetamine were
necessary to maintain prelesion rates of responding for self-
administration (Glick, 1973; Glick and Cox, 1978; Glick and Marsanico,
1975). Conversely, lesions of the caudate or substantia nigra shifted the
dose response curve to the left, so that lower doses of these drugs were
needed to maintain prelesion rates of intake (Glick, 1974; Glick et al.,
1975). A number of similar sites produced negative results for both
opiates and psychostimulants (Glick and Cox, 1977; 1978). This type of
convergence of effects suggests that the neuroanatomical substrates of
intravenous self-administration are similar for these drug classes.

Table 1. Effects of Electrolytic Lesions on Intravenous Self-
Administration.

Structure	Heroin or morphine	Cocaine or amphetamine	Reference
Anterior cingulate	D		Trafton and Marques '71
Frontal cortex	D	I	Glick and Cox, '78
			Glick '73
			Glick and Marsanico, '75
Hippocampus	D		Glick and Cox, '78
Septum		N	
Medial raphe	D		Glick and Cox, '77
Caudate nucleus	I		Glick '74
Substantia nigra	I		Glick and Cox, '77
Locus coeruleus	N		Glick and Cox, '77
Medial forebrain bundle	I		Glick '74
Amygdala	N		Glick and Cox, '78
Nucleus accumbens	N		Glick and Cox, '78
Olfactory tubercle	N		Glick and Cox, '78

I=Increased sensitivity, D=Decreased sensitivity, N=No effect.

Many of these studies made use of dose effect relationships to draw conclusions about the effects of lesions on self-administration. This is a practice that has unfortunately not been utilized in some recent studies. Interpretation of increases and decreases in responding is difficult without this type of information. Behavioral changes may result from changes in response capability or in stimulus properties of the self-administered drug, in addition to changes in reinforcing efficacy. If multiple doses of the self-administered drug are tested across conditions or other control procedures utilized, more meaningful conclusions can be drawn.

Electrolytic lesions are rather non-specific, since they destroy all neuronal elements, e.g., fibers and cell bodies, within a given brain region. It is difficult, therefore, to ascribe changes in self-administration behavior to a specific neuroanatomical system or to a particular neurochemical mechanism. Some of these problems may be obviated by the use of neurotoxins which can affect precise neuronal elements or populations within a brain region.

A major focus of neurotoxin lesion studies has been to investigate the role of dopamine in the nucleus accumbens. Lesions of the accumbens that deplete dopamine levels by 90% or more produce long-lasting reductions in cocaine and amphetamine maintained self-administration behavior. These decreases in self-administration behavior would be difficult to interpret as a motivational deficit rather than a response deficit, if similar results had not been obtained with a progressive ratio procedure (Koob et al., 1987). In contrast, comparable dopaminergic depletions in the accumbens do not have a significant effect on heroin self-administration (Pettit et al., 1984). In a study of morphine self-administration in physically dependent rats, however, smaller dopaminergic depletions in the range of 20% caused a significant shift in the dose-effect curve to the right (Smith et al., 1985). The fact that dependent rats were studied may be the source of the discrepancy between these studies, although the degree of dopaminergic depletion may also be a significant variable. Serotonin depletions of the nucleus accumbens with the neurotoxin, 5,7-dihydoxytryptamine, in rats self-administering morphine also shifted the dose-effect curve to the right suggesting a decreased sensitivity to morphine, but such lesions had little effect on amphetamine self-administration (Lyness et al., 1980). Although serotonin and dopamine may play different roles in opiate and psychostimulant self-administration, other similarities in the neural substrates have been observed. In a study by Zito and her colleagues (1985) in which kainic acid lesions of the nucleus accumbens were made, the destruction of the cell bodies of interneurons within the nucleus and efferents from the nucleus disrupted both cocaine and heroin self-administration. Despite some neurochemical differences, then, neuroanatomically the substrates of opiate and psychostimulant self-administration are similar in that processing within the accumbens is involved in both.

Studies of the connections of the nucleus accumbens have revealed other similarities in the neuroanatomical circuitry subserving opiate and psychostimulant self-administration. Lesions of the ventral pallidum with ibotenic acid, which like kainic acid destroys cell bodies without damaging fibers, depressed responding for both heroin and cocaine (Hubner and Koob, 1989).

Studies of other forebrain structures have yielded somewhat contradictory results. The role of the medial prefrontal cortex, for example, remains controversial. Because this brain region supports responding maintained by direct local infusions of cocaine, it has been

Table 2. Effects of Neurotoxin Lesions on Intravenous Self-
Administration.

Structure	Neurotoxin	Heroin or Morphine	Cocaine or Amphetamine	Reference
Nucleus accumbens	6-OHDA	I,N	D	Smith et al. '85 Pettit et al. '84 Roberts et al. '80 Lyness et al. '79
Ventral tegmental area	6-OHDA		D	Roberts and Koob '82
Medial prefrontal cortex	6-OHDA		N	Martin-Iverson et al.'86
Caudate	6-OHDA		N	Koob et al. '87
Nucleus accumbens	5,7 DHT	I	N	Smith et al. '87 Lyness et al. '80
Nucleus accumbens	AF64A	I		
Nucleus accumbens	Kainic acid	D	D	Zito et al. '85
Ventral pallidum	Ibotenic acid	D	D	Hubner and Koob '90

I=Increased responding; D=Decreased responding; N=No effect

hypothesized that the medial prefrontal cortex is an important substrate of cocaine self-administration (see below). Depletion of dopamine in the medial prefrontal cortex, produced by 6-hydroxydopamine, however, did not alter intravenous self-administration of cocaine (Martin-Iverson et al., 1986). This failure to demonstrate an effect may relate to the fact that there may potentially be multiple sites where cocaine interacts with receptors to initiate activity in reinforcement circuits. Activation of a site such as the medial prefrontal cortex may be sufficient, but not necessary for the mediation of cocaine self-administration.

Verification of the specificity of the involvement of some of these brain regions in self-administration behavior has been investigated with neurotoxin lesions of components of these circuits in rats where responding on separate levers was maintained by the presentation of food, water or intravenous morphine with 24 hour access. Bilateral kainate (Dworkin et al., 1988c) and 5,7-dihydroxytryptamine (Dworkin et al., 1988a) lesions of the nucleus accumbens lower and generally flatten the dose-intake function for morphine self-administration without altering the intake of food or water. In contrast, 6-OHDA or sham-vehicle lesions have no effect (Dworkin et al., 1988b). These data indicate that serotonergic innervation and efferents from and/or interneurons in the nucleus accumbens are necessary for the intravenous self-administration of morphine by rats. Neurotoxin lesions of the ventral tegmental area, ventral pallidum and nucleus accumbens appear to attenuate processes necessary for intravenous amphetamine, cocaine, heroin and morphine to maintain self-administration. However, the specificity of some of these effects are yet to be determined. The lesion studies in the rats with drug, food and water access suggest that the neuronal systems responsible for the maintenance of food and water intake are separate and\or more generally distributed than those responsible for drug intake.

In considering these studies, the conclusion that seems the most apparent is that there is no single neurotransmitter substrate of self-administration behavior, but that multiple neurochemical systems must be involved. The single neurotransmitter approach to reinforcement is too simplistic to account for all of the available data, as well as all of the variables involved in self-administration behavior. It is necessary to consider multiple neurochemical systems and entire circuits to explain intravenous self-administration.

INTRACRANIAL SELF-ADMINISTRATION

Another method that can provide information about the neuronanatomical and neurochemical substrates of drug reinforcement is intracranial self-administration in which performance of a task results in the direct infusion of a drug into a discrete brain region. With intracranial self-administration it is possible to demonstrate that a given brain area is involved in the initiation of neuronal activity in systems mediating the reinforcing actions of a drug. It is also an important technique because it allows the circumvention of some of the problems of non-specific drug effects that occur with systemic drug administration.

Amphetamine has been shown to be reliably self-administered into the nucleus accumbens at a rate significantly above vehicle in rats (Hoebel et al., 1983) and into the orbitofrontal cortex by monkeys (Phillips and Rolls, 1981). The self-administration into the nucleus accumbens was dose dependent with 1.0 n molar injections maintaining the highest and most stable rates of intake, while 10.0 n molar doses were necessary in the orbitofrontal cortex. The dose range used in these studies could potentially have exceeded the concentrations necessary for specific binding. Therefore, without pharmacological blockade studies, it is difficult to demonstrate the specificity of these behaviors.

The intracranial self-administration of cocaine into the medial prefrontal cortex but not into the ventral tegmental area or nucleus accumbens by rats has been demonstrated by Goeders and his colleagues (Goeders and Smith, 1983; Goeders et al., 1986). The self-administration occurs significantly above vehicle and was dose dependent with inverted U shaped dose-intake relationships. Optimum doses were in the 90-120 pmol range while doses to 3000 pmol were ineffective in the nucleus accumbens and ventral tegmental area. Since cocaine is not thought to directly bind to dopamine receptors the location of the site of action was investigated with neurotoxin lesions. Removal of dopamine innervations of the self-administration site with 6-OHDA lesions attenuated intake to the level and patterns of that seen with the vehicle (Goeders and Smith, 1983). These data are consistent with the hypothesis that cocaine initiates reinforcing neuronal activity in the medial prefrontal cortex by a direct action on dopamine nerve endings since removal results in a loss of reinforcing efficacy.

Opiate agonists have been shown to be self-administered into the nucleus accumbens, hypothalamus and ventral tegmental area by rats. Morphine was self-administered at rates significantly above vehicle into the ventral tegmental region at 150 pmol doses (Bozarth and Wise, 1981), the lateral hypothalamus at 300 pmol doses (Olds, 1979), and into the nucleus accumbens at 700 pmol doses (Olds, 1982). Naloxone attenuated this self-administration demonstrating activation of opiate receptors to be necessary. Methionine enkephalin was self-administered into the nucleus accumbens (Goeders et al., 1984) and D-Ala-methionine enkephalinamide into the posterior hypothalamus (Olds and Williams, 1980) in a dose dependent manner with 500 pmol of the former and 150 pmol of the latter maintaining highest rates of intake. This self-administration was also dependent on the interaction of these substances with opiate receptors since naloxone attenuated intake. These data demonstrate these opiate agonists to initiate reinforcing neuronal activity through receptor mediated processes at several forebrain sites.

NEUROCHEMICAL MEASUREMENTS

Lesion studies of intravenous self-administration and intracranial self-administration studies are particularly useful for the identification of specific brain regions involved in the mediation of the reinforcing effects of drugs of abuse. It is important, however, not only to identify a single critical structure, but also to identify the inputs that influence activity in that structure, as well as the output pathways from that structure through which central actions are eventually translated into behavioral responses. One way the circuits involved in drug self-administration can be described is through the measurement of the neurochemical changes that accompany this behavior.

This type of approach has been utilized by Smith and his colleagues in order to identify the neural circuits which mediate opiate self-administration. In these studies (Smith et al., 1982; 1984a; 1984b) concurrent measurements of dopamine, norepinephrine, serotonin, aspartate, glutamate, glycine, and gamma-aminobutyric acid (GABA) were made in eleven brain regions of rats self-administering morphine. These rates were compared to turnover rates in littermate controls receiving yoked-morphine and yoked-saline infusions. The passive infusion of morphine resulted in elevated turnover rates of dopamine and GABA in the caudate and in the nucleus accumbens. Changes in norepinephrine turnover were found in the amygdala and glutamate utilization was increased in the septum. A similar pattern of changes in transmitter turnover rates was seen as a consequence of the contingent administration of morphine, but the changes were generally much larger. In addition, in animals self-administering morphine increased glutamate and GABA turnover were found in the hippocampus, amygdala, and septum. In the nucleus accumbens, in contrast to the increases in dopamine turnover observed in the yoked morphine group, dopamine along with norepinephrine and serotonin turnover rates were significantly decreased in self-administering animals. The contingent presentation of morphine appears to activate pathways distinct from those that are activated by the response independent presentation of the drug.

The differences in the distribution of turnover rate changes in self-administering and yoked-morphine animals are not the result of differences in morphine dosing, since both groups had identical histories of morphine administration. Nor does it appear that the disparities between these groups are the result of differences in the response requirements of the schedule of morphine reinforcement, because the lever presses emitted by the self-administering animals actually represented a very small proportion of the total motor activity. The critical variable was the control of reinforcer presentation specific to the self-administering group. The neurochemical turnover changes that are unique to the contingent presentation of morphine represent the activation of pathways mediating the positive reinforcing properties of this drug.

344

In separate experiments turnover rates of acetylcholine were also assessed in rats self-administering morphine and yoked controls (Smith et al., 1984a; 1984b). Turnover was increased in the diagonal band of Broca and in the septum of both groups of animals receiving morphine. Increased turnover of acetylcholine was found in the frontal cortex only in self-administering animals, suggesting that the changes in this area are specific to opiate reinforcement rather than to the pharmacological effects of morphine.

The results of these studies taken together with data from other studies with other methodologies suggest two circuits involved in intravenous morphine self-administration (Fig. 1). The first is a hippocampal circuit consisting of pathways from hippocampus to nucleus accumbens to amygdala to entorhinal cortex to hippocampus. This circuit may be modulated by dopaminergic inputs from the ventral midbrain and by serotinergic afferents from the raphe nuclei, both to the nucleus accumbens or to other components of this circuit. A second circuit that may be involved in reward processes includes a frontal cortex to dorsal striatum to globus pallidus to frontal cortex loop. Again monoamines from the brainstem may modulate activity in this circuit through afferent inputs to the striatum and frontal cortex. Opiate interactions with these circuits could occur at a number of sites such as the nucleus accumbens which has been shown to be a positive site for intracranial self-administration of met-enkephalin or the hippocampus where opiates have local excitatory actions.

These circuits may represent general reinforcement pathways and not act exclusively as substrates of opiate reward processes. This is supported by recent studies in which neurotransmitter turnover rates of biogenic amines were measured in animals lever pressing for brain stimulation reward to the ventral tegmental area (Smith et al., 1990). Comparisons of these rates were made to those measured in animals receiving electrical brain stimulation non-contingently and to those not receiving stimulation. A preliminary analysis of these data showed increases in dopamine and decreases in serotonin turnover in the nucleus accumbens only in animals lever pressing for brain stimulation reward. In other areas changes were also seen that were consistent with the results obtained in animals self-administering morphine. Despite differences in the form of the reinforcer then, the circuits critical to reinforcement appear to be similar.

METABOLIC MAPPING

Metabolic mapping with the 2-[^{14}C]deoxyglucose method (Sokoloff et al., 1977) is another useful procedure with which the neuronal circuitry underlying a given behavioral process can be delineated. This method can measure the changes in functional activity throughout the central nervous system that accompany a behavioral, pharmacological, or physiological manipulation. This is accomplished through the measurement of rates of glucose utilization in specific brain structures with quantitative autoradiography. Because glucose is the sole substrate of energy metabolism in the brain under normal physiological conditions, changes in local rates of glucose utilization are a direct reflection of changes in energy consumption. In turn, the amount of energy used is in proportion to the amount of neuronal or functional activity in a particular brain region. Changes in glucose utilization then are a measure of functional changes in the central nervous system.

Although there have been a number of studies mapping the changes in functional activity that accompany the administration of a variety of abused drugs, these have generally not utilized paradigms in which the reinforcing effects of these drugs could be isolated. In some cases however, experimental conditions have been chosen which approximate the conditions under which the drug is abused. For example, we have measured rates of glucose utilization in freely-moving rats following the intravenous administration of cocaine (Porrino et al., 1988) and methamphetamine (Pontieri et al., in press) in doses within the range of those self-administered by rats. The intravenous route is a frequently used route of self-administration, and its pharmacokinetics closely resemble those of the intranasal and intrapulmonary routes of ' administration. In these studies functional activity was selectively elevated in portions of the mesocorticolimbic system including the nucleus accumbens, prefrontal cortex and olfactory tubercle, particularly at doses which support self-administration. Although these studies are suggestive they do not directly address the question of which circuits are activated during self-administration.

In order to map the distribution of alterations in local functional activity associated with the reinforcing properties of cocaine, the 2-[^{14}C]deoxyglucose method was applied to rats self-administering the drug during the experimental session. Rates of glucose utilization in these rats were compared to littermate control rats receiving yoked infusions of either cocaine or saline. Preliminary findings indicate that the

Fig. 1. The neuronal circuits identified in morphine self-administering rats that appear to mediate the reinforcing effects of opiates include two forebrain circuits that likely have the ability to reverberate activity. Feedforward pathways from brain stem nuclei likely can potentiate or antagonize activity in these circuits depending on which subset of inputs are activated. Feedback pathways from these two forebrain circuits likely modulate activity in these feedforward pathways, thus fine tuning the mediating networks. The actual neuronal pathways identified in these morphine self-administering animals are:

A. Neuronal circuits mediating effects of reinforcing actions of morphine in self-administering rats. These include a frontal cortex-caudate nucleus circuit involving glutamate and acetylcholine pathways with aspartate and GABA interneurons, and a hippocampal-entorhinal cortex-hippocampus circuit also involving glutamate and acetylcholine pathwyas and aspartate and GABA interneurons in the nucleus accumbens and hippocampus.

B. Feedforward brainstem pathways modifying activity in the two forebrain circuits. Dopaminergic innervations of the caudate nucleus-putamen from the substantia nigra and of the nucleus accubmens and septum from the ventral tegmental area may modulate activity in these structures, while norepinephrine may do the same in the septum and serotonin in the nucleus accumbens.

C. Feedback forebrain pathways modulating activity in brain stem regions of origin of feedforward pathways. GABA pathways from the caudate nucleus to the substantia nigra and from the nucleus accumbens to the ventral tegmental area may modulate activity in these areas that send innervations to these forebrain circuits.

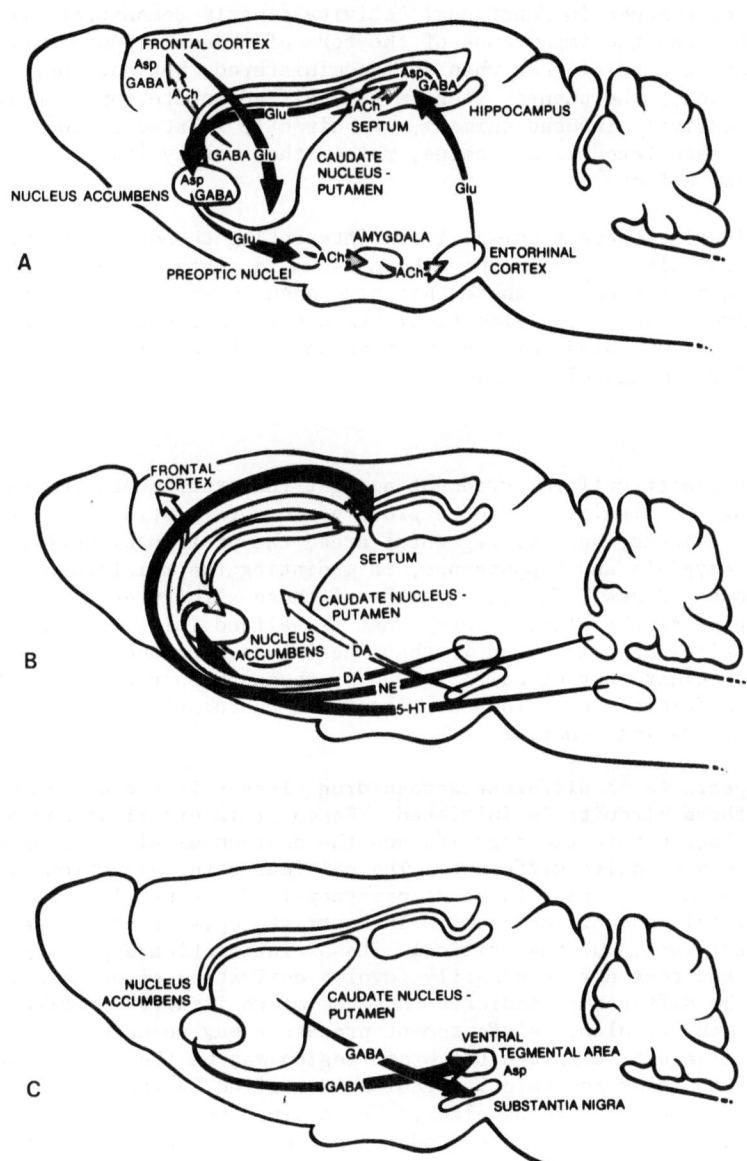

contingent administration of cocaine (self-administration) produced large widespread elevations in glucose utilization throughout the mesocorticolimbic system including the prefrontal cortex, nucleus accumbens, olfactory tubercle, and amygdala, when compared to yoked saline controls. Increases were also evident in the extrapyramidal motor system. Passive cocaine infusion resulted in increases in functional activity as compared to yoked saline animals in portions of the extrapyramidal motor system, but not in forebrain limbic structures. The differences in the two patterns of changes in functional activity clearly demonstrate the differences between the importance of the form of presentation of the drug. Cocaine is a reinforcer when self-administered, but not when passively infused. The changes seen in the self-administering animals, but not the passively infused animals, are directly related to the reinforcement associated with cocaine, rather than merely its pharmacological effects.

In considering these studies, it is interesting to note that once again the neural circuits which have been shown to subserve cocaine self-administration correspond to those that have been shown to mediate other forms of reinforcement, e.g. electrical brain stimulation reward (Porrino et al., 1984) and morphine self-administration (Smith et al., 1982) regardless of the methodology used.

CONCLUSIONS

There is almost uniform agreement amongst the experiments reviewed here regarding the significance of a group of forebrain structures, e.g., the nucleus accumbens, ventral tegmental area, the prefrontal cortex, and possibly the amygdala and hippocampus, in mediating the reinforcing effects of abused drugs. The involvement of these structures does not appear to depend on the class of drug that is self-administered, nor on the methodological approach used in the experiments. On the basis of this evidence it is clear that circuits formed by these brain regions and their afferent and efferent connections are important neuroanatomical substrates of self-administration behavior.

What appears to be different across drug classes is the way in which activity in these circuits is initiated. Receptor interactions may occur at different loci within the circuits and the neurochemical bases of these interactions may be quite different. The critical sites of action of cocaine, for example, appear to be at presynaptic dopaminergic reuptake sites in striatal and cortical loci. In contrast, opiates also appear to act at striatal loci, but the critical site of interaction appears to be postsynaptic and does not necessarily involve activation of dopaminergic systems. These differences indicate that the microcircuitry involved in opiate and psychostimulant reinforcement processes may be unique to the pharmacological agent, although the brain region may be the same. The importance of the neuroanatomical circuits identified by the studies reviewed here, however, remains clear. These circuits may not only be involved in the rewarding effects produced by abused drugs, but they may constitute more general reinforcement systems mediating the effects of other reinforcers.

ACKNOWLEDGEMENTS

Supported in part by USPHS Research Grant DA-01999, DA-03628, DA-03832, DA-03631, and DA K05 00114.

REFERENCES

Bozarth, M.A. and Wise, R.A., 1981, Intracranial self-administration of morphine into the ventral tegmental area in rats, Life Sci., 28:551.

Dworkin, S.I., Goeders, N.E., Guerin, G.F., Co, C., and Smith, J.E., 1988a, Effects of 5,7-dihydoxytrptamine lesions of the nucleus accumbens in rats responding on a concurrent schedule of food, water, and intravenous morphine self-administration, in: "Problems of drug dependence," L. Harris, ed., NIDA Research Monograph No. 81, Washington, D.C., p. 138.

Dworkin, S.I., Guerin, G.F., Co, C., Goeders, N.E., and Smith, J.E., 1988b, Lack of an effect of 6-hydroxydopamine lesions of the nucleus accumbens on intravenous morphine self-administration, Pharmacol. Biochem. Behav., 30:1051.

Dworkin, S.I., Guerin, G.F., Goeders, N.E., and Smith, J.E., 1988c, Kainic acid lesions of the nucleus accumbens selectively attenuate morphine self-administration, Pharmacol. Biochem. Behav., 29:175.

Glick, S.D., 1973, Impaired tolerance to the effects of oral amphetamine intake in rats with frontal cortex ablations, Psychopharmacology, 28:363.

Glick, S.D., 1974, Changes in sensitivity to the rewarding property of morphine following lesions of the medial forebrain bundle or caudate nucleus in rats, Arch. Int. Pharmacodyn. Therap., 212:214.

Glick, S.D. and Cox, R.D., 1977, Changes in morphine self-administration after brainstem lesions in rats, Psychopharmacology, 52:151.

Glick, S.D. and Cox, R.D., 1978, Changes in morphine self-administration after tel-diencephalic lesions in rats, Psychopharmacology, 57:283.

Glick, S.D. and Marsanico, R.G., 1975, Time-dependent changes in amphetamine self-administration following frontal cortex ablations in rats, J. Comp. Physiol. Psychol., 88:355.

Glick, S.D., Cox, R.D., and Crane, A.M., 1975, Changes in morphine self-administration and morphine dependence after lesions of the caudate nucleus in rats, Psychopharmacology, 41:219.

Goeders, N.E. and Smith, J.E., 1983, Cortical dopaminergic involvement in cocaine reinforcement, Science, 221:773.

Goeders, N.E., Dworkin, S.I., and Smith, J.E., 1986, Neuropharmacological assessment of cocaine self-administration into the medial prefrontal cortex, Pharmacol. Biochem. Behav., 24:1429.

Goeders, N.E., Lane, J.D., and Smith, J.E., 1984, Intracranial self-administration of methionine enkephalin into the nucleus accumbens, Pharmacol. Biochem. Behav., 20:451.

Hoebel, B.G., Monaco, A.P., Hernandez, L., Aulisi, E.F., Stanly, B.G., and Lenard, L., 1983, Self-injection of amphetamine directly into the brain, Psychopharmacology, 81:158.

Hubner, C.B. and Koob, G.F., 1990, The ventral pallidum plays a role in mediating cocaine and heroin self-administration in the rat, Brain Res., 508:20.

Koob, G.F. and Goeders, N.E., 1989, Neuroanatomical substrates of drug self-administration, in: "The Neuropharmacological Basis of Reward," J.M. Liebman and S.J. Cooper, ed., Clarendon Press, Oxford, p. 214.

Koob, G.F., Vaccarino, F.J., Amalric, M., and Bloom, F.E., 1987, Positive reinforcement properties of drugs: search for neural substrates, in: "Brain reward systems and abuse," J.Engel and L. Oreland, ed., Raven, New York, p. 35.

Lyness, W.H., Friedle, N.M., and Moore, K.E., 1979, Destruction of dopaminergic nerve terminals in nucleus accumbens: effect on d-amphetamine self-administration, Pharmacol. Biochem. Behav., 11:553.

Lyness, W.H., Friedle, N.M., and Moore, K.E., 1980, Increased self-administration of d-amphetamine after destruction of 5-hydroxytryptaminergic neurons, Pharmacol. Biochem. Behav., 12:937.

Martin-Iverson, M.T., Szostak, C., and Fibiger, H.C., 1986, 6-Hydroxydopamine lesions of the medial prefrontal cortex fail to influence intravenous self-administration of cocaine, Psychopharmacology, 88:310.

Olds, M.E., 1979, Hypothalamic substrate for the positive reinforcing properties of morphine in the rat, Brain Res., 168:351.

Olds, M.E., 1982, Reinforcing effects of morphine in the nucleus accumbens, Brain Res., 237:429.

Olds, M.E. and Williams, K.N., 1980, Self-administration of D-Ala^2-Met-enkephalinamide at hypothalamic self-stimulation sites, Brain Res., 194:155.

Pettit, H.O., Ettenberg, A., Bloom, F.E., and Koob, G.F., 1984, Destruction of dopamine in the nucleus accumbens selectively attenuates cocaine but not heroin self-administration in rats, Psychopharmacology, 84:167.

Phillips, A.G. and Rolls, E.T., 1981, Intracranial self-administration of amphetamine by rhesus monkeys, Neurosci. Lett., 24:81.

Pontieri F.E., Crane, A.M., Seiden, L.S., Kleven, M.S., and Porrino, L.J., Metabolic mapping of the effects of intravenous methamphetamine administration in freely-moving rats. Psychopharmacology, in Press.

Porrino, L.J., Domer, F.R., Crane, A.M., and Sokoloff, L., 1988, Selective alterations in cerebral metabolism within the mesocorticolimbic dopaminergic system produced by acute cocaine administration in rat, Neuropsychopharmacology, 1:109.

Porrino, L.J., Esposito, R.U., Seeger, T.F., Crane A.M., Pert, A., and Sokoloff, L., 1984, Metabolic mapping of the brain during rewarding self-stimulation, Science, 224:306.

Roberts, D.C.S. and Koob, G.F., 1982, Disruption of cocaine self-administration following 6-hydroxydopamine lesions of the ventral tegmental area in rats, Pharmacol. Biochem. Behav., 17:901.

Roberts, D.C.S., Koob, G.F., Klonoff, P., and Fibiger, H.C., 1980, Extinction and recovery of cocaine self-administration following 6-hydroxydopamine lesions of the nucleus accumbens, Pharmacol. Biochem. Behav., 12:781.

Smith, J.E., Co., and Lane, J.D., 1984a, Limbic acetylcholine turnover rates correlated with rat morphine-seeking behaviors, Pharmacol. Biochem. Behav., 20:429.

Smith, J.E., Co, C., and Lane, J.D., 1984b, Limbic muscarinic and benzodiazepine receptor changes with chronic intravenous morphine and self-administration, Pharmacol. Biochem. Behav., 20:443-450.

Smith, J.E., Co, C., Freeman, M.E., and Lane, J.D., 1982, Brain neurotransmitter turnover correlated with morphine-seeking behavior of rats, Pharmacol. Biochem. Behav., 16:509.

Smith, J.E., Dworkin, S.I., Co, C., and Porrino, L.J., 1990, Concurrent brain glucose utilization and biogenic monoamine neurotransmitter turnover rates with VTA brain stimulation reinforcement. Soc. Neurosci. Abstr. 19, in press.

Smith, J.E., Guerin, G.F., Co, C., Barr, T.S., and Lane, J.D., 1985, Effects of 6-OHDA lesions of the central medial nucleus accumbens on rat intravenous morphine self-administration, Pharmacol. Biochem. Behav., 23:843.

Smith, J.E., Shultz, K., Co, C., Goeders, N.E., and Dworkin, S.I., 1987, Effects of 5,7-dihydrotryptamine lesions of the nucleus accumbens on rat intravenous morphine self-administration, Pharmacol. Biochem. Behav., 26:607.

Sokoloff, L., Reivich, M., Kennedy, C., DesRosiers, M.H., Patlak, C.S., Pettigrew, K.D., Sakurada, O., and Shinohara, M., 1977, The [^{14}C]deoxyglucose method for the measurement of local cerebral glucose utilization: theory, procedure and normal values in the conscious and anesthetized albino rat. J. Neurochem. 28:897.

Trafton, C.L. and Marques, P.R., 1971, Effects of septal area and cingulate cortex lesions on opiate addiction behavior in rats, J. Comp Physiol. Psychol., 75:227.

Weeks, J.R., 1962, Experimental morphine addiction: method for automatic intravenous injections in unrestrained rats, Science, 138:143.

Zito, K.A., Vickers, G., and Roberts, D.C.S., 1985, Disruption of cocaine and heroin self-administration following kainic acid lesions of the nucleus accumbens, Pharmacol. Biochem. Behav., 23:1029.

BASAL FOREBRAIN CHOLINERGIC SYSTEM: A FUNCTIONAL ANALYSIS

David Olton, Alicja Markowska, Mary Lou Voytko,

Ben Givens, Linda Gorman, Gary Wenk

Department of Psychology
The Johns Hopkins University
Baltimore, MD 21218

INTRODUCTION

Describing the function of the basal forebrain cholinergic system (BFCS) is complicated by the fact that both anatomical and neurochemical criteria define the system. Ideally, functions would be described for only the cholinergic cells in only this brain area. As will be indicated below, for technical reasons, functional analyses of the BFCS have not often achieved this goal. For example, studies examining the behavioral consequences of lesions in the nucleus basalis magnocellularis (NBM) and medial septal area (MSA), two areas of the BFCS, have often used a neurotoxin such as ibotenic (IBO) acid to produce the lesion. Although IBO clearly destroys many cholinergic cells, it is not specific for just the cholinergic system, and destroys noncholinergic cells also. Determining the extent to which the behavioral consequences of these lesions depend upon the destruction of the cholinergic cells, as compared to the noncholinergic cells, is very difficult. Likewise, measures of sodium dependent high affinity choline uptake (HACU) in the frontal cortex (FC), a projection area for cholinergic cells in the NBM, is complicated by the fact the FC has intrinsic cholinergic neurons. Again, determining the extent to which the behaviorally-induced changes in HACU are due to activity in the NBM cells as compared to the intrinsic cells is complicated. In short, the manipulations and measurements discussed in this chapter clearly alter and reflect, respectively, activity in the BFCS. However, they are not always specific for just the cholinergic cells in the BFCS, with the result that conclusions about the role of cholinergic nerve cells in the basal forebrain must be considered carefully.

Functional analyses of the BFCS have been guided by both its anatomical and neurochemical characteristics. Anatomically, two of the important target sites of the BFCS,

The Basal Forebrain, Edited by T.C. Napier *et al.*
Plenum Press, New York, 1991

the hippocampus (HIP) and the frontal cortex (FC), have had extensive functional analyses. The conceptual frameworks and experimental procedures used to examine HIP and FC have been applied to the MSA and NBM, respectively, as a first guide to making predictions about their function (Fuster, 1989; Kolb, 1984; O'Keefe, and Nadel, 1978; Olton et al., 1979). Neurochemically, the cholinergic system has had extensive analysis through pharmacological manipulations, which indicate the behavioral impairments produced by drugs that disrupt cholinergic function, and the occasional behavioral improvements produced by drugs that enhance cholinergic function (Bartus et al., 1985a).

The data reviewed in this chapter lead to the following conclusions. First, interference with the activity of cells in the NBM and the MSA can produce behavioral impairments similar to those produced by lesions of their respective target areas, the FC and HIP. This pattern of results is particularly striking in rats. Less information is available in primates, and the currently available information about the effects of lesions in primates does not fit into any easily identifiable pattern yet. Second, the cells in the NBM and MSA are activated (as measured by HACU) in many different kinds of tasks. The psychological demands that influence the activity of NBM and MSA cells remain to be fully characterized, but they probably include many different types of psychological processes. Together, these data suggest that cholinergic cells in the BFCS have an important influence on a variety of cognitive processes, particularly ones that involve the HIP and FC.

Many of the general statements in this chapter describing the BFCS and the effects of lesions in the NBM and MSA have been reviewed in detail elsewhere (Olton and Wenk, 1987; Olton, Wenk, and Markowska, in press; Wenk and Olton, 1987). Consequently, the present chapter will not provide detailed references for statements that are generally supported in these previous reviews, and the reader is referred to them for a complete list of references. Statements about particular articles in the present chapter are followed by the appropriate references.

NEUROTOXIC LESIONS

The logic of a lesion analysis of function is the same as that used to assess and interpret malfunction in any system. Empirically, diagnostic behavioral tests indicate the ways in which the behavior of the animal is altered following the lesion. Theoretically, the pattern of results is used to infer the kinds of functions that are disrupted by the lesion and those that remain intact. The logic of this approach has been summarized in detail elsewhere (Olton, 1986), and is applied to the functional analysis of the BFCS in the same way that it has been applied to other neural systems.

The ideal lesion is one that involves all of the cholinergic cells in a given area, and is limited to just these cells; it does not involve noncholinergic cells or fibers of passage in the same area, nor does it involve any cells in any other area.

Unfortunately, the technique to produce this ideal lesion does not yet exist. Most lesions have been made with ibotenic (IBO) acid. When used appropriately, IBO can destroy cell bodies in a circumscribed area, and spare fibers of passage through that area. However, IBO is not selective for cholinergic cells. Thus, unlike the ideal lesion, IBO destroys noncholinergic as well as cholinergic cells. In some areas and at some concentrations, IBO is lethal (Wenk et al., 1984). Consequently, restricted amounts of IBO in restricted locations have been used, with the result that not all of the cholinergic cells in the desired area are destroyed.

IBO's effectiveness in destroying cholinergic cell bodies in the BFCS is usually assessed by measuring choline acetyltransferase (ChAT) activity in the appropriate projection area. Because the HIP has few if any intrinsic cholinergic nerve cells, a completely effective MSA lesion should eliminate ChAT. In reality, IBO lesions of MSA typically reduce ChAT to 30%-50% of normal. Because the FC has intrinsic cholinergic neurons, which account for approximately 30% of the ChAT activity in the FC, a complete NBM lesion should reduce ChAT to approximately 30% of normal. In reality, IBO lesions have reduced ChAT to 50%-70% of normal.

The "good news" from these data is that IBO does produce substantial damage to cholinergic nerve cells in selective portions of the BFCS. The "bad news" is composed of two parts. First, the cholinergic damage is not complete. Second, the damage is not limited to the cholinergic system.

The "good news" means that appropriate behavioral tests should be sensitive to the loss of cholinergic nerve cells in the BFCS. The "bad news" means that the data must be interpreted with some caution. First, the absence of an effect might result from an incomplete lesion, which spares enough of the intended area so that it can support normal behavior in the test being given. Second, the presence of a behavioral change might result from damage to noncholinergic cells rather than from damage to cholinergic cells.

Rats: Lesions Produced by IBO

MSA lesions produced behavioral effects similar to those following HIP lesions, although the magnitude and the duration of the effect were often smaller. For example, consider the pattern of results obtained from a test of spatial working memory in a T-maze (Hepler et al., 1985). For the forced run, one arm of the maze was blocked, forcing the rat to enter the other arm. For the subsequent choice run, both arms were available, but reinforcement was only in the arm not entered on the immediately preceding forced run. Because the arm that was available for the forced run of each trial changed randomly from trial to trial, the rat had to remember the arm that was entered on the immediately preceding forced run in order to choose the correct arm on the choice run. Normal rats learned this task quickly, and performed it well, with choice accuracy consistently above 90%.

Lesions of the HIP or any of its extrinsic fiber pathways produced a severed and enduring impairment of performance in this type of task in a T-maze (Markowska et al., 1989; Olton et al., 1979). Choice accuracy was at approximately the level expected by chance, and showed little improvement in choice accuracy in spite of extensive postoperative testing. MSA lesions, during immediate postoperative testing, also produced a substantial impairment of choice accuracy (Hepler et al., 1985). With continued testing, however, this impairment gradually decreased so that at the end of postoperative testing, the choice accuracy in this basic test procedure was normal. Choice accuracy in other tasks that require working memory was also consistently impaired by MSA lesions.

The magnitude and duration of the MSA impairment must depend upon a number of variables, many of which remain to be specified. At least one relevant variable is the demand for recent memory. The longer the delay interval during which information had to be remembered, the greater the magnitude and/or duration of the impairment (Hepler et al., 1985; Bartus et al., 1985b). The size of the lesion should also be relevant. Although systematic manipulations of the size of an NBM or MSA lesion have not been reported yet, the pattern of results following MSA lesions is markedly similar to that following partial fornix lesions (Raffaelle and Olton, 1988). Like complete fornix lesions, partial lesions produced an impairment of choice accuracy. However, the magnitude and duration of the impairment produced by the partial lesions was less than that produced by the complete lesions. Nonlinear functions relating the size of the lesion to the behavioral and neural consequences of it indicate the caution that ought to be taken when drawing conclusions from incomplete lesions (Messing, 1986; Messing and Aanosen, 1989; Gerbec et al., 1988). The length of the postoperative recovery time, the amount of preoperative training, and the amount of postoperative testing should all contribute to the recovery of the impairment (Davis and Volpe, in press; Handelmann & Olton, 1981; Olton and Markowska, 1989; Volpe et al., 1989). Changes in cholinergic and noncholinergic systems may also be important (Wenk et al., 1989; Wenk and Olton, 1984).

NBM lesions also produced impairments in tests of recent memory. These were of similar magnitude and duration to those produced by MSA lesions.

The number of different tasks used to assess performance following NBM lesions is larger than that used to assess behavior following MSA lesions. This relative emphasis on the NBM may reflect the initial focus on this area in Alzheimer's Disease, and the relatively little information that was available about the functions of the NBM, as compared to those of the MSA. In any case, NBM lesions produced behavioral impairments in tasks that do not require recent memory. For example, place discrimination in a water maze was substantially impaired by NBM lesions (Whishaw et al., 1985). A circular tank was filled with water. A platform was placed in a constant location in the tank so the surface of the platform was just below the level of the water, and not visible to the rat. For each trial, the rat was placed in one of four locations at the edge of the tank, and allowed to swim to the hidden platform. In this type of task, rats use spatial stimuli around the tank to identify the location of the platform, and form a spatial cognitive map of the topological relationship between the platform and the stimuli. Normal rats acquired this task quickly, reducing the length of the path and the time required to go from the edge of the tank to the platform. NBM lesions impaired performance. During the initial test sessions, the time taken to swim from the edge of the tank to the platform was increased. Throughout behavioral testing, the deviation from the most direct path to reach the platform remained substantially higher. This pattern of results occurred for both postoperative acquisition of the task, and postoperative retention of the preoperatively learned discrimination.

Lesions of the NBM, and also of the MSA, have often produced temporary impairments in general sensory-motor skills. For a few days postoperatively, rats often failed to eat, groom, and show normal reflexes. In extreme cases, some intubation was necessary to provide nutritional support for several days. These behavioral changes typically disappeared within a few days, and were not present at the time of behavioral testing in the experiments described above.

Rats: Functional Dissociations Between NBM and MSA

As indicated above, NBM and MSA lesions produced similar behavioral impairments in some tasks. This similarity is not surprising because HIP and FC lesions in rats produce the same pattern of behavioral impairments. Thus, the absence of an empirical dissociation in those tasks should not be taken to indicate the absence of a functional dissociation. Indeed, in the few behavioral tests that did produce behavioral dissociations following FC and HIP lesions in rats, similar dissociations were produced by NBM and MSA lesions.

One series of experiments demonstrated three dissociations in temporal discriminations, a pattern of results that was identical to dissociations produced by lesions of the fornix and the FC (Meck et al., 1984, 1987; Olton et al., 1987, 1988). These experiments used variations of an operant procedure that trained rats to perform different types of temporal discriminations. The following dissociations were found: (1) The memory for the expected time of reinforcement was decreased by MSA lesions and increased by NBM lesions. (2) The working memory for the duration of a previous stimulus was impaired by MSA lesions and unaffected by NBM lesions. (3) Divided attention for timing two stimuli simultaneously was impaired by NBM lesions but was unaffected by MSA lesions.

Primacy and recency effects were also dissociated following small MSA and NBM lesions (Kesner, 1988; Kesner et al., 1987, 1989). Primacy produces better memory for information at the beginning of the list than at the end of the list. Partial MSA lesions impaired primacy; partial NBM lesions had no effect. Recency produces better memory for information at the end of the list than at the beginning of the list. Partial NBM lesions disrupted recency; partial MSA lesions had no effect.

These dissociations demonstrate that the functions of the NBM and MSA are not equivalent and that appropriate behavioral tests can discriminate between lesions in these two areas. The fact that MSA lesions produced behavioral changes similar to those of HIP lesions, whereas NBM lesions produced behavioral effects similar to those of FC lesions, suggests that the functions of these two basal forebrain areas are closely linked to the functions of their target regions. Indeed, this summary of the lesion data suggests that the projections from the NBM to the FC are necessary for normal function of the FC, and the projections from the MSA to the HIP are necessary for normal function of the HIP. Consequently, one of the functions of the NBM and MSA is to permit their respective projection areas, the FC and HIP, to perform their functions. The functions of the NBM are probably not limited to this category of FC functions. The NBM has projections to many other brain areas, and appropriate behavioral analyses should be able to demonstrate the behavioral relevance of these projections.

Rats: Cholinergic and Non-cholinergic Contributions

As mentioned above, IBO does not destroy all cholinergic cells in a desired area and spare all noncholinergic cells. The extensive damage to the cholinergic system produced by IBO is likely to contribute to the behavioral impairments produced by these lesions. However, the noncholinergic damage may also influence behavior. The behavioral effects produced by another neurotoxin, quisqualic (QUIS) acid, suggest that some of the impairments produced by IBO may result from IBO's damage to noncholinergic neurons.

QUIS, as compared to IBO, when injected into the NBM, produced a greater loss of ChAT activity in the FC, suggesting that QUIS destroyed more cholinergic cells in the NBM than IBO. Immunocytochemical identification of cholinergic cells in the NBM confirmed this impression, although the comparison was not quantitative.

If the behavioral effects produced by IBO were due only to its cholinergic pathology, then QUIS, because it produces more extensive cholinergic pathology, should produce a larger (or at least equivalent) behavioral deficit. Such is not the case. Although QUIS did produce a behavioral impairment, the duration and magnitude of this impairment was less than that produced by IBO. (See Table 1 in Sarter and Dudchenko for a summary; Dunnett et al., 1987; Dunnett, in press; Etherington et al., 1987; Markowska et al., in press; Robbins et al., 1989; Wenk et al., 1989).

The fact that QUIS, like IBO, produced a behavioral impairment during immediate postoperative testing is consistent with the conclusion that the cholinergic projections from the NBM to the FC are required for normal performance in the tasks that were tested. The fact that

QUIS produced a behavioral impairment that was smaller in magnitude and shorter in duration than that produced by IBO complicates any interpretation of the behavioral data in terms of the cholinergic portions of the basal forebrain system.

An adequate interpretation of the pattern of results from IBO and QUIS will depend upon more extensive information about the neural effects of these two neurotoxins. Here are some possible resolutions. IBO may destroy one or more noncholinergic systems that are spared by QUIS, and this additional destruction may be responsible for the greater behavioral impairments (Coffey et al., 1988; Sarter and Dudchenko, in press; Wenk et al., 1989). Alternatively, the current measures of cholinergic pathology may lead to erroneous conclusions about the functional effects of these two neurotoxins on cholinergic activity. Direct measurements of acetylcholine release in the FC might help to resolve this last issue.

In the context of the current discussion, IBO and QUIS are tools, and not the phenomena of primary concern. Consequently, the development of more selective neurotoxins may eliminate the importance of understanding the mechanisms of action of IBO and QUIS. For the moment, however, these are the tools that have been used most extensively to make lesions in the NBM and MSA.

In summary, these data demonstrate that neurotoxic lesions in the NBM and MSA can produce impairments in many behavioral tasks that require psychological functions mediated by the FC and HIP, respectively. Attributing the functional impairments to specific neurotransmitter systems is currently difficult; however a case can be made for the involvement of both cholinergic and noncholinergic neurons in this area.

Primates: Lesions Produced By IBO

Relatively little information is available about the effects of BFCS lesions in primates. The available data suggest that basal forebrain lesions made by IBO produce less dramatic results in primates than in rats. The reasons for these apparent species differences are not yet immediately obvious, and any suggestions must be made with caution given the limited amount of data currently available from primates.

The results from experiments with primates are similar to those from experiments with rats, with the exception that the magnitude and duration of the impairment is often considerably less. The experiments reviewed below have three general outcomes. First, BFCS lesions produced moderate behavioral impairments during immediate postoperative testing. Second, continued postoperative testing often led to substantial recovery of function so that the performance of the lesion group often returned to normal. Third, the lesion group had greater sensitivity to the anticholinergic drug scopolamine, even when its performance was normal without the drug.

Combined lesions of the MSA and NBM produced a small impairment in visual delayed nonmatch-to-sample with long delays and long list lengths (Algner, et al., 1990, submitted). In this task, one or more sample stimuli were presented. A delay interval followed. Then each sample was presented simultaneously with a novel stimulus. The correct response was to choose the novel stimulus. With short delays, the lesions had no effect on choice accuracy. With the longest delay (approximately 2 minutes) and a list of 10 sample stimuli prior to the first choice, these lesions produced a slight but significant decrease in choice accuracy. The magnitude of this impairment was generally less than that produced by combined lesions of the hippocampus, amygdala, and temporal lobe cortex. This pattern of results suggest that BFCS lesions disrupted temporal lobe function sufficiently to produce a mild impairment. The pattern of this impairment, growing in magnitude as task demand (delay length, list length) increased, suggests that the lesions impaired working memory. This impairment was transient. At the end of four months, the lesion group performed as well as controls.

Selective lesions of the NBM did not impair choice accuracy in this same visual delayed nonmatch-to-sample task (Aigner et al, 1987). However, scopolamine, a cholinergic antagonist, produced a greater impairment in monkeys with NBM lesions than in control monkeys. This increased sensitivity of the monkeys with lesions suggests that although the behavioral test procedures did not detect the functional significance of the NBM lesion, nonetheless, the brain was compromised. Because the scopolamine injections were intramuscular, they must have affected the entire brain, rather than being restricted to just the BFCS. Thus, determining the locus of the increased sensitivity of the brain to the anticholinergic drug is not directly possible. Likewise, the increased sensitivity to scopolamine cannot be used to indicate that the cholinergic system, as compared to other transmitter systems, was selectively compromised. No information is available about sensitivity to drugs that affect other neurotransmitter systems. Consequently, the increased sensitivity to scopolamine might just reflect a compromised brain that is more susceptible to disruption by any kind of interference. This difficulty in interpreting the extent to which sensitivity to scopolamine following BFCS lesions reflects damage specifically to the cholinergic system can be reduced by dose-response curves for functional antagonists of many different neurotransmitter systems. If the dose-response curve to only scopolamine has a leftward shift, and the dose-response curves for the antagonists of other neurotransmitter systems are normal, this pattern of results would suggest selective damage to the cholinergic system. If, however, similar leftward shifts appear following antagonists for other neurotransmitter systems, the leftward shift of the dose-response curve to scopolamine may not reflect specific damage to the cholinergic system. Both of these interpretations require additional assumptions about the role of different neurotransmitter systems in these tasks, assumptions which have yet to be fully tested.

Marmosets with lesions of the NBM had immediate postoperative impairments on several visual discriminations, a recovery of function, and a continued sensitivity to scopolamine (Ridley et al, 1985). Preoperatively, each marmoset was trained on a visual discrimination and several reversals. When the preoperative discrimination was tested postoperatively, the NBM group took more trials to reach criterion than the control group. The lesion group showed normal performance, however, during subsequent discrimination reversals and the acquisition of a new visual discrimination. A low dose of scopolamine (.03 mg/kg, IM) differentially impaired the choice accuracy of the NBM group in serial reversal performance.

A similar pattern of results was produced in the acquisition of visual discriminations (Ridley et al, 1986). NBM lesions produced a severe impairment in initial postoperative retention of a preoperatively learned discrimination. Subsequent postoperative testing included acquisition of new visual discriminations. The NBM lesions produced impairments in acquisition, but the magnitude of these impairments gradually decreased until acquisition was normal. Postoperative retention of discriminations learned postoperatively was unaffected by NBM lesions. A low dose of scopolamine (.03 mg/kg IM) had a differential effect in the lesion group, markedly increasing the number of trials to reach criterion in the acquisition of a new discrimination.

Lesions of the vertical limb of the diagonal band of Broca also produced significant impairments (Ridley et al, 1989). Impairments occurred on tests of serial reversal discrimination and on conditional discriminations. Normal performance occurred for postoperative retention of preoperatively learned discriminations, and postoperative acquisition of visual discriminations. This pattern of performance was interpreted in terms of impairments in rule-based behavior as distinguished from association with stimuli.

Squirrel monkeys with lesions of the NBM had impairments in a large set of tasks: visual reversal, concurrent object discriminations, delayed nonmatch-to-sample, and a discrimination of visual angles. The impairment in nonmatch-to-sample was present at all delays, even the shortest. This pattern of performance, in addition to the impairment in the discrimination of visual angles, suggests that these lesions produced a substantial disruption of general sensory and cognitive processes. The particular behavioral procedures used here had significant variations from the ones described previously, making direct comparisons complicated. However, the fact that the monkeys received the lesions prior to any behavioral testing may have been an important variable influencing the severity of their impairment (Irle and Markowitsch, 1987).

A large set of behavioral tasks is being used to assess the affects of BFCS lesions in our own laboratory (Voytko et al, 1990). The systematic set of tasks given both preoperatively and postoperatively included the following: visual delayed nonmatch-to-sample, delayed response, and visual and spatial discriminations. Visual delayed nonmatch-to-sample and delayed response tasks were given with a greater task demand (longer delay intervals for both, longer list length for nonmatch-to-sample), and successive reversals of visual discriminations were included.

The test apparatus had three panels in front of the monkey. For the visual discrimination and delayed nonmatch-to-sample tasks, slides of objects were projected on the back of the appropriate panel. For the delayed response task and spatial discriminations, the appropriate panels were illuminated with white light. This series of tests was designed to assess many different psychological functions, and incorporated procedures that are sensitive to lesions of the prefrontal cortex, temporal lobe cortex, and hippocampus. Each monkey was trained on a set of the tasks preoperatively. The lesions were produced by injections of IBO throughout the entire BFCS. Magnetic resonance imagery (MRI) guided the stereotaxic placement of the injections, and confirmed the lesion's effectiveness. Although histological and neurochemical data are not yet available from behaviorally tested monkeys, postsurgical MRI data suggest that the lesions produced major destruction of the NBM, but may have spared some of the ascending limb of the diagonal band and the MSA.

The preliminary data (two monkeys with lesions, two control monkeys) are striking; the lesions had no significant effect on performance in any task. This lack of impairment is puzzling, especially given the fact that the variations in task demand for each of the tasks produced systematic variation of performance in the control monkeys. Consequently, the absence of an impairment in the monkeys with lesions is not due to insensitive behavioral test procedures. Likewise, the lack of impairment is not likely due to a failure to assess the appropriate psychological function. The number of behavioral tests was substantial, and each was designed to require a different aspect of memory. Final assessment of the lesions must wait for histological and neurochemical analyses. However, the currently available data from the MRI of the behaviorally tested monkeys, and the pathological data from other monkeys with similar lesions, suggest that the lesions produced major disruption of the CH3 and CH4 portions of the BFCS, even if they didn't reach all of the MSA and vertical limb of the diagonal band.

Taken together, the general pattern of results from the experiments with primates is similar to that from rats, with the qualification that the magnitude and duration of the impairment in some cases is reduced.

HIGH AFFINITY CHOLINE UPTAKE

High affinity choline uptake (HACU) is a measure of the activity of cholinergic neurons (Durkin, 1989; Wenk et al, 1984). If certain psychological functions involve cholinergic neurons, then behavioral tasks that engage those functions should increase HACU in those neurons. Because HACU can be measured in small regions of the brain, this technique can provide information that is anatomically and neurochemically specific. Because almost all of the cholinergic innervation of the HIP comes from the MSA, HACU in the HIP can be taken as a good index of the activity of neurons in the MSA. Because the FC has cholinergic input from cells other than in the NBM, HACU in the FC is not as precise a measure of the activity of projecting cells from the NBM.

HACU in the HIP of rats was increased by performance in many different maze tasks (Wenk et al, 1984). The rats were first trained to perform the discrimination. HACU was measured immediately after a day's test. Three different procedures produced a significant increment in HACU in the HIP; spatial working memory on a radial arm maze, left-right discrimination on a T-maze, reversal of a left-right discrimination on a T-maze. The increased HACU persisted for at least 20 days after testing. The acquisition of two-way active avoidance also increased HACU.

HACU in the FC was generally unaffected by any of the behavioral manipulations which increased HACU in the HIP. None of the behaviors altered HACU in the frontal cortex.

These results demonstrate that one or more components of the behavioral experience stimulated HACU in HIP. The contribution of each of the different components of the experience remains to be established. In the experiments reviewed above (Wenk et al, 1984), two comparison groups were included. The cage group remained in the home cage and received no behavioral experience, the treadmill group walked for a similar amount of time on a treadmill. HACU in these two groups was not substantially different, suggesting that the general sensory-motor experience of walking was not sufficient to stimulate HACU. Likewise, the difference between the treadmill control group and the discrimination groups described above was significant, suggesting that the associative processes involved in each of the discrimination tasks was sufficient to activate HACU. However, more systematic analyses of the quantitative relationship between a given task demand and the rate of HACU are necessary to provide a quantitative analysis of the relative importance of different types of experiences in altering HACU.

In mice, behavioral experience also changed HACU in the HIP (Toumane et al., 1988, 1989). Mice were trained in two spatial discriminations on a radial arm maze. For the working memory discrimination, each arm had only one pellet of food so that the optimal strategy was to visit each arm once. For the reference memory discrimination, some arms never had food so that the optimal strategy was never to visit these arms.

HACU in the HIP after the first day of testing was substantially increased immediately after the day's test experience, remained elevated for several hours after that experience, and then returned to baseline levels one day later. Reversal of the reference memory discrimination after 9 days of training produced no additional increment in HACU.

HACU in the FC was also elevated after the first day's testing. The absolute amount of the increase in FC and the duration of the increase were less than those in the HIP.

The duration of the increased HACU following a day's test decreased as the number of test days increased. After the first day's testing, HACU in the HIP and FC remained substantially elevated one hour after the test session. After the ninth day of testing, HACU had already returned to baseline levels at this time. The exact reason for the change in the duration of HACU remains to be specified. From the first day to the ninth day of training, choice accuracy in both discrimination tasks improved markedly, which was associated with changes in many aspects of behavior (a reduction in the amount of time spent on the maze, the number of arms entered, etc.).

The specific experiential factor responsible for the changes in HACU in the experiments with mice cannot yet be identified because the single control group was a group that remained in the cage, and received no behavioral experience. Thus, any component of the maze experience might have been responsible for the increase in HACU.

Together, these data demonstrate that one or more components of the behavioral experience of performing spatial discriminations can increase HACU in the HIP, and perhaps in the FC. No particular conclusions can be drawn about the type of cognitive experience that engages the choline uptake system. The data from the treadmill control task of Wenk, Hepler, and Olton (1984) suggest that the experience of walking by itself is not sufficient to increase HACU, at least after extensive experience in the apparatus. The fact that HACU in the HIP was increased following performance in a well-learned left-right discrimination indicates that the HACU system can be engaged by tasks that do not involve working memory. Finally, the increase in HACU in the HIP cannot be necessary for accurate performance of the left-right discrimination task in the T-maze because hippocampal lesions do not impair performance in this type of task (O'Keefe and Nadel, 1978, page 461).

MICROINFUSIONS

The activity of cells in the BFCS can be transiently and reversibly altered by direct microinfusion into the basal forebrain of specific neurotransmitter agonists or antagonists through an intracranial cannula. The functional effects of these microinfusions can be examined by testing animals before the infusion, and at different time points after the microinfusion. If the microinfusion interferes with the activity of cholinergic neurons, and this activity is required for normal function of the projection area of those neurons, then the microinfusion should produce behavioral changes similar to those following lesions of the BFCS.

A variety of studies have investigated the mnemonic effects of microinfusions into the BFCS. Microinfusions into the NBM impaired choice accuracy in a spatial working memory discrimination on the radial arm maze (Robinson et al, 1988; Majchrzak et al, 1990). Microinfusions into the MSA produced a specific impairment in spatial working memory (Mizumori et al, 1989), and altered conditioning (Powell et al, 1985), inhibitory avoidance (Nagel and Huston, 1988), spatial memory (Bostock et al, 1988), and latent inhibition (Gallagher et al, 1987). The heterogeneity of neurochemical input to the BFCS is reflected in the diversity of different neuroactive substances that have been used in these studies.

Cholinergic and GABAergic systems have important connections in the MSA (Leranth and Frotscher, 1989). The MSA cholinergic neurons that project to the HIP receive inhibitory synapses from GABA afferents. Consequently, muscimol, a GABA agonist, should inhibit the activity of these MSA cholinergic cells and produce behavioral impairments similar to those caused by MSA lesions. These same MSA cholinergic neurons have excitatory synapses from cholinergic afferents, presumably arising from local collaterals. Consequently, scopolamine, should block these excitatory afferents, reduce the activation of cholinergic neurons, and produce behavioral impairments similar to those following MSA lesions.

These predictions were assessed with an experiment in which saline, scopolamine, or muscimol were infused into the MSA, theta was recorded from the HIP to confirm the electrophysiological effectiveness of the neurochemical infusions, and the rats were tested in a spatial working memory test in a T-maze (Givens and Olton, in press). A limited dose-response curve was obtained for each substance. At a dose that disrupted hippocampal theta (muscimol, 30 ng; scopolamine, 15 mg), these compounds produced a substantial impairment in choice accuracy 10 minutes after the infusion. Both the electrophysiological and behavioral effects were transitory; 90 minutes after the infusion, hippocampal theta and choice accuracy had returned to baseline levels. The electrophysiological and behavioral effects of the infusions were highly correlated (r=.78); the greater the disruption of the hippocampal theta, the greater the impairment of choice accuracy (Mitchel et al, 1982).

These data provide strong evidence that the cholinergic and/or GABAergic cells in the MSA make an important contribution to accurate working memory performance. Single unit analyses are currently being conducted to determine exactly what component of working memory stimulates the activity of these MSA neurons. If the dissociations following infusions into the MSA are similar to those following MSA lesions, this pattern of results would indicate that the MSA neurons contribute to the associative processes of recent memory.

This technique also offers the possibility of improving choice accuracy by enhancing cholinergic function in the MSA. In the study just mentioned (Givens and Olton, in press), injections of carbachol, which should activate MSA cholinergic neurons, produced a slight (but statistically insignificant) increase in the power of hippocampal theta, and a slight (and also statistically insignificant) increase in choice accuracy. The absence of a statistically significant effect might have occurred for at least two different reasons. First, the MSA cholinergic system might function at its optimal level in normal young rats. Consequently, any change in activity, even an increase, might have detrimental effects. Alternatively, the carbachol may have produced a real improvement in MSA cholinergic function, but the test parameters made the procedure insensitive to these improvements. In the baseline control conditions, choice accuracy was almost perfect, making an improvement very difficult to detect. If this latter explanation is correct, decreasing choice accuracy (by increasing task demand) in the control condition should unmask a beneficial effect of the carbachol injection.

Preliminary data from microinfusions into the MSA of rats treated with scopolamine and into aged rats suggest that beneficial effects can follow intraseptal microinfusions (Markowska et al, 1990). Scopolamine was administered systemically at one of two doses (0.2 mg/kg or 0.4 mg/kg), and saline (0.5 μl) or carbachol (1 μg) was microinfused into the MSA. Scopolamine reduced choice accuracy. Saline in the MSA had no effect on choice accuracy. Carbachol (as compared to saline) in the MSA improved choice accuracy in rats treated with the lower dose of scopolamine. Preliminary data from aged rats, which also have lower levels of choice accuracy, suggest that activation of MSA neurons with the cholinergic agonist oxotremorine (12 μg and 6 μg) infused into the MSA can partially alleviate memory impairments present in these rats. Manipulation of GABAergic function may also have significant therapeutic potential (Sarter et al, 1990).

SUMMARY

This chapter has been organized empirically, focusing on the types of approaches that have been taken to understand BFCS function. This approach reflects the state of our knowledge about the behavioral and psychological functions of the BFCS. Considerable information has been gathered in the very short time that the BFCS has been the object of intense investigation.

The results from the neurotoxic lesions and from the HACU studies provide some points of consistency and some puzzling differences. Both approaches to the study of basal forebrain function suggest that the MSA is involved in tasks that require spatial working memory; MSA lesions impaired choice accuracy, and HACU in the HIP was increased after performance. The pattern of results in simpler tasks is more difficult to interpret. In a left-right reference memory discrimination in a T-maze, MSA lesions did not impair acquisition or performance, whereas HACU in the HIP was activated during performance. This pattern of results suggests that although the MSA is engaged during this type of task, its activity is not necessary for normal performance.

These, and other comparisons indicate the need for a systematic analysis of task demand (Olton, 1989b). Parametric manipulations of different task demands in a systematic fashion can indicate the extent to which the BFCS is involved in the function associated with each parametric manipulation.

Ultimately, of course, the organization of this material should focus on particular psychological functions, rather than the techniques and procedures used to gather the information. Achieving this goal is going to require careful attention to the design of behavioral experiments so that definitive conclusions can be made about the extent to which the BFCS is involved in a given psychological function. A systematic application of task analysis can achieve this goal (Olton, 1986, 1989a, 1989b). For example, BFCS lesions in rats impair choice accuracy in spatial working memory tasks, and performance in these tasks engages the HACU system, at least in the HIP. If the spatial functions of this task involve the BFCS, then a nonspatial version of the task should produce a different pattern of results. If the spatial nature of the task is unimportant for BFCS function, then a nonspatial version of the task should produce the same results. By systematically changing one characteristic of the task at a time, the contribution of each component can be assessed.

A combination of microinfusions in the BFCS, measurement of cholinergic activity by HACU or microdialysis in the appropriate projection area, and behavioral testing can be a powerful approach to the analysis of BFCS function. The microinfusions can provide anatomical and neurochemical specificity, the measurements of cholinergic activity can provide empirical information about the effects of these manipulations on activity of the cholinergic system, and appropriate behavioral testing can indicate the types of functions that are altered following these alterations of cholinergic activity.

In summary, the data reviewed here indicate the kinds of behavioral procedures that involve the BFCS. The next step is to take these data and design diagnostic experiments to indicate the kinds of psychological functions that are associated with the activity of the BFCS. Progress in this direction can be assisted markedly by manipulations and measurements that are anatomically and neurochemically specific, and by the use of conceptual analyses that have been applied to the functions of the hippocampus and the frontal cortex.

REFERENCES

Aigner, T., Mitchell, S., Aggleton, J., et al. (1987). Effects of scopolamine and physostigmine on recognition memory in monkeys with ibotenic-acid lesions of the nucleus basalis of Meynert. Psychopharmacology, 92, 292-300.
Algner, T. G., Mitchell, S. J., Aggleton, J. P., et al. (1990). Transient impairment of recognition memory following ibotenic-acid lesions of the basal forebrain in macaques. Manuscript Submitted.
Bartus, R. T., Flicker, C., Dean, R. L., Pontecorvo, M., Figueiredo, J. C., & Fisher, S. K. (1985a). Selective memory loss following nucleus basalis lesions: Long

term behavioral recovery despite persistent cholinergic deficiencies. <u>Pharmacol. Biochem. Behav.</u>, <u>23</u>, 125-135.

Bartus, R.T., Reginald, L.D., Pontecorvo, M.J., & Flicker, C. (1985b). The cholinergic hypothesis: a historical overview, current perspective, and future directions. In D. S. Olton, E. Gamzu, & S. Corkin. (Eds.), <u>Memory dysfunctions: An integration of animal and human research from preclinical and clinical perspectives</u> (pp. 332-358). New York: The New York Academy of Sciences.

Bostock, E., Gallagher, M., & King, R.A. (1988). Effects of opioid microinjections into the medial septal area on spatial memory in rats. <u>Behavioral Neuroscience</u>, <u>102</u>, 643-652.

Coffey, P. J., Perry, V. H., Allen, Y., Sinden, J., & Rawlins, J. N. P. (1988). Ibotenic acid induced demyelination in the central nervous system: A consequence of a local inflammatory response. <u>Neuroscience Letter</u>, <u>84</u>, 171-184.

Davis, H.P., & Volpe, B.T. (In Press). Memory performance after ischemic or neurotoxin damage of the hippocampus. In L. R. Squire, & E. Lindenlaub. (Eds.), <u>The Biology of Memory. Symposia Medica Hoechst 23</u> New York: Stuttgart.

Dunnett, S. B., Whishaw, I. Q., Jones, G. H., & Bunch, S. T. (1987). Behavioural, biochemical and histochemical effects of different neurotoxic amino acids injected into nucleus basalis magnocellularis of rats. <u>NEUROSCIENCE.</u>, <u>20</u>, 653.

Dunnett, S. B. (In Press). Comparison of short-term memory deficits in animal models of ageing using an operant delayed response task in rats. <u>The Biology of Memory</u>.

Durkin, T. (1989). Central cholinergic pathways and learning and memory processes: presynaptic aspects. <u>Comp. Biochem. Physiol.</u>, <u>93A(1)</u>, 273-280.

Etherington, R., Mittleman, G., & Robbins, T. W. (1987). Comparative effects of nucleus basalis and fimbria-fornix lesions on delayed matching and alternation tests of memory. <u>Neurosci. Res. Communic.</u>, <u>1</u>, 135-143.

Fuster, J.M. (1989). <u>The prefrontal cortex: Anatomy, physiology, and neuropsychology of the frontal lobe</u> (2nd ed.). New York:Raven Press.

Gallagher, M., Meagher, M.W., & E. Bostock. (1987). Effects of opiate manipulations on latent inhibition in rabbits: Sensitivity of the medial septal region to intracranial treatments. <u>Behavioral Neuroscience</u>, <u>101</u>, 315-324.

Gerbec, E. N., Messing, R. B., & Sparber, S. B. (1988). Parallel changes in operant behavioral adaptation and hippocampal corticosterone binding in rats treated with trimethyltin. <u>Brain Research</u>, <u>460</u>, 346-351.

Givens, B. S., & Olton, D. S. (In Press). Modulation of medial septal area alters working memory: GABA and Ach may influence memory through a common mechanism in the medial septal area. <u>Behavioral Neuroscience</u>.

Handelmann, G. E., & Olton, D. S. (1981). Recovery of function after neurotoxic damage to the hippocampal CA3 region: Importance of postoperative recovery interval and task experience. <u>Behav. Neural Biol.</u>, <u>33</u>, 453-464.

Hepler, D. J., Olton, D. S., Wenk, G. L., & Coyle, J. T. (1985). Lesions in nucleus basalis magnocellularis and medial septal area of rats produce qualitatively similar memory impairments. The Journal of Neuroscience, 5(4), 866-873.

Irle, E., & Markowitsch, H. J. (1987). Basal forebrain-lesioned monkeys are severely impaired in tasks of association and recognition memory. Annals of Neurology, 22, 735-743.

Kesner, R. P., Adelstein, T., & Crutcher, K. A. (1987). Rats with nucleus basalis magnocellularis lesions mimic mnemonic symptomatology observed in patient with dementia of the Alzheimer's type. Behav. Neurosci., 101(4), 451-456.

Kesner, R. P. (1988). Reevaluation of the contribution of the basal forebrain cholinergic system to memory. Neurobiology of Aging, 9, 609-616.

Kesner, R. P., Crutcher, K. A., & Beers, D. R. (1989). Serial position curves for item (spatial location) information: Role of the dorsal hippocampal formation and medial septum. Brain Research, 454, 219-226.

Kolb, B. (1984). Functions of the frontal cortex of the rat: A comparative review. Brain Research Review, 8, 65-98.

Leranth, C. & Frotscher, M. (1989). Organization of the septal region in the rat: Cholinergic-Gabaergic interconnections and the termination of hippocampo-septal fibers. Journal of Comparative Neurology, 289, 304-314.

Majchrzak, M., Brailowsky, S., & Will, B. (1990). Chronic infusion of GABA and saline into the nucleus basalis magnocellularis of rats: II. Cognitive impairments. Behavioral Brain Research, 37, 45-56.

Markowska, A. L., Olton, D. S., Murray, E. A., & Gaffan, D. (1989). A comparative analysis of the role of the fornix and cingulate cortex in memory: Rats. Experimental Brain Research, 74, 255-269.

Markowska, A.L., Givens, B., & Olton, D.S. (1990). Cholinergic activation of medial septal area can restore working memory in old rats and in scopolamine-treated young rats. Society for Neuroscience Abstracts.

Markowska, A. L., Wenk, G. L., & Olton, D. S. (In Press). Nucleus basalis magnocellularis and memory: Differential effects of two neurotoxins. Behavioral and Neural Biology.

Meck, W. H., Church, R. M., & Olton, D. S. (1984). Hippocampus, time, and memory. Behavioral Neuroscience, 98, 3-22.

Meck, W. H., Church, R. M., Wenk, G. L., & Olton, D. S. (1987). Nucleus basalis magnocellularis and medial septal area lesions differentially impair temporal memory. Journal of Neuroscience, 7(11), 3505-3511.

Messing, R. B., & Aanonsen, L. M. (1989). Trimethyltin (TMT) effects on hippocampal phencyclidine receptors: Dose-dependent decreases in CA and increases in dentate. Society for Neuroscience Abstracts, 1351.

Messing, R. B., & Sparber, B. S. (1986). Increased forebrain adrenergic ligand binding induced by trimethlytin. Toxicology Letters, 32, 107-112.

Mitchel, S. J., Rawlins, J. A., Steward, O., & Olton, D. S. (1982). Medial septal area lesions disrupt theta rhythm and cholinergic staining in medial entorhinal cortex, and produce impaired radial arm maze behavior in rats. The Journal of Neuroscience, 2, 292-300.

Mizumori, S.J., McNaughton, B.L., Barnes, C.A. & Fox, K.B. (1989). Preserved spatial coding in hippocampal CA1 pyramidal cells during reversible suppression of CA3c output: evidence for pattern completion in the hippocampus. Journal of Neuroscience, 9, 3915-3928.

Nagel, J.A., & Huston, J.P. (1988). Enhanced inhibitory avoidance learning produced by post-trial injections of substance P into the basal forebrain. Behavioral and Neural Biology, 49, 374-385.

O'Keefe, J., & Nadel, L. (1978). The Hippocampus as a cognitive map Oxford:Clarendon Press.

Olton, D.S. (1986). Interventional approaches to memory: Lesions. In J. L. Martinez, & R. Kesner. (Eds.),Learning and Memory: A Biological View (pp. 379-397). New York: Academic Press.

Olton, D.S. (1989a). Mnemonic functions of the hippocampus: single unit analyses in rats. In V. Chan-Palay, & C. Kohler. (Eds.),Neurology and neurobiology. The Hippocampus New Vistas (pp. 411-424). New York: Alan R. Liss, Inc..

Olton, D.S. (1989b). Dimensional mnemonics. In G. H. Bower. (Ed.), (pp. 1-23). San Diego: Academic Press.

Olton, D. S., Becker, J. T., & Handelmann, G. E. (1979). Hippocampus, space and memory. The Behavioral and Brain Sciences, 2, 313-322.

Olton, D.S., & Wenk, G.L. (1987). Dementia:Animal models of the cognitive impairments produced by degeneration of the basal forebrain cholinergic system. In H. Y. Meltzer. (Ed.),Psychopharmacology: The third generation of progress (pp. 941-953). New York: Raven Press.

Olton, D. S., Meck, W. H., & Church, R. M. (1987). Separation of hippocampal and amygdaloid involvement in temporal memory dysfunctions. Brain Research, 404, 180-188.

Olton, D. S., Wenk, G. L., Church, R. M., & Meck, W. H. (1988). Attention and the frontal cortex as examined by simultaneous temporal processing. Neuropsychologia, 26(2), 307-318.

Olton, D.S., & Markowska, A.L. (1989). The effects of preoperatiave experience upon postoperative performance of rats following lesions of the hippocampal systems. In J. Schulkin. (Ed.),Preoperative events (pp. 151-173). Hillsdale, New Jersey: Lawrence Erlbaum Associates, Publishers.

Olton, D.S., Wenk, G.L., & Markowska, A.M. (In Press). Basal forebrain, memory, and attention. In R. Richardson. (Ed.),Activation to acquisition: Functional aspects of the basal forebrain Boston: Birkhauser.

Powell, D.A., Hernandez, L., & Buchanan, S.L. (1985) Intraseptal scopolamine has differential effects on Pavlovian eye blink and heart rate conditioning. Behavioral Neuroscience, 99, 75-87.

Raffaele, K. C., & Olton, D. S. (1988). Hippocampal and amygdaloid involvement in working memory for nonspatial stimuli. Behavioral Neuroscience, 102, 349-355.

Ridley, R. M., Baker, H. F., Drewett, B., & Johnson, J. A. (1985). Effects of ibotenic acid lesions of the basal forebrain on serial reversal learning in marmosets. Psychopharmacology, 60, 438-443.

Ridley, R. M., Murray, T. K., Johnson, J. A., & Baker, H. F. (1986). Learning impairment following lesion of the basal nucleus of Meynert in the marmoset: Modification by cholinergic drugs. Brain Research, 376, 108-116.

Ridley, R. M., Aitken, D. M., & Baker, H. F. (1989). Learning about rules but not about reward is impaired following lesions of the cholinergic projection to the hippocampus. Brain Research, 502, 306-318.

Robbins, T. W., Everitt, B. J., Ryan, C. N., Marston, H. M., Jones, G. H., & Page, K. J. (1989). Comparative effects of quisqualic and ibotenic acid-induced lesions of the substantia innominata and the globus pallidus on the acquisition of a conditional visual discrimination: Differential effects on cholinergic mechanisms. Neurosci., 28, 337-352.

Robinson, S.E., Hambrecht, K.L. & Lyeth, B.G. (1988). Basal forebrain carbachol injection reduces cortical acetylcholine turnover and disrupts memory. Brain Research, 445, 160-164.

Sarter, M., Bruno, J. P., & Dudchenko, P. (1990). Activating the damaged basal forebrain cholinergic system: tonic stimulation versus signal amplification. Psychopharmacology, 101, 1-17.

Sarter, M., & Dudchenko, P. (In Press). Dissociative effects of ibotenic and quisqualic acid-induced basal forebrain lesions on cortical AChE-positive fiber density and cytochrome oxidase activity. Neuroscience.

Toumane, A., Durkin, T., Marighetto, A., Galey, D., & Jaffard, R. (1988). Differential hippocampal and cortical cholinergic activation during the acquisition, retention, reversal and extinction of a spatial discrimination in an 8-arm radial maze by mice. Behavioral Brain Research, 30, 225-234.

Toumane, A., Durkin, T., Marighetto, A., & Jaffard, R. (1989). The durations of hippocampal and cortical cholinergic activation induced by spatial discrimination testing of mice in an eight-arm radial maze decrease as a function of acquisition. Behavioral and Neural Biology, 52, 279-284.

Volpe, B. T., Davis, H. P., & Colombo, P. J. (1989). Preoperative training modifies radial maze performance in rats with ischemic hippocampal injury. Stroke, 20, 1700-1706.

Voytko, M. L., Olton, D. S., Richardson, R. T., Wenk, G. L., & Price, D. L. (1990). Lack of memory impairments following basal forebrain lesions in monkeys. Society of Neuroscience.

Wenk, G. L., & Olton, D. S. (1984). Recovery of neocortical choline acetyltransferase activity following ibotenic acid injection in the nucleus basalis of Meynert in rats. Brain Research, 293, 184-186.

Wenk, G., Hepler, D., & Olton, D. (1984). Behavior alters the uptake of [3H]choline into acetylcholinergic neurons of the nucleus basalis magnocellularis and medial septal area. Behavioural Brain Research, 13, 129-138.

Wenk, G. L., Cribbs, B., & McCall, L. (1984). Nucleus basalis magnocellularis: Optimal coordinates for selective reduction of choline acetyltransferase in frontal neocortex by ibotenic acid injections. <u>Exp. Brain Res.</u>, <u>56</u>, 335-340.

Wenk, G.L., & Olton, D.S. (1987). Basal forebrain cholinergic neurons and Alzheimer's disease. In J. T. Coyle. (Ed.), <u>Animal models of dementia</u> (pp. 81-101). New York: Alan R. Liss, Inc.

Wenk, G. L., Markowska, A. L., & Olton, D. S. (1989). Basal forebrain lesions and memory: Alterations in neurotensin, not acetylcholine, may cause amnesia. <u>Behav. Neurosci.</u>, <u>103</u>, 765-769.

Whishaw, I. Q., O'Connor, W. T., & Dunnett, S. B. (1985). Disruption of central cholinergic systems in the rat by basal forebrain lesions or atropine: Effects on feeding, senosrimotor behaviour, locomotor activity and spatial navigation. <u>Behavioural Brain Research</u>, <u>17</u>, 103-115.

CHOLINE-INDUCED SPATIAL MEMORY FACILITATION CORRELATES WITH

ALTERED DISTRIBUTION AND MORPHOLOGY OF SEPTAL NEURONS

Rebekah Loy*, D. Heyer*, C.L. Williams†, and W.H. Meck†

*Department of Neurology
University of Rochester
435 East Henrietta Road
Rochester, NY 14620

†Departments of Psychology
Barnard College and
Columbia University
New York, NY 10027

INTRODUCTION

Within the basal forebrain is a population of magnocellular neurons that provides cholinergic innervation to the hippocampus, olfactory bulb, amygdala and cortex. Extensive lesion and pharmacological studies in animals and humans support the hypothesis that these basal forebrain cholinergic neurons, in particular the medial septal nucleus (MSN) and the vertical nucleus of the diagonal band of Broca (DBv), and the nucleus basalis (NB), which project to the hippocampus and neocortex, respectively, have an important role in memory function. Memory loss associated with Alzheimer's disease correlates highly with degeneration of NB neurons and with reduction of cortical acetylcholinesterase and choline acetyltransferase (ChAT) activities (Whitehouse et al., 1982; Bartus et al., 1985). In addition, lesions of MSN/DBv and/or NB in rodents, or selective blockade of cholinergic activity by muscarinic receptor antagonists in either animal or human subjects, elicit learning and memory deficits (Davies, 1985; Meck et al., 1987).

In order to increase the synthesis and release of acetylcholine, we and others have provided supplemental dietary choline, and have found that this improves memory function in adult animals (Bartus et al., 1980; Mizumori et al., 1985; Meck and Church, 1987).   Recently, we have also found that either pre- or postnatal treatment with choline results not only in long-lasting facilitation of spatial memory (Meck et al., 1988; Meck and Williams, 1990), but in decreased ChAT activity and increased muscarinic receptor binding in hippocampus and cortex (Meck et al., 1989). While these studies suggest that perinatal choline treatment causes an organizational change in cholinergic function, the nature of the underlying mechanism is unclear. In order to define the origin of these changes, which presumably underlie the improvement in spatial memory, we have examined the morphology and distribution of basal forebrain neurons, using as a marker immunoreactivity for nerve growth factor receptor (NGFRir). This marker is useful, as cells immunoreactive for NGFR correspond closely to the cholinergic neurons in MSN/DBv and NB, and cell bodies are reliably stained in their entirety (Johnson et al., 1987; Springer et al., 1987; Koh and Loy, 1989).

The Basal Forebrain, Edited by T.C. Napier *et al.*
Plenum Press, New York, 1991

MATERIALS AND METHODS

Materials

Choline chloride was provided by Syntex Agricultural. Monoclonal antibody 192 IgG (Chandler et al., 1984) was purified and provided by P.S. DiStefano (Abbott Labs, Abbott Park, IL). This antibody has been fully characterized (Taniuchi and Johnson, 1985) and shown to bind specifically to both high- and low-affinity rat NGF receptors. Affinity-purified biotinylated horse anti-mouse IgG (preabsorbed to remove cross-reactivity for rat IgG), normal horse serum (NHS), and the avidin-biotin-peroxidase complex were purchased from Vector Lab (Burlingame, CA). Other chemical were obtained from Sigma (St. Louis, MO).

Animals

Sprague Dawley (Charles River) rats were mated in the laboratory and treated from the day of conception with 300 mg/kg/day choline chloride delivered in a 0.05M saccharin solution, which served as the only source of drinking water. Prenatally choline-treated and control pups were marked by toe clipping. Litters were standardized to 5 each male and female on the day after birth (PD1), evenly divided between prenatally choline-treated and control pups. All pups were raised by dams which had received no choline chloride during pregnancy. Prenatally choline-treated pups were treated postnatally from PD 1 through PD 30 with subcutaneous injections of 150 mg/kg choline chloride. Rats received restricted food intake as part of 12-arm radial maze behavioral testing from approximately 3 months to 7 months of age.

Immunocytochemistry

At 7 months of age, rats were anesthetized with sodium pentobarbital (50 mg/kg, i.p.) and perfused transcardially with 100 ml 0.1 M phosphate buffer containing 0.9% sodium chloride, pH 7.4 (PBS), followed by 200 ml 4% paraformaldehyde in PBS. The brains were removed and transferred to 30% sucrose in PBS for cryoprotection. Frozen sections were cut in the horizontal plane at 40 µm, collected serially in a cryoprotectant solution containing 30% ethylene glycol and 30% sucrose in PBS, and stored at -20°C. Every tenth section was rinsed in PBS and pretreated in PBS containing 3% hydrogen peroxide and 10% methanol to remove endogenous peroxidase activity. Sections were then preincubated in 0.4% Triton-X100 in PBS for 30 min at room temperature, then in 0.1% Triton-X100 with 3% NHS in PBS for 1 hr at room temperature. Sections were incubated in 1 µg/ml 192-IgG in PBS containing 0.1% Triton-X100 and 3% NHS for 48 hr at 4°C, washed, and incubated in affinity-purified, biotinylated horse anti-mouse antibody, 1:500 in PBS, overnight at 4°C. Following washing, sections were incubated in avidin-biotin-peroxidase complex for 3 hr at room temperature, then reacted with 0.04% 3,3-diaminobenzidine tetrahydrochloride containing 0.005% hydrogen peroxide and 2.5% nickel sulfate in acetate-imidazole buffer, pH 7.2, for 5 min at room temperature. Sections were washed, mounted on slides, dehydrated and coverslipped with permount. Control sections were run with no primary antibody.

Morphometry and Statistics

Sections at -6.8, -7.4 and -8.1 mm ventral to bregma (Paxinos and Watson, 1986) were selected for analysis of number, size, shape and distribution of immunoreactive cell profiles. The landmark defining the -6.8 level is the decussation of the anterior commissure, rostral to which is the DBv and MSN. Level -7.4 is defined by the presence of the rostral component of the anterior commissure, medial to which the DBv is still fused along the midline. At level -8.1 the DBv is clearly separated into left and right components, yet the nuclear boundaries are still nearly circular, not having fused with the cells of the horizontal limb nucleus of the diagonal band-substantia innomminata. Morphometric analyses were made using a Zeiss Image-based Analysis System (IBAS) and Leitz microscope at 10x. Images were digitized and immunoreactive profiles selected based on their density. Cells obviously overlapping were interactively separated, major dendrites truncated at the cell body, and profiles less than 30 μm^2 eliminated from the fields. Each level was analyzed in four fields, rostral and caudal from each side, which at 10x magnification included all but the most rostral cells scattered along the midline. Within these fields minimum and maximum profile diameter, profile area and number of profiles were measured. In addition, "form factor" was determined by the slope of a linear regression line fit to the minimum and maximum cell diameters for individual subjects. Lower estimates of slope indicate the degree to which cells become progressively rounder as they increase in size. Group means were compared using Students t test and analysis of variance.

RESULTS

Cell Number and Distribution

Comparison of the number of NGFRir cell profiles in the whole MSN/DBv reveals no significant change in the choline-treated groups relative to controls (Figure 1). However, the rostral-caudal distribution of MSN/DBv cell bodies is altered (Figure 2). That is, more NGFRir cells occur rostrally while fewer cells occur caudally in choline-treated basal forebrain relative to controls (p<.05).

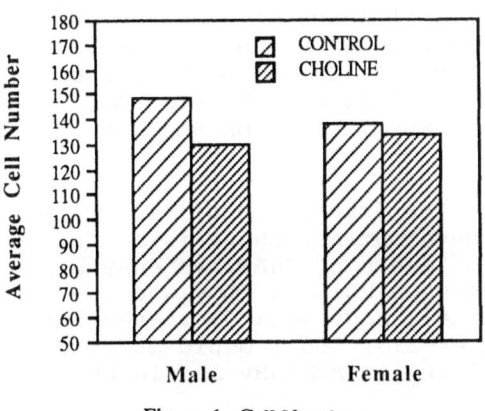

Figure 1. Cell Number

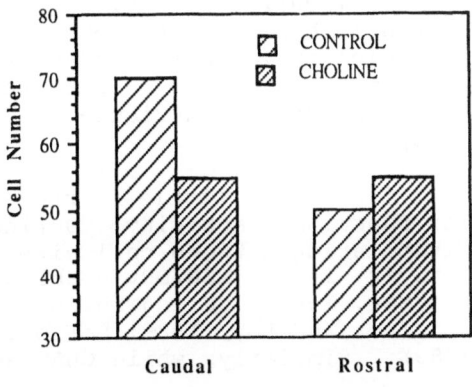

Figure 2. Regional Distribution

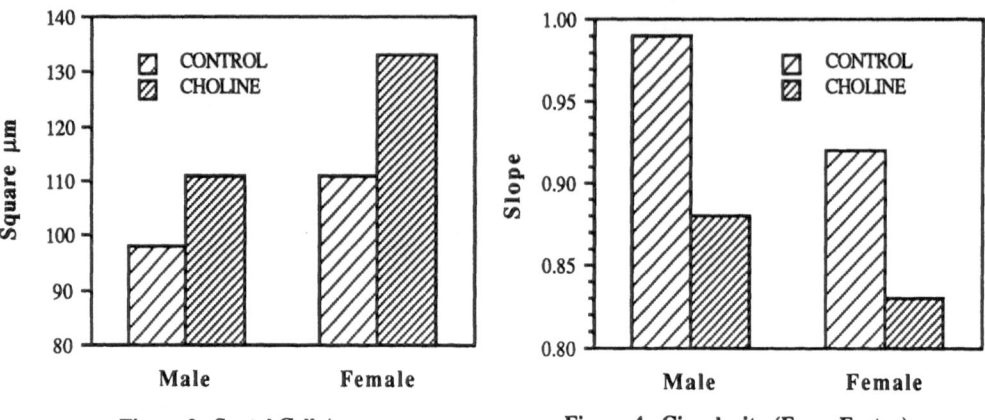

Figure 3. Septal Cell Area

Figure 4. Circularity (Form Factor)

Cell Size

The average area of NGFRir somata in the MSN/DBv of choline-treated rats is 10% larger than in control rats. This is true for the nucleus as a whole when all levels are included (p<.05), although the effect is most robust in the dorsal and rostral regions (Figure 3). Both male and female choline-treated rats show a similar increase in cell area. While there is a trend at some levels for cells to be larger in untreated females compared to males, this comparison in the nucleus as a whole did not reach significance, perhaps due to the small female sample available and relatively large variance in this preliminary analysis.

Circularity

Relative to controls, choline-treated rats of both sex are not only larger, but are rounder and have greater uniformity (p<.05), as indicated by a decrease in the "form factor" (Figure 4).

DISCUSSION

In the present study medial septal cell bodies in choline-treated rats were larger, rounder and more uniform than controls. While cell shape can influence the ability of growth factors to mobilize intracellular calcium in fibroblasts (Tucker et al., 1990), the physiological significance of a change in neuronal shape is not known. It may, however, reflect a retention of immature characteristics, as during development these neurons appear rounder and they lack the thick primary dendrites which contribute to their asymmetry (Koh and Loy, 1989).

Similarly, little is known of the importance of cell size for functional activity. During development, there is a dissociation between cell size and levels of ChAT, as enzyme activity in the MSN/DBv doubles between PD18-90 (Loy and Sheldon, 1987), during which time cell body area decreases 35-40% (Koh and Loy, 1989; Sofroniew et al., 1987; Gould et al., 1989). Similarly, while ChAT activity in MSN/DBv and in the dentate gyrus is greater in females between PD 18-90 (Loy and Sheldon, 1987), and muscarinic receptor binding in the hippocampus is greater in males (Loy et al., 1985), we found no significant sex difference in septal cell area, although NGFRir basal forebrain cells are steroid-sensitive (Toran-Allerand and MacLusky, 1989) and the expression of NGF mRNA is sexually dimorphic (Kornack et al., 1989). Somatic hypertrophy in some systems, however, accompanies axonal sprouting (Goldschmidt and Steward, 1980; Hendrickson and Dineen, 1982); this may be the basis for the observed increase in NB cell size following contralateral cortex ablation or corpus callosum section (Pearson et al., 1984; 1985).

Several observations suggest that basal forebrain cell size correlates with memory ability. In a subpopulation of old rats, basal forebrain cells showed continued shrinkage at 30 months compared to 10-month young adult rats (Koh and Loy, 1988). On a spatial water maze task (Morris, 1984) the aged rats as a group showed longer latencies over the course of training and poorer retention. Behavioral performance also correlates significantly with the number of NGFRir profiles in both the MSN/DBv and NB (Koh et al., 1989). In man, NGFRir cells from NB of Alzheimer patients show a 30% decrease in cell area relative to age- and sex-matched controls (Rinne et al., 1987; Loy et al., 1989; Kordower et al., 1989; Mufson et al., 1989; Vogels et al., 1990). Cell shrinkage also follows transection of basal forebrain axons (Springer et al., 1987; Daitz and Powell, 1954), and both anatomical and behavioral dysfunctions can be reversed by intraventricular injection of NGF (Fischer et al., 1987; Mandel et al., 1989).

The mechanisms controlling the reduction in somal size during development, aging and following axotomy are not known. Gould et al. (1989) showed that these cells undergo considerable dendritic remodeling over the first 3 postnatal weeks. It is possible that there is also an axonal remodeling with significant collateral withdrawal, and concommitant somal shrinkage. In support of this, other regressive changes take place in the septohippocampal cholinergic system beginning around postnatal day 15. Septal terminals within the hippocampus initially show a diffuse distribution, then become laminated in the adult pattern, presumably by withdrawal of collaterals, between postnatal days 14 and 21 (Milner et al., 1983). Muscarinic receptors, initially diffuse, become segregated by postnatal day 17 (Ben-Barak and Dudai, 1979; Rotter et al., 1979). Nicotinic receptors also initially show a more widespread distribution than in adults, with binding declining after day 15 (Ben-Barak and Dudai, 1979; Hunt and Schmidt, 1978 and 1979). While ChAT and acetylcholinesterase activities mature at about postnatal day 20 (Nadler et al., 1974; Loy and Sheldon, 1987), high affinity uptake of choline in the hippocampus declines between days 16 and 30 (Shelton et al., 1979). Together these changes suggest a remodeling and likely a withdrawal of axonal collaterals during the development of the septohippocampal system. We propose that one effect of perinatal choline treatment may be to partially prevent the naturally-occuring reduction in somal cell size. This may occur by stabilization of axon collaterals, as acetylcholine can inhibit neurite growth and synaptic plasticity in vitro (Lipton et al., 1988; Lipton and Kater, 1988).

Since ChAT activity and basal forebrain cell growth are sensitive to NGF (Hefti, 1986; Hefti and Werner, 1986), some of these regressive developmental events may be related to levels of NGF, which decline in hippocampus over approximately the same postnatal timecourse (Large et al., 1986; Whittemore et al., 1986; Auburger et al., 1987). In the periphery, NGF causes a hypertrophy of sympathetic neurons (Levi-Montalcini and Booker, 1960); and in the CNS, basal forebrain ChAT activity as well as cell size, survival and neurite outgrowth are regulated by NGF (Hefti and Werner, 1986; Higgins et al., 1989; and others). The reported effects of choline on basal forebrain morphology and cholinergic expression are not likely to be regulated solely by NGF, however, as NGF _elevates_ basal forebrain ChAT activity as well as increases cell size in sensitive neurons.

Following perinatal treatment with choline, the caudal-rostral distribution of DB/MSN cell bodies, as well as their size and circularity, is altered. This unexpected finding could indicate an effect of perinatal choline on basal forebrain cytogenesis (Ashkenazi et al., 1989) or on patterns of migration. Alternately, this may reflect a differential sensitivity to choline, which could lead to selective withdrawal or maintenance of collaterals, allowing survival of neurons in subregions of the MSN/DBv. The most choline-sensitive region analyzed, the dorsal-rostral area, for instance, contains relatively more MSN neurons, and these project to dorsal-rostral regions of the hippocampal formation (Milner et al., 1983). In order to determine the potential relevance of these or other regressive events in the influence of perinatal choline treatment, further experiments will be necessary to demonstrate that collaterals in the septohippocampal system are withdrawn, that this is accompanied by a reduction in cell size, or by an apparent redistribution of cells within the nucleus, and that choline treatment prevents such regressive events.

Indeed, it is not known if the effects of perinatal choline are specific to cholinergic neurons or if choline acts through a more general anabolic action on neuronal or glial membranes. In relation to this, we have recently found that MSN/DBv neurons immunoreactive for ChAT show a larger choline-elicited increase in cell size (30%, data not shown) than those immunoreactive for NGFR. Since the NGFR-positive population includes some GABAergic neurons (Arimatsu and Miyamoto, 1989; Dreyfus et al., 1989), it will be interesting to determine if this latter subpopulation shows any choline-induced change in cell size.

ACKNOWLEDGEMENTS: NGF receptor antibody 192-IgG was kindly provided by Peter S. DiStefano, Abbott Labs. Supported by PHS grant #NS25169.

REFERENCES

Arimatsu, Y., and Miyamoto, M., 1989, Colocalization of NGF receptor immunoreactivity and [^{3}H]GABA uptake activity in developing rat septum/diagonal band neurons in vitro, Neurosci. Lett., 99:39.
Ashkenazi, A., Ramachandran, J., and Capon, D.J., 1989, Acetylcholine analog stimulates DNA synthesis in brain-derived cells by a specific muscarinic receptor subtype. Nature, 340:146.
Auburger, G., Heumann, R., Hellweg, R., Korsching, S., and

Thoenen, H., 1987, Developmental changes of nerve growth factor and its mRNA in the rat hippocampus: comparison with choline acetyltransferase, Dev. Biol., 120:322.

Bartus, R.T., Dean, R.L. III., Goas, J.A., and Lippa, A.S., 1980, Age-related changes in passive avoidance retention: Modulation with dietary choline, Science, 209:301.

Bartus, R.T., Dean, R.L., Pontecorvo, M.J. and Flicker, C., 1985, The cholinergic hypothesis: A historical overview, current perspective, and future directions, Ann. N.Y. Acad. Sci., 444:332.

Ben-Barak, J. and Dudai, Y., 1979, Cholinergic binding sites in rat hippocampal formation: properties and ontogenesis. Brain Res., 166:245.

Chandler, C.E., Parsons, L.M., Hosang, M., and Shooter, E.M., 1984, A monoclonal antibody modulates the interaction of nerve growth factor with PC 12 cells, J. Biol. Chem. 259:6882.

Daitz, H.M., and Powell, T.P.S., 1954, Studies of the connexions of the fornix system, J. Neuro. Neurosurg. Psychiat., 17:75.

Davies, P., 1985, A critical review of the role of the cholinergic system in human memory and cognition. Ann. N.Y. Acad. Sci., 444:212.

Dreyfus, C.F., Bernd, P., Martinez, J.H., Rubin, S.J., and Black, I.B., 1989, GABAergic and cholinergic neurons exhibit high-affinity nerve growth factor binding in rat basal forebrain, Exper. Neurol., 104:181.

Fischer, W., Wictorin, K., Bjorklund, A., Williams, L.R., Varon, S., and Gage, F.H., 1987, Amelioration of cholinergic neuron atrophy and spatial memory impairment in aged rats by nerve growth factor, Nature, 329:65.

Goldschmidt, R.B., and Steward, O., 1980, Time course of increases in retrograde labeling and increases in cell size of entorhinal cortex neurons sprouting in response to unilateral entorhinal lesions, J. Comp. Neurol., 189:359.

Gould, E., Farris, T.W., and Butcher, L.L, 1989, Basal forebrain neurons undergo somatal and dendritic remodeling during postnatal development: a single-section Golgi and choline acetyltransferase analysis, Devel. Brain Res., 46:297.

Hefti, F., 1986, Nerve growth factor promotes survival of septal cholinergic neurons after fimbrial transections, J. Neurosci., 6:2155.

Hefti, F., and Werner, W.J., 1986, Nerve growth factor and Alzheimer's disease, Ann. Neurol., 20:275.

Hendrickson, A., and Dineen, J.T., 1982, Hypertrophy of neurons in dorsal lateral geniculate nucleus following striate cortex lesions in infant monkeys, Neurosci. Lett., 30:217.

Higgins, G.A., Koh, S., Chen, K.S., and Gage, F.H., 1989, NGF induction of NGF receptor gene expression and cholinergic neuronal hypertrophy within the basal forebrain of the adult rat, Neuron, 3:247.

Hunt, S., and Schmidt, J., 1978, Some observations on the binding patterns of α-bungarotoxin in the central nervous system of the rat, Brain Res., 157:213.

Hunt, S., and Schmidt, J., 1979, The relationship of α-bungarotoxin binding activity and cholinergic termination within the rat hippocampus. Neurosci., 4:585.

Johnson, E.M.Jr., Taniuchi, M., Clark, H.B., Springer, J.E., Koh, S., Tayrien, M.W., and Loy, R., 1987, Demonstration of the retrograde transport of nerve growth factor receptor in the peripheral and central nervous system, J. Neurosci., 7:923.

Koh, S., Chang, P., Collier, J.T., and Loy, R., 1989, Loss of NGF receptor immunoreactivity in basal forebrain neurons of aged rats: correlation with spatial memory impairment, Brain Res., 498:397.

Koh, S., and Loy, R., 1988, Age-related loss of nerve growth factor sensitivity in rat basal forebrain neurons, Brain Res., 440:396.

Koh, S., and Loy, R., 1989, Development of NGF-sensitive rat basal forebrain neurons and their afferent projections to hippocampus and cortex, J. Neurosci., 9:2999.

Kornack, D.R., Lu, B., and Black, I.B., 1989, Sexually dimorphic expression of the NGF receptor gene in rat brain. Soc. Neurosci. Abs., 15:954.

Kordower, J.H., Gash, D.M., Bothwell, M., Hersh, L., and Mufson, E.F., 1989, Nerve growth factor receptor and choline acetyltransferase remain colocalized in the nucleus basalis (Ch4) of Alzheimer's patients, Neurobiol. Aging, 10:67.

Large, T.H., Bodary, S.C., Clegg, D.O., Weskamp, G., Otten, U., and Reichardt, L.F., 1986, Nerve growth factor gene expression in the developing rat brain, Science, 234:352.

Levi-Montalcini, R., and Booker, R., 1960, Excessive growth of the sympathetic ganglia evoked by a protein isolated from mouse salivary glands, Proc. Natl. Acad. Sci., 46:373.

Lipton, S.A., Frosch, M.P., Phillips, M.D., Tauck, D.L., and Aizenman, E., 1988, Nicotinic antagonists enhance process outgrowth by rat retinal ganglion cells in cultures. Science , 239:1293.

Lipton, S.A., and Kater, S.B., 1988, Neurotransmitter regulation of neuronal outgrowth, plasticity and survival. Trends in Neurosci., 12:265.

Loy, R., Heyer, D., Clagett-Dame, M., and DiStefano, P.S., 1989, Localization of NGF receptor in human basal forebrain using two new monoclonal antibodies: comparison of normal and Alzheimer brains, Soc. Neurosci. Abs., 15:1364.

Loy, R., and Sheldon, R.A., 1987, Sexually dimorphic development of cholinergic enzymes in the rat septohippocampal system. Dev. Brain Res., 34:156.

Loy, R., Tayrien, M.W., and Springer, J., 1985, Sexual dimorphism in regional muscarinic receptor binding in rat hippocampus, Anat. Rec., 211:100A.

Mandel, R.J., Gage, F.H., and Thal, L.J., 1989, Spatial learning in rats: correlation with cortical choline acetyltransferase and improvement with NGF following NBM damage, Exp. Neurol., 104:208.

Meck, W.H., and Church, R.M., 1987, Nutrients that modify the speed of internal clock and memory storage processes, Behav. Neurosci., 101:465.

Meck, W.H., Church, R.M., Wenk, G.L., and Olton, D.S., 1987, Nucleus basalis magnocellularis and medial septal area lesions differentially impair temporal memory, J. Neurosci., 7:3505.

Meck, W.H., Smith, R.A., and Williams, C.L., 1988, Pre- and postnatal choline supplementation produces long-term facilitation of spatial memory, Devel. Psychobiol., 21:339.

Meck, W.H., Smith, R.A., and Williams, C.L., 1989, Organizational changes in cholinergic activity and enhanced spatial memory as a function of pre- and/or postnatal choline supplementation, Behav. Neurosci, 103:1234.

Meck, W.H., and Williams, C.L., 1990, Choline supplementation during pre- and postnatal development eliminates proactive interference in spatial memory, Devel. Brain Res., in press.

Milner, T.A., Loy, R., and Amaral, D.G., 1983, An anatomical study of the development of the septo-hippocampal projection in the rat, Dev. Brain Res., 8:343.

Mizumori, S.J., Patterson, T.A., Sternberg, H., Rosenzweig, M.R., Bennett, E.L., and Timiras, P.S., 1985, Effects of dietary choline on memory and brain chemistry in aged mice. Neurobiol. Aging, 6:51.

Morris, R.G.M., 1984, Development of a water-maze procedure for studying spatial learning in the rat, J. Neurosci. Meth., 11:47.

Mufson, E.J., Bothwell, M.A., and Kordower, J.H., 1989, Loss of nerve growth factor receptor-containing neurons in Alzheimer's disease: a quantitative analysis across subregions of the basal forebrain, Exper. Neurol., 105:221.

Nadler, J.V., Matthews, D.A., Cotman, C.W., and Lynch, G.S., 1974, Development of cholinergic innervation in the hippocampal formation of the rat. II. Quantitative changes in choline acetyltransferase and acetylcholinesterase activities, Develop. Biol., 36:142.

Paxinos, G., and Watson, C., 1986, The Rat Brain in Stereotaxic Coordinates, Second Edition. Academic Press, Orlando, FL.

Pearson, R.C.A., Sofroniew, M.V., and Powell, T.P.S., 1984, Hypertrophy of immunohistochemically identified cholinergic neurons of the basal nucleus of Meynert following ablation of the contralateral cortex in the rat, Brain Res., 311:194.

Pearson, P.C.A., Sofroniew, M.V., and Powell, T.P.S., 1985, Hypertrophy of cholinergic neurones of the rat basal nucleus following section of the corpus callosum, Brain Res., 338:337.

Rinne, J.O., Paljarvi, L., and Rinne, U.K., 1987, Neuronal size and density in the nucleus basalis of Meynert in Alzheimer's disease, J. Neurol. Sci., 79:67.

Rotter, A., Field, P.M., and Raisman, G., 1979, Muscarinic receptors in the central nervous system of the rat. III. Postnatal development of binding of [3H]propylbenzilylcholine mustard, Brain Res.,Rev., 1:185.

Shelton, D.L., Nadler, J. V., and Cotman, C.W., 1979, Development of high affinity choline uptake and associated acetylcholine synthesis in the rat fascia dentata, Brain Res., 163:263.

Sofroniew, M.V., Pearson, R.C.A., and Powell, T.P.S., 1987, The cholinergic nuclei of the basal forebrain of the rat: normal structure, development and experimentally induced degeneration, Brain Res., 411:310.

Springer, J.E., Koh, S., Tayrien, M.W., and Loy, R., 1987, Basal forebrain magnocellular neurons stain for nerve growth factor receptors: correlation with cholinergic cell bodies and effects of axotomy, J. Neurosci. Res., 17:111.

Taniuchi, M., and Johnson, E.M. Jr., 1985, Characterization of the binding properties and retrograde axonal transport of a monoclonal antibody directed against the rat nerve growth factor receptor, J. Cell Biol, 101:1100.

Toran-Allerand, D., and MacLusky, N.J., 1989, Co-localization of NGF and estrogen receptors: implications for the basal forebrain, <u>Soc. Neurosci. Abs</u>, 15:954.

Tucker, R.W., Meade-Cobun, K., and Ferris, D., 1990, Cell shape and increased free cytosolic calcium $[Ca^{2+}]_i$ induced by growth factors, <u>Cell Calcium</u>, 11:201.

Vogels, O.J.M., Broere, C.A.J., Ter Laak, H.J., Ten Donkelaar, H.J., Nieuwenhuys, R., and Schulte, B.P.M., 1990, Cell loss and shrinkage in the nucleus basalis of Meynert complex in Alzheimer's disease, <u>Neurobiol. Aging</u>, 11:3.

Whitehouse, P.J., Price, D.L., Struble, R.G., Clark, A.W., Coyle, J.T., and DeLong, M.R., 1982, Alzheimer's disease and senile dementia: loss of neurons in the basal forebrain, <u>Science</u> , 215:1237.

Whittemore, S.R., Ebendal, T., Lårkfors, L., Olson, L., Seiger, Å., Strømberg, I., and Persson, H., 1986, Developmental and regional expression of β-nerve growth factor messenger RNA and protein in the rat central nervous system, <u>Proc. Natl. Acad. Sci</u>., 83:817.

THE PHARMACOLOGY OF BASAL FOREBRAIN INVOLVEMENT IN COGNITION

James J. Chrobak, T. Celeste Napier,
Israel Hanin and Thomas J. Walsh[*]

Department of Pharmacology and Experimental Therapeutics,
Loyola University Chicago, Stritch School of Medicine
Maywood, IL and Department of Psychology, Rutgers
University, New Brunswick, NJ

Magnocellular basal nucleus (MBN) neurons, a majority of which are
cholinergic, directly influence cortical neurophysiology. These neurons
appear to be part of a diffuse cortically projecting system that includes
brain stem monoaminergic neurons as well as the MBN column. The entire
system appears to modify cortical excitability in relation to an animal's
behavioral state (see Saper, 1987). By regulating activity within the
entire cortical mantle, this system can influence a broad range of
cognitive phenomena (sensory processing, attention, motivation, memory).
The restrictive terminal distribution of individual MBN neurons (Bigl et
al., 1982: Price and Stern, 1983; Nyakas et al., 1987) however, as
compared to monoaminergic projections, may allow for a more discrete
modulation of cortical activity.

MBN neurons follow a rough topographical distribution such that
those within the medial septum/vertical limb of the diagonal band
(MS/vlDB) innervate the hippocampal and entorhinal cortices
[septohippocampal (SH)]; those within the horizontal limb DB innervate
the olfactory tubercle; and those within the rostrocaudal extent of the
ventral pallidum (VP) throughout the extended amygdala (Alheid and
Heimer, 1988), or nucleus basalis (NB), innervate the remaining cortical
mantle, including the basolateral amygdala [basocortical (BC)] (Rye et
al., 1984; Gaykema et al., 1990). While the majority of MBN neurons are
cholinergic, a portion of the SH projection (approximately 40-50%) and BC
projection is GABAergic (Kohler et al., 1984; Fisher et al., 1988) and
may also release neuroactive peptides.

A variety of neurotransmitter systems converge on MBN neurons (see
Zaborsky, this book), with GABAergic inputs being the most predominant
(Ingham, Bolam and Smith, 1988; Zaborsky et al., 1986; Leranth and
Frotscher, 1989; Kohler and Chan-Palay, 1983). While extensive
comparisons of distinct portions of the MBN column have not been
described, certain similarities exist between the afferent inputs and
morphology of the basalis and septal pole of the MBN (Dinopoulos et al.,
1988; Alheid and Heimer, 1989). Microinfusion of pharmacological agents
into the NB or MS produce similar alterations in cortical cholinergic
indices (Costa et al., 1983; Wood and McQuade, 1983). Similarly,
putative MBN neurons within the NB and MS respond comparably to

The Basal Forebrain, Edited by T.C. Napier *et al.*
Plenum Press, New York, 1991

iontophoretic drug administration (Lamour and Epelbaum, 1988).

The major source of GABAergic input to MBN neurons within the MS is the lateral septum (Kohler and Chan-Palay, 1983: Leranth and Frotscher, 1989). MBN neurons within the VP and NB appear to receive prominent GABAergic inputs from the nucleus accumbens (see Ingham et al., 1988; and Figure 1). These inputs may serve to close a feedback loop whereby descending corticolimbic inputs (Raisman, 1966; Swanson and Cowan, 1979: Nauta et al., 1978; Alonso and Kohler, 1984; Swanson and Kohler, 1986) can influence MBN neurons. Within the MS, the GABAergic input serves as a common pathway for the regulation of MBN activity by descending corticolimbic and ascending aminergic influences (Costa et al., 1983). The degree to which this scenario reflects inputs to the VP and NB is unclear, although these MBN neurons receive predominant GABAergic contacts and a paucity of synaptic contacts in general (Ingham et al., 1988; Zaborsky et al., 1986).

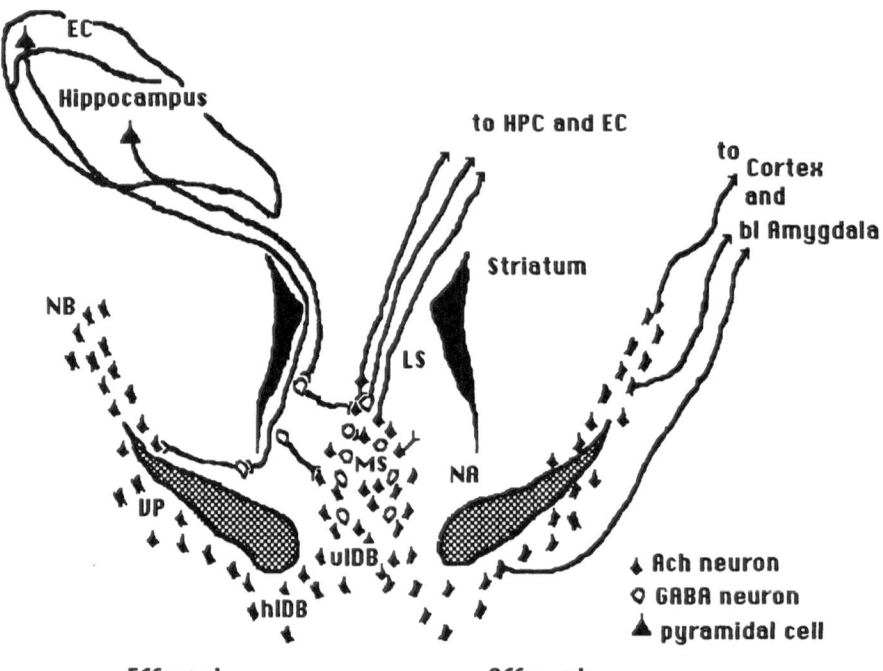

Fig. 1.   A diagram of the MBN neurons embedded within the basal
            forebrain. Septohippocampal neurons project via the
            fimbria-fornix to innervate the hippocampus (HPC)
            and entorhinal cortex (EC); basocortical neurons
            project to the remaining cortical targets including
            the basolateral amygdala. Medial septal (MS) and
            vertical limb of the diagonal band (vlDB) MBN neurons
            receive GABAergic projections from the lateral septum
            (LS) which are in turn influenced by hippocampal
            efferents. A similar relationship exists whereby EC
            projections influence GABAergic neurons within the
            nucleus accumbens (NA), which then influence MBN
            neurons in the ventral pallidum (VP) and possibly the
            remaining nucleus basalis (NB). hlDB = horizontal
            limb of the diagonal band.

Our interest in the MBN derives from its strategic position to influence cortical functions, as well as the utility of modulating MBN activity as a means to deconstruct component cognitive processes. A variety of findings indicate that particular memory processes are disrupted following various forms of corticolimbic insult in mammals, while other memory processes remain intact . A number of theoretical models have attempted to define the precise nature of preserved and impaired functions (see Squire, 1987). Based upon this literature, Sherry and Schacter (1987, pg. 446) have suggested a "distinction between a memory system that supports gradual or incremental learning and is involved in the acquisition of habits and skills and a system that supports one trial learning and is necessary for forming memories that represent specific situations and episodes" (see also Tulving, 1983). These authors, recognizing the "functional incompatibility" of these systems within an information processing perspective, have avoided the idiosyncratic nature of various models by referring to a declarative/episodic system as Memory System I, and a procedural/habit system as Memory System II. We agree that there are at least two memory systems in mammals and that a distinct memory system provides for the representation of episodic events. Further, we suggest that it is information extracted from unique temporal and spatial references (episodes) that is predominantly susceptible to the disruptive effects of corticolimbic insult, in particular the disruption of hippocampal, entorhinal and frontal cortical function. Given that a temporally graded retrograde impairment can occur in many amnesic syndromes, it is likely that this memory system participates in the maintenance of "memories" over a protracted time frame and in the consolidation of memory into a neurophysiologically distinct, less mutable, state (see Squire 1987).

This dissociation has been observed in a variety of mammalian models (see Sherry and Schacter, 1987; Walsh and Chrobak, 1990, for references). Although the tasks and sensory stimuli vary, mammals with corticolimbic insults are noticeably impaired in tasks that require them to maintain, over a retention interval, representations of recently acquired information. We have referred to this system as working/episodic (W/E) memory (Chrobak et al., 1988; see Roitblat, 1982; Olton, 1986; Mishkin, 1982; and Pribram, 1986 for discussion), and have been utilizing a delayed-non-match-to-sample radial arm maze (DNMTS-RAM) spatial task to examine W/E memory and its neural substrate in the rat (see Figure 2).

A disruption of W/E memory may occur despite a sparing of the ability to acquire and perform other tasks that are characterized by slow, incremental learning, where specific responses are contingently reinforced over repeated trials. This compartmentalization of memory systems is one of several similar theoretical frameworks that attempt to explain the dissociation in memory impairments observed in a variety of human, primate, and rodent amnesic syndromes (see also Gaffan, 1985; Thomas and Gash, 1986; and Weiskrantz, 1982). While experimentally observed dichotomies in memory performance may artifactually misrepresent a unified memory system, these dichotomies serve as critical conceptual frameworks that guide further experimental analysis of brain-behavior relationships.

Over the last decade increased attention has been focused upon the neurodegeneration of the basal forebrain cholinergic system in patients with Alzheimer's Disease. This disorder typically presents with a marked loss of memory for recent events (i.e, episodic memory) and progresses toward widespread cognitive dysfunction (Van Hoesen and Damasio, 1987). Considerable evidence indicates that hypofunctional activity within forebrain cholinergic neurons contributes to the cognitive deficits.

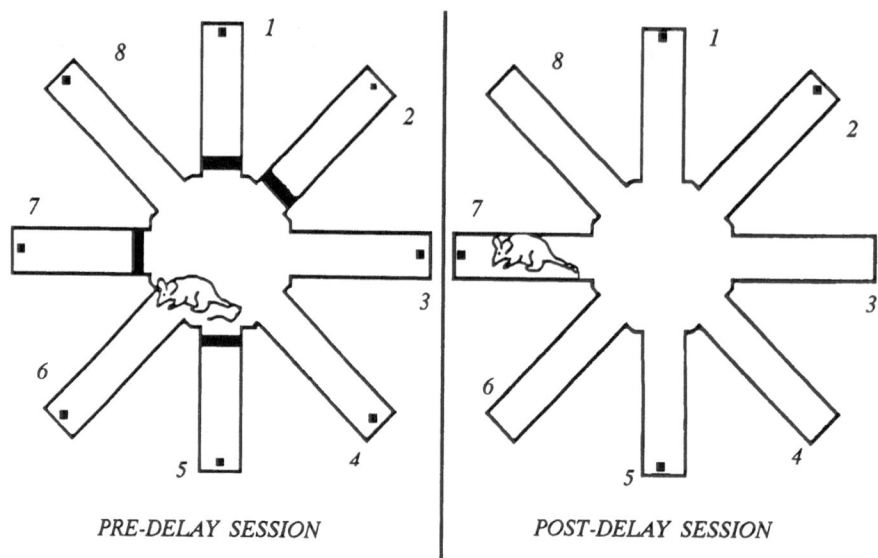

PRE-DELAY SESSION   POST-DELAY SESSION

Fig. 2. The DNMTS-RAM is a unique tool for studying working/
episodic memory and its neural substrate in the rat.
In the performance of this task, rats are capable of
identifying several unique spatial locations and
remembering over the course of a trial which arms
have or have not been chosen. When a delay is inserted between choices, rats are capable of retaining
information concerning choice events for as long as
24 hours (Beatty and Shavalia, 1980). In this task
the rat is allowed access to only four of eight maze
arms during a pre-delay session, i.e., 3,4,6,8 (see
left panel). The pattern of choices is varied daily.
Varying retention intervals can then be imposed.
Following the delay, the animal is returned to the
maze and is allowed access to all eight arms. During
the post-delay session only those arms not visited
during the pre-delay session contain food, i.e., 1,2,
5,7. Various post-delay performance indices include:
the number of CORRECT CHOICES in the first four post-
delay choices (range = 0-4); the number of POST-DELAY
ERRORS (range = 0-10); RETROACTIVE ERRORS, the number
of post-delay errors that were entries into pre-delay
chosen arms (range = 0-4); and PROACTIVE ERRORS, the
number of post-delay errors that were repeat entries
into post-delay chosen arms (range = 0-9).

Several strategies have been used to examine the role of MBN and/or
its cholinergic component in cognition in normal and pathological states
(see Hagan and Morris, 1988; Durkin, 1989), including neurotoxic lesions
of the MS or NB, and local pharmacological manipulation within these
respective nuclei. Analysis of the behavioral impairments resulting from
chronic damage to the MBN have supported the involvement of this system
in cognitive processes. Researchers utilizing excitatory amino acid
analogues, such as ibotenic acid, to damage neurons within the MS and NB
have shown that such lesions can impair the acquisition of several
learning tasks and disrupt performance of preoperatively acquired
cognitive tasks (Hepler et al., 1985; see Olton and Wenk, 1987).

These insults predominantly impair performance in tasks that require the use of W/E memory processes, without necessarily disrupting performance in reference/habit memory tasks (Hepler et. al., 1985; see Durkin, 1989). This dissociation has been more thoroughly demonstrated for insults to the SH system, while insult to BC system has led to conflicting results (see Robbins et al., 1989; Murray and Fibiger, 1985). Based on the foregoing discussion, one might suggest that insult to the ventrocaudal pole of the MBN (i.e., VP and NB) could produce a broader range of cognitive impairment, based upon its innervation of a vast area of the cortical mantle, including sensory areas. Thus, differences in the cognitive impairments associated with such insults would reflect the disruption of cortical processing in various areas, producing a broad range of cognitive deficits including sensory, attentional, associative, and mnemonic deficits. Insult to the SH system, primarily affecting the innervation of the hippocampal and entorhinal cortices, produces a more circumscribed impairment due to the polymodal associative nature of the sensory information processed in this area of cortex and its unique involvement in representing the context of that information (i.e., an episodic event).

Damage to the MS/VLDB or its major efferent pathway, the fimbria-fornix, produces impairment in the performance of W/E memory tasks. These lesions clearly disrupt SHC indices, yet they also invariably disrupt either non-cholinergic fibers traversing the fimbria-fornix, including the GABAergic SH projection, septal non-cholinergic cell populations, as well as cortico-septal and cortico-accumbens fibers that travel via the fimbria-fornix. Thus it is difficult to ascribe the observed deficits to the selective disruption of SHC activity.

A number of laboratories have been examining the effect of ethylcholine aziridinium ion (AF64A) as a tool to induce a selective cholinergic hypofunction (see Hanin, et al., 1987; Hanin, 1990; Walsh and Chrobak, 1990 for review). AF64A is an analog of choline containing a reactive aziridinium ring that can akylate nucleophilic sites. Within restrictive dose ranges the transport of this toxin by the high affinity choline uptake (HAChU) carrier can result in a site-directed insult to cholinergic neurons (Pittel, Fisher and Heldman, 1987, Uney and Marchbanks, 1987; Curti and Marchbanks, 1984; Rylett and Colhoun, 1980; Potter, et al. 1989). Administration of this toxin into the lateral cerebral ventricles can produce a regionally selective alteration of SHC indices without any effect on striatal or cortical cholinergic indices (see above reviews for references).

Intraventricular administration of AF64A can produce cognitive impairments that are associated with persistent reductions in hippocampal cholinergic indices (Chrobak, et al., 1986; 1988; 1989a; Gower, et al., 1989; Jarrard et al., 1984; Pope et al., 1987; Tedford et al., 1990; Walsh et al., 1984). We have reported that intraventricular AF64A (3nmol/side) can produce a deficit in the performance of W/E memory tasks, without necessarily affecting the performance or acquisition of reference/habit memory tasks (Chrobak et al., 1986; 1988). Thus, the cognitive deficit induced by this treatment is consistent with the deficits observed following other insults to the SH system. The fact that this deficit can be blocked by the prior administration of the potent HAChU inhibitor, hemicholinium-3, demonstrates that the impairment associated with this treatment is a selective consequence of the compound's direct cholinotoxic properties, or to alterations that occur secondarily to an AF64A-induced cholinergic hypofunction (Chrobak et al., 1989a). Further, they indicate that a selective alteration of SHC activity is sufficient to disrupt W/E memory processes, and suggest that the cholinergic innervation of the HPC is critically involved in the physiological activity of the HPC that subserves such memory processes.

Analysis of the mnemonic deficits induced by graded insult to the SHC by lower doses of AF64A (1.5 and 0.75nmol/side) indicates that the degree of impairment can vary as a function of the retention interval in a DNMTS-RAM task (Chrobak and Walsh, submitted). Thus, animals treated with 1.5nmol/ventricle of AF64A did not reveal significant deficits until delays of at least 1 hr were imposed between choices 4 and 5 in a DNMTS-RAM task. Likewise, deficits were not induced in animals treated with 0.75nmol/ventricle until delays of 2 hrs were introduced. Neither group of AF64A-treated animals were significantly impaired in a standard RAM task when no delay was inserted between choices 4 and 5. Analysis of the post-delay errors indicates that only retroactive errors were significantly affected (see Figure 3). This finding suggests that the deficit observed in these animals may be related specifically to their inability to maintain representations of episodic events over an extended retention interval.

Such findings suggest that the cognitive deficits observed in animals with a graded insult to the SHC system are dependent upon the mnemonic load associated with task performance, with no impairment observed with short retention intervals ("easier"), and increasing impairments as the retention interval is lengthened ("more difficult"). It is important to note that such deficits were observed at retention intervals where control animals post-delay performance was unaffected. These data indicate that selective disruption of the cholinergic portion of the MBN that innervates the hippocampal and entorhinal cortices is sufficient to disrupt cognitive performance, and that this disruption is associated with an impairment in W/E memory processes, in encoding and/or retrieving representations of episodic events.

While considerable emphasis has been placed on depicting the cognitive insult resulting from neurotoxic lesions of MBN neurons, few studies have examined the effects of direct pharmacological manipulations

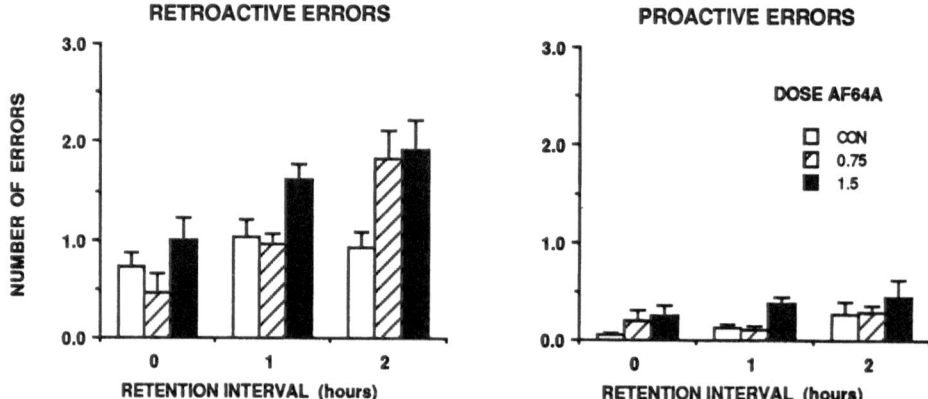

Fig. 3. Effects of intraventricular AF64A treatment on the number of retroactive and proactive errors during the post-delay session of a DNMTS-RAM task. CON = control artificial cerebrospinal fluid infusions; 0.75 and 1.5 refer to nanomole doses of AF64A/ventricle.

within this region. Such acute non-toxic manipulations have several advantages over chronic insult to these neurons. Particularly, one may examine the effects of repeated manipulations within a single subject. This benefit, coupled with the advantages of a DNMTS-RAM task that allows for repeated examination of memory performance as well as delayed post-training injections, provides a powerful tool with which to explore the neurobiology of memory.

Pharmacological manipulations within the basal forebrain can affect the acquisition, retention, or extinction of learned responses (see Table 1). Many of these manipulations are known to alter cortical indices of dynamic cholinergic activity (HAChU and acetylcholine turnover). It is important to consider, however, that such treatments would also affect non-cholinergic cortical projections, projections to other targets, as well as communication between intrinsic neurons within the basal forebrain.

With respect to the interpretation of cognitive deficits following intracranial infusion, it is important to recognize that basal forebrain infusions can produce alterations in motor and primarily locomotor activity levels, (Allen and Crawford, 1984; Breese et al 1984; Mogenson, 1987; Napier et al., 1987; 1988; Kalivas, this book). Mogenson (1987) has proposed that hippocampal and amygdala influences on the ventral pallidum, via the nucleus accumbens, modulate locomotor activity. These findings must be considered when examining the effects of intracranial infusions on cognitive performance. While pre-training drug manipulation may disrupt cognitive processing, impairments may be a consequence of altered motoric function. The advantages of post-training procedures have been carefully elucidated (see McGaugh, 1989) and may be very important in light of the aforementioned concerns.

Huston and colleagues have reported that post-training administration of substance P into either the MS or the NB facilitates, whereas administration of muscimol (50 ng) into the NB disrupts, retention of a passive avoidance task. It is important to note their finding that vehicle injections into the NB can alter performance relative to control animals, thus the facilitatory effect of substance P occurs despite a vehicle-associated deficit, and the impairment produced by muscimol is significantly greater than that of the vehicle (Nagel & Huston, 1988).

Recently, we have attempted to examine the effects of direct infusions into the VP in order to compare the effects of muscimol infusions into this brain region with those observed in the MS (see below). However this area of the basal forebrain appears to be very sensitive to insult, including vehicle infusions. Infusion of physiological saline as well as artificial CSF solutions produced a significant impairment in the performance of a DNMTS-RAM task on the day following administration (Chrobak and Napier, 1989). Animals' performance returned to baseline within two or three days of testing, although it was characterized by increased variability. We observed no impairment in responding with sham injection, however, in which the injector was lowered directly into tissue. This, coupled with a repeated injection procedure, would seem to suggest that physical damage surrounding the injection site was insufficient to induce the deficit. Such findings are consistent with previous reports concerning chronic, and acute, vehicle-induced impairments in cognitive performance (Altman et al., 1985; Nagel and Huston, 1988) that appear to manifest in more difficult cognitive testing paradigms (Elrod and Buccafusco, 89). We also observed no deficit in performance in a standard RAM paradigm in which animals received bilateral saline injections into the VP five

TABLE 1. ACUTE PHARMACOLOGICAL MANIPULATIONS WITHIN
         THE BASAL FOREBRAIN ON COGNITIVE PERFORMANCE
         IN THE RAT

MEDIAL SEPTUM	DRUG and BEHAVIORAL EFFECT
Staubli & Huston, (80)	Post-infusion of Substance P facilitates passive avoidance.
Blaker et al., (84)	Pre-infusion of muscimol increases extinction responding in a lever-pressing paradigm.
Bostock et al., (88)	Post-infusion of naloxone facilitates, and B-endorphin impairs, RAM acquisition in a novel environment. Pre- and post-infusion of B-endorphin has no effect on RAM performance.
Chrobak et al., (89b)	Post-infusion of muscimol impairs performance of a DNMTS-RAM task.
Chrobak & Napier, (90)	Post-infusion of bicuculline impairs performance of a DNMTS-RAM task.
Brioni et al., (90)	Pre-infusion of muscimol impairs acquisition of a water maze task.

NUCLEUS BASALIS	DRUG and BEHAVIORAL EFFECT
Kafetzopoulos et al., (86)	Post-infusion of Substance P facilitates passive avoidance.
Nagel & Huston (88)	Post-infusion of Substance P, muscimol and vehicle impairs passive avoidance.
Robinson et al., (88)	Pre-infusion of carbachol impairs RAM performance.
Chrobak & Napier, (89)	Post-infusion of vehicle impairs performance of DNMTS-RAM task. Pre-infusion of vehicle has no effect on performance of standard RAM task.

Pre-infusion refers to studies in which drugs were administered immediately prior to testing in any particular paradigm. Post-infusion refers to studies in which drugs were administered subsequent to the event(s) to be remembered.

minutes prior to RAM testing. These observations highlight the importance of uninjected controls in protocols that involve infusions within the MBN and suggest that the current emphasis on the understanding the neurobiology of basocortical systems should extend to examining the peculiar sensitivity of this region to insult.

Post-training intraseptal administration of the GABAergic agonist muscimol and the GABAergic antagonist bicuculline can also produce deficits in the performance of the DNMTS-RAM task. Administration of 0.5ug of bicuculline immediately following the pre-delay session produced a significant decrease in the number of correct choices (Chrobak and Napier, 1990) and an increase in retroactive errors during the post-delay session at both a one and four hour retention interval (see Figure 4).

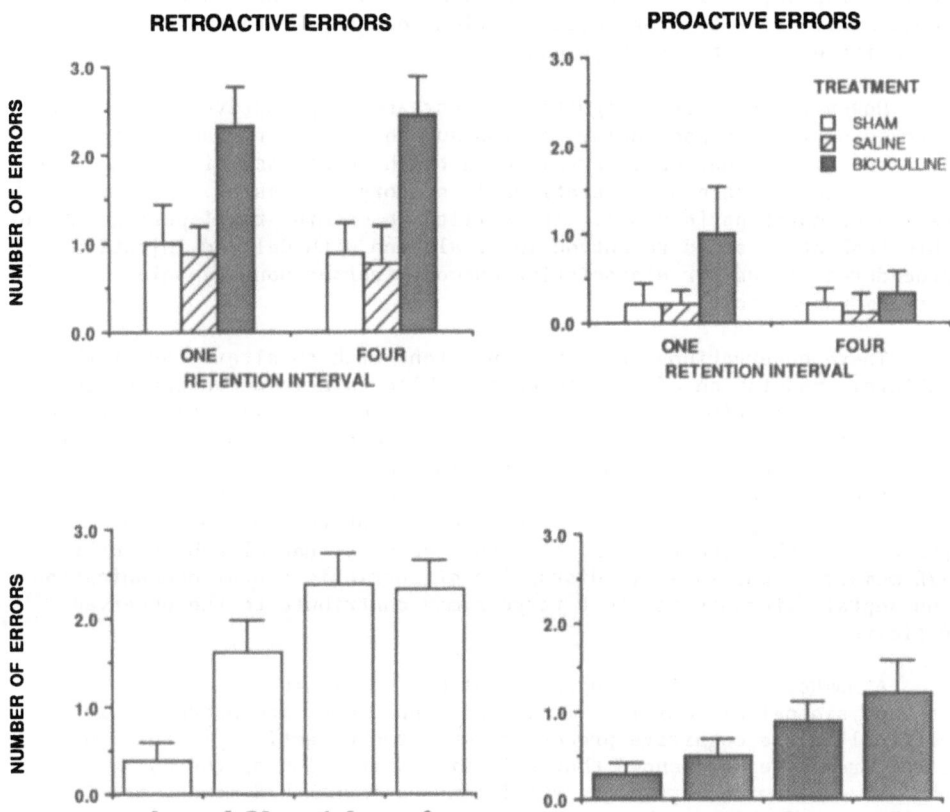

Fig. 4. Effects of post-training infusion of GABAergic agents [bicuculline (0.5ug), top; muscimol, bottom] into the medial septum on the number of retroactive and proactive errors during the post-delay session of a DNMTS-RAM task.

Administration of the drug two hours following the pre-delay session, however, did not significantly affect performance. Recently, we have observed significant impairments in post-delay performance with doses as low as 0.05ug and which vary as a function of the retention interval (Chrobak and Napier, in preparation). The lack of effect with the delayed injection procedure would seem to indicate a time-dependent effect on memory processes. Alternatively, phasic changes in the activity of septohippocampal neurons or intraseptal GABAergic neurons may alter the effectiveness of drug treatments during the delay interval.

Intraseptal administration of muscimol between choices 4 and 5 of the DNMTS-RAM task also produces a deficit in post-delay performance (Chrobak et al., 1989b). Thus, a dose-related decrease in correct choices in the first four post-delay choices and an increase in errors occurred following this GABAergic agonist. Examination of the post-delay errors, as shown in Figure 4, illustrates that the deficit in post-delay performance can occur as a consequence of mnemonic and/or non-mnemonic effects. Thus, with post-training administration of both muscimol and bicuculline, the animal's performance at one and four hours following intraseptal drug administration is characterized by an increase in errors to pre-delay chosen arms. This is indicative of an inability to respond in a discriminative manner to arms entered prior to drug administration. This incapacity to recall those events occurring immediately prior to drug administration contrasts with the absence of proactive errors at most drug doses.

However, there is a significant increase in proactive errors (arms entered during the post-delay session subsequent to drug administration) at a one hour retention interval, with the highest dose of muscimol, that may be a consequence of alterations in sensory processing, attention, memory or motor performance. The ability to examine rats' performance of this task at extended retention intervals and with delayed-injection procedures allows for dissociating mnemonic *versus* non-mnemonic performance deficits.

These observations would be consistent with an alteration of the GABAergic modulation of SH efferents. While we have demonstrated that a selective alteration of SHC is sufficient to disrupt W/E memory, these pharmacological manipulations would appear to disrupt both cholinergic as well as GABAergic neurotransmission. Indeed, disruption of either, based upon their connectivity to hippocampal circuitry (see Freund and Antal, 1988), would appear to be sufficient to disrupt the physiological processes within the temporal lobe that serve as neural substrates for W/E memory. Additionally, disruption of intrinsic septal communication and septal efferents to other targets may contribute to the observed deficits.

A number of findings suggest that MBN modulation of neurophysiological activity within the hippocampus and neocortex is critical to the cognitive processes subserved by activity within these structures. We have shown that selective disruption of the MS cholinergic neurons system is sufficient to impair W/E memory processes. Modulation of the septohippocampal projection *via* intraseptal infusion of GABAergic agents can likewise produce an acute impairment in W/E memory. These studies support observations that chronic and acute disruption of MBN activity is manifested as a cognitive impairment that can vary as a function of information processing demands.

Anatomical description of the potential inputs to the MBN suggest that these neurons receive direct and indirect input from a circumscribed set of corticolimbic structures, yet influence activity within the entire cortical mantle. Topographically organized afferents from the hippocampal formation to the lateral septum appear to modulate the activity of SH afferents (Meibach and Siegal, 1977; Swanson and Cowan, 1979) and a similar anatomical circuitry exists for the ventral striatum to modulate the more ventral extent of the MBN. While this description may not be unique to MBN neurons, the input and output of the MBN projection, as compared to other ascending systems, appears to allow for much greater topographical specificity in modulating cortical activity. Such a system could thus provide for spatial and temporal modulation of this projection. One could hypothesize that the importance of such modulation critically depends upon the demands made upon the system, i.e., its critical value increasing with greater information processing demands. In this regard ganglioside-induced behavioral recovery following SHC insult was particularly sensitive to task demands. Thus, this treatment completely ameliorated standard RAM deficits, but was totally ineffective in altering the deficit induced by imposition of minimal delays between arm choices in the DNMTS version of the task (Emerich and Walsh, 1990).

Future research will need to evaluate the role of descending cortical influences on MBN neurons, and the relationship between increasing task demands and the topographical specificity of MBN neuronal activity. The MBN projection may represent a feedback circuit whereby descending cortical projections may "backtalk", or modulate activity within other cortical areas. Our ability to determine the validity of this hypothesis will be critically important in our attempts to maintain or alleviate cognitive dysfunction resulting from MBN pathology. If MBN neurons are critically influenced by descending cortical influences, then the degree to which one might be able to literally amplify signal processing in this pathway, without degrading the pattern, may limit the efficacy of particular pharmacological compounds, notably cholinomimetics, and likewise may explain discrepancies in the human and animal literature concerning the effects of these compounds. The greater the information processing demands placed upon this system by task complexity, the more one may require increased specificity in the signals (fine tuning) conducted by the MBN afferents; specificity in the signal presumably being the product of the spatiotemporal pattern of activity. Constraints on our ability to amplify signal processing in SH and BC projection systems, and to alleviate cognitive dysfunction associated with MBN pathology, will be better understood by an increased understanding of the neurobiology of this system.

ACKNOWLEDGEMENTS

Funding for original studies described in this manuscipt were provided by a Basic Science Research Grant to TJW, NIMH grant #MH42572 to IH, the Potts Estate and NIMH grant #MH45180 to TCN, and the Illinois Department of Public Health to JJC.

REFERENCES

Alheid G.F., and Heimer, L., 1988, New perspectives in basal forebrain organization of special relevance for neuropsychiatric disorder: the striatopallidal, amygdaloid and corticopetal components of substantia innominata, Neuroscience, 27:1.

Allen, C.A., and Crawford, I.L., 1984, GABAergic agents in the medial septal nucleus affect hippocampal theta rhythm and acetylcholine utilization, Brain Res., 322:261.

Alonso, A., and Kohler, C., 1984, A study of the reciprocal connections between the septum and the entorhinal area using anterograde and retrograde axonal transport methods in the rat brain, J. comp. Neurol., 225:327.

Altman, H.J., Crosland, R.D., Jenden, D.J., and Berman, R.F., 1985, Further characterization of the nature of the behavioral and neurochemical effects of lesions to the nucleus basalis of Meynert in the rat, Neurobiol. Aging, 6:125.

Beatty, W.W., and Shavalia, D.A., 1980, Rat spatial memory: resistance to retroactive interference at long retention intervals, Anim. Learn. Behav., 8:550.

Bigl, V., Woolf, N.J., and Butcher, L., 1982, Cholinergic projections from the basal forebrain to frontal, parietal, temporal, occipital, and cingulate cortices: a combined fluorescent tracer and acetylcholinesterase analysis, Brain Res. Bull., 8:727.

Blaker, W.D., Peruzzi, G., and Costa, E., 1984, Behavioral and neurochemical differentiation of specific projections in the septal-hippocampal cholinergic pathway of the rat, Proc. Natl. Acad. Sci., 81:1880.

Bostock, E., Gallagher, M., and King, R.A., 1988, Effects of opioid microinjections into the medial septal area on spatial memory in rats, Behavioral Neuroscience, 102:643.

Breese, G.R., Frye, G.D., McCown, T.J., and Mueller, R.A., 1984, Comparisons of CNS effects induced by TRH and bicuculline after microinjection into medial septum, substantia nigra and inferior colliculus: absence of support for a GABA antagonist action of TRH, Pharmacol. Biochem. Behav., 21:145.

Brioni, J.D., Decker, M.W., Gamboa, L.P., Izquierdo, I., and McGaugh, J.L., 1990, Muscimol injections in the medial septum impair spatial learning, Brain Research, 522:227.

Chrobak, J.J. and Napier, T.C., 1989, Vehicle infusion into the basal forebrain produces task-specific cognitive deficits in the rat, Soc. Neurosci. Abs. 20.

Chrobak, J.J., and Napier, T.C., 1990, Intraseptal administration of bicuculline produces working memory impairments in the rat, Behav. Neural Bio., in press.

Chrobak, J.J., and Walsh, T.J., Dose and delay dependent working/episodic memory impairments following intraventricular administration of AF64A, submitted.

Chrobak, J.J., Hanin, I., and Walsh, T.J., 1986, AF64A (ethylcholine aziridinium ion), a cholinergic neurotoxin, selectively impairs working memory in a multiple component T-maze task, Brain Research, 414:14.

Chrobak, J.J. Hanin, I., Schmechel, D.E., and Walsh, T.J., 1988, AF64A-induced working memory impairment: behavioral, neurochemical and histological correlates, Brain Research, 463:107.

Chrobak, J.J. Spates, M., Stackman, R.W., and Walsh, T.J., 1989a, Hemicholinium-3 prevents the working memory impairments and the cholinergic hypofunction induced by ethylcholine aziridinium ion (AF64A), Brain Research, 504:269.

Chrobak, J.J., Stackman, R.W., and Walsh, T.J., 1989b, Intraseptal administration of muscimol produces dose-dependent memory impairments in the rat, Behav. Neural Bio., 52:357.

Costa, E., Panula, P., Thompson, H.K., and Cheney, D.L., 1983, The transynaptic regulation of the septal-hippocampal cholinergic neurons, Life Sciences, 32:165.

Curti, D., and Marchbanks, R.M., 1984, Kinetics of irreversible inhibition of choline transport in synaptosomes by ethylcholine mustard aziridinium, J. Membrane Biol., 82:259.

Dinopoulos, A., Parnavelas, J.G., Uylings, H.B.M., and Van Eden, C.G., 1988, Morphology of neurons in the basal forebrain nuclei of the rat: a golgi study, J. comp. Neurol., 272:461.

Durkin, T., 1989, Central cholinergic pathways and learning and memory processes: presynaptic aspects, Comp. Biochem. Phsyiol., 93:273.

Elrod, K., and Buccafusco, J.J., 1988, Microinjection of vehicle into the nucleus basalis magnocellularis results in task-specific impairment of passive avoidance responding, Res. Comm. Psychol. Psychiat. Behav., 13:271.

Emerich, D.F., and Walsh, T.J., 1990, Ganglioside AGF2 promotes task-specific recovery and attenuates the cholinergic hypofunction induced by AF64A, Brain Research, 527:299.

Fisher, R.S., Buchwald, N.A., Hull, C.D., and Levine, M.S., 1988, GABAergic basal forebrain neurons project to the neocortex: the localization of glutamic acid decarboxylase and choline acetyltransferase in feline corticopetal neurons, J. comp. Neurol., 272:489.

Freund, T.F., and Antal, M., 1988, GABA-containing neurons in the septum control inhibitory interneurons in the hippocampus, Nature, 336:170.

Gaffan, D., 1985, Hippocampus: memory, habit and voluntary movement, Philos. Trans. R. Soc. London Ser. B, 308:87.

Gaykema, R.P.A., Luiten, P.G.M., Nyakas, C., and Traber, J., 1990, Cortical patterns of the medial septum-diagonal band complex, J. comp. Neurol., 293:103.

Gower, A.J., Rousseau,D, Jamsin, P., Gobert, J., Hanin, I., and Wulfert, E., 1989, Behavioural and histological effects of low concentrations of intraventricular AF64A, Eur. J. Pharmacol., 166:271.

Hagan, J.J., and Morris, R.G.M., 1988, The cholinergic hypothesis of memory: a review of animal experiments, in "Handbook of Psychopharmacology, Volume 20, Psychopharmacology of the Aging Nervous System," L.L. Iversen, S.D. Iversen, and S.H. Snyder, eds., Plenum Press, New York, pp. 237-323.

Hanin, I., 1990, AF64A-induced cholinergic hypofunction, in "Cholinergic Neurotransmission: Functional and Clinical Aspects," S-M. Aquilonius and P.G.Gillberg , eds., Elsevier Science Publishers. B.V., Amsterdam, pp. 289-299.

Hanin, I., Fisher, A., Hortnagl, H., Leventer, S.M., Potter, P.E., and Walsh, T.J., 1987, Ethylcholine mustard aziridinium (AF64A; ECMA) and other potential cholinergic neuron-specific neurotoxins, in "Psychopharmacology- The Third Generation of Progress," H.Y. Meltzer, ed., Raven, New York, pp. 341-349.

Hepler, D.J., Olton, D.S., Wenk, G.L., and Coyle, J.T., 1985, Lesions in nucleus basalis magnocellularis and medial septal area of rats produce qualitatively similar memory impairments, J. Neurosci., 5:866.

Ingham, C.A., Bolam, J.P., and Smith, A.D., 1988, GABA-immunoreactive synaptic boutons in the rat basal forebrain: comparison of neurons that project to the neocortex with pallidosubthalamic neurons, J. comp. Neurol., 273:263.

Jarrard, L.E., Kant, G.J., Meyerhoff, J.C.,and Levy, A., 1984, Behavioral and neurochemical effects of intraventricular AF64A administration in rats, Pharmacol. Biochem. Behav., 21:273.

Kafetzopoulos, E., Holzhauer, M.S., and Huston, J.P., 1986, Substance P injected into the region of the nucleus basalis magnocellularis facilitates performance of an inhibitory avoidance task, Psychopharm., 90:281.

Kohler, C.,and Chan-Palay, V., 1983, Distribution of gamma aminobutyric acid containing neurons and terminals in the septal area, Anat. Embryol., 167:53.

Kohler, C., Chan-Palay, V., and Wu, J-Y., 1984, Septal neurons containing glutamic acid decarboxylase immunoreactivity project to the hippocampal region in the rat brain, Anat. Embryol., 169:41.

Lamour, Y. and Epelbaum, J., 1988, Interactions between cholinergic and peptidergic systems in the cerebral cortex and hippocampus, Progress in Neurobiology, 31:109.

Leranth, C., and Frotscher, M., 1989, Organization of the septal region in the rat brain: cholinergic-gabaergic interconnections and the termination of hippocampo-septal fibers, J. comp. Neurol., 289:304.

McGaugh, J.L., 1989, Dissociating learning and performance: drug and hormone enhancement of memory storage, Brain Res. Bull., 23:339.

Meibach, R.C., and Siegal, A., 1977, Efferent connections of the hippocampal formation in the rat, Brain Res., 124:197.

Mishkin, M., 1982, A memory system in the monkey, Philos. Trans. R. Soc. London Ser. B., 298:85.

Mogenson, G.J., 1987, Limbic and motor integration, in "Progress in Psychobiology and Physiological Psychology, Volume 12," J.N. Sprague and A.N. Epstein, eds., Academic Press, New York, pp. 117-170.

Murray, C.L., and Fibiger, H.C., 1985, Learning and memory deficits after lesions of the nucleus basalis magnocellularis: reversal by physostigmine, Neuroscience, 14:1025.

Nagel, J.A., and Huston, J.P., 1988, Enhanced inhibitory avoidance learning produced by post-trial injections of substance P into the basal forebrain, Behav. Neural Biol., 49:374.

Napier, T.C., and Marx, K., 1987, Enkephalin unilaterally microinjected into the ventral pallidal/nucleus basalis induces circling, Neurosci. Abstr., 13:445.

Napier, T.C., An, D., Austin, M.C., and Kalivas, P.W., 1988, Opiates microinjected into the ventral pallidum/substantia innominata (VP/SI) produce locomotor responses that involve dopaminergic systems. Neurosci. Abstr., 15:1173

Nauta, W.J.H., Smith, G.P., Gaull, R.L.M., and Domesick, V.B., 1978, Efferent connections and nigral afferents of the nucleus accumbens septi in the rat, Neurosci., 3:385.

Nyakas, C., Luiten, P.G.M., Spencer, D.G., and Traber, J., 1987, Detailed projection patterns of septal and diagonal band efferents to the hippocampus in the rat with emphasis on innervation of CA1 and dentate gyrus, Brain Res. Bull., 18:533.

Olton, D.S., 1986, Hippocampal function and memory for temporal context in "The Hippocampus, Volume 4," R.L. Isaacson and K. Pribram, eds., Plenum Press, New York, pp. 316-348.

Olton, D.S., and Wenk, G.L., 1987, Dementia: animal models of the cognitive impairments produced by degeneration of the basal forebrain cholinergic system, in "Psychopharmacology: The Third Generation of Progress," H.Y. Meltzer (ed), Raven Press, New York, pp. 941-953.

Pittel, Z., Fisher, A., and Heldman, E., 1987, Reversible and irreversible inhibition of high-affinity choline transport caused by ethylcholine aziridinium ion, J. Neurochem., 49:468.

Potter, P.E., Tedford, C.E., Kindel, G.H., and Hanin, I., 1989, Inhibition of high affinity choline transport attenuates both cholinergic and non-cholinergic effects of ethylcholine aziridinium (AF64A), Brain Res., 13:283.

Pope, C.N., Ho, B.T., and Wright, A.A., 1987, Neurochemical and behavioral effects of N-ethyl-acetylcholine aziridinium chloride in mice, Pharmacol. Biochem. Behav., 26:365.

Pribram, K., 1986, The hippocampal system and recombinant processing, in "The Hippocampus, Volume 4," R.L. Isaacson and K. Pribram, eds., Plenum Press, New York, pp. 329-369.

Price, J.L., and Stern, R., 1983, Individual cells in the nucleus basalis-diagonal band complex have restricted axonal projections to the cerebral cortex in the rat, Brain. Res., 269:352.

Raisman, G., 1966, The connections of the septum, Brain, 9:317.

Robbins, T.W., Everitt, B.J., Marston, H.M., Wilkinson, J., Jones, G.H., and Page, K.J., 1989, Comparative effects of ibotenic acid- and quisqualic acid-induced lesions of the substantia innominata on attentional function in the rat: further implications for the role of the cholinergic neurons of the nucleus basalis in cognitive processes, Behav. Brain Res., 35:221.

Robinson, S.E., Hambrecht, K.L., and Lyeth, B.G., 1988, Basal forebrain carbachol injection reduces cortical acetylcholine turnover and disrupts memory, Brain Res., 445:160.

Roitblat, H.L., 1982, The meaning of representations in animal memory, Behav. Brain Sci., 5:353.

Rye, D.B., Wainer, B.H., Mesulam, M-M, Mufson, E.J., and Saper, C.B., 1984, Cortical projections arising from the basal forebrain: a study of cholinergic and noncholinergic components employing combined retrograde tracing and immunohistochemical localization of choline acetyltransferase, Neurosci., 13:627.

Rylett, R.J., and Colhoun, E.H., 1980, Kinetic data on the inhibition of high-affinity choline transport into rat forebrain synaptosomes by choline-like compounds and nitrogen mustard analogues, J. Neurochem., 34:713.

Saper, C.B., 1987, Diffuse cortical projection systems: anatomical organization and role in cortical function, in "Handbook of Physiology, Section 1, The Nervous System, Vol. 5, Part 1," V.B. Mountcastle, F. Plum, and S.R. Geiger, eds., Am. Physiol. Soc., Bethesda, pp. 169-210.

Sherry, D.F., and Schacter, D.L., 1987, The evolution of multiple memory systems, Psych. Rev., 98:439.

Staubli, U., and Huston, J.P., 1980, Facilitation of learning post-trial injection of substance P into the medial septal nucleus, Behav. Br. Res., 1:245.

Swanson, L.W., and Kohler, C., 1986, Anatomical evidence for direct projections from the entorhinal area to the entire cortical mantle in the rat, J. Neurosci., 6:3010.

Swanson, L.W., and Cowan, W.M., 1979, The connections of the septal region in the rat, J. comp. Neurol., 186:621.

Squire, L.R., 1987, "Memory and Brain," Oxford Univ. Press, London and New York.

Tedford, C.E., Lorens, S.A., Corey, J.C., Lokhorst, D., Kindel, G., Wulfert, E., and Hanin, I., Behavioral and neurochemical effects of AF64A in young and old fisher-344 male rats, in "Alzheimer's and Parkinson's Diseases: Basic and Therapeutic Strategies," M. Yoshida, A. Fisher and T. Nagatsu, eds., Plenum Press, NY (in press).

Thomas, G.J., and Gash, D.M., 1986, Differential effects of posterior septal lesions on dispositional and representational memory, Behav. Neurosci., 100:712.

Tulving, E., 1983, "Elements of Episodic Memory", Clarendon Press, Oxford, England.

Uney J.B., and Marchbanks, R.M., 1987, Specificity of ethylcholine mustard aziridinium as an irreversible inhibitor of choline transport in cholinergic and noncholinergic tissue, J. Neurochem., 48:1673.

Van Hoesen, G.W., and Damasio, A.R., 1987, Neural correlates of cognitive impairment in Alzheimer's disease, in "Handbook of Physiology, Section 1, The Nervous System, Vol. 5, Part 1," V.B. Mountcastle, F. Plum, and Geiger, S.R., eds., Am. Physiol. Soc., Bethesda, pp. 871-898.

Walsh, T.J., and Chrobak, J.J., 1990, Animal models of Alzheimer's disease: role of hippocampal cholinergic system in working memory, in "Current Topics in Animal Learning: Brain, Emotion & Cognition," L. Dachowsky and C. Flaherty, eds., Erlbaum, Hillsdale, New Jersey, pp. 347-379.

Walsh, T.J., Tilson, H.A., DeHaven, D.L., Mailman, R.B., Fisher, A., and Hanin, I., 1984, AF64A, a cholinergic neurotoxin, selectively depletes acetylcholine in hippocampus and cortex, and produces long-term passive avoidance and radial-arm maze deficits in the rat, Brain Research, 321:91-102.

Weiskrantz, L., 1982, Comparative aspects of studies of amnesia, Philos. Trans. R. Soc. London Ser. B, 298:97.

Wood, P.L., and McQuade, P., 1986, Substantia innominata - cortical cholinergic pathway: regulatory afferents, in "Dynamics of Cholinergic Function," I. Hanin, ed., Plenum Press, New York, pp. 999-1006.

Zaborsky, L., Heimer, L., Eckenstein, F., and Leranth, C., 1986, GABAergic input to cholinergic forebrain neurons: an ultrastructural study using retrograde tracing of HRP and double immunolabeling, J. comp. Neurol., 250:282.

BEHAVIORAL PHARMACOLOGY AND BIOCHEMISTRY OF CENTRAL

CHOLINERGIC NEUROTRANSMISSION

Hans C. Fibiger, Geert Damsma, and Jamie C. Day

Division of Neurological Sciences
Department of Psychiatry
University of British Columbia
Vancouver, Canada

CHOLINERGIC MECHANISMS IN LEARNING AND MEMORY:
PHARMACOLOGICAL STUDIES

Although there is no question that muscarinic receptor antagonists such as atropine and scopolamine can have deleterious effects on the acquisition and post-acquisition performance of a broad spectrum of learned behaviors, at present there is no consensus concerning the psychological mechanisms underlying these antimuscarinic-induced deficits and it has not been possible to characterize them in a unitary theoretical framework. For reasons that are discussed below, this should not be regarded as surprising.

Since the early work of Carlton (1963) who proposed that a central cholinergic system antagonizes behavioral activation, particularly that associated with non-reinforcement, there have been ongoing attempts to characterize the function of central cholinergic neurons in terms of single, comprehensive constructs. According to Carlton (1963) the level of central activation "could be viewed as controlling the tendency for *all* responses to occur, whereas an inhibitory cholinergic system would act to antagonize this action on non-reinforced responses. The net result of this interaction would be that changes in activation would result in changes in the likelihood of occurrence of only a *few* responses, those that were reinforced. A cholinergic system thus appeared to be involved in the necessary and obviously adaptive 'steering' function assumed to be lacking in activation" (p. 27). Although many other hypotheses regarding central cholinergic function have been proposed in the intervening decades, it is interesting that the most recent attempt at a unitary characterization of central cholinergic function has much in common with Carlton's early speculations. Specifically, Whishaw and Petrie (1988) have recently proposed that cholinergic systems are involved in the selection of strategies and movements that are a prerequisite for learning.

The "selection" function of Whishaw and Petrie (1988) is not dissimilar to the "steering" function of Carlton (1963). To a certain extent, therefore, attempts to characterize the function of central cholinergic systems have come full circle in 25 years. This less than notable achievement raises questions about the assumptions upon which much of this research and thinking are based.

There have been many other attempts to characterize the functions of central cholinergic systems on the basis of the behavioral effects of systemically administered antimuscarinic agents. Wagman and Maxey (1969) concluded that scopolamine (0.5-1.0 mg/kg) impairs interoceptive cue discrimination and improves discrimination of exteroceptive stimuli. Brown and Warburton (1971) and Warburton and Brown (1971) found that scopolamine (0.0625-0.75 mg/kg) impaired performance on a DRL 15-second schedule of reinforcement and suggested that the drug reduces stimulus sensitivity (i.e. the signal-to-noise ratio). These results were inconsistent with response disinhibition hypotheses and were interpreted as indicating that the drug impairs attentional mechanisms. Other findings have also been viewed as being consistent with an attentional hypothesis. Cheal (1981) concluded that scopolamine (0.01-1.0 mg/kg) produces a deficit in the maintenance of attention in gerbils. Furthermore, her results were inconsistent with memory, response inhibition or stimulus selection hypotheses of cholinergic function.

The radial maze and the Morris water maze have been used extensively to analyse the nature of the antimuscarinic drug-induced deficits in learning and memory in rats. Eckerman et al. (1980) observed that post-acquisition administration of scopolamine (0.17-1.0 mg/kg) decreased the accuracy of performance in an 8-arm radial maze. Watts et al. (1981) found that scopolamine (0.5 mg/kg) impaired the acquisition of an 8-arm radial maze task. However, animals using non-spatial strategies were relatively resistant to the drug's deleterious effects. The results were taken as indicating that scopolamine interferes preferentially with working memory. Using a different task, Heise and coworkers (1976) also obtained evidence that scopolamine may have more potent actions on working than on reference memory measures. Thus, scopolamine treated rats (0.125-1.0 mg/kg) were unimpaired in a cued alternation task (a test of reference memory) but showed reduced response accuracy in non-cued spatial alternation which requires both working and reference memory for its successful execution. Using an 8-arm radial maze in which only half of the arms were ever baited, Wirsching and colleagues (1984) also concluded that scopolamine (0.1-0.8 mg/kg) can selectively impair measures of working memory (re-entries into baited arms) without having significant effects on reference memory (as indicated by entries into arms that were never baited). Buresova et al. (1986) studied the effects of scopolamine (0.1-1.0 mg/kg) on a number of Morris water maze and radial water maze tasks and concluded that the disruptive effect of scopolamine is proportional to the degree to which performance of the task requires working memory. Overtrained mapping strategies were much less affected by the drug. Using a Maier three table apparatus, Ellen and coworkers (1986) found that atropine (50 mg/kg) produced a selective impairment on a spatial integration task; rule learning and visual discrimination learning were not affected. These results were interpreted as indicating that central cholinergic blockade disrupts working memory for spatial information, rather than producing a general impairment in working memory.

While the above studies have emphasized the selective vulnerability of working memory to central antimuscarinic blockade, many other studies using similar procedures and drug doses have reached substantially different conclusions. Okaichi and Jarrard (1982) studied the effect of scopolamine (0.5-1.5 mg/kg) on several tasks in a radial arm maze and found that the drug did not differentially affect measures of working and reference memory. Both showed significant dose-related deficits. Sutherland et al. (1982) reported that atropine (50 mg/kg) impaired the acquisition of a reference memory task in a Morris water maze; in addition, unlike controls, the atropine-treated animals failed to utilize a spatial mapping strategy. Hagan and coworkers (1986) studied the effects of atropine (10 or 50 mg/kg) and confirmed the finding of Sutherland et al. (1982) that blockade of muscarinic receptors impairs the acquisition of a spatial navigation task in a Morris water maze. The acquisition of a nonspatial, two-platform discrimination task in the same apparatus was also disrupted. Inasmuch as neither of these tasks has a working memory component and because the spatial navigation task is a reference memory problem without a working memory component, these results provide strong evidence that the effects of muscarinic blockade are not restricted to working memory. Similar conclusions were reached by Spencer and colleagues (1985) who used a substantially different task (lever pressing on a variable intertrial interval continuous nonmatching to sample schedule or an analogous discrimination schedule) to conclude that scopolamine (0.125-0.50 mg/kg) interferes not only with certain aspects of working memory but also possibly with the "retrieval of response rules from reference memory". In a similar vein, on the basis of a literature review, Spencer and Lal (1983) concluded that the predominant effect of central muscarinic blockade is to disrupt encoding and retrieval processes in reference memory.

Other investigators have provided evidence that antimuscarinic drugs reduce stimulus discriminability or alter the sensory processing of conditioned stimuli. Kirk et al. (1988), for example, used a delayed conditional discrimination procedure to show that scopolamine (0.005-0.375 mg/kg) affected initial discriminability of auditory stimuli at much lower doses than were required to increase the rate of forgetting. The authors therefore suggested that the effects of scopolamine on working memory tasks could be more the result of changed stimulus processing than of impaired memorial processes. In contrast, Spencer and coworkers (1985) found little evidence that scopolamine (0.125-0.50 mg/kg) interfered with visual stimulus perception. Perhaps the effects of antimuscarinic agents on stimulus discrimination are determined by the sensory mode under investigation (i.e. visual vs. auditory). This does not appear to be the case, at least with respect to classical conditioning. Harvey and coworkers (1983) found that scopolamine (0.005-1.6 mg/kg) significantly retarded the rate of acquisition and the final asymptotic performance of conditioned nictitating membrane responses to both tone and light-conditioned stimuli. These authors concluded that scopolamine retards the acquisition of this classically conditioned response by blocking the "excitatory properties" of conditioned auditory stimuli or "altering the sensory processing of stimuli used as conditioned stimuli".

The pharmacological studies briefly reviewed above serve to illustrate the considerable divergence of opinion that exists about the psychological or psychophysical mechanisms through which antimuscarinic agents produce their disruptive effects on the acquisition and maintenance of learned behaviors. The mechanisms that have been proposed include behavioral inhibition, movement and strategy selection, working memory, retrieval from reference memory, attention, decisional processes and sensory processing. A problem with a substantial portion of this work is the absence of dose-response analyses. To some extent this is no doubt due to the complicated and time-consuming nature of the behavioral tests used in many of these studies. Nevertheless, this shortcoming severely limits the conclusions that can be drawn from many of these experiments. A limitation of other portions of this literature is the frequent failure to use methylated forms of scopolamine and atropine as controls for the potent consequences of peripheral muscarinic blockade. However, even if studies lacking dose-response analyses and methylated drug controls are ignored, it is difficult to detect conceptually monolithic effects of central muscarinic receptor blockade. For anatomical and neuropharmacological reasons, this is not surprising and might even have been predicted.

ANATOMY OF CENTRAL CHOLINERGIC SYSTEMS

The great majority of cholinergic neurons in the central nervous system can be conveniently subdivided into 4 major groupings. (1) Neurons in the cholinergic basal nuclear complex are distributed in a continuum across several classically defined nuclei in the basal forebrain from the medial septal nucleus rostrally to the substantia innominata and globus pallidus caudally (Schwaber et al., 1987). They have widespread telencephalic projections, including the cerebral cortex, hippocampus and olfactory bulbs. (2) Cholinergic interneurons in dopamine-rich regions such as the caudate-putamen, nucleus accumbens and olfactory tubercle form another distinct grouping. (3) Neurons of the mesopontine tegmental cholinergic group are located in the tegmentum of the midbrain and pons and also appear to form a continuum that is delimited rostrally just caudal to the substantia nigra and caudally by the locus coeruleus (Satoh et al., 1983). The major projections of these cells include the thalamus, pretectal area, lateral hypothalamus, superior colliculus, lateral preoptic area, magnocellular basal forebrain regions and medial prefrontal cortex (Semba and Fibiger, 1989). (4) The somatic motoneurons of the cranial motor nuclei and the spinal cord form another convenient grouping of cholinergic neurons. Although neurons in these 4 groups account for the majority of cholinergic activity in the central nervous system, in some species small populations of cholinergic neurons lying outside these groups have been identified in the medial habenula, lateral hypothalamus, cerebral cortex and hippocampus (Semba and Fibiger, 1989). Despite the fact that some actions of acetylcholine released from central cholinergic neurons are mediated by nicotinic receptors, muscarinic receptors are distributed so diffusely in the brain (Rotter et al., 1979a,b; Mash and Potter, 1986) that a substantial portion of the cholinergic transmission arising from the cholinergic basal nuclear complex, from striatal interneurons, and from the mesopontine tegmental group must be mediated by muscarinic receptors.

THE FALLACY OF UNITARY FORMULATIONS OF CENTRAL CHOLINERGIC FUNCTION

Given the near complete innervation of the neuraxis by cholinergic neurons and the widespread distribution of central muscarinic receptors, attempts at unitary formulations concerning the effects of systemically administered antimuscarinic drugs cannot be taken seriously because they are based on an anatomy that does not exist. Why then do systemically administered muscarinic receptor antagonists disrupt the performance of learned behaviors? It seems certain that they must do so for a variety of reasons, probably including many of those mentioned above. Given the diffuse anatomy of cholinergic transmission one would not expect otherwise.

Neuropharmacological considerations may further account for the diverse conclusions that have been reached on the basis of psychopharmacological experiments that have utilized different environmental conditions and behavioral demands. In theory, the degree of functional blockade produced by a given dose of a competitive antagonist will depend in part on the concentration of the transmitter in the vicinity of the receptor. The greater the concentration of transmitter, the less the antagonist-induced blockade. We have recently obtained evidence indicating that under different environmental and behavioral conditions, the amount of ACETYLCHOLINE (ACh) releaseD can vary both within and between different brain structures (see below). For example, under condition A, ACh release in the hippocampus might be higher than that in the cortex, while in condition B cortical ACh release might exceed that in the hippocampus. It should be noted that conditions A and B need not refer only to different environmental circumstances; they could also refer to two groups of animals having different levels of training on the same behavioral task. In either case, the administration of a given dose of a muscarinic receptor antagonist would be predicted to have substantially different consequences in conditions A and B. Specifically, in condition A, cortical function would be more severely impacted than would processes dependent on cholinergic mechanisms in the hippocampus because the net level of muscarinic blockade would be greater in the cortex than in the hippocampus. In condition B, the opposite would apply. Even though other factors such as differing receptor affinities would also influence the differential regional actions of systemically applied antagonists, the above example serves to illustrate the logical fallacy of attempting to characterize the effects of systemically administered anticholinergic agents within a unitary theoretical framework. Under some testing conditions, cholinergic blockade in one brain region would be expected to result in the most significant functional consequences; under others, behavioral deficits mediated by cholinolytic effects in the other brain structures might be most conspicuous. Viewed in this context, it could be predicted that behavioral experiments utilizing systemically applied antimuscarinics would generate a broad spectrum of seemingly inconsistent results. As indicated above, the literature on this subject is quite consistent with this prediction.

CHOLINERGIC MECHANISMS IN LEARNING AND MEMORY: LESION STUDIES

An alternative to studying the role of cholinergic systems in learning and memory by pharmacological means is to lesion selected populations of cholinergic neurons with axon-sparing excitotoxins and to investigate the effects of such lesions on the acquisition and post-acquisition performance of learned behaviors. In the past 5 years, this has become a frequently used technique in attempts to define the contributions of cholinergic mechanisms to learning and memory (Bartus et al., 1985; Dubois et al., 1985, Everitt et al., 1987; Flicker et al., 1984; Hepler et al., 1985a,b; Knowlton et al., 1985; Miyamoto et al., 1987; Murray and Fibiger, 1985; Murray and Fibiger, 1986; Peternel et al., 1988). For reasons discussed below, however, this will probably prove to be a short-lived approach to the problem. The most serious potential confound with the lesion paradigm is that wherever they are found in the forebrain, cholinergic neurons are always intermingled with non-cholinergic cells. Thus, when an excitotoxin such as ibotenic acid is applied by stereotaxic injection into the nucleus basalis magnocellularis (nBM), both cholinergic and non-cholinergic neurons are inevitably damaged and the lesions also typically spread to damage adjacent structures such as the globus pallidus, amygdala and some thalamic nuclei (Murray and Fibiger, 1985; Dunnett et al., 1987; Everitt et al., 1987; Arbogast and Kozlowski, 1988). Any resultant effects of these lesions on measures of learning and memory must be interpreted with these facts in mind. One method of indirectly addressing these limitations has been to determine if lesion-induced deficits can be reversed either by drugs that increase central cholinergic tone (*i.e.* acetylcholinesterase (AChE) inhibitors or muscarinic receptor agonists) or by transplants of foetal, cholinergic neuron-containing grafts into the cholinergically denervated target structures. Both strategies have been used successfully to reverse some of the behavioral deficits produced by lesions of the nBM or the fornix-fimbria (Murray and Fibiger, 1985; 1986; Low et al., 1982; Dunnett et al., 1982; 1985; Fine et al., 1985). Nevertheless, as Murray and Fibiger (1985) pointed out, physostigmine-induced amelioration of nBM lesion-induced learning and memory impairments suggests, *but does not prove*, that these behavioral deficits are due to damage to cholinergic neurons in the nBM rather than to non-cholinergic neurons in the same region. It remains possible, for example, that the lesion deficit and the pharmacological reversal of that deficit are due to actions on different neural substrates that summate to produce effects on behavior, rather than reflecting effects of these manipulations on a common cholinergic mechanism. As subsequent events have made clear, this cautionary note was well-founded.

A major clue that the learning and memory deficits produced by ibotenic acid lesions of the nBM were not primarily due to damage to cortically-projecting cholinergic neurons was obtained by Etherington and colleagues (1987). These investigators compared the effects of ibotenic and quisqualic acid lesions of the nBM on the performance of an operant delayed matching task and a delayed alternation task in rats. Earlier, Dunnett et al. (1987) had found that quisqualic acid injected into the nBM produced less non-specific cell damage than ibotenic acid when the toxins were injected in concentrations that produced equivalent decreases in cortical choline acetyltransferase (ChAT) activity. Etherington et al. (1987) reported that only the ibotenate lesioned animals showed significant impairments in accuracy in the performance of these tasks; the quisqualate group did not differ from controls. However, when cortical ChAT activity was measured in these groups, the quisqualate lesions reduced ChAT activity by 51%, whereas the decrease was only 17% in the ibotenate lesioned group. These results cast severe doubt on the hypothesis that the ibotenic acid lesion-induced deficit was cholinergic in nature. In subsequent studies this group of investigators confirmed and extended these preliminary observations. Thus, Robbins et al., (1989) reported that quisqualate lesions of the nBM failed to significantly affect the acquisition of a visual conditional discrimination task, despite producing substantial (44%) reductions in cortical ChAT activity 10 months after the lesions. Again, however, ibotenic acid lesions did result in a behavioral deficit but produced smaller decreases (27%) in cortical ChAT activity 3-5 months after the surgery. As observed previously, therefore, there was a dissociation between the ability of these excitotoxins to produce deficits in conditional learning and the extent to which they decreased cortical ChAT activity. Using a different behavioral task, such a dissociation has recently been confirmed by Wenk et al. (1989). These results strongly suggest that the many deficits in the acquisition and performance of learned behaviors that have been demonstrated after ibotenate lesions of the nBM cannot be attributed alone to the loss of cortically-projecting cholinergic neurons. Clearly, ibotenate-induced damage to non-cholinergic cells in the vicinity of the nBM must contribute importantly to the memory impairments following nBM cell loss. While this evidence indicates that damage to cholinergic neurons in the nBM is not alone sufficient to account for the memory impairments observed after ibotenic acid lesions of this structure, whether concomitant damage to cholinergic neurons is necessary for the demonstration of these deficits remains to be determined.

Dunnett and co-workers (1989) have recently studied the effects of nBM quisqualate lesions on short-term memory capacities of the rat as assessed by delayed matching to position and delayed non-matching to position tasks. These lesions reduced cortical ChAT activity by 55% but did not affect the performance of previously trained animals on either task. However, the lesions selectively impaired the acquisition of the non-matching contingency and disrupted reversal to the non-matching task in animals previously trained on the matching task. While these are potentially important observations, it should be emphasized that as Dunnett et al. (1989) pointed out, even though quisqualic acid clearly has advantages over ibotenic acid with respect to cholinergic neurotoxicity in the basal forebrain, at present there is no reason to believe that this excitotoxin does not also damage non-cholinergic neurons in the vicinity of the cholinergic cells. Thus, it would not be appropriate to conclude that quisqualate-induced deficits have a cholinergic basis. Given the lesson of ibotenic acid, investigators are now obliged to exercise extreme caution in interpreting behavioral results based even on quisqualate lesions. In fact, inasmuch as firm conclusions about selective cholinergic mechanisms cannot be achieved with excitotoxin lesions, the utility of this approach as a tool for studying central cholinergic function is doubtful.

Studies of central cholinergic systems would be greatly advanced by the availability of a specific neurotoxin for cholinergic neurons. Unfortunately, although early studies indicated that AF64A might be such an agent, subsequent work has demonstrated that this compound has limited selectivity for cholinergic neurons (Levy et al., 1984; McGurk et al., 1987; Until such an agent becomes available perhaps the best approach for studying the regional functions of cholinergic systems is by the local injection of cholinergic agonists and antagonists into discrete regions of the central nervous system. At present, only a few investigators have used this approach and, despite its limitations (Myers, 1971), it appears capable of producing worthwhile results (Brito et al., 1983). In addition, the use of brain microdialysis to monitor neurotransmitter release in awake, freely moving animals holds considerable promise as a technique for elucidating the behavioral correlates of regional ACh release (Damsma et al., 1987).

CHOLINERGIC TRANSMISSION IN THE HIPPOCAMPUS AND FRONTAL CORTEX STUDIED BY IN VIVO MICRODIALYSIS

In the past decade, brain microdialysis (Johnson and Justice, 1983; Ungerstedt, 1984) has increasingly been applied in the neurosciences as a technique to sample interstitial fluid from laboratory animals. Pharmacological investigations have validated the use of microdialysis to study the release of neurotransmitters (Imperato and Di Chiara, 1984; Westerink et al., 1987). ACh collected during microdialysis is thought to reflect synaptic overflow of neuronal release because dialysate ACh is calcium dependent, tetrodotoxin-sensitive, and responds to pharmacological manipulations in the predicted manner (Damsma et al., 1987, 1988).

In vivo brain microdialysis is based on the principle that molecules can diffuse from the interstitial fluid into a surgically implanted semipermeable tube. This dialysis tube is continuously perfused with a physiological solution and the collected dialysate is assayed for ACh using HPLC separation, a post-column enzyme reactor, and electrochemical detection (ED). The dialysis probe is small (outer diameter approximately 250m) and is implanted stereotaxically, allowing specific measurement from defined brain regions. Following recovery from surgery (typically for 24-48 hours), the dialysis tube outlet can be connected by flexible tubing to automated analytical equipment allowing the animal free movement and minimal disturbance by the experimenter.

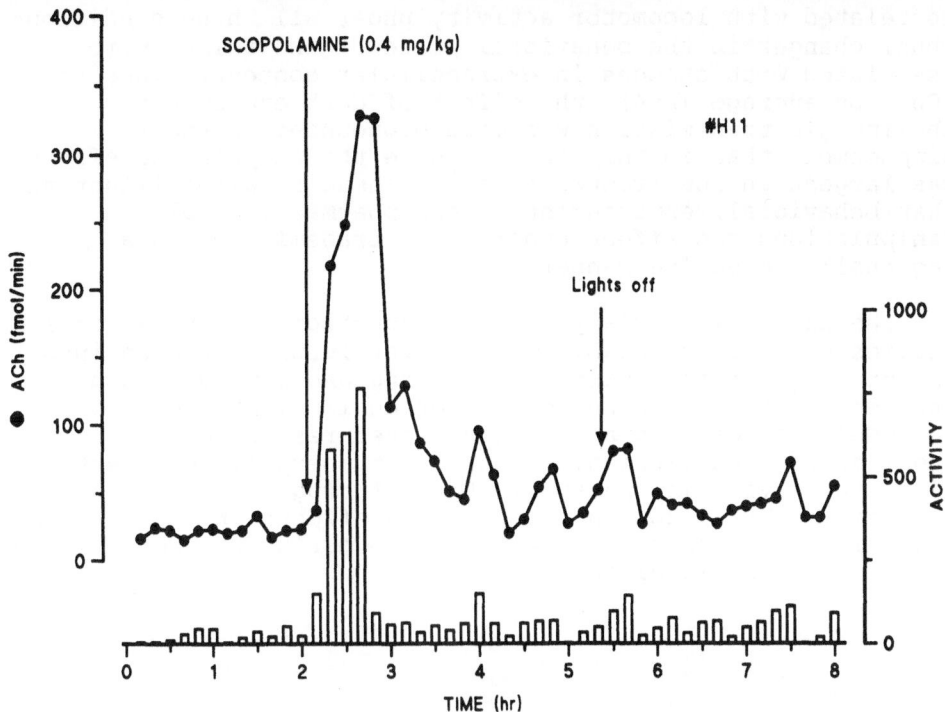

Fig. 1. Dialysate concentrations of ACh and locomotor activity
during spontaneous behavior, scopolamine
administration and exposure to dark. A dialysis
membrane implanted in the dorsal hippocampus of a rat
was perfused at 5 μl/min with a solution containing
0.1 μM neostigmine. ACh was analyzed on-line by HPLC-
EC and locomotion was counted as the number of
photocell beam breaks in 10 min intervals. Lights off
occurred at the usual time in the rats daily cycle.

To date, limited microdialysis investigations have been
employed to study the neurochemical correlates of behavior.
For example, the effects of feeding (Radhakishun et al., 1988)
and sexual behavior (Damsma et al., in press) on mesolimbic
dopamine, and the effect of stress on striatal dopamine
(Abercrombie, 1989) and hippocampal norepinephrine
(Abercrombie et al., 1988) have been reported. Recently, ACh
dialysis has also been employed in behavioral studies.
Nilsson (1990), for example, has reported an increase in
hippocampal ACh during a 15 min period of handling by the
experimenter. Also, scopolamine-induced behavioral activation
and concomitant changes in cholinergic transmission have been
measured in the hippocampus and frontal cortex (Toide, 1989)
and in the striatum (Watanabe and Shimizu, 1989). Preliminary
studies (below) further demonstrate the feasibility of
combining ACh microdialysis with behavioral measurements, both
under drug and drug-free conditions.

We have recently measured locomotor activity of rats
during perfusion of either the frontal cortex or hippocampus
(Day et al., submitted). Figure 1 shows dialysate ACh
concentrations and motor activity of two subjects during
spontaneous behavior, scopolamine administration and exposure
to dark. On an individual basis, ACh dialysate concentrations

correlated with locomotor activity under all three conditions. Thus, changes in the behavioral state of the animal were associated with changes in extracellular concentrations of ACh. On average (n=4), the effect of dark exposure on cholinergic transmission was more pronounced in the hippocampus than in the cortex, while the scopolamine effect was largest in the frontal cortex. These findings illustrate that behavioral, environmental, and pharmacological manipulations can affect cholinergic transmission in a regionally selective manner.

The use of microdialysis in conjunction with behavioral studies holds considerable promise for future research into the cholinergic mechanisms of learning and memory. Such studies could explore the effect of environmental cues and well-defined learning and memory procedures on regional cholinergic transmission. Also, a yet unexploited potential of microdialysis is the application of drugs through the dialysis membrane to study local drug effects on cholinergic transmission in specific brain regions during the performance of various behavioral tasks.

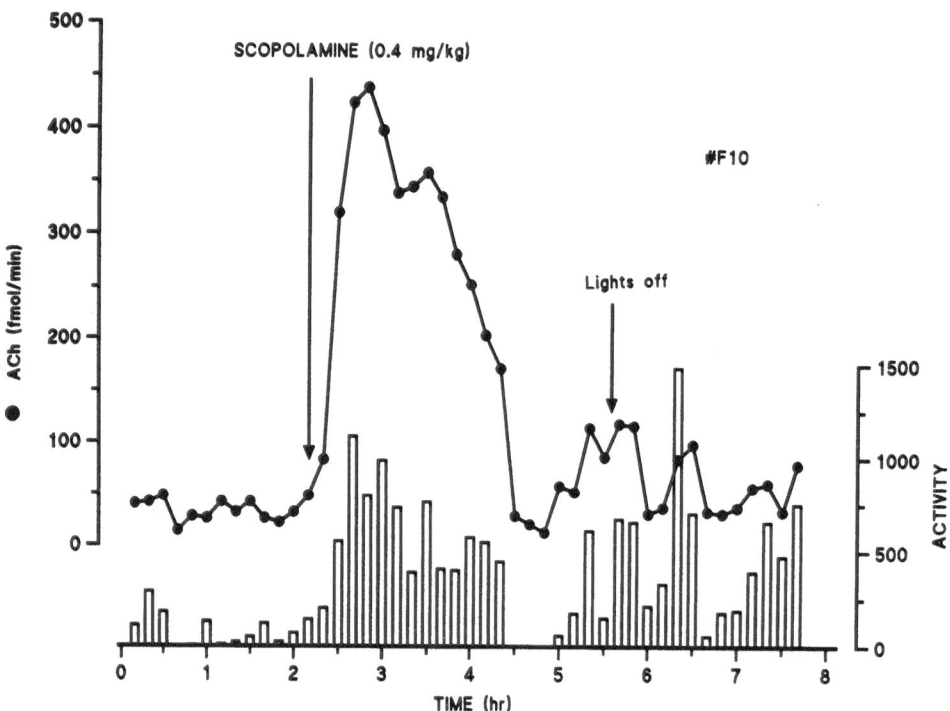

Fig. 2. Dialysate concentrations of ACh and locomotor activity during spontaneous behavior, scopolamine administration and exposure to dark. A dialysis membrane implanted in the frontal cortex of a rat was perfused at 5 µl/min with a solution containing 0.1 µM neostigmine. ACh was analyzed on-line by HPLC-EC and locomotion was counted as the number of photocell beam breaks in 10 min intervals. Lights off occurred at the usual time in the rats daily cycle.

SUMMARY

Systemically administered cholinergic (muscarinic) receptor antagonists can impair the acquisition and post-acquisition performance of a variety of learned behaviors. At present, there is no consensus about the psychological mechanisms underlying these deficits. Behavioral inhibition, working (short-term) memory, reference (long-term) memory, attention, movement and strategy selection, and stimulus processing are among the constructs that have been proposed as underlying the effects of muscarinic receptor blockade. On the basis of neuroanatomical and neuropharmacological considerations it is contended that debates about the nature of the mediating events are pointless because they are on an anatomy that does not exist. Specifically, given that cholinergic neurons innervate almost the entire neuraxis and that muscarinic cholinergic receptors are distributed throughout the central nervous system, it is virtually certain that systemically applied antimuscarinic drugs will influence a broad spectrum of brain functions. In addition, the nature of the deficits produced by scopolamine and atropine, which are competitive antagonists, will depend on the regional endogenous rate of acetylcholine release, which may in turn be influenced by the particular environment and/or level of training imposed on the animal. As the literature seems to indicate, therefore, the effects of competitive antagonists will vary as a function of both the behavioral test and the level of training. Accordingly, attempts at unitary formulations of central cholinergic function are ill-conceived and illusory.

Another approach to understanding central cholinergic function has been based on the use of local injections of excitotoxins into brain regions such as the basal forebrain that contain cholinergic neurons. Recent published reports indicate, that many of the behavioral deficits observed after ibotenic acid lesions of the basal forebrain are due primarily to the loss of non-cholinergic neurons. The inherent limitations of the excitotoxin lesion approach for unravelling the functions of central cholinergic systems are such that they cannot produce definitive information and might best, therefore, be abandoned. At present, a reliable selective toxin for cholinergic neurons is not available and urgently required. Until such a compound is identified, local intracerebral applications of antimuscarinic agents may be the preferred procedure for studying the behavioral correlates of regional blockade of cholinergic activity. Brain microdialysis in freely moving animals also holds considerable promise with respect to defining the circumstances under which acetylcholine is released in discrete regions of the central nervous system. At present, the function of central cholinergic systems and the possible role of each in learning and memory remain poorly understood.

REFERENCES

Abercrombie, E.D., Keefe, K.A., DiFrischia, D.S. and Zigmond, M.J., 1989, Differential effect of stress on in vivo dopamine release in striatum, nucleus accumbens, and medial frontal cortex, J. Neurochem., 52:1655.

Abercrombie, E.D., Keller, Jr., R.W., and Zigmond, M.J., 1988, Characterization of hippocampal norephinephrine release as measured by microdialysis perfusion: pharmacological and behavioral studies, <u>Neuroscience</u>, 27:987.

Arbogast, R.E., and Kozlowski, M.R., 1988, Quantitative morphometric analysis of the neurotoxic effects of the excitotoxin, ibotenic acid, on the basal forebrain, <u>Neurotoxicology</u>, 9:39.

Bartus, R.T., Flicker, C., Dean, R.L., Pontecorvo, M., Figueriredo, J.C., and Fisher, S.K., 1985, Selective memory loss following nucleus basalis lesions: long term behavioral recovery despite persistent cholinergic deficiencies, <u>Pharmacol. Biochem. Behav.</u>, 23:125.

Brito, G.N.O., Davis, B.J., Stopp, L.C., and Stanton, M.E., 1983, Memory and the septo-hippocampal cholinergic system in the rat, <u>Psychopharmacology</u>, 81:315.

Brown, K., and Warburton, D.M., 1971, Attenuation of stimulus sensitivity by scopolamine, <u>Psychonomic Sci.</u>, 22:297.

Buresova, O., Bolhuis, J.J., and Bures, J., 1986, Differential effects of cholinergic blockade on performance of rats in the water tank navigation task and in a radial water maze, <u>Behav. Neurosci.</u>, 100:476.

Carlton, P.L., 1963, Cholinergic mechanisms in the control of behavior by the brain, <u>Psychol. Rev.</u>, 70:19.

Cheal, M, 1981, Scopolamine disrupts maintanance of attention rather than memory processes, <u>Behav. Neural Biol.</u>, 33:163.

Damsma, G., Pfaus, J.G., Nomikos, G.G., Wenkstern, D.G., Blaha, C., Phillips, A.G., and Fibiger, H.C., 1990, Sexual behavior enhances central dopamine transmission in the male rat, <u>Brain Res. Rev.</u> (in press).

Damsma, G., Westerink, B.H.C., de Vries, J.B., Van Den Berg, C.J., and Horn, A.S., 1987, Measurement of acetylcholine release in freely moving rats by means of automated intracerebral dialysis, <u>J. Neurochem.</u>, 48:1523.

Damsma, G., Westerink, B.H.C., de Boer, P., de Vries, J.B., and Horn, A.S., 1988, Basal acetylcholine release in freely moving rats detected by on-line trans-striatal dialysis: pharmacological aspects, <u>Life Sci.</u>, 43:1161.

Day, J.C., Damsma, G., and Fibiger, H.C., 1990, Acetylcholine release in the hippocampus, cortex and striatum of rats correlates with locomotor activity: an in vivo microdialysis study, Submitted.

Dubois, B., Mayo, W., Agid, Y., LeMoal, M., and Simon, H., 1985, Profound disturbances of spontaneous and learned behaviors following lesions of the nucleus basalis magnocellularis in the rat, <u>Brain Res.</u>, 338:249.

Dunnett, S.B., Low, W.C., Iversen, S.D., Stenevi, U., and Bjorklund, A., 1982, Septal transplants restore maze learning in rats with fornix-fimbria lesions, <u>Brain Res.</u>, 251:335.

Dunnett, S.B., 1985, Comparative effects of cholinergic drugs and lesions of nucleus basalis or fimbria-fornix on delayed matching in rats, <u>Psychopharmacology</u>, 87:357.

Dunnett, S.B., Toniolo, G., Fine, A., Ryan, C.N., Bjorklund, A., and Iversen, S.D., 1985, Transplantation of embryonic ventral forebrain neurons to the neocortex of rats with lesions of nucleus basalis magnocellularis. II. Sensorimotor and learning impairments, <u>Neuroscience</u>, 16:787.

Dunnett, S.B., Whishaw, I.Q., Jones, G.H., and Bunch, S.T.,

1987, Behavioural, biochemical and histochemical effects of different neurotoxic amino acids injected into nucleus basalis magnocellularis of rats, Neuroscience, 20:653.

Dunnett, S.B., Rogers, D.C., and Jones, G.H., 1989, Effects of nucleus basalis magnocellularis lesions in rats on delayed matching and non-matching to position tasks, Eur. J. Neurosci., 1:395.

Eckerman, D.A., Gordon, W.A., Edwards, J.D., MacPhail, R.C., Gage, M.I., 1988, Effects of scopolamine, pentobarbital, and amphetamine on radial arm maze performance in the rat, Pharmacol. Biochem. Behav., 12:595.

Ellen, P., Taylor, H.S., and Wages, C., 1986, Cholinergic blockade effects on spatial integration versus cue discrimination performance, Behav. Neurosci., 100:720.

Etherington, R., Mittleman, G., and Robbins, T.W., 1987, Comparative effects of nucleus basalis and fimbria-fornix lesions on delayed matching and alternative tests of memory, Neurosci. Res. Commun., 1:135.

Everitt, B.J., Robbins, T.W., Evenden, J.L., Marston, H.M., Jones, G.H., and Sirkia, T.E., 1987, The effects of excitotoxic lesions of the substantia innominata, ventral and dorsal globus pallidus on the acquisition and retention of a conditional visual discrimination: implications for cholinergic hypotheses of learning and memory, Neuroscience, 22:441.

Fine, A., Dunnett, S.B., Bjorklund, A., and Iversen, S.D., 1985, Cholinergic ventral forebrain grafts into the neocortex improve passive avoidance memory in a rat model of Alzheimer disease, Proc. Natl. Acad. Sci. USA, 82:5227.

Flicker, C., Ferris, S.F., Bartus, R.T., and Crook, T., 1984, Effects of aging and dementia upon recent visuospatial memory, J. Neurobiol. Aging, 5:75.

Hagan, J.J., Tweedie, F., and Morris, R.G.M., 1986, Lack of task specificity and absence of posttraining effects of atropine on learning, Behav. Neurosci., 100:483.

Harvey, J.A., Gormezano, I., and Cool-Hauser, V.A., 1983, Effects of scopolamine and methylscopolamine on classical conditioning of the rabbit nictitating membrane response, J. Pharmacol. Exp. Ther., 225:42.

Heise, G.A., Conner, R., and Martin, R.A, 1976, Effects of scopolamine on variable intertrial interval spatial alternation and memory in the rat, Psychopharmacology, 49:131.

Hepler, D.J., Olton, D.S., Wenk, G.L., and Coyle, J.T., 1985a, Lesions in nucleus basalis magnocellularis and medial septal area of rats produce qualitatively similar memory impairments, J. Neurosci., 5:866.

Hepler, D.J., Wenk, G.L., Cribbs, B.L., Olton, D.S., and Coyle, J.T., 1985b, Memory impairments following basal forebrain lesions, Brain Res., 346:8.

Imperato, A., and Di Chiara, G., 1984, Trans-striatal dialysis coupled to reverse phase high performance liquid chromatography with electrochemical detection: a new method for the study of the in vivo release of endogenous dopamine and metabolites, J. Neurosci., 4:966.

Johnson, R.D., and Justice, J.B., 1983, Model studies for brain dialysis, Brain Res. Bull., 10:567.

Kirk, R.C., White, K.G., and McNaughton, N., 1988, Low dose scopolamine affects discriminability but not rate of forgetting in delayed conditional discrimination, Psychopharmacology, 96:541.

Knowlton, B.J., Wenk, G.L., Olton, D.S., and Coyle, J.T., 1985, Basal forebrain lesions produce a dissociation of trial-dependent and trial-independent memory performance, Brain Res., 345:315.

Levy, A., Kant, G.J., Meyerhoff, J.L., and Jarrard, L.E., 1984, Non-cholinergic neurotoxic effects of AF64A in the substantia nigra, Brain Res., 305:169.

Low, W.C., Lewis, P.R., Bunch, S.T., Dunnett, S.B., Thomas, S.R., Iversen, S.D., Bjorklund, A., and Stenevi, U., 1982, Function recovery following neural transplantation of embryonic septal nuclei in adult rats with septo-hippocampal lesions, Nature, 300:260.

Mash, D.C., and Potter, L.T., 1986, Autoradiographic localization of M1 and M2 muscarine receptors in the rat brain, Neuroscience, 19:551.

McGurk, S.R., Hartgraves, S.L., Kelly, P.H., Gordon, M.N., and Butcher, L.L., 1987, Is ethylcholine mustard aziridinium ion a specific cholinergic neurotoxin?, Neuroscience, 22:215.

Miyamoto, M. Kato, J., Narumi, S., and Nagaoka, A., 1987, Characteristics of memory impairment following lesioning of the basal forebrain and medical septal nucleus in rats, Brain Res., 419:19.

Murray, C.L., and Fibiger, H.C., 1985, Learning and memory deficits after lesions of the nucleus basalis magno-cellularis: reversal by physostigmine, Neuroscience, 14:1025.

Murray, C.L., and Fibiger, H.C., 1986, Piolcarpine and physostigmine attenuate spatial memory impairments produced by lesions of the nucleus basalis magnocellularis, Behav. Neurosci., 100:23.

Myers, R.D., 1971, Methods for chemical stimulation of the brain in "Methods in Psychobiology, Vol. 1,": R.D. Myers, ed., Academic Press, New York, p. 247.

Nilsson, O.G., Kalen, P., Rosengren, E., and Bjorklund, A., 1990, Acetylcholine release in the rat hippocampus as studied by microdialysis is dependent on axonal impulse flow and increases during behavioural activation, Neuroscience, (in press).

Okaichi, H., and Jarrard, L.E., 1982, Scopolamine impairs performance of a place and cue task in rats, Behav. Neural Biol., 35:319.

Peternel, A., Haughey, D., Wenk, G., and Olton, D., 1988, Basal forebrain and memory: neurotoxic lesions impair serial reversals of a spatial discrimination, Psychobiology, 16:54.

Radhakishun, F.S., Van Ree, .M., and Westernik, B.H.C., 1988, Scheduled eating increases dopamine release in the nucleus accumbens of food-deprived rats as assessed with on-line brain dialysis, Neurosci. Lett., 85:351.

Robbins, T.W., Everitt, B.J., Ryan, C.N., Marston, Marston, H.M., Jones, G.H., and Page, K.J., 1989, Comparative effects of quisqualic and ibotenic acid-induced lesions of the substantia innominata and globus pallidus on the acquisition of a conditional visual discrimination: differential effects on cholinergic mechanisms, Neuroscience, 28:337.

Rotter, A., Birdsall, N.J.M., Burgen, A.S.V., Field, P.M., Hulme, E.C., and Raisman, G., 1979a, Muscarinic receptors

in the central nervous system of the rat. I. Technique
for autoradiographic localization of the binding of
[^{3}H]propylbenzilylcholine mustard and its distribution in
the forebrain, Brain Res. Rev., 1:141.
Rotter, A., Birdsall, N.J.M., Burgen, A.S.V., Field, P.M.,
Hulme, E.C., Raisman, G., 1979b. Muscarinic receptors in
the central nervous system of the rat. II. Distribution
of binding of [^{3}H]propylbenzilylcholine mustard in the
midbrain and hindbrain, Brain Res. Rev., 1:167.
Satoh, K., Armstrong, D.M., and Fibiger, H.C., 1983, A
comparison of the distribution of central cholinergic
neurons as demonstrated by acetylcholinesterase pharmaco-
histochemistry and choline acetyltransferase
immunohistochemistry, Brain Res. Bull., 11:693.
Schwaber, J.S., Rogers, W.T., Satoh, K., and Fibiger, H.C.,
1987, Distribution and organization of cholinergic neurons
in the rat forebrain demonstrated by computer-aided data
acquisition and three-dimensional reconstruction, J. Comp.
Neurol., 263:309.
Semba, K., and Fibiger, H.C., 1989, Organization of central
cholinergic systems, Prog. Brain Res., 79:37.
Spencer, Jr., D.G., and Lal, H., 1987, Effects of
anticholinergic drugs on learning and memory, Drug
Development Res., 3:489-502, (1987).
Spencer, Jr., D.G. Pontecorvo, M.J., and Heise, G.A., 1985,
Central cholinergic involvement in working memory:
effects of scopolamine on continuous nonmatching and
discrimination performance in the rat, Behav. Neurosci.,
99:1049.
Sutherland, R.J., Whishaw, I.Q., and Regehr, J.C., 1982,
Cholinergic receptor blockade impairs spatial localization
by use of distal cues in the rat, J. Comp. Physiol.
Psychol., 96:563.
Toide, K., 1989, Effects of scopolamine on extracellular
acetylcholine and choline levels and on spontaneous motor
activity in freely moving rats measured by brain dialysis,
Pharmacol. Biochem. Behav., 33:109.
Ungerstedt, U, 1984, Measurement of neurotransitter release
by intracranial dialysis, "Measurement of Neurotransmitter
Release In Vivo," C.A. Marsden, ed., John Wiley & Sons
Ltd., New York, pp. 81.
Wagman, W.D., and Maxey, G.C., 1969, The effects of
scopolamine hydrobromide and methyl scopolamine
hydrobromide upon the discrimination of interoceptive and
exteroceptive stimuli, Psychopharmacologia, 15:280.
Warburton,, D.M., and Brown, K., 1971, Attenuation of stimulus
sensitivity induced by scopolamine, Nature, 230:126.
Watanabe, H., and Shimizu, H., 1989, Effect of anticholinergic drugs on striatal acetylcholine release and
motor activity in freely moving rats studied by brain
microdialysis, Jpn. J. Pharmacol., 51:75.
Watts, J., Stevens, R., and Robinson, C., 1981, Effects of
scopolamine on radial maze performance in rats, Physiol.
Behav., 26:845.
Wenk, G.L., Markowski, A.L., and Olton, D.S., 1989, Basal
forebrain lesions and memory: alterations in neurotensin,
not acetylcholine, may cause amnesia, Behav. Neurosci.,
103:1624.
Westerink, B.H.C., Damsma, G., Rollema, H., De Vries, J.B.,
and Horn, A.S., 1987, Scope and limitations of in vivo

brain dialysis: a comparison of its application to various neurotransmitter systems, <u>Life Sci.</u>, 41:1763.

Whishaw, I.Q., and Petrie, B.F., 1988, Cholinergic blockade in the rat impairs strategy selection but not learning and retention of nonspatial visual discrimination problems in a swimming pool, <u>Behav. Neurosci.</u>, 102:662.

Wirsching, B.A., Beninger, R.J., Jhamandas, K., Boegman, R.J., and El-Defrawy, S.R., 1984, Differential effects of scopolamine on working and reference memory of rats in the radial maze, <u>Pharmacol. Biochem. Behav.</u> 20:659.

IN VITRO CELL CULTURES AS A MODEL OF THE BASAL FOREBRAIN

B.H. Wainer,[1,2,4] H.J. Lee[1], J.D. Roback[2] and D.N. Hammond[3,4]

Departments of Pharmacological and Physiological Sciences[1], Pathology[2], Neurology[3], and Pediatrics[4], The University of Chicago, Chicago, Illinois, 60637

SUMMARY

The basal forebrain has attracted considerable attention because of its putative role in complex functions such as learning, memory and behavioral state control as well as its vulnerability in neurological disorders such as Alzheimer's Disease (AD). The finding that nerve growth factor provides trophic support for the cholinergic basal forebrain neurons has stimulated further interest in understanding trophic interactions of basal forebrain neurons as well as in possible trophic factor therapeutic stategies for disease states. Our laboratory has utilized primary cell cultures and developed immortalized central nervous system cell lines to study the trophic interactions that establish and maintain the septohippocampal pathway, a basal forebrain component which plays an essential role in cognitive function and is prominently affected in AD. The results of our primary cell culture studies have demonstrated the importance of trophic signals elaborated by the hippocampus in mediating the development of septal cholinergic neurons. Nerve growth factor plays an important role in this process, but it cannot account for all of the trophic signals elaborated by authentic hippocampal target cells. The development by this laboratory of clonal cell lines of septal and hippocampal lineage offers the prospect of investigating both the response to and elaboration of neural trophic signals at a more precise level of resolution than can be achieved with primary cultures. The technology and information that is generated from the engineering of such cell lines will also serve as a strategy to study trophic interactions in other brain circuits in future years, and to investigate possible changes or dysfunctions that occur neurological diseases.

INTRODUCTION

The mechanisms utilized by central neurons to establish and maintain synaptic connections with specific target cell populations remain poorly understood. Knowledge gained from the study of such mechanisms has important implications not only for elucidating the principles of neuronal development, but also for investigating the pathogenesis of certain nervous system diseases (Price, 1986; Saper, 1988). AD is the most common neurodegenerative disease and as schematically depicted in Fig. 1, it involves the progressive degeneration of cortico-cortical association pathways, elements of the entorhinal-hippocampal circuitry, and subcortical diffuse ascending systems such as the cholinergic basal forebrain cell groups (Price, 1986; Saper et.al.1987; Wainer, 1989b). All of these pathways are synaptically linked and contribute to cognitive function. Recent therapeutic strategies have focused on enhancing cholinergic transmission either by providing precursors of acetylcholine (ACh), or prolonging ACh degradation with anticholinesterases (Becker and Giacobini, 1988). These strategies have not produced dramatic results clinically (Hefti and Weiner, 1986), and do not address the cause of degeneration in both cholinergic and noncholinergic pathways. One pathogenetic mechanism to explain the observed pattern of neuronal degeneration is a dysfunction of essential trophic interactions, mediated by target cell-derived macromolecular signals

(Hsiang et.al.1987; Hsiang et.al.1988; Purves, 1988) that maintain the viability of particular neural circuits (Appel, 1981; Hefti and Weiner, 1986; Saper et.al.1987; Wainer, 1989b). This possibility has recently received substantial attention with the observation that NGF, a well-characterized neurotrophic factor in the peripheral nervous system (PNS), is also likely to function as a trophic factor for central cholinergic neurons that degenerate in AD (Thoenen et.al.1987; Whittemore and Seiger, 1987). It is probable that other growth factors, likewise initially isolated in the periphery, are active in the brain (Werner et.al.1988; Lindholm et.al.1987; Morrison et.al.1987; Anderson et.al.1988) and that additional substances in the central nervous system (CNS) are yet to be identified (Walicke, 1989). This theoretical construct raises the possibility of a new era in therapeutic management that would consist of the delivery of growth factors to the nervous system in order to repair or prevent the degeneration of neural pathways (Rosenberg et.al.1988). The efficacy of this approach has already been demonstrated in rodents and non-human primates, and in fact NGF therapy for patients suffering from AD is likely to be tested clinically in the near future (Phelps et.al.1989; Marx, 1990).

The purpose of the present review is to discuss work carried out in this laboratory to: **1)** Investigate trophic interactions that establish and maintain one of the pathways at risk in AD, the septohippocampal projection; and **2)** To present a CNS cell immortalization strategy that we have developed to-identify and study neural trophic factors. For further discussions of approaches to study neural trophic interactions and growth factors, the reader is referred to other recent reviews (Thoenen et.al.1987; Whittemore and Seiger, 1987; Walicke, 1989; Hefti et.al.1989; Snider and Johnson, 1989).

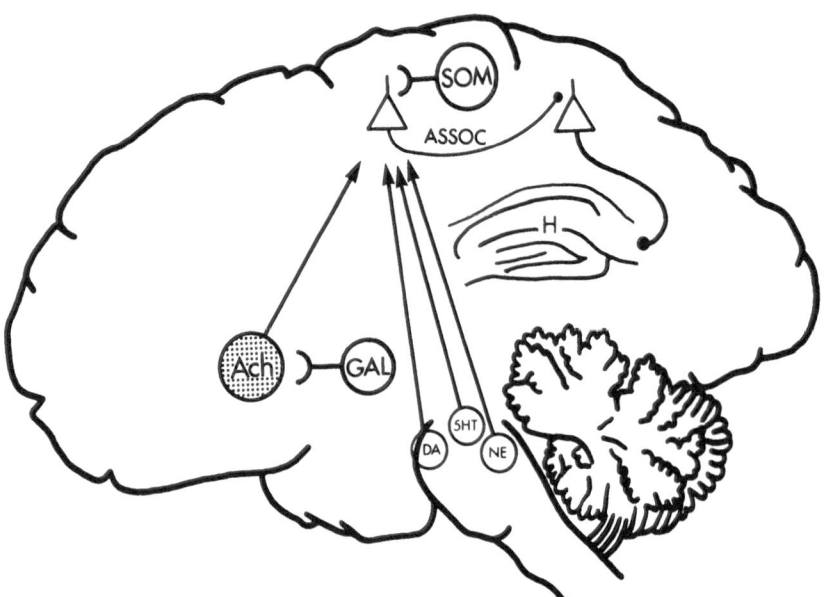

Fig. 1. Schematic diagram illustrating some of the major neural connections that are at risk in Alzheimer's Disease (Price, 1986; Wainer, 1989b). Assoc, cortico-cortico association pathways; H, hippocampal circuitry, including entorhinal cortical inputs (i.e. perforant pathway); SOM, cortical somatostatin-containing neurons (Davies et.al.1980); ACh, the magnocellular basal forebrain cholinergic projections to cortex (including septohippocampal); GAL; galanin inputs to basal forebrain cholinergic neurons (Chan Palay, 1988). Other diffuse ascending projections at risk include: DA, Dopaminergic; 5HT, Serotoninergic; NE, Noradrenergic. Diagram modified from Saper et al.(1987).

PRIMARY CELL CULTURE STUDIES OF BASAL FOREBRAIN NEURONS

Our laboratory has studied the detailed anatomical principles of organization of central cholinergic pathways (Wainer et.al.1984; Rye et.al.1984; Rye et.al.1987; Wainer and Mesulam, 1990). The septohippocampal pathway contains a substantial cholinergic component, is part of a contiguous basal forebrain system (Rye et.al.1984; Wainer et.al.1985; Wainer and Mesulam, 1990), and plays an important role in memory (Meck et.al.1987; Zola Morgan et.al.1986). Although this pathway is not the only locus of vulnerability in AD (Price, 1986; Saper, 1988), it is severely affected and serves as a useful model system to examine the factors that establish and sustain neural connectivity. Information that is obtained from a careful study of this system may be directly applied as a therapeutic strategy in AD, and may serve as a rationale for the study and approach to both the repair and prevention of degeneration in other pathways which are at risk.

The use of *in vitro* cell culture techniques represents a powerful tool to investigate neural trophic interactions because cells from areas of interest can be studied under controlled conditions which are more amenable to experimental manipulation than *in situ* investigations. We have employed reaggregating cell cultures to examine the mechanisms that instruct developing septal cholinergic neurons to establish and maintain stable synaptic contacts with hippocampal target cells (Hsiang et.al.1987; Hsiang et.al.1988; Wainer et.al.1989a). In this system, the brain areas of interest (hippocampus and/or septum) are dissected from mouse embryos, "dissociated" into a single cell suspension, and cultured under conditions which promote the reassociation, or "reaggregation", of these single cells into spheres, or "reaggregates" of approximately 200 μm in diameter (Garber and Moscona, 1972). Within these reaggregates, cells are surrounded by, and may interact with, neighboring cells in three-dimensions. This situation closely approximates the environment *in vivo*. In fact, this environment is essential for some physiological interactions involved in the development of neural systems. Glutamine synthetase, an enzyme found in the Mueller glia of the retina, is induced by hydrocortisone if the intact retina is grown in explant culture (Piddington and Moscona, 1967), or if the dissociated retina is grown in reaggregating culture (Morris and Moscona, 1970). However, if the dissociated cells are plated in monolayer culture, where the cell-cell interactions are more restricted, there is no steroidal induction (Morris and Moscona, 1970). Work by Heller and colleagues (Hoffmann et.al.1983) has demonstrated an increased survival of the developing dopaminergic neurons of the substantia nigra when they are coaggregated with a normal target area, the corpus striatum, but not with a non-target area, the tectum. No such survival effects are observed in monolayer cultures (Prochiantz et.al.1979).

Initially, we studied the effects of hippocampal target cells on the development of septal cholinergic neurons in septal-hippocampal (S-H) reaggregate co-cultures derived from embryonic day 15 (E15) mouse brains (Hsiang et.al.1987). As a control, nontarget cells from the cerebellum were coaggregated with the septal cells (S-Cb coaggregates). The results demonstrated a significant increase in the numbers of cholinergic cells in S-H coaggregates (relative to S-Cb), and proliferation of numerous fibers with axon-like morphologies (Hsiang et.al.1987; Wainer et.al.1989a). In septal-cerebellar coaggregates, the cells and fibers appeared to be degenerating by light microscopic observation and cholinergic fibers were excluded from regions of the coaggregates occupied by the nontarget cerebellar cells . In a subsequent ultrastructural study, it was determined that cholinergic axons form synaptic contact with hippocampal target cells in this culture system (Hsiang et.al.1988). In contrast, in the presence of nontarget cerebellar cells, the cholinergic neurons develop initially, but subsequently undergo a process of cell degeneration and death. These results demonstrate that hippocampal cells elaborate trophic signals that are essential for the development, survival, and synaptogenic potential of septal cholinergic neurons.

Because NGF is synthesized by cells in the hippocampus (Roback et.al.1990), we next investigated its role as a putative hippocampal-derived cholinergic trophic factor (Hsiang et.al.1989). In the presence of exogenous NGF, septal reaggregates demonstrated increases in choline acetyltransferase (ChAT) activity, cholinergic cells, and fiber density. In contrast, when NGF antibodies were added to S-H coaggregates to neutralize endogenous NGF, there was only a partial loss of cholinergic markers, and no qualitative changes in cholinergic cells or fibers at the light microscopic level (Fig. 2, A-G), or synapse formation at the ultrastructrual level (Fig. 2H) (Hsiang et.al.1989). These findings contrast with previous *in vivo* studies demonstrating destruction of sympathetic and sensory neurons following anti-NGF treatment in the PNS (Levi-Montalcini and Booker, 1960; Johnson et.al.1980), and antagonism of sympathetic sprouting in the CNS (Springer and Loy, 1985). Although considerable attention has focused on NGF as the primary cholinergic neural trophic factor in the CNS, evidence for such a hypothesis is not conclusive as summarized in Table 1. This point has important implications when considering the development of trophic factor treatment strategies for AD. If attention is singularly focused on NGF, a lack of success in clinical trials may lead to an untimely abandonment of such strategies; whereas a consideration of other factors which may act in concert with NGF may provide a more effective therapeutic modality.

We conclude that NGF mediates cholinergic development and maintenance in the septohippocampal system in concert with an unknown number of additional target cell-associated cues. These conclusions are supported by the observations of other groups: Hofer and Barde (1988) have shown that both exogenous NGF and brain-derived neurotrophic factor prevent naturally occurring cell death in quail dorsal root ganglion neurons. Appel and co-workers. (Ojika and Appel, 1984; Bostwick et.al.1987) and Wise et al. (Emerit et.al.1989) have provided evidence for non-NGF molecule(s) influencing basal forebrain cholinergic neurons. Hefti and colleagues, utilizing an *in vitro* explant culture system, have demonstrated a requirement for NGF in the formation of the septohippocampal pathway (Gahwiler et.al.1987) while Levitt and co-workers have demonstrated a similar requirement for a non-NGF limbic system-associated membrane glycoprotein in an identical culture system (Keller et.al.1989). More recent evidence suggests that the cytolymphokine,

Table 1. Is NGF the Primary Basal Forebrain Cholinergic Neurotrophic Factor?

Evidence For:

i). Exogenous NGF increases ChAT activity in the basal forebrain and striatum both *in situ* and *in vitro* (Mobley et.al.1985; Gnahn et.al.1983; Hsiang et.al.1989; Hartikka and Hefti, 1988a).

ii). Exogenous NGF "rescues" septal cholinergic neurons after axotomy (Williams et.al.1986).

iii). NGF supports survival of basal forebrain cholinergic neurons in low-density monolayer cultures (Hartikka and Hefti, 1988b).

Evidence Against:

i). Infusion of anti-NGF antibodies *in situ* decreases basal forebrain ChAT activity but does not lead to death of cholinergic neurons (Vantini et.al.1989).

ii). Antibodies against NGF do not completely neutralize the neurotrophic effects of hippocampal target cells on septal cholinergic neurons *in vitro* (Hsiang et.al.1989).

iii). Antibodies against LAMP disrupt the formation of the cholinergic septohippocampal projection *in vitro* (Keller et.al.1989).

iv). bFGF and IL-3 also "rescue" septal cholinergic neurons after axotomy (Anderson et.al.1988; Kamegai et.al.1990).

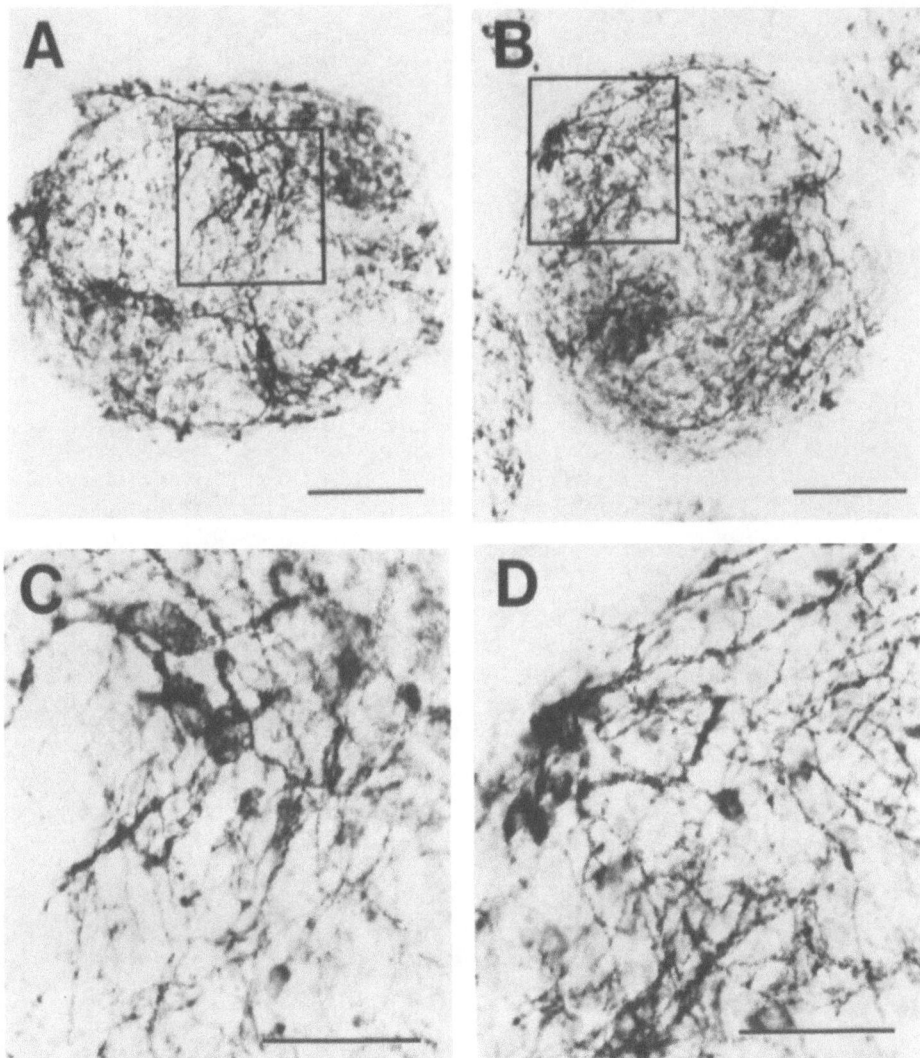

Fig. 2. Lack of Effect of Anti-NGF Antiserum on Septal-Hippocampal (S-H) coaggregates: Photomicrographs of sections of 21-day old Acetylcholinesterase (AChE)-stained S-H coaggregates cultured in media alone, media containing non-immune serum, or media containing Anti-NGF antiserum beginning at day 2 in culture. **A**: S-H coaggregate grown in culture medium alone. AChE histochemistry reveals a typical target cell-induced pattern of fibers and cells similar to that described in previous studies (Hsiang et.al.1987). The boxed area is shown at higher magnification in C. **B**: S-H coaggregate cultured in the presence of non-immune serum . AChE histochemistry reveals a pattern of cells and fibers similar to those in A. The boxed area is shown at higher magnification in D. **C**: Higher-power photomicrograph of boxed area of AChE-positive cells and fibers in A. These fine caliber fibers are well-defined with extensive arborizations and varicosities. **D**: Higher-power photomicrograph of (Legend cont'd. on subsequent page).

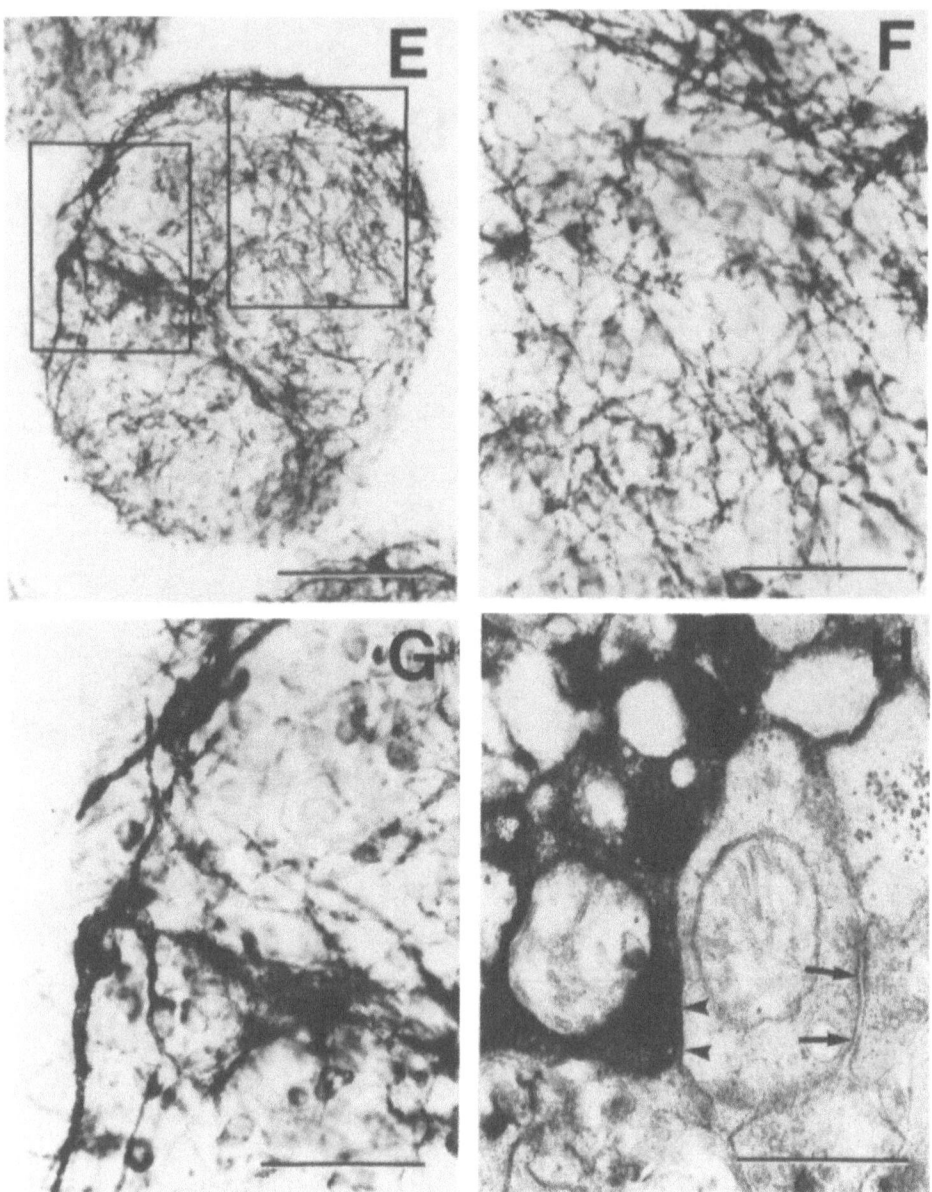

Fig. 2 cont'd. boxed area of AChE-positive cells and fibers in B. The appearance of these cells and fibers does not differ from those shown in C. **E:** S-H coaggregate cultured in the presence of Anti-NGF antiserum (1:200 dilution). AChE histochemistry reveals fibers and cells similar to those in A. The boxed areas are shown at higher magnification in F and G. **F & G:** Higher-power photomicrographs of boxed areas of AChE-positive cells (G) and fibers (F) in E. The appearance of these cells and fibers are similar to those shown in A and B suggesting that NGF antibodies do not inhibit cholinergic fiber proliferation in S-H coaggregates. **H:** High-power electron micrograph showing AChE-labeled synapse (arrowheads) and unlabeled synapse (arrows) in a 21-day old S-H coaggregate which was cultured in the presence of Anti-NGF antiserum beginning on the second day of culture. The identification of labeled synapses in this case suggests that anti-NGF does not inhibit the process of synaptogenesis between septal cholinergic terminals and hippocampal target cells. Scale Bars: A, B, E = 200 μm; C, D, F, G = 50 μm; H = 0.5 μm. From Hsiang et al. (1989).

Table 2. Central Nervous System Neurotrophic Factors

NGF-like Family:

Nerve Growth Factor (NGF) supports the neurochemical and morphological differentiation of basal forebrain and striatal cholinergic neurons (Hartikka and Hefti, 1988a; Mobley et.al.1985).

Brain-Derived Neurotrophic Factor (BDNF) is synthesized in the brain and supports the survival of retinal ganglion cells (Johnson et.al.1990).

Neurotrophin-3 (NT-3) bears extensive sequence similarity to both NGF and BDNF and is synthesized in the brain (Maisonpierre et.al.1990; Hohn et.al.1990).

Insulin-like Family:

Insulin and **Insulin-like Growth Factors I** and **II (IGF-I & II)** stimulate the neurochemical differentiation of selected central nervous system cholinergic and dopaminergic neurons (Knusel et.al.1990).

Peripheral Growth Factors:

Epidermal Growth Factor (EGF) increases survival and process outgrowth of subneocortical telencephalic neurons (Morrison et.al.1987).

Basic and **Acidic Fibroblast Growth Factors (bFGF & aFGF)** support the survival and fiber proliferation of diverse populations of central nervous system neurons (Walicke, 1988).

Other:

Ciliary Neurotrophic Factor (CNTF) is synthesized by brain astrocytes and exerts trophic actions on multiple peripheral neuronal populations (Lillien et.al.1988; Barbin et.al.1984).

Interleukin 3 (IL-3) enhances fiber outgrowth and stimulates neurochemical differentiation of basal forebrain cholinergic neurons (Kamegai et.al.1990).

Limbic System Associated Membrane Glycoprotein (LAMP) is important for the formation of the septohippocampal pathway *in vitro* (Keller et.al.1989).

IL-3, may also provide trophic support for septal cholinergic neurons (Kamegai et.al.1990). Taken together, this evidence suggests that the trophic mechanisms mediating the formation and maintenance of this single pathway may be the result of the interaction of a number of trophic influences which are possibly interactive. A partial list of putative neural trophic factors is presented in Table 2.

CNS CLONAL CELL LINES

The challenge of identifying novel trophic factors is complex because of the cellular heterogeneity of the brain and the likelihood that such factors are expressed in extremely minute quantities. Although a trophic "activity" may be detected in a crude brain extract or through the use of primary cell cultures, purification and characterization of the bioactive molecule(s) are still quite formidable. Alternatively, the utilization of clonal cell lines provides an approach that can circumvent the problems cited above (Banker and Goslin, 1988; Lendahl and McKay, 1990). In fact, the original identification of NGF was achieved by virtue of its expression in a rat sarcoma cell line (Levi-Montalcini, 1987). Theoretically, cell lines might be isolated either from spontaneously arising neural tumors (DeLellis et.al.1973; Warren and Chute, 1972) or from spontaneous "immortalization" of cultured primary cells (Ronnett et.al.1990). For example PC12 cells, derived from a rat pheochromocytoma, have been employed extensively for studies on the function of NGF and its receptor (Tishler and Greene, 1975; Guroff, 1985). Although spontaneous immortalization of primary neurons in culture is an exceedingly infrequent event (Bottenstein, 1985), considerable attention has surrounded the recent report of a "neuronal" cell line that has been isolated from human cerebral cortex (Ronnett et.al.1990). Such strategies have several limitations. First, cell lines derived from neural tumors carry with them the malignant features of their tumor cell lineage. Second, most of the tumor cell lines arise in the peripheral nervous system and extrapolation of the properties of such cells to processes occuring in specific central nervous system pathways may not be entirely justified. This fact is further complicated by the extensive heterogeneity present even with specific brain regions. For example, the properties of hippocampal pyramidal cells vary considerably between various sectors of Ammon's Horn (i.e. CA1-4) (Prince and Connors, 1986). Finally, the strategy of isolating spontaneously "immortalized" cell lines does not rule out the possibility of inherent cellular abnormalities. In the recent report by Ronnett et al. (1990) the particular cell line of interest, HCN-1A, was derived from the brain of a patient with megalencephaly, a pathological condition characterized by abnormal growth of immature neuron-like cells. The observation that this cell line expresses five distinct putative neurotransmitters is in contrast to what has been observed in normal cortical neurons, which suggests that the HCN-1A cell line may reflect a more primitive pluripotential stage of neuronal development.

An alternative approach is to modify cells genetically for purposes of immortalization. Using this technology, permanent cell lines could be established from specific brain regions that either elaborate or respond to trophic signals which participate in the formation and maintenance of the synaptic circuitry of those regions. To date, little work has been done in this area. Two general strategies are available for engineering such cell lines. The first approach is to utilize retroviral transduction of oncogenes to immortalize primary brain cells (Cepko, 1988; Cepko, 1989; Lendahl and McKay, 1990). An example of this approach was the use of a temperature sensitive mutant of the SV-40 viral T antigen to immortalize embryonic rat brain cells (Frederiksen et.al.1988). Several cell lines were generated that expressed varying degrees of neuronal or glial differentiation and which proliferated at the permissive temperature, 33°C, or could be induced to "differentiate" at the nonpermissive temperature, 37-39°C. A hippocampal cell line generated through this approach, HT4, has been reported to exhibit an immature neuronal phenotype and also to express NGF. A preliminary report by Whittemore and co-workers (1988) has demonstrated that the HT4 cells can be grafted into the adult rat brain and the cells do not appear to proliferate. In addition, when the recipient rats receive lesions of the septohippocampal pathway, the grafted cells partially prevent retrograde degeneration of septal cholinergic neurons possibly through the elaboration of NGF. While this approach is attractive, there are also some disadvantages. Retroviral transduction is only effective with cells that retain the capacity to replicate DNA. In addition, once a cell is immortalized, it tends be remain locked within a particular developmental window. This phenomenon has been exploited by immunologists to study the stages of lymphocyte differentiation (Paige and Wu, 1989; Alt et.al.1987). Therefore, while viral gene transduction might yield cell lines to study the early stages of neuronal development, it is less likely to provide cell lines that express the phenotypic repertoire of mature neurons which are almost invariably post-mitotic.

A second approach utilized previously by our laboratory employs somatic cell fusion techniques in which primary brain cells are fused via polyethylene glycol to a neuroblastoma cell line as illustrated schematically in Fig. 3 (Hammond et.al.1986; Hammond et.al.1990b; Lee et.al.1990b; Lee et.al.1990a; Wainer et.al.1989a). The neuroblastoma, N18TG2, is deficient in hypoxanthine phosphoribosyl transferase (HPRT⁻) and can be selectively eliminated in media containing hypoxanthine, aminopterin, and thymidine (HAT). This is precisely the approach that has been used to generate monoclonal antibody-producing hybridomas (Kohler and Milstein, 1975). The fusion technique allows one to immortalize cell populations that are postmitotic and therefore more likely to express highly differentiated neuronal phenotypes. Once neural cell lines are established, the probability of identifying one which expresses interesting trophic activities is increased because of the clonal nature of the cell lines. Finally, the likelihood of successfully purifying and characterizing the molecules responsible for a "trophic activity", or investigating the mechanism of a trophic

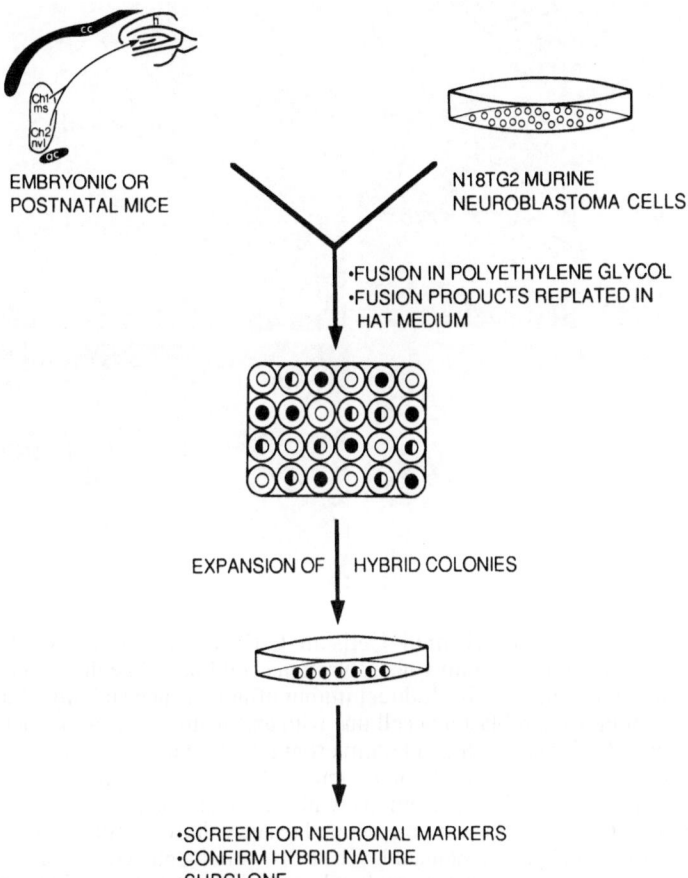

Fig. 3. Immortalization of Septal and Hippocampal Neurons. Schematic diagram outlining the protocol used to generate septum x neuroblastoma or hippocampus x neuroblastoma hybrid cell lines from C57BL/6 mice and the N18TG2 neuroblastoma (Hammond et.al.1986; Lee et.al.1990a). To facilitate adherence of primary cells to neuroblastoma cells, the primary cells are incubated with phytohemagglutinin (PHA) prior to fusion. Somatic cell fusion is mediated by polyethylene glycol. The fusion products are grown in hypoxanthine, aminopterin, and thymidine (HAT) media to select for hybrid cells. Aminopterin blocks de novo purine biosynthesis and is used to select against parent N18TG2 neuroblastoma cells which are deficient in hypoxanthine-guanine-phosphoribosyl transferase, an enzyme which is essential for salvage pathway purine biosynthesis requiring exogenous hypoxanthine and thymidine. Abbreviations; Ch1/ms, cholinergic neurons of medial septal nucleus; Ch2/nvl, cholinergic neurons of nucleus of vertical limb of diagonal band of Broca; ac, anterior commissure; cc, corpus callosum; h, hippocampus.

response, is enhanced since the cell lines represent an infinite source of starting material. In order to establish cell lines that exhibit the capacity to respond to target cell-mediated trophic signals and cell lines that elaborate such signals, we have employed somatic cell fusion to develop both septal and hippocampal cell lines.

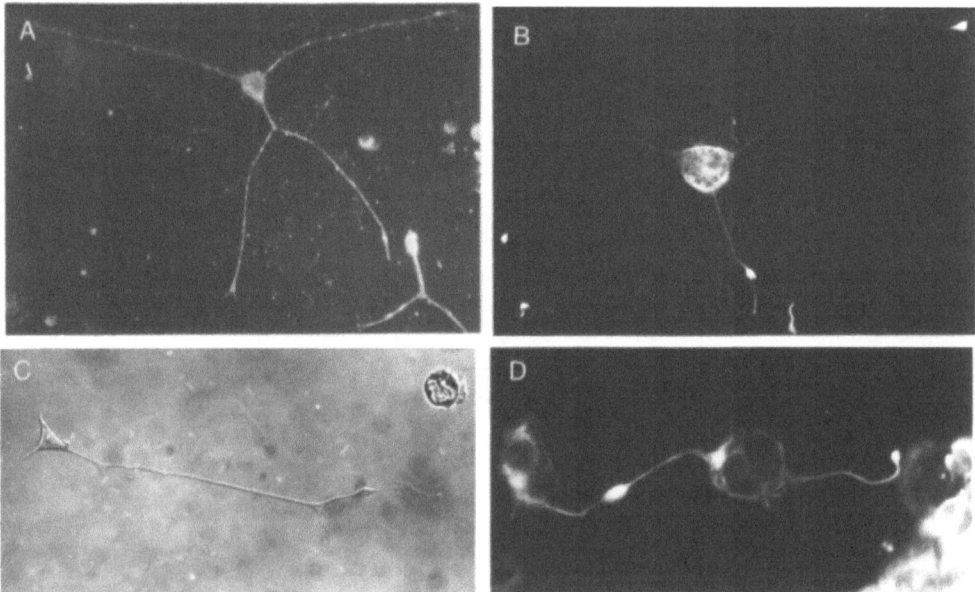

Fig. 4. Neuronal Features of Hybrid Cells in Culture. A. Phase contrast photomicrograph of a septum x neuroblastoma cell line. Note the elaboration of very long, thin neurites.. **B.** Indirect immunofluorescence staining of a postnatal day 21 septum x neuroblastoma cell line with anti-neurofilament monoclonal antibodies. **C.** Phase contrast photomicrograph of a hippocampus x neuroblastoma cell line. **D.** Indirect immunofluorescence staining of a hippocampus x neuroblastoma cell line with anti-neurofilament monoclonal antibodies. Cells were grown on untreated tissue culture plastic and then fixed with 2% formaldehyde in methanol. The cells were incubated with a primary antibody solution containing monoclonal antibodies to the low, middle, and high molecular weight neurofilaments proteins. Immunoreactivity was visualized with a fluorescein-conjugated secondary antibody.From Wainer et al. and Lee et al. (1989a; 1990a).

Septum x Neuroblastoma Cell Lines: Initially, we generated cell lines from embryonic septum (Hammond et.al.1986; Hammond et.al.1990b), and have since established techniques to generate cell lines from young adult animals (Lee et.al.1990b). Fig. 3 schematically illustrates the somatic cell fusion procedure. The hybrid nature of the cell lines has been documented by culturing in selective medium (i.e HAT), karyotyping, and

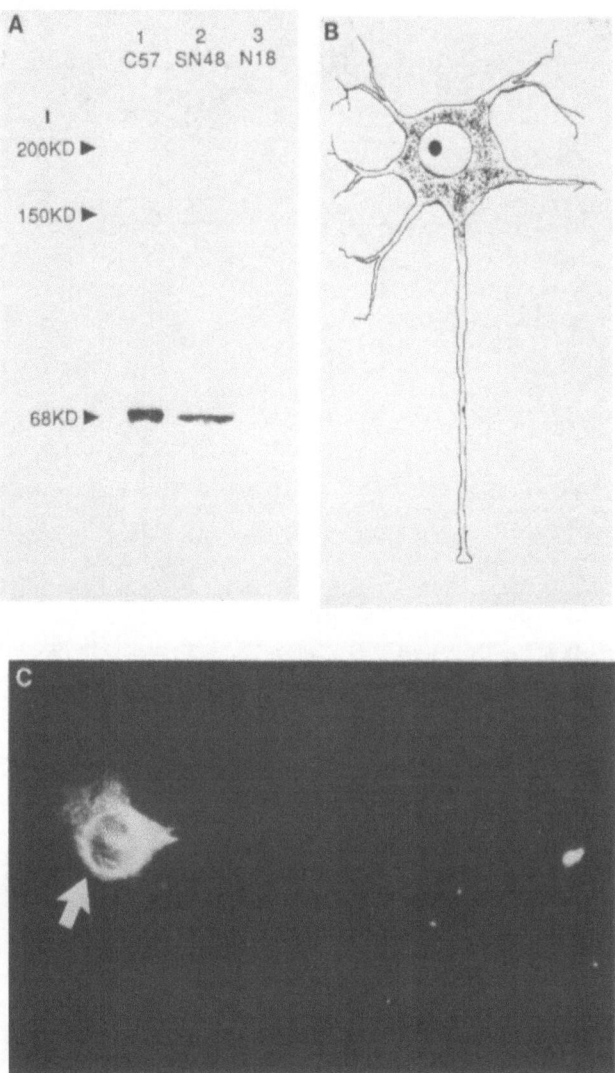

Fig. 5. Septal hybrid cell line: expression of the low molecular weight neurofilament protein (NF-L). **A.** Immunoblot analysis shows that a septal hybrid cell line (SN48.B12.2.C6; lane 2) synthesizes NF-L at a molecular size comparable to mouse spinal cord (lane 1). The parent neuroblastoma (N18TG2; lane 3) does not synthesize this protein. **B and C.** NF-L is preferentially distributed to the cell body of mature neurons (NF-L illustrated as stipling in B) and the septal hybrid cell line (NF-L localized by immunofluorescence in C). In the septal hybrid cell line, the protein is also present at low levels in processes (arrowhead) and high levels in growth cones. The SN48.B12.C6 cells were grown in the absence of differentiating agents and specialized substrates. Magnification = 1390x. From Lee et al. (1989).

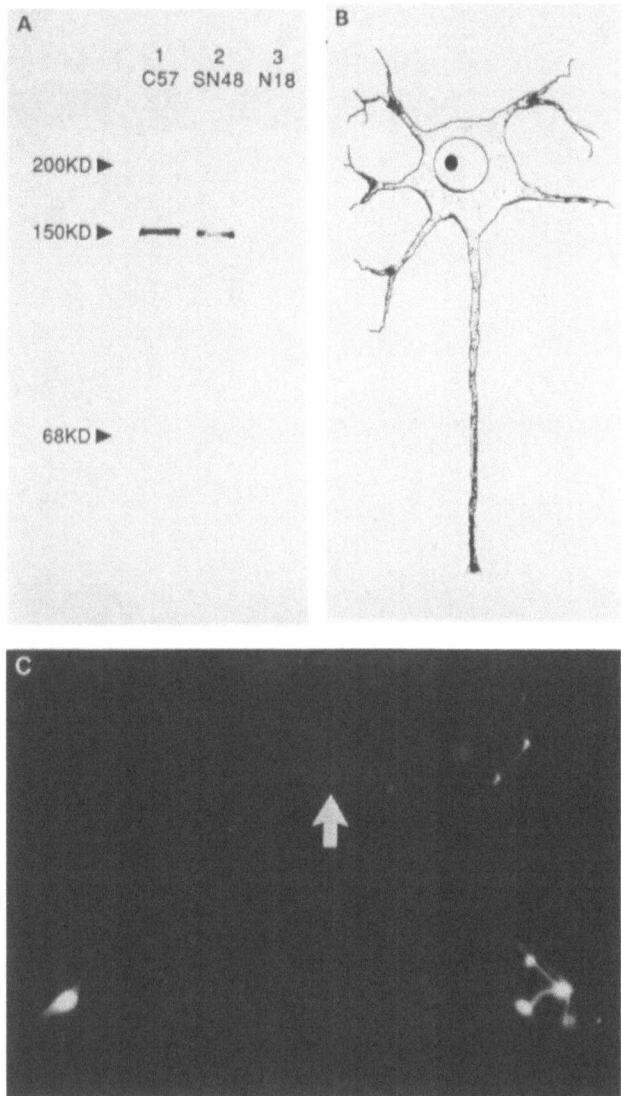

Fig. 6. Septal hybrid cell line: expression of the middle molecular weight neurofilament protein (NF-M) in its intermediately phosphorylated state [P++]. A. Immunoblot analysis shows that a septal hybrid cell line (SN48.B12.2.C6; lane 2) synthesizes NF-M[P++] at a molecular size comparable to mouse spinal cord (lane 1). The parent neuroblastoma (N18TG2; lane 3) does not synthesize this protein. **B and C.** NF-M [P++] is preferentially distributed to the distal processes of mature neurons (NF-M[P++] illustrated as stipling in B) and the septal hybrid cell line (NF-M[P++] localized by immunofluorescence in C). In the septal hybrid cell line, the protein is also present at low levels in the cell body (arrowhead) and proximal processes. The SN48.B12.C6 cells were grown in the absence of agents and specialized substrates. Magnification = 1390x. From Lee et al. (1989).

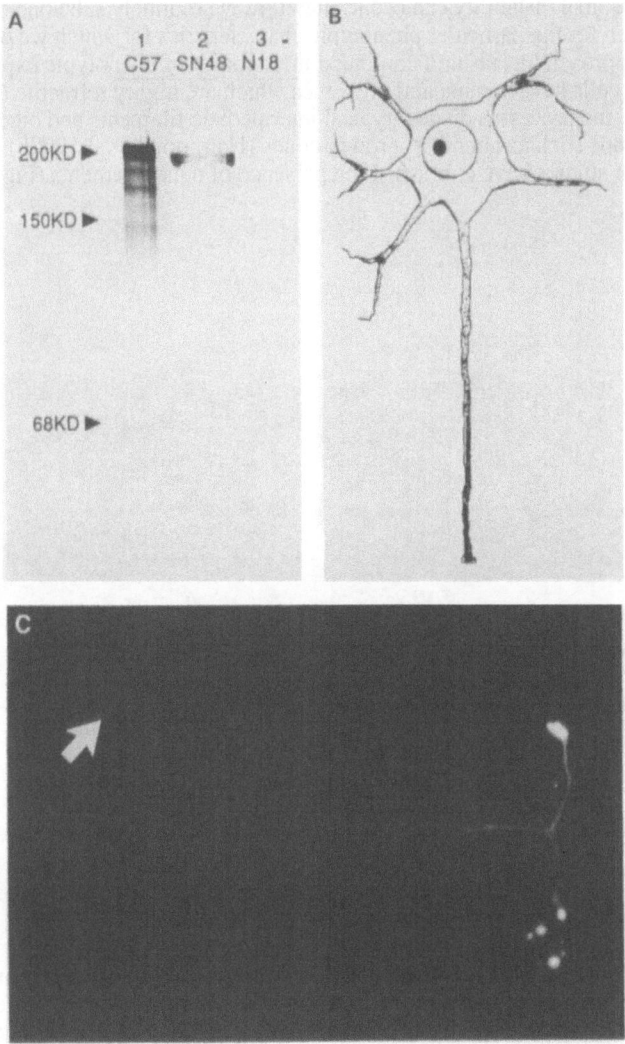

Fig. 7. Septal hybrid cell line: expression of the high molecular weight neurofilament protein (NF-H) in its most highly phosphorylated state [P+++]. A. Immunoblot analysis shows that a septal hybrid cell line (SN48.B12.2.C6; lane 2) synthesizes NF-H[P+++] at a molecular size comparable to mouse spinal cord (lane 1). The parent neuroblastoma (N18TG2; lane 3) does not synthesize this protein. **B and C.** NF-H[P+++] is preferentially distributed to the distal processes of mature neurons (NF-H[P+++] illustrated as stipling in B) and the septal hybrid cell line (NF-H[P+++] localized by immunofluorescence in C). Note the similarity of the distribution of NF-H[P+++] and NF-M[P++] (Appendix 8). In the septal hybrid cell line, the protein is also present at low levels in the cell body (arrowhead) and proximal processes. The SN48.B12.C6 cells were grown in the absence of differentiating agents and specialized substrates. Magnification = 1390x. From Lee et al. (1989).

isozyme analysis. To date, 60 septal hybrid lines (SN) have been generated from E14, 15, and 18 and P21 tissues. The neuronal phenotypes of these cell lines have remained stable under continuous culture conditions for over one year (tissue culture plastic, Dulbecco's modified Eagle medium (DMEM) supplemented with 10% fetal calf serum). As with any cell line, chromosomal instability can occur; therefore we routinely subclone cell lines of interest and screen for the particular phenotypic characteristics for which we originally selected. This approach has insured continued the stability of phenotypic expression in these lines. The cells have somata and processes which are highly refractile (Fig. 4A). Ultrastructurally, the processes display typical intermediate filaments and other features of developing neurons such as abundant growth cones (Hammond et.al.1990b). Immunochemical studies have confirmed the presence of neurofilaments (Fig. 4B), and

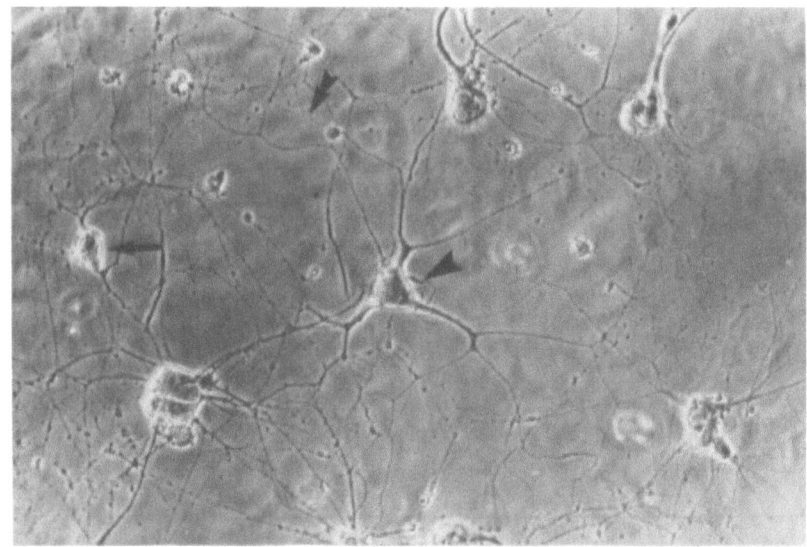

Fig. 8. Septal Hybrid Cell Line Co-Cultured with Primary Hippocampal Neurons. SN56.B5.G4 cells (arrowhead) morphologically differentiate and extend varicose processes (double arrowhead) toward E16 hippocampal target neurons (arrow) after 28 days in culture. Prior to culturing, both populations were labeled with different fluorescent markers and were easily identified at the end of the culture period (data not shown). Note that SN56.B5.G4 have not formed colonies and thus appear to be amitotic. Magnification = 400x.

demonstrate that the cytoskeletal features of some lines are sufficiently differentiated to express the low molecular weight neurofilament subunit as well as differentially phosphorylated isoforms of the middle and high molecular weight subunits, a cytoskeletal array characteristic of mature neurons (Figs. 5-7) (Lee et.al.1989; Carden et.al.1987). The neurofilament-positive septal lines are unique among other neuronal cell lines in that they 1) do not contain abnormal perinuclear aggregations of the low molecular weight neurofilament and 2) contain highly phosphorylated or "mature" neurofilament isoforms (Trojanowski,

1987). Lines cultured continuously for over 1 year still express all neurofilament subunit isoforms and, presumably, the various kinases and phosphatases required for regulation of the neurofilament phosphorylation state. Intermediate filaments characteristic of non-neuronal cells, such as cytokeratins or GFAP, are not detected in SN lines expressing neurofilaments (Lee et.al.1989; Wainer et.al.1989a). The cells also form aggregates under rotation-mediated culture conditions wheareas the N18TG2 neuroblastoma parent cell line does not (Hammond et.al.1986).

One important question relating to experimentally engineered cell lines is the extent to which they can interact with normal neuronal counterparts. Such studies are in progress with the hybrid cell lines previously generated. As illustrated in Fig. 8, septal cell lines can be co-cultured with primary hippocampal neurons and differentiate to an extent to which they are almost indistinguishable from the primary neurons. Although not shown in the Fig., the septal cell line was prelabeled with fluorescent microspheres to permit definitive identification. These cultures are currently being evaluated ultrastructurally to assess whether synaptogenesis occurs.

Several septal hybrid lines stably express neurochemical markers typical of septal cholinergic neurons such as ChAT and AChE (Hammond et.al.1986; Hammond et.al.1990b; Lee et.al.1990b; Blusztajn and Wurtman, 1983). In collaboration with Dr. J K. Blusztajn, we have determined that at least one postnatal septal line, SN56.B5.G4, exhibits three additional cholinergic properties: synthesis of ACh; high-affinity sodium-dependent choline uptake; and ACh release following depolarization with high potassium (unpublished observations). There have been no previous reports of cell lines which express all components of normal ACh turnover. A preliminary examination of the electrophysiological properties of SN6.10.2 has been performed in collaboration Drs. D.A. Brown and J. A. Sim (unpublished observations). This line expresses neurofilament proteins and high levels of ChAT activity. Following differentiation with retinoic acid, these cells exhibit resting membrane potentials of -30 to -60 mV. The cells also display sodium and potassium (K^+) currents necessary for generating action potential.

Septal cholinergic neurons express the NGF receptor and are responsive to NGF (Hefti et.al.1985; Hartikka and Hefti, 1988b; Hsiang et.al.1989). In preliminary studies selected septal lines were screened for either morphological or neurochemical (ChAT activity) responses to NGF(Hammond et.al.1990b; Lee et.al.1990b). These particular cell lines did not respond with any observable effects, however, it should be noted that the assays were performed under culture conditions that promote cell proliferation (i.e. 10% serum supplementation). More recently we have determined that some of the septal cell lines express NGF receptor mRNA as well as high and low affinity binding sites (H.C. Palfrey, unpublished observations). These studies have been carried out under conditions that promote cell differentiation. As cited above, a recent report by Tabira and co-workers has described trophic responses of septal cholinergic neurons to IL3 (Kamegai et.al.1990). In this study one of our septal cell lines, SN6.10.2.2 was evaluated and was found to respond to IL3 with significant elevations of ChAT activity. These findings strongly suggest that the existing septal cell lines will be quite useful for studying neuronal responses to neural trophic signals.

Hippocampus x Neuroblastoma Cell Lines To date 48 hippocampal hybrid cell lines (HN) have been generated from E18 and P21 tissues (Lee et.al.1990a). Many of these lines exhibit neuronal morphologies (Fig. 4C) and cytoskeletal features (Fig 4D), as well as excitable properties typical of hippocampal neurons (Fig. 9). HN25.2G10 also expresses the middle and high molecular weight isoforms of neurofilament proteins. Several HN lines express high levels of NGF mRNA and NGF protein (Fig 10). The N18TG2 neuroblastoma expresses neither NGF protein nor mRNA. Several hippocampal cell lines were additionally screened for muscarinic receptors. A radioligand binding assay for M_1-type receptors was employed utilizing N-methyl scopolamine and pirenzepine as ligands. HN33 expressed 8 femtomoles of M_1-type receptors per million cells or 4800 receptors per cell; 5 times as many as N18TG2 neuroblastoma cells and comparable to the number found on NGF-activated PC12 cells (unpublished observations). RNA extracts from these cell lines are presently being probed for the presence of specific muscarinic receptor subtype mRNA species. In monolayer co-cultures with primary septal cells, hippocampal cell lines differentiate and elaborate processes that appear to contact their primary septal counterparts. Ultrastructural studies to confirm the presence of morphological identifiable synapses are in progress. When cultured under rotation conditions, HN10, HN25.2.G10, and HN25.2.C11 cells each form aggregates similar in size and shape to those formed by primary hippocampal cells, and they also coaggregate with primary septal cells. These HN lines thus express the molecules required for the cell-cell interactions involved in aggregate formation.

<u>Trophic Activities of HN Cell Lines</u>: The production of NGF by these cell lines suggested that they may express non-NGF trophic factors. We thus developed a microculture bioassay to screen either conditioned media or subcellular fractions of hippocampal cell lines for the presence of molecules which affect cholinergic neurons (Lee et.al.1990a). Studies employing this bioassay system indicate that it is sensitive to effects of NGF in a dose-response fashion (Wainer et.al.1989a). We have screened 22 HN cell lines, as well as

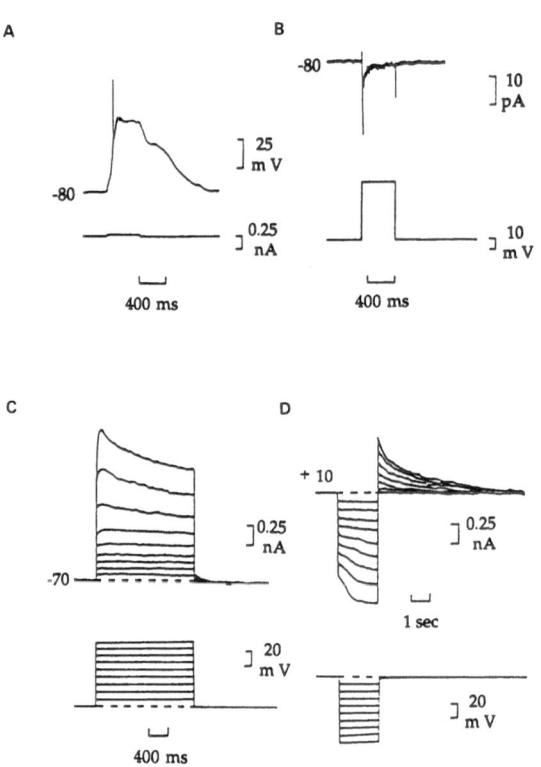

Fig. 9. **Membrane Properties and Ionic Conductances expressed in Retinoic Acid Differentiated Hippocampal Hybrid Cells. A.** An action potential and depolarizing afterpotential(upper trace) evoked by a depolarizing current injection (lower trace) and **B.**, an inward current recorded under voltage-clamp (upper trace) induced by a depolarizing voltage step (lower trace) in the same cell held at -80 mV. **C** and **D.** A series of current (top traces) and voltage (lower traces) recorded in a voltage-clamped hybrid cell held at -70 mV and +10 mV, respectively. **C.** Sustained and depolarizing voltage steps from a holding potential of -70 mV while in **D.**, inactivating outward currents were recorded from +10 mV following 2 second prepulses to various potentials between -20 and -100 mV. Note also in **D** the activation of a time-dependent inwardly rectifying current with cell hyperpolarization. From Lee et al. (1990a).

N18TG2 cells. Most of the lines have little effect (less than 75% increase in ChAT). However, conditioned medium from one line, HN10, promotes a 4-fold increase in ChAT activity (Fig. 11). We have found more recently that the HN10 bioactivity can be salt-extracted from a membrane preparation with a 5-6 fold increase in specific activity (Hammond et.al.1989; Hammond et.al.1990a). We have completed preliminary studies of the HN10 activity which show the following: 1) the addition of the anti-NGF monoclonal antibody, 23C4, to primary septal cells blocks the effect of exogenous NGF, but not the effect of the HN10 membrane extract (Wainer et.al.1989a); 2) epidermal growth factor (EGF) or basic fibroblast growth factor (bFGF) which have been shown to be neurotrophic factors are not likely to be responsible for the HN10 effect (Fig. 11) (Lee et.al.1990a); 3) loss of virtually all HN10 activity occurs when the membrane extract is heated to 90°C for 20 minutes; and 4) loss of activity also occurs when the HN10 membrane extract is prepared in the absence of protease inhibitors. In preliminary unpublished work, gel filtration studies reveal a sharp band of activity in the approximate molecular size range of 40-44 Kd and in a

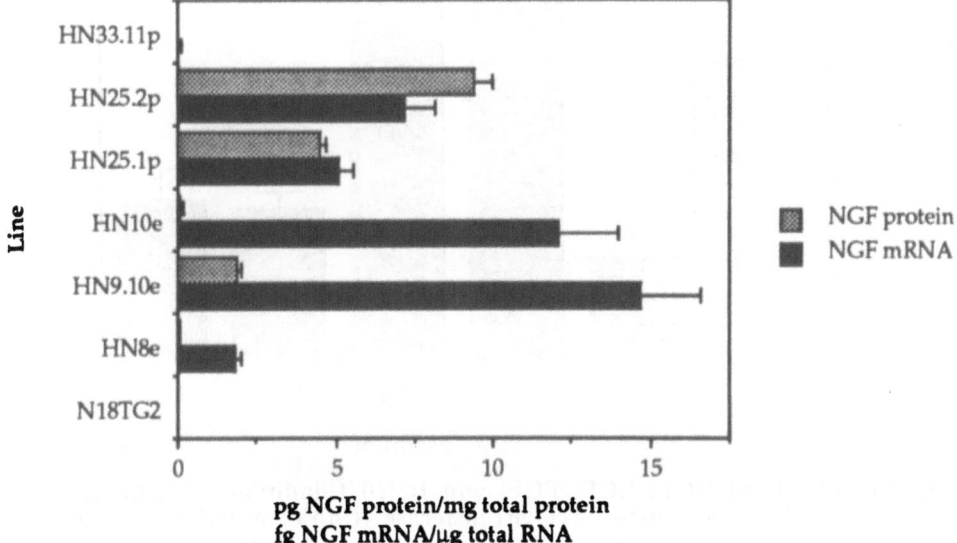

Fig. 10. **NGF Protein and mRNA Content of Hippocampus x Neuroblastoma (HN) cell lines.** HN cell lines were grown in 10% fetal calf serum in DMEM, harvested, and assayed. NGF protein was measured by a 2-site ELISA specific for ß-NGF. NGF mRNA was quantitated by northern blot analysis. The DNA probe was prepared from the 0.9 kb *Pst* I fragment of the NGF cDNA comprising a portion of the coding sequence as well as part of the 5' untranslated region. From Lee et al. (Lee et.al.1990a).

region of the chromatogram that is relatively low in protein content. In addition, the activity can be released from crude membrane fractions with a marked enhancement of specific activity by treatment with low concentrations of trypsin (10 µg/ml). These findings are consistent with the possibility that intact protein(s) or protein domain(s) are required for the HN10 activity, and with the possibility that the activity resides specifically in an extramembranous domain. These observations are preliminary and considerable work will be required to determine if this HN10 activity is not attributable to a previously identified growth factor and to further characterize it. However, the fact that the activity has been identified in a clonal cell line makes such studies more feasible because of the unlimited source of starting material and the clonal nature of the cells expressing it.

CONCLUSIONS AND FUTURE DIRECTIONS

The results of the studies described above strongly suggest that the trophic interactions that influence both the development and maintenance of cholinergic basal forebrain neurons are likely to be mediated by more than one macromolecule such as NGF. The strategy

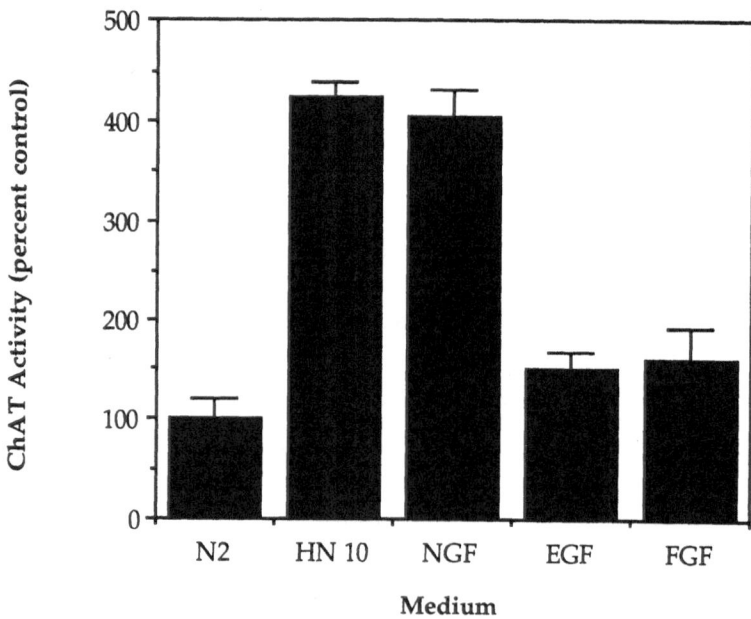

Fig. 11. Effects of NGF, EGF, FGF, and HN10 Conditioned Media on Septal Cell Choline Acetyltransferase (ChAT) Activity. Primary septal cells were cultured in the presence of N2 medium only (control), in the presence of NGF at 100 ng/ml, in the presence of epidermal growth factor (EGF) at 1 nM final concentration, in the presence of basic fibroblast growth factor (FGF) at 1 nM final concentration, and in the presence of 100 μg/ml of HN10 conditioned medium. The ChAT activity of the septal cells treated with each of the various reagents is expressed as percent of control (N2 medium alone). HN10 conditioned medium exhibited a significant increase in ChAT activity compared to the control medium (p = 0.001). NGF resulted in an increase in ChAT approximately (p = 0.001) equal to that of the HN10 conditioned medium. EGF and FGF did not have a significant effect on ChAT activity. From Lee et al. (1990a).

outlined for the generation of region-specific clonal cell lines represents a powerful approach for elucidating the molecular basis of such interactions. In addition, similar strategies can be applied to study the trophic mechanisms sustaining neural connections in other neural pathways. Ultimately, this research may lead to meaningful and potentially useful therapeutic approaches in AD and other neurological diseases, involving trophic factor therapy or neural grafting strategies. Cell lines expressing appropriate trophic factors may provide a ready source of this material. Importantly, such a therapeutic approach would promote the survival and normal function of neurons, rather than simply ameliorate symptoms as is the case with approaches which substitute exogenous compounds for a deficient neurotransmitter. Beneficial effects of trophic factor therapy on cell survival and function may occur even if factor deficit is not the primary cause of disease.

ACKNOWLEDGEMENTS

This research is supported by PHS **RO1 NS25787,** The Alzheimer's Disease and Related Disorders Association, The Brain Research Foundation of The University of Chicago (BHW); PHS **5 KO8** NS01244, The French Foundation (DNH); and PHS **5 T32 HD070009** (HJL & JDR). The authors express gratitude for the excellent technical support of Ms. L. Sherman, and Mr. S.D. Price.

REFERENCES

Alt, F., Blackwell, K. and Yancopoulos, G.D. (1987) Development of the primary antibody repertoire. Science 238:1079.

Anderson, K.J., Dam, D., Lee, S. and Cotman, C.W. (1988) Basic fibroblast growth factor prevents death of lesioned cholinergic neurons in vivo. Nature 332:360.

Appel, S.H. (1981) A unifying hypothesis for the cause of amyotrophic lateral sclerosis, Parkinsonism, and Alzheimer disease. Ann. Neurol. 10:499.

Banker, G. and Goslin, K. (1988) Developments in neuronal cell culture. Nature. 336:185.

Barbin, G., Manthorpe, M. and Varon, S. (1984) Purification of the chick eye ciliary neuronotrophic factor. J. Neurochem. 43:1468.

Becker, R.E. and Giacobini, E. (1988) Mechanisms of cholinesterase inhibition in senile dementia of the Alzheimer type: Clinical, pharmacological, and therapeutic aspects. Drug Dev. Res. 12:163.

Blusztajn, J.K. and Wurtman, R.J. (1983) Choline and cholinergic neurons. Science 221:614.

Bostwick, J.R., Appel, S.H. and Perez-Polo, J.R. (1987) Distinct influences of nerve growth factor and a central cholinergic trophic factor on medial septal explants. Brain Res. 422:92.

Bottenstein, J.E. (1985) Growth and differentiation of neural cells in defined media. In:" Cell Culture in the Neurosciences."eds:Bottenstein, J.E. and Sato, G.,Plenum Press, New York, pp. 3.

Carden, M.J., Trojanowski, J.Q., Schlaepfer, W.W. and Lee, V.M.-Y. (1987) Two-stage expression of neurofilament polypeptides during rat neurogenesis with early establishment of adult phosphorylation patterns. J. Neurosci. 7:3489.

Cepko, C. (1988) Immortalization of neural cells via oncogene transduction. Trends Neurosci. 11:6.

Cepko, C.L. (1989) Immortalization of neural cells via retroviral-mediated oncogene transduction. Ann. Rev. Neurosci. 12:47.

Chan Palay, V. (1988) Galanin hyperinnervates surviving neurons of the human basal nucleus of Meynert in dementias of Alzheimer's and Parkinson's disease: a hypothesis for the role of galanin in accentuating cholinergic dysfunction in dementia. J Comp.Neurol. 273:543.

Davies, P., Katzman, R. and Terry, R.D. (1980) Reduced somatostatin-like immunoreactivity in cerebral cortex from cases of Alzheimer disease and Alzheimer senile dementia. Nature 288:279.

DeLellis, R.A., Merk, F.B., Deckers, P., Warren, S. and Balogh, K. (1973) Ultrastructure and in vitro growth characteristics of a transplantable rat pheochromocytoma. Cancer 32:227.

Emerit, M.B., Segovia, J., Alho, H., Mastrangelo, M.J. and Wise, B.C. (1989) Hippocampal membranes contain a neurotrophic activity that stimulates cholinergic properties of fetal rat septal neurons cultered under serum-free conditions. J. Neurochem. 52:952.

Frederiksen, K., Jat, P.S., Valtz, N., Levy, D. and McKay, R. (1988) Immortalization of precursor cells from the mammalian cns. Neuron 1:439.

Gahwiler, B.H., Enz, A. and Hefti, F. (1987) Nerve growth factor promotes development of the rat septo-hippocampal cholinergic projection in vitro. Neurosci.Lett. 75:6.

Garber, B.B. and Moscona, A.A. (1972) Reconstruction of brain tissue from cell suspensions. I. Aggregation of patterns of cells dissociated from different regions of the developing brain. Dev. Biol. 27:217.

Gnahn, H., Hefti, F., Heumann, R., Schwab, M.E. and Thoenen, H. (1983) NGF-mediated increase of choline acetyltransferase (ChAT) in the neonatal forebrain: Evidence for a physiological role of NGF in the brain?. Dev. Br. Res. 9:45.

Guroff, G. (1985) PC12 Cells as a model of neuronal differentiation. In:" Cell Culture in the Neurosciences." eds:Bottenstein, J.E. and Sato, G., Plenum Press, New York, pp. 245-272.

Hammond, D.N., Wainer, B.H., Tonsgard, J.H. and Heller, A. (1986) Neuronal properties of clonal hybrid cell lines derived from central cholinergic neurons. Science 234:1237.

Hammond, D.N., Lee, H.J. and Wainer, B.H. (1989) A membrane-associated factor influences central cholinergic neuron development. Ann. Neurol. 26:445.

Hammond, D.N., Lee, H.J. and Wainer, B.H. (1990a) A hippocampal cell line expresses a membrane-associated cholinergoc trophic activity not attributable to nerve growth factor. In:" Alzheimer's and Parkinson's Disease II: Basic and therapeutic strategies." eds:Nagatsu, T., Fisher, A. and Yoshida, M.,Plenum Press, New York, in press.

Hammond, D.N., Lee, H.J., Tonsgard, J.H. and Wainer, B.H. (1990b) Development and characterization of clonal cell lines from septal cholinergic neurons. Brain Res. 512:190.

Hartikka, J. and Hefti, F. (1988a) Comparison of nerve growth factor's effects on development of septum,striatum, and nucleus basalis cholinergic neurons in vitro. J. Neurosci. Res. 21:352.

Hartikka, J. and Hefti, F. (1988b) Development of septal cholinergic neurons in culture: Plating density and glial cells modulate effects of NGF on survival, fiber growth, and expression of transmitter-specific enyzmes. J. Neurosci. 8:2967.

Hefti, F., Hartikka, J., Eckenstein, F., Gnahn, H., Heumann, R. and Schwab, M. (1985) Nerve growth factor increases choline acetyltransferase but not survival or fiber outgrowth of cultured fetal septal cholinergic neurons. Neuroscience 14:55.

Hefti, F. and Weiner, W.J. (1986) Nerve growth factor and Alzheimer's Disease. Ann. Neurol. 20:275.

Hefti, F., Hartikka, J. and Knusel, B. (1989) Function of neurotrophic factors in the adult and aging brain and their possible use in the treatment of neurodegenerative diseases. Neurobiol.Aging 10:515.

Hofer, M.M. and Barde, Y.-A. (1988) Brain-derived neurotrophic factor prevents neuronal death in vivo. Nature 331:261.

Hoffmann, P.C., Hemmendinger, L.M., Kotake, C. and Heller, A. (1983) Enhanced dopamine cell survival in reaggregates containing telencephalic target cells. Brain Res. 274:275.

Hohn, A., Leibrock, J., bailey, k. and Barde, Y.-A. (1990) Identification and characterization of a novel member of the nerve growth factor/brain-derived neurotrophic factor family. Nature 344:339.

Hsiang, J., Wainer, B.H., Shalaby, I.A., Hoffmann, P.C., Heller, A. and Heller, B.R. (1987) Neurotrophic effects of hippocampal target cells on developing septal cholinergic neurons in culture. Neuroscience 21:333.

Hsiang, J., Price, S.D., Heller, A., Hoffmann, P.C. and Wainer, B.H. (1988) Ultrastructural evidence for hippocampal target cell-mediated trophic effects on septal cholinergic neurons in reaggregating cell cultures. Neuroscience 26:417.

Hsiang, J., Heller, A., Hoffmann, P.C., Mobley, W.C. and Wainer, B.H. (1989) The effects of nerve growth factor on the development of septal cholinergic neurons in reaggregate cell cultures. Neuroscience 29:209.

Johnson, E.M.Jr., Gorin, P.D., Brandeis, L.D. and Pearson, J. (1980) Dorsal root ganglion neurons are destroyed by in utero exposure to maternal antibody to nerve growth factor. Science 210:916.

Johnson, J.E., Barde, Y.-A., Schwab, M. and Thoenen, H. (1990) Brain-derived neurotrophic factor supports the survival of cultured rat retinal ganglion cells. J. Neurosci. 6:3031.

Kamegai, M., Niijima, K., Kunishita, T., Nishizawa, M., Ogawa, M., Araki, M., Ueki, A., Konishi, Y. and Tabira, T. (1990) Interleukin 3 as a trophic factor for central cholinergic neurons in vitro and in vivo. Neuron 4:429.

Keller, F., Rimvall, K., Barbe, M.F. and Levitt, P. (1989) A membrane glycoprotein associated with the limbic system mediates formation of the septo-hippocampal pathway in vitro. Neuron 3:551.

Knusel, B., Michel, P.P., Schwaber, J.S. and Hefti, F. (1990) Selective and nonselective stimulation of central cholinergic and dopaminergic development in vitro by nerve growth factor, basic fibroblast growth factor, epidermal growth factor, insulin and the insulin-like growth factors I and II. J. Neurosci. 10:558.

Kohler, G. and Milstein, C. (1975) Continuous cultures of fused cells secreting antibody of defined specifity. Nature 256:495.

Lee, H.J., Elliot, G.J., Hammond, D.N., Lee, V.M.-Y. and Wainer, B.H. (1989) Expression of the mature array of neurofilament protein isoforms by a clonal cell line from the cns. Soc. Neurosci. Abs. 15:327.

Lee, H.J., Hammond, D.N., Large, T.H., Sim, J.A., Brown, D.A., Otten, U.H. and Wainer, B.H. (1990a) Neuronal properties and trophic activities of immortalized hippocampal cells from embryonic and young adult mice. J. Neurosci. 10:1779.

Lee, H.J., Hammond, D.N., Large, T.H. and Wainer, B.H. (1990b) Immortalized young adult cholinegic neurons from the medial septal region: Production and characterization. Dev. Brain Res. 52:219.

Lendahl, U. and McKay, R.D.G. (1990) The use of cell lines in neurobiology. Trends Neurosci. 13:132.

Levi-Montalcini, R. and Booker, B. (1960) Destruction of the sympathetic ganglia in mammals by an antiserum to a nerve-growth protein. Proc. Natl. Acad. Sci. USA 46:384.

Levi-Montalcini, R. (1987) The nerve growth factor 35 years later. Science 237:1154.

Lillien, L.E., Sendtner, M., Rohrer, H., Hughes, S.M. and Raff, M.C. (1988) Type-2 astrocyte development in rat brain cultures is initiated by a CNTF-like protein produced by type-1 astrocytes. Neuron 1:485.

Lindholm, D., Heumann, R., Meyer, M. and Thoenen, H. (1987) Interleukin-1 regulates synthesis of nerve growth factor in non-neuronal cells of rat sciatic nerve. Nature 330:658.

Maisonpierre, P.C., Belluscio, L., Squinto, S., Ip, N.Y., Furth, M.E., Lindsay, R.M. and Yancopoulos, G.D. (1990) Neurotrophin-3: A neurotrophic factor related to NGF and BDNF. Science 247:1446.

Marx, J. (1990) NGF and Alzheimer's: Hopes and Fears. Science 247:408.

Meck, W.H., Church, R.M., Wenk, G.L. and Olton, D.S. (1987) Nucleus basalis magnocellularis and medial septal area lesions differentially impair temporal memory. J Neurosci. 7:3505.

Mobley, W.C., Rutkowski, J.L., Tennekoon, G.I., Buchanan, K. and Johnston, M.V. (1985) Choline acetyltransferase activity in striatum of neonatal rats increased by nerve growth factor. Science 229:284.

Morris, J.E. and Moscona, A.A. (1970) Induction of glutamine synthetase in embryonic retina: Its dependence on cell interactions. Science 167:1736.

Morrison, R.S., Kornblum, H.I., Leslie, F.M. and Bradshaw, R.A. (1987) Trophic stimulation of cultured neurons from neonatal rat brain by EGF. Science 238:72.

Ojika, K. and Appel, S.H. (1984) Neurotrophic effects of hippocampal extracts on medial septal nucleus in vitro. Proc. Natl. Acad. Sci. USA 81:2567.

Paige, C.J. and Wu, G.E. (1989) The B cell repertoire. FASEB J. 3:1818.

Phelps, C.H., Gage, F.H., Growdon, J.H., Hefti, F., Harbaugh, R., Johnston, M.V., Khachaturian, Z.S., Mobley, W.C., Price, D.L., Raskind, M., Simpkins, J., Thal, L.J. and Woodcock, J. (1989) Potential use of nerve growth factor to treat Alzheimer's Disease. Neurobiol. Aging 10:205.

Piddington, R. and Moscona, A.A. (1967) Precocious induction of retinal glutamine synthetase by hydrocortisone in the embryo and in culture. Age-dependent differences in tissue response. Biochim. Biophys. Acta 141:429.

Price, D.L. (1986) New perspectives on Alzheimer's disease. Annu.Rev.Neurosci. 9:489.

Prince, D.A. and Connors, B.W. (1986) Mechanisms of interictal epileptogenesis. In:" Basic Mechanisms of the Epilepsies, Molecular and Cellular Approaches: Advances in Neurology Vol. 44." eds:Delgado-Escueta, A.V., Woodbury, D.M., Ward Jr., A.A. and Porter, R.J., Raven Press, New York, pp. 275-299.

Prochiantz, A., diPorzio, U., Kato, A., Berger, B. and Glowinski, J. (1979) In vitro maturation of mesencephalic dopaminergic neurons from mouse embryos is enhanced in the presence of their striatal target cells. Proc. Natl. Acad. Sci. USA 76:5387.

Purves, D. (1988)"Body and Brain: A Trophic Theory of Neural Connections." Harvard University Press, Cambridge, Massachusetts.

Roback, J.D., Large, T.H., Otten, U. and Wainer, B.H. (1990) Nerve growth factor expression in the isolated hippocampus in vitro. Dev. Biol. 137:451.

Ronnett, G.V., Hester, L.D., Nye, J.S., Connors, K. and Snyder, S.H. (1990) Human cortical neuronal cell line: Establishment from a patient with unilateral megalencephaly. Science 248:603.

Rosenberg, M.B., Friedmann, T., Robertson, R.C., Tuszynski, M., Wolff, J.A., Breakefield, X.O. and Gage, F.H. (1988) Grafting genetically modified cells to the damaged brain restorative effects of ngf expression. Science 242:1575.

Rye, D.B., Wainer, B.H., Mesulam, M.-M., Mufson, E.J. and Saper, C.B. (1984) Cortical projections arising from the basal forebrain: a study of cholinergic and noncholinergic components employing combined retrograde tracing and immunohistochemical localization of choline acetyltransferase. Neuroscience 13:627.

Rye, D.B., Saper, C.B., Lee, H.J. and Wainer, B.H. (1987) Pedunculopontine tegmental nucleus of the rat: cytoarchitecture, cytochemistry, and some extrapyramidal connections of the mesopontine tegmentum. J Comp.Neurol. 259:483.

Saper, C.B., Wainer, B.H. and German, D.C. (1987) Axonal and transneuronal transport in the transmission of neurological disease: potential role in system degenerations, including Alzheimer's disease. Neuroscience 23:389.

Saper, C.B. (1988) Chemical neuroanatomy of Alzheimer's disease. In:" Handbook of Psychopharmacology Vol. 20." eds:Iversen, L.L., Iversen, S.D. and Snyder, S.H., Plenum, New York, pp. 131-156.

Snider, W.D. and Johnson, E.M.Jr. (1989) Neurotrophic molecules. Ann.Neurol. 26:489.

Springer, J.E. and Loy, R. (1985) Intrahippocampal injections of antiserum to nerve growth factor inhibit sympathohippocampal sprouting. Brain Res.Bull. 15:629.

Thoenen, H., Bandtlow, C. and Heumann, R. (1987) The physiological function of nerve growth factor in the central nervous system: comparison with the periphery. Rev. Physiol. Biochem. Pharmacol. 109:146.

Tishler, A.S. and Greene, L.A. (1975) Nerve growth factor-induced process formation by cultured rat pheochromocytoma cells. Nature 258:341.

Trojanowski, J.Q. (1987) Neurofilament proteins and human nervous system tumors. J. Histochem. Cytochem. 35:999.

Vantini, G., Schiavo, N., Di Martino, A., Polato, P., Triban, C., Callegaro, L., Toffano, G. and Leon, A. (1989) Evidence for a physiological role of nerve growth factor in the central nervous system of neonatal rats. Neuron 3:267.

Wainer, B.H., Levey, A.I., Mufson, E.J. and Mesulam, M.-M. (1984) Cholinergic systems in mammalian brain identified with antibodies against choline acetyltransferase. Neurochem. Int. 6:163.

Wainer, B.H., Levey, A.I., Rye, D.B., Mesulam, M.-M. and Mufson, E.J. (1985) Cholinergic and non-cholinergic septohippocampal pathways. Neurosci.Lett. 54:45.

Wainer, B.H., Lee, H.J., Roback, J.D. and Hammond, D.N. (1989a) The use of reaggregating cell cultures and immortalized central nervous system cells to study cholinergic trophic mechanisms. In:" Novel Approaches to the Treatment of Alzheimer's Disease." eds:Meyer, E.M. and Simpkins, J., Plenum Press, New York, pp. 71-94.

Wainer, B.H. (1989b) Neurotrophic factors and Alzheimer's disease. Neurobiol. Aging 10:540.

Wainer, B.H. and Mesulam, M.-M. (1990) Ascending cholinergic pathways in the rat brain. In:" Brain Cholinergic Systems." eds:Steriade, M. and Biesold, D., Oxford University Press, New York, in press.

Walicke, P.A. (1988) Basic and acidic fibroblast growth factors have trophic effects on neurons from multiple cns regions. J. Neurosci. 8:2618.

Walicke, P.A. (1989) Novel neurotrophic factors, receptors, and oncogenes. Ann. Rev. Neurosci. 12:103.

Warren, S. and Chute, R.N. (1972) Pheochromocytoma. Cancer 29:327.

Werner, M.H., Nanney, L.B., Stoscheck, C.M. and King, L.E. (1988) Localization of immunoreactive EGF receptors in human nervous system. J. Histochem. Cytochem. 36:81.

Whittemore, S.R. and Seiger, A. (1987) The expression, localization and functional significance of ß-nerve growth factor in the central nervous system. Brain Res. Rev. 12:439.

Whittemore, S.R., Holets V.R., Gonzalez-Carvajal, M. and Levy, D. (1988) Transplantation of a hippocampally-derived immortal, temperature-sensitive, neuronal cell line into adult rat cns. Soc. Neurosci. Abs. 14:586.

Williams, L.R., Varon, S., Peterson, G.M., Wictorin, K., Fischer, W., Bjorklund, A. and Gage, F.H. (1986) Continuous infusion of nerve growth factor prevents basal forebrain neuronal death after fimbria fornix transection. Proc.Natl.Acad.Sci.USA. 83:9231.

Zola Morgan, S., Squire, L.R. and Amaral, D.G. (1986) Human amnesia and the medial temporal region: enduring memory impairment following a bilateral lesion limited to field CA1 of the hippocampus. J Neurosci. 6:2950.

THE EPIDEMIOLOGY OF DEMENTIA IN PATIENTS

WITH PARKINSON'S DISEASE

Karen Marder and Richard Mayeux

Departments of Neurology and Psychiatry
College of Physicians and Surgeons
Columbia University, New York, New York

Though in his 1817 monograph on the 'Shaking Palsy', James Parkinson
believed in "the absence of any injury to the senses and to the
intellect", it is now widely accepted that dementia is often seen in
idiopathic Parkinson's disease. Prevalence estimates of cognitive
impairment in Parkinson's disease have ranged from under 10% to over 80%
(Brown and Marsden, 1984), attesting to the variable methodologies
employed and populations studied.

Determination of the prevalence of dementia, the number of people
with Parkinson's disease who are demented at a given point in time is
critical for both practical and theoretical reasons. Demented patients
with Parkinson's disease are more likely to develop adverse side effects
to the medications used in the treatment of Parkinson's disease.
Anticholinergics are more likely to cause confusion, delirium, and
drug-induced hallucinations in demented Parkinson's patients. Levodopa
and other dopamine agonists are capable of causing hallucinations and
delirium, particularly in demented patients. No medications have been
causally implicated in the development of dementia. There is also a
suggestion that the development of dementia may decrease survival
(Mindham et al., 1982; Mayeux et al., 1990).

If there truly is an association between idiopathic Parkinson's
disease and dementia, several intriguing questions can be posed. The
most central question is whether the dementia is unique to Parkinson's
disease or whether it is due to coexistent Alzheimer's disease. If the
dementia is due to Alzheimer's disease is this a chance association? Or,
is Parkinson's disease a form of Alzheimer's disease or visa versa? If
dementia is more common in the setting of Parkinson's disease than in an
age-matched population without Parkinson's disease, does Parkinson's
disease in some way predispose to the development of dementia? Perhaps
dementia is part of the natural history of idiopathic Parkinson's disease
and will occur as an inevitable consequence in all patients at a certain
time? Lastly, are the risk factors for the development of dementia
genetic or environmental, and are they similar to those known for
Alzheimer's disease?

In this review, the epidemiology of dementia in Parkinson's disease
will be discussed in terms of the prevalence, incidence and risk factors

The Basal Forebrain, Edited by T.C. Napier *et al.*
Plenum Press, New York, 1991

Table 1. Prevalence of Dementia in Parkinson's Disease

INVESTIGATOR	YEAR	NUMBER PATIENTS	% WITH DEMENTIA	SAMPLE
Lewy	1923	70	64	?
Mjones	1949	194	40	population survey
Pollock & Hornabrook	1966	131	20	population survey
Celesia & Wanamaker	1972	153	40	inpatient
Martin et al.	1973	100	81	?
Marttila & Rinne	1976	444	29	population survey
Lieberman et al.	1979	520	32	outpatient
Sroka et al.	1981	93	28	inpatient
Mindham et al.	1982	40	20	outpatient
Lees et al.	1985	48	15	outpatient
Sutcliffe	1985	226	11.9	population survey
Taylor et al.	1985	100	8	outpatient
Rajput et al.	1987	118	6	case records
Mayeux et al.	1988	339	10.9	hospital based
Girotti et al.	1988	147	14	outpatient

for dementia in Parkinson's disease. Methodological problems in these studies will be enumerated, and suggestions on further avenues of investigation will be offered.

PREVALENCE OF DEMENTIA IN PARKINSON'S DISEASE

Numerous prevalence surveys have attempted to answer the question of how often dementia occurs in idiopathic Parkinson's disease. The prevalence rate is the proportion of a defined group having a condition at any time within a stated period. In this case, prevalence surveys were conducted to determine the number of Parkinson's patients who were demented among all patients with Parkinson's disease over a certain time period. The wide range of estimates (Table 1) stems from a lack of a uniform definition of dementia. The DSM-III R definition of dementia, formulated to describe patients with Alzheimer's disease has been adopted by many investigators as a reasonable, working definition of dementia. Without a biological marker to identify demented subjects or an accurate means of determining disease onset since the patient may be incapable of providing any historical information, a widely recognized definition incorporating some measure of disease severity is imperative. The DSM-III R definition requires a "loss of intellectual abilities of sufficient severity to interfere with social or occupational functioning", and stipulates impairment in memory and either impairment in abstract thinking, judgement, disturbance in higher cortical function, or personality change. (DSM III-R APA 1987). This definition, though perhaps the best available, is difficult to apply to Parkinson's patients whose profound physical disabilities limit their social and occupational function to a great extent even before the development of cognitive impairment. In addition, impaired performance on visuospatial and perceptual motor tasks, and difficulty in planning and modulating consecutive motor behavior are well described in Parkinsonian patients who are not demented. There is a continuum of cognitive decline and the definition of dementia must be clearly delineated.

The problems inherent in the application of different definitions of dementia to prevalence estimates are illustrated in the studies by Martin et al. (1973) and by Mjones (1949). Martin and coinvestigators (1973), using a combination of an interview and a mental state exam, found that

81 of 100 patients with idiopathic Parkinson's disease were demented. Mild impairment, that would probably not meet DSM-III R criteria for dementia was seen in 58% of the sample, leaving 23% with a moderate to severe dementia. Mjones (1949) included all forms of "mental change" which he labeled reactive or organic in his definition of dementia. Since depression can occur in 20 to 90% of patients with Parkinson's disease (Patrick and Levy, 1922; Mindham, 1970), reactive cases of "mental changes" should probably have been eliminated from the 40% estimate of dementia. Marttila and Rinne (1976), Lieberman et al. (1979), and Celesia and Wanamaker (1972) all used a similar definition of dementia and determined prevalence rates of 29%, 32% and 40% respectively. Each of these investigators classified dementia by stages. Stage 1, mild slowing with impairment in abstract reasoning, concept formation and calculation, would not satisfy current DSM-III-R criteria for dementia. If the prevalence were recalculated excluding stage 1, the prevalence of dementia in these three studies would be 15% (Mayeux, 1988).

In a thoughtful review of the studies in which the prevalence of dementia in idiopathic Parkinson's Disease (IPD) has been addressed over the past 60 years, Brown and Marsden (1984) calculated that 35.1% of 2530 patients enrolled in 17 studies were demented. They argued that this was probably an overestimate and that the true prevalence of dementia was more likely 20%. In addition to lack of uniformity of definition of dementia, they cited a number of other methodological problems. The inclusion of patients with arteriosclerotic parkinsonism with patients with idiopathic Parkinson's disease in some studies may have led to overestimates of dementia (Martilla and Rinne, 1976; Pollock and Hornabrook, 1966). Sampling variability may also have played some role in the different reported estimates, depending on whether surveys were population based or hospital based.

Taylor et al. (1985), echoed the lower prevalence figures suggested by Brown and Marsden (1984). In a group of 100 patients, 8% were moderately to severely demented while 7% were confused, possibly secondary to medications, and were a possible 'at risk' group for dementia. Lees (1985), recalculated the prevalence of dementia in 8 studies and found that 7-15% were demented using DSM-III criteria.

Perhaps one of the most important factors accounting for the disparity in estimates of dementia in Parkinson's disease is the different ages of the subjects studied. Numerous studies have shown that dementia clearly increases with age. Martilla and Rinne (1976) found the prevalence of dementia in those over age 70 to be almost twice that of those under 60.

In a retrospective analysis of 339 patients with idiopathic Parkinson's disease, seen over an 18 month period, 10.9% were found to be demented using DSM-III criteria for dementia (Mayeux et al., 1988). Demented patients were significantly older than non-demented patients and the prevalence of dementia for those 70 or older was 25.5% as compared to 10.5% in those under 70. The odds of becoming demented if the motor manifestations of Parkinson's disease began after age 70 were almost 4 times greater than if Parkinson's disease began at a younger age, suggesting that the age of onset of PD might influence disease course (Mayeux et al., 1988). In a survey of 147 consecutive patients, Girotti et al. (1988), estimated that 14.3% were demented. The percentage of demented subjects clearly rose with age, with 7.1% of those under 50 demented as compared to 16% of those 60-69 and 25% of those over 70 demented by DSM-III criteria.

The studies by Mayeux et al.(1988) and Girotti et al.(1988), calculated age specific prevalence rates of dementia among those with IPD and demonstrated that much of the variability in reported rates might be due to the differences in the ages of the patient groups studied. Higher dementia rates would be expected in samples with older patients.

INCIDENCE OF DEMENTIA IN PARKINSON'S DISEASE

Incidence, rather than prevalence can provide a direct measure of the risk or probability of developing a disease or symptom such as dementia. Incidence may be defined as the number of new cases of a disease in a population during a specified period of time. In this instance, incidence rate would be defined as the number of previously nondemented patients with Parkinson's disease who become demented during a specific time period. High prevalence of a disease does not necessarily imply a high incidence of the disease since prevalence reflects both disease duration and incidence. High prevalence may reflect a process that is of low incidence but of long duration. In contrast, low prevalence may reflect a disease with a high incidence that is rapidly fatal or quickly cured. For example, an influenza epidemic may affect many people, but since it resolves so quickly, the prevalence of influenza never really reflects the high incidence. Therefore, to truly estimate disease risk, incidence rates are fundamental.

Only two studies of the incidence of dementia among patients with Parkinson's disease have been undertaken. Mindham et al. (1982) assessed a cohort of 40 patients with Parkinson's disease and 40 psychiatric outpatients, 3 years after the initial evaluation. Four patients in the Parkinson's group developed dementia in the index period as compared to one patient in the control group. The overall prevalence at the initial assessment was 20%. Four of sixteen demented Parkinsonians died during the three year interval while none of the demented subjects in the control group died during the same interval.

Mayeux et al.(1990) assessed a cohort of 249 patients, almost 5 years after the initial review yielded a point prevalence of 10.9%. During the reevaluation, an additional 15 cases of dementia were identified during the prevalence period, and the prevalence estimate was recalculated to be 15.9%. Sixty five new cases of dementia were identified. Using the sum of the number of person-years for each case as the denominator, an overall incidence of 69 per 1000 person-years of observation was calculated. This was almost 6 times that expected when compared to the rate of dementia in an age-matched cohort. As age increased, the cumulative incidence of dementia in this cohort increased so that by age 85, the risk of dementia was over 65%.

This extremely high incidence rate despite a relatively low prevalence rate may imply that disease duration is shortened by the development of dementia. Of the 249 records reviewed, excluding those who were previously demented, 41 subjects died during the 5 year period; 11.9% of the nondemented group and 29.2% of the demented group. The fact that the demented subjects were older, had more severe manifestations and were significantly more disabled than the nondemented patients may have contributed to their reduced survival (Marder et al., 1990[a]). Numerous studies have suggested that cognitive impairment parallels physical impairment, and that the more demented subjects have more severe motor manifestations. (Celesia and Wanamaker 1972; Marttila and Rinne 1976; Mindham et al., 1982; Pirozzolo et al., 1982; Lieberman et al., 1987; Rajput et al., 1987; Mayeux et al., 1988)

The cumulative incidence of dementia of 65% at age 85 may reflect the natural history of Parkinson's disease or it may reflect a chance association of two diseases of the elderly.

If the dementia seen in idiopathic Parkinson's disease is actually concommitant Alzheimer's disease, then the risk factors for dementia should be the same in both cases. The known risk factors for Alzheimer's diseaseare age, family history of dementia, particularly in a sibling, family history of Down's syndrome, head injury and history of thyroid disease. Each of these risk factors were queried in addition to questions about toxin exposure, personal habits such as smoking, alcohol consumption and sports participation, medication use and family history of dementia in first degree relatives, Parkinson's disease and lymphoma or leukemia. When a structured interview was administered to surrogates of 17 demented subjects with IPD and 54 nondemented subjects, two factors emerged as possible risks for dementia. Age and family history of dementia in a first degree relative were significantly associated with the presence of dementia in the Parkinson's patients surveyed. Demented patients were older than nondemented patients although the duration of symptoms was similar. A family history of dementia was present in 30% of the demented group and 5.6% of the nondemented group and this was most often reported among siblings. The first degree relatives of demented patients with Parkinson's disease had over 6 times the risk of dementia than nondemented patients with Parkinson's disease. No other differences were seen on any of the other risk factors (Marder et al., 1990[b]). A further suggestion that there may be a shared etiology between Alzheimer's disease and Parkinson's disease emerges from a study of 198 Dutch patients with Alzheimer's disease who were compared to age and sex-matched healthy controls. Patients with Alzheimer's disease were almost three times more likely to have a relative with Parkinson's disease than were controls, (Hofman et al., 1989).

This familial clustering does not necessarily argue for a genetic etiology. Common environmental exposures such as dietary factors or toxins may also account for familial clustering. The cycad plant, suggested as the toxin causing the Guam amyotrophic lateral sclerosis Parkinson's complex provides evidence that an environmental exposure can give the impression of familial inheritance (Spencer et al., 1987).

OTHER EVIDENCE FOR PARKINSON'S DISEASE AND CONCOMMITANT ALZHEIMER'S DISEASE

In addition to the risk factor information, neuropathological and neurochemical data have been cited to support or refute the contention that the dementia seen in Parkinson's disease is due to Alzheimer's disease.

Considering the neuropathological data, the results are inconclusive. Some, (Hakim, 1979) suggest that there is concommitant Alzheimer's disease (AD) because of the neuropathological changes consistent with AD such as plaques and tangles present in 33 of 34 brains of patients with Parkinson's disease. Dementia was more common in the Parkinson's patients but there was no neuropathological difference seen between demented and nondemented Parkinson's patients. In a study of 36 patients with idiopathic Parkinson's disease, (Boller et al., 1980), neuropathological changes seen in AD such as plaques and tangles were seen in all 9 severely demented patients, but they were not specific, since 3/13 Parkinson's patients with normal mental status also showed changes consistent with Alzheimer's disease. Gaspar and Gray (1984) also found that pathological changes of AD were more severe in the more severely demented Parkinson's patients, although there was a severely demented subject without these changes. A number of smaller studies (Heston 1981; Perry et al., 1983; Chui et al., 1986) also failed to confirm a strong association between the presence of Alzheimer's lesions in the cortex and dementia in Parkinson's disease. Neuropathological evidence of Parkinson's disease may be present in the setting of Alzheimer's disease, even if

Parkinson's disease is not clinically diagnosed. In 20 pathologically confirmed Alzheimer's brains, there was pathological evidence of Parkinson's disease in 55%. Rigidity was present in 80% of the cases with PD pathology but no tremor was seen (Ditter and Mirra, 1987). There seems to be substantial overlap between Alzheimer's disease and Parkinson's disease or parkinsonism.

Recent evidence has suggested that demented Parkinsonians may have greater loss of medial substantia nigra than nondemented patients even when patients with plaques and tangles are removed from the analysis (Rinne et al, 1989). The selective loss in the medial substantia nigra in demented subjects might argue for the dementia being unique to Parkinson's disease. Parallel decline in motor and cognitive function also suggests a common etiology, possibly dopaminergic, for the dementia of Parkinson's disease.

Neither biochemical nor neuropathological evidence have settled the question of whether the dementia seen in patients with Parkinson's disease is due to coincident Alzheimer's disease, or is unique to Parkinson's disease. In part, this may be because rigorous clinical assessments were not always performed on those cases on whom neuropathological or neurochemical analysis was available. Prospective studies of patients with idiopathic Parkinson's disease on whom careful neurological and neuropsychological assessments are done prior to the onset of dementia will provide invaluable information on the natural history, risk factors and ultimately the neuropathological substrate of dementia in Parkinson's disease.

Acknowledgement

This work was supported by Federal grants AG-02802-08, AG-07232-02, M01-RR00645-1945, and the Parkinson's Disease Foundation.

References

American Psychiatric Association, 1987,"Diagnostic and Statistical Manual of Mental Disorders, Third Edition-Revised," Washington, D.C.

Boller F, Mizutani T, Roessmann U, Gambetti P: Parkinson's disease, dementia, and Alzheimer disease, clinicopathological correlations, 1980, Annals of Neurology, 7:329-335.

Brown RG, and Marsden CD, How common is dementia in Parkinson's disease?, 1984, Lancet, 1:1262-1265.

Celesia GG, and Wanamaker WM, Psychiatric Distrubances in Parkinson's disease, 1972, Dis Nerv Syst,33:577-583.

Chui, H, Mortimer, J, Slager, U, Zarow, C, Bondareff, W, Webster, D, Pathologic correlates of dementia in Parkinson's disease, 1986, Arch Neurol, 43:991-995.

Ditter, S, and Mirra, S. Neuropathologic and clinical features of Parkinson's disease in Alzheimer's patients, 1987, Neurology, 37:754-760.

Gaspar, P, and Gray, F, Dementia in idiopathic Parkinson's diease,1984, Acta Neuropathol, 64:43-52.

Girotti F, Soliveri P, Carella F, Piccolo I, Caffarra P, Musicco M, Caraceni T, Dementia and cognitive impairment in Parkinson's disease,1988, J Neuro Neurosurg Psychiatry, 51:1498-1502.

Hakim A, and Mathieson G, Dementia and Parkinson disease: a neuropathologic study, 1979, Neurology, 29:1209-1214.

Heston,L. 1981, Genetic studies of dementia: with emphasis on Parkinson's disease and Alzheimer neuropathology, in, "Epidemiology of Dementia", Mortimer J, Schuman, L, ed. Oxford Univ Press, New York.

Hofman A, Shulte W, Tanja T, van Duijn C, Haaxma R, Lameris A, Otten V, Saan R. History of dementia and Parkinson's disease in 1st degree relatives of patients with Alzheimer's disease 1989, Neurology, 39:1589-1592.

Lees A, Parkinson's disease and dementia, 1985, Lancet, 1:43-44.

Lewy, F, 1923, "Die lehre von tonus und der bewegung zugleich systematiche untersuchiner sur klinik, physiologie, pathologie und pathogenese der paralysis agitans," Springer, Berlin.

Lieberman A, Dziatolowski M, Kupersmith M, Serby M, Goodgold A, Korein J, Goldstein M, Dementia in Parkinson disease, 1979, Ann Neurology, 6:355-359.

Marder K, Mirabello E, Chen J, Bell K, Dooneief G, Cote L, Stern Y, Mayeux R, Death rates among demented and nondemented patients with Parkinson's disease, to be presented at Fourth Annual Parkinson Study Group Symposium, Atlanta, Georgia, October 1990[a].

Marder K, Flood P, Cote L, Mayeux R, A pilot study of risk factors for dementia in Parkinson's disease, 1990[b], Movement Disorders, 5:156-161.

Martin W, Loewenson R, Resch J, Baker A, Parkinson's disease. A clinical analysis of 100 patients, 1973, Neurology, 23:783-790.

Marttila, R, and Rinne, U, Dementia in Parkinson's disease, 1976, Acta Neurol Scand, 54:431-441.

Mayeux R, The epidemiology of major forms of dementia in the United States, 1988, Progress in Clinical Neurosciences, 1:335-347.

Mayeux, R, Stern Y, Rosenstein, R, Marder, K, Hauser A, Cote L, Fahn S, An estimate of the prevalence of dementia in idiopathic Parkinson's disease, 1988, Arch Neurol, 45:260-262.

Mayeux R, Chen, J, Mirabello, E, Marder, K, Bell, K, Dooneief, G, Cote, L, Stern, Y, An estimate of the incidence of dementia in idiopathic Parkinson's disease, (in press) Neurology.

Mindham R, Ahmed, S, Clough C., A controlled study of dementia in Parkinson's disease, 1982, J Neurol Neurosurg Psychiat, 45:969-974.

Mindham, R, Psychiatric syndromes in Parkinsonism, 1970, Journal Neurol, Neurosurgery, Psychiatry, 30:88-191.

Mjones H, Paralyis agitans, 1949, Acta Psychiat Neurol,54: 1-195.

Parkinson, J., 1817, "An Essay on the Shaking Palsy," Sherwood, Neely, and Jones, London. p 1.

Patrick, H, and Levy, D, Parkinson's disease: a clinical study of 146 cases, 1922, Arch Neurol, 7:711-720.

Perry, R, Tomlinson, B, Candy, J, Blessed, G, Foster, J, Bloxham, C, Perry, E, Cortical cholinergic deficit in mentally impaired parkinsonian patients, 1983, Lancet, 789-790.

Pirozzolo, F, Hansch, E, Mortimer, J, Webster, D, Kuskowski, M, Dementia in Parkinson disease: a neuropsychological analysis, 1982, Brain Cogn., 1:71-83.

Pollock, M, and Hornabrook, R, The prevalence, natural history and dementia of Parkinson's disease, 1966, Brain, 89: 429-448.

Rajput, A, Offord, K, Beard, C, Kurland, L, A case-control study of smoking habits, dementia, and other illnesses in idiopathic Parkinson's disease, 1987, Neurology, 37:226-232.

Rinne J, Rummukainen, J, Paljarvi, L, Rinne, U, Dementia in Parkinson's disease is related to neuronal loss in the medial substantia nigra, 1989, Ann Neurol, 26:47-50.

Spencer, P, Nunn, P, Hugon, J, Ludolph, A, Ross, S, Robertson, R. Guam amyotrophic lateral sclerosis-parkinsonism-dementia linked to a plant excitant neurotoxin, 1987, Science, 237: 517-522.

Sroka, H, Elizan T, Yahr, M, Burger, A, Mendoza, M, Organic mental syndrome and confusional states in Parkinson's disease: relationship to computerized tomographic signs of cerebral atrophy, 1981, Arch Neurol, 38: 339-342.

Sutcliffe, R, Parkinson's disease in the district of the Northhampton Health Authority, United Kindom. A study of prevalence and disability, 1985, Acta Neurol Scand, 72: 363-379.

Taylor, A, Saint-Cyr, Lang A, Dementia prevalence in Parkinson's disease, 1985, Lancet, 1: 1037.

PATHOLOGY IN THE CHOLINERGIC BASAL FOREBRAIN:

IMPLICATIONS FOR TREATMENT

Peter J. Whitehouse

Alzheimer Center,and Department of Neurology
University Hospitals and Case Western Reserve
University
Cleveland, Ohio

INTRODUCTION

The development of effective treatments for the cognitive impairment characteristic of dementing illnesses represents a major challenge to neuroscientists in the future. Success in this endeavor will depend on understanding the biological basis of normal cognition. In order to make scientific progress in understanding the neural basis of cognition and the consequences of dysfunction in these systems in dementia, we must isolate components or subsystems in order to study them. As we make simplifying assumptions, we must remember the dangers of such simplication and the importance of eventually modelling the entire network of interacting systems that support normal thought and that are impaired in dementia.

In this paper, we will review aspects of pathology that occur in one system, the cholinergic basal forebrain system (ChBF) in several dementias, particularly Alzheimer's Disease (AD) and Parkinson's Disease (PD). Short and long term implications of this pathology for development of therapies will then be considered.

CHOLINERGIC HYPOTHESIS--A CRITIQUE

In the 1960's, alterations in acetylcholinesterase (AChE) were found in neocortex in AD brains. Descriptions of reductions in choline acetyltransferase (ChAT), a more specific marker for cholinergic neurons, followed in the 1970's. Finally, in the early 1980's, the anatomical substrate of the loss of cholinergic markers in tele-encephalon was identified; namely, loss of cells in the medial septum, diagonal band of Broca and nucleus basalis of Meynert, now called the cholinergic basal forebrain (ChBF) (reviews see Bartus, Whitehouse).

The evidence that this system plays a role in normal cognition and that it is affected in AD and related dementias is

The Basal Forebrain, Edited by T.C. Napier *et al.*
Plenum Press, New York, 1991

considerable and will not be reviewed here extensively. In humans, anticholinergic drugs are known in humans to cause delirium and associated memory difficulties, and preliminary evidence has suggested that cholinomimetics of one kind or another can be of limited benefit in some dementias. In AD and PD, autopsy studies have found positive correlations between loss of cholinergic markers and premortem severity of dementia (Perry, Ruberg). Moreover, animal studies have consistently demonstrated that lesioning basal forebrain leads to cognitive difficulties, which can be partially alleviated by cholinergic drugs (Olton).

The concept of a "cholinergic hypothesis" of dementia is widely discussed yet poorly defined and understood. The cholinergic hypothesis is often considered in discussions of the possible etiology of AD, along with genetic or environmental toxin hypotheses. This is not appropriate, since this hypothesis does not address etiology, rather it relates to the neural underpinnings of at least a portion of the cognitive impairment that occurs in AD and some related diseases. The cholinergic hypothesis is often set up as a "straw man." Rarely, have investigators claimed that the cholinergic hypothesis explains all--or even a major part--of the cognitive impairment in the disease. Only future studies will be able to demonstrate how much of the cognitive impairment in AD can be accounted for by pathology in this system and how effective therapeutic interventions will be that focus exclusively on this particular system.

It is also important to consider that non-cholinergic cells are also present in the basal forebrain and may be damaged in AD. Several studies have recently suggested that some of the cognitive impairment associated with pathology in the basal forebrain may in fact be related to dysfunction of cells that are noncholinergic. These lesion studies in rats showed that a neurotoxin (quinolinic acid), which produces profound reductions of ChAT in cortex when injected into the cholinergic basal forebrain, may not cause as severe learning and memory problems as a toxin which produces less ChAT loss (ibotenic acid) (Wenk).

TYPES OF PATHOLOGY IN THE BASAL FOREBRAIN

The first observations of pathology in the nucleus basalis of Meynert, the major portion of ChBF, in degenerative disease occurred in the first part of this century when loss of cells was reported in Parkinson's Disease. (For review see Whitehouse.) In fact, Lewy originally described the intracellular inclusion which now bears his name not in the substantia nigra, but in the substantia nominata which includes the nucleus basalis of Meynert. Hassler may have been the first to associate pathology in this structure with specific clinical syndrome, when he suggested that bradyphrenia, or slowness of mentation, may be associated with pathology in this structure in PD. Early observations were also made that neurofibrillary tangles occurred with great frequency in this structure in AD (Ishii).

It was not until the 1980's, however, that the full extent of the pathology in this system was described, and the relationship between loss of cells in this structure and the loss of cholinergic markers in cortex and hippocampus was established. Most studies have confirmed that cell loss occurs in AD and some related disorders, although other authors have focused on the fact that shrinkage may also occur prior to cell death. Alterations in several neurotransmitter markers have been reported in the basal forebrain, which presumably reflect death of nerve cells and alterations in afferent connections. For example, a marked reduction of nicotinic receptors (Shimohammi) has been reported as well as alterations in bioaminergic markers.

In addition to cell loss and alterations in neurotransmitter markers, studies using *in situ* hybridization have described abnormalities in the expression of genes in these neurons. Younkin and his colleagues, for example, reported that neurons in the nucleus basalis of Meynert overexpress the messenger RNA for certain forms of the amyloid precursor protein (Palmert). Earlier work had suggested that dysfunction of cells in this structure may contribute to the abnormal neurites in some of the cortical plaques which have acetylcholinesterase associated with them (Struble). However, it has now become clear that the neurites surrounding senile plaques include neurotransmitter markers of many different neural systems.

DISEASES IN WHICH NEUROPATHOLOGY OCCURS IN THE BASAL FOREBRAIN

Table 1 lists dementias in which pathology has been described in the basal forebrain. The most common illnesses are Alzheimer's Disease and Parkinson's Disease; however, other rarer diseases, although not as well studied, demonstrate pathology in this structure as well. Not all neurodegenerative diseases, however, manifest cell loss in this structure. Table 2 represents the few disorders that have been studied in which this system appears to be spared (Clarke).

TABLE 1
DISEASES WITH PATHOLOGY IN THE BASAL FOREBRAIN

Alzheimer's Disease
Chronic Alcoholism
Creutzfeldt-Jakob Disease
Dementia Pugilistica
Diffuse Lewy Body Disease
Down's Syndrome
Hallervordin Spatz Disease
Multi-infarct dementia/hypoxia
Olivopontine cerebellar atrophy
Parkinson's Disease
Parkinson in Dementia Complex of Guam
Progressive Supranuclear Palsy
Subacute Sclerosing Panencephalitis
Wernicke-Korsakoff Syndrome

TABLE 2

DISEASES WITHOUT CONSISTENT PATHOLOGY IN THE BASAL FOREBRAIN

Pick's Disease
Cortical degeneration associated
with motor neuron disease
Huntington's Disease

IMPLICATIONS FOR TREATMENT

Short Term

Programs for the development of cholinergic treatments probably account for the single largest investment in a specific drug development strategy in dementia. For years, studies of cholinesterase inhibitors, such as physostigmine, have provided tantalizing hints, but no conclusive evidence, that treatment with cholinomimetic agents can alleviate some of the symptoms of patients with these disorders (Drachman, Davis). Most attention has been focused recently on Tacrine, a cholinesterase inhibitor that was originally reported to be dramatically effective in certain patients with AD (Summers). The design of the original study makes interpreting this work difficult, and a multi-center study is underway in this country, with almost 200 subjects enrolled. Recently, a New Drug Application (NDA) was submitted to the FDA for Tacrine by Warner Lambert. Smaller studies, using somewhat different designs, e.g. cross-over rather than parallel designs, have been inconsistent with some reporting negative and others positive results (Gauthier). Another related compound, HP-029 (Hoechst-Roussel), is also being tested and is apparently showing encouraging results.

In addition to cholinesterase inhibitors, considerable energies are being placed on developing drugs that act selectively at cholinergic receptor sites. Such drugs could stimulate post synaptic cells without depending on the integrity of the pre-synaptic cholinergic neurons which are affected by the disease process. Several types of muscarinic receptors have been defined using different methods--genetic, pharmacological and physiological. The controversy in defining muscarinic subtypes has made interpretation of this literature difficult, however. Claims have been made that two major subtypes, M-1, which are post-synaptic, and M-2, which are pre-synaptic, exist in human brain, and that the M-2 pre-synaptic site is selectively reduced in the disease (Mash, et al.). Such data would suggest that finding an M-1 agonist or and M-2 antagonist might be effective strategies to block the pre-synaptic autoreceptors. Our work demonstrating that the most significant alteration in cholinergic receptors in Alzheimer's Disease and Parkinson's Disease is reductions in nicotinic receptors (Whitehouse with Kellar) has been confirmed by others. (Shimohama, Nordberg, Perry) Preliminary attempts to treat the cognitive symptoms in AD with nicotinic compounds have not been successful, however.

Long-Term Strategies

Ultimately, effective biological treatment for AD should be based on an understanding of the cause of the disease. In the absence of such understanding of the etiology, intermediate strategies can be based on preventing the cell death, which is characteristic of the disease. Nerve growth factor, an approximately 120 amino acid protein, is known to promote the viability of cholinergic basal forebrain cells and to be a member of a class of related trophic factors including brain-derived neurotrophic factor and neurotropin 3 (Barde). In animals, the intraventricular infusion of NGF can prevent cell death and improve some of the cognitive dysfunction associated with pathology in the cholinergic basal forebrain (Gage). More animal studies are needed prior to considering the introduction of this therapy with human subjects. Nevertheless, these approaches offer hope for the not-too-distant future of interventions which may actually slow the progression of the disease (Heft).

CONCLUSION

Tremendous advances have been made in our understanding of the basic biology of the cholinergic basal forebrain. (See this Volume.) The fact that this structure is damaged in many degenerative dementias and that strong evidence exists that it has a role in the cognitive deficits of these diseases, suggest that appropriate attention is being paid to the therapeutic implications of dysfunction in this system. Nevertheless, since the time Alzheimer described the disease in 1907, it is clear that AD is a multi-system disorder, and as a result it is not yet clear how much therapeutic impact can be made by focusing exclusively on any one system. Some animal studies suggest that combining bioamminergic and cholinergic agents may be more effective; human studies using combination treatments have been relatively disappointing, however (Davis). As we narrow our focus and simplify our approaches to the study of AD in order to apply scientific methods, it is important to remember advice concerning such simplification from two notable individuals. Alfred N. Whitehead suggested that one should "seek simplicity but distrust it." Albert Einstein suggested that one should "simplify as far as possible, but no further." Whether therapeutic interventions in AD and related dementias based on cholinergic mechanisms can be effective or not remains to be seen.

REFERENCES

Barde, Y-A, 19 , Trophic factors and neuronal survival, Neuron.
Bartus, R.T., Dean, R.L., Beer, B., Lippa, A.S., 1982, The cholinergic hypothesis of geriatric memory dysfunction, Science, 217:408-417.
Gage, F.H., Chen, K.S., Buzsaki, G., Armstrong, D., 1988, Experimental approaches to age related cognitive impairments, Neurobiol. Aging, 9:645.
Gauthier, S., Bouchard, R., Lamontagne, A., Bailey, P., Bergman, H., Ratner, J., Tesfaye, Y., Saint-Martin, M., Bacher, Y., Carrier, L., Charbonneau, R., Clarfield, A.M., Collier, B., Dastoor, D., Gauthier, L., Germain, M., Kissel, C., Krieger, M., Kushnir, S., MAsson, H., Morin, J., Nair, V., Neirinck, L., and Suissa, S., 1990, Tetrahydroaminoacridine-Lecithin combination treatment in patients with intermediate-stage Alzheimer's disease, N Engl J Med 322:1272.
Hefti, F., Hartikka, J., Knusel, B., 1989, Function of neurotrophic factors in the adult and aging brain and

their possible use in the treatment of neurodegenerative diseases, <u>Neurobiol. Aging</u>, 10:515-533.

Ishii, T., 1966, Distribution of Alzheimer's neurofibrillary changes in the brain stem and hypothalamus of senile dementia, <u>Acta. Neuropathol.</u> 6:181.

Mash, D.C., Flynn, D.D., Potter, L.T., 1985, Loss of M2 muscarinic receptors in the cerebral cortex in Alzheimer's and experimental cholinergic denervation, <u>Science</u>, 228:1117.

Olton, D.S., Gamzu, E., Corkin, S., 1985, Memory Dysfunctions: An Integration of Animal and Human Research from Preclinical and Clinical Perspectives. Annals of the New York Academy of Sciences Vol. 444, 1985.

Palmert, M.R., Golde, T.E., Cohen, M.L., Kovacs, D.M., Tanzi, R.E., Gusella, J.F., Usiak, M.F., Younkin, L.H., Younkin, S.G., 1990, Amyloid protein precursor messenger RNAs: Differential expression in Alzheimer's disease, <u>Science</u> 241:1080.

Perry, E.K., Tomlinson, B.E., Blessed, G., Bergmann, K., Gibson, P.H. and Perry, R.H., 1976, Correlation of cholinergic abnormalities with senile plaques and mental test scores in senile dementia, <u>Br. Med. J.</u> 2:457.

Ruberg, M., Ploska, A., Javoy-Agid, F., and Agid, Y., 1982, Muscarinic binding and choline acetyltransferase activity in parkinsonian subjects with reference to dementia, <u>Brain Res.</u> 232:129.

Shimohama, S., Taniguchi, T., Fujuwara, M. et al., 1985, Biochemical characterization of the nicotinic cholinergic receptors in human brain: Binding of (-)-[^{3}H]nicotine, <u>J. Neurochem.</u> 45:604.

Sunderland, T., Tariot, P.N., Newhouse, P.A., 1988, Differential responsivity of mood, behavior, and cognition to cholinergic agents in elderly neuropsychiatric populations, <u>Brain Res</u> 13:371.

Whitehouse, P.J., Struble, R.G., Hedreen, J.C., Clark, A.W., Price, D.L., 1985, Alzheimer's disease and related dementias: Selective involvement of specific neuronal systems, <u>CRC Crit. Rev. Clin. Neurobiol.</u> 1(4):319.

HISTORICAL ASPECTS OF THE CHOLINERGIC TRANSMISSION

Alexander G. Karczmar

Department of Pharmacology, Loyola University Chicago,
Stritch School of Medicne, Maywood, IL and Research
Services, Hines V.A. Hospital, Hines, IL

My long-held personal opinion that the transmission phenomena are
more exciting than other physiological and pharmacological events may be
subjective and biased, but there cannot be any doubt that the synaptic or
junctional phenomena generally, and the cholinergic transmission
specifically, constitute most rewarding processes, both intellectually
and aesthetically. Nevertheless, I will try to prove this truism.

Part of this excitement is generated, of course, by the original
mystery, today almost solved, of the neuron-to-neuron or neuron-to-non-
nervous tissue communication; and by the link, forged early in this
century, between cholinergic transmission and behavioral function,
including mentation. To me, the elation comes primarily from the long
sequence of heuristic steps that characterize the scientific progress of
cholinergic research, and, no less, from the courageous missteps that
occurred quite frequently in the course of this progress. Furthermore, it
is not a coincidence that this progress was so brilliant and yet so
unpredictable or even erratic; the persons involved were brilliant and
ready to take risk, whether for the sake of science or for their own
sake... To prove then what needs no proof I will describe in this chapter
some of the pertinent people, their successes, and even their goofs.

MISCONCEPTIONS AND RECOVERIES

Let me present the matter of misconceptions or "goofs" and the
recoveries which frequently followed the former, in the form of a quiz.
Accordingly, I will refer to the perpetrators in question by their
initials (see Tables 1 and 2 and REFERENCES FOR THE QUIZ); please
identify the individuals!

As these misconceptions or, indeed errors in scientific thinking are
being described, it must be emphasized that these are brilliant and
"poetic" errors. Indeed, some of them, such as the notions of DN as to
the role of acetylcholine (ACh) and acetylcholinesterase (AChE) in axonal
conduction, were generated, I believe, by the artistic sense of the
perpetrators of the misconceptions in question. Thus, DN was driven by

The Basal Forebrain, Edited by T.C. Napier *et al.*
Plenum Press, New York, 1991

Table 1. Goofs and Booboos[a] (a sample only).

DN (1960)	"You can kill me - you can't kill the theory" or "To get incredible results you must use incredible doses"
JCE (1941-1946)	"Presynaptic action current (is) responsible for......excitatory action at the neuromuscular junction and sympathetic ganglion"
CH (1955)	"Les deux actions du DFP, l' inactivation des cholinesterases et les actions d'excitation parasympathique, ne sont pas dependantes l'une de l'autre".
MH (1969)	"Les deux etats du sommeil.....sont.... modules par des mecanismes monaminergiques" ... "very close...to catecholamines."
FFW (1968)	"Recurrent inhibition from major axone collaterals to motoneurones is a monosynaptic pathway" (there is no Renshaw cell).
DK,MRR & ELB (1965-67)	"Difference of..3% in the total cortical...2% in the total subcortical AChE...correlates.. with..adaptive behavior... and...learning"
WFR, Jr.(1963)	"The data strongly implicate muscle as the principal tissue of ACh origin, and therefore..in.. neuromuscular structure, formation and release of ACh may not be correlated to synaptic events".
AGK (1946)	"Trophic maintenance by the nerves"
PdC (1854)	"TEPP tastes so sweet". Actually, PDC lived to be 91.
AOZ (1967)	"Identity of....cholinesterases with cholinoreceptors"

[a] For references see text and the REFERENCES FOR THE QUIZ.

his[1] belief in the unity of all biochemical energy-related events and the need to maintain a mathematical balance between these events, a belief based on the influence of his teachers and men he admired - Otto Meyerhof, Keith Lucas and A. V. Hill. Indeed, her studies on this matter are a pleasure to read (eg., DN, 1961, 1963a); and one feels like exclaiming: "Si non e vero e ben trovato"! Yet, DN was a stubborn scientist. Symbolically, this may be illustrated by the following story: in the course of the 1960 Rio de Janeiro Symposium on Bioelectrogenesis (Chagas and Paes de Carvalho, 1961) DN fell asleep in a beach chair; he woke up as he was being wheeled by JCE and Harry Grundfest into the ocean and exclaimed: "You can kill me, but you cannot kill the theory!" Less symbolically, many investigators believe that undue commitment of DN to the AChE and ACh hypothesis of conduction and continued draining of her energy in unconvincing experimentation cost DN the Nobel prize. She was also unduly fond of using excessive concentrations of drugs in her experiments - such as $10^{-3}M$ concentrations of ACh or diisopropyl-fluorophosphate (DFP) - to prove her point (Table 1).

Nor will I omit to show that many of those who fumbled recovered brilliantly.

Thus, DN compensated for her erroneous notion (Tables 1 and 2) with respect to the role of AChE in conduction (for this story, see AGK, 1967) by discovering, characterizing and elucidating the role of choline acetylase, as she and Machado called the enzyme since then referred to officially as choline o-acetyl transferase (CAT; E.C.2.3.1.6.); he was also one of the first to crystallize AChE and to attempt the chemical identification and purification of the cholinoceptive (cholinergic) receptor. I will return again to this individual, so brilliant and so difficult to understand.

Table 2[a]. Recoveries

JCE, PF & KK	"Cholinergic...Synapses in a Central Nervous System Pathway"
DN & ALM	"On addition of ATP to... brain... and electric organs...extracts, the first enzymatic acetylation of choline was achieved".
CH & EN	"The Chemical Control of Respiration".
DK, MRR & ELB	"Relate biochemical activities in the cortex to adaptive behavior pattern... plasticity".
MJ	"Potent inhibitory action of atropine...upon PS and PGO spikes".

[a] For references, see text and the REFERENCES FOR THE QUIZ.

[1] I use the personal pronouns - "his", "her" - in their non-gender sense; thus, in this quiz I will refer sometimes to "her" and sometimes to "him", with no gender significance intended, and so as better to mislead those quizzed! Besides, it is so clumsy to refer to him/her and she/he!

Then, there is the JCE heresy (see JCE, 1964) and JCE's subsequent recovery (Tables 1 and 2) which ultimately earned him the Nobel Prize for a series of studies beginning with the JCE, PF and KK papers of 1953 and 1954. JCE commented subsequently (JCE, 1987) that the latter became the fastest published MS in her experience! Both the heresy and the recovery were based on a series of brilliant experiments...

Similarly, while CH, a somewhat supercilious patrician raconteur and gourmet, may be faulted for believing that organophosphorus (OP) drugs including di-isopropylfluorophosphonate exert all of their actions at the ganglia and at the other synapses and junctions independently of their inhibition of AChE (Tables 1 and 2), yet CH managed to earn the Nobel Prize for his demonstration of the chemical control of respiration via the chemo- and related receptors of the carotid body and sinus (CH, 1955). In fact, she stressed in her studies of these centers their exquisite sensitivity to ACh; actually, Landgren (see Landgren et al., 1952) proposed a neurotransmitter role for ACh at those sites, a role which, unfortunately, could not be ultimately justified (CH, o.c.).

In the same context, while MJ claimed initially that the catecholamines and serotonin are the sole transmitters regulating the balance between the slow and the REM sleep, he was probably the first to find that atropine blocks the latter (MJ, 1967, 1972; Tables 1 and 2). She also was among the first to "play the monamine game" in regulation of a brain center, and she was a pioneer in the studies of establishing the multitransmitter nature of the pathways controlling the sleep system, studies which led ultimately to the revelation of the cholinergic role in the control of the REM sleep by Hobson, McCarley and their associates (see Baghdoyan, 1987).

In experimental terms, MRR, DK and ELB and their associates (DK, MRR and ELB, 1966; MRR et al., 1958 and 1967) never "recovered" from their heresy of imputing functional and learning significance to 2 and 3% differences in brain AChE and to the complex formulae concerning the ratios between butyrylcholinesterase (BuChE) and AChE, as they seem to have stuck to this notion for good (see AGK, 1969, and discussion statement by ELB); however, the notion posited by this brilliant, interdisciplinary team of the flexibility and plasticity of the brain as illustrated by appropriate biochemical parameters constituted a heuristic speculation posited long before its time had arrived (Tables 1 and 2).

Still less "recovery" seems to characterize AOZ's (1967; Table 1) claim that cholinesterases are identical with the cholinergic receptors; however, in terms of the state of the art at the time when her data were obtained, her hypothesis was tenable and exciting at the time of the early, vigorous search for the identity of the receptors. As to the "goof" of PdC (Table 1) who imbued in 1854 the potent organophosphate antiChE tetraetylphosphate (TEPP) and should have perished for his error but did not (Holmstedt, 1972; AGK, 1972), this "goof" may be considered, again, as a state of the art error since one of the Edinburgh pioneers of the studies of the "truth" (Calabar) bean, Robert Christison (see Holmstedt, 1972) was similarly rash in consuming some 10 mg worth of physostigmine of the bean without first testing in animals the appropriate pharmacodynamics of the concoction - yet he survived. Second, both PdC and Christison redeemed themselves as they lived to be 91 and 85, respectively.

Similarly, FFW (1968; Table 1) who denied the presence of the Renshaw cell some 15 years after the publication of JCE's papers (JCE, PF and KK, 1953 and 1954) and a few years after its almost definitive anatomical identification (JCE, 1969; Karczmar, 1891), redeemed himself by surviving, apparently in silence, JCE's (o.c.) deadly critique of his study which JCE referred to as "a most audacious attack" on the Renshaw cell and on the existence of the synaptic transmission at this site (JCE, 1969).

Another, strange "goof" which was never retracted was that of AGK as she had a strange notion in the forties of suggesting a trophic and, indeed, morphogenetic role for the nerves, whether cholinergic or otherwise (AGK, 1946; see also AGK, 1963 and 1967)!

Now, we come to the story of WFR, Jr., FGS, and their associates (Table 1) which - via P.T. Feng as an intermediary - is linked with the "recovery" by JCE from his early heresy of the neuromyal and ganglionic transmission being generated by "electric action potential". The full story may be - probably! - reconstituted as follows. P. T. Feng and his collaborators were engaged in the thirties and forties at the old Peiping (ancient spelling) Union Medical College in the pharmacological studies which included the use of physostigmine, guanidine, veratrine, etc.; among other phenomena they demonstrated that retrograde discharge can be evoked from the toad motor nerve endings (see, for example, Dun and Feng, 1940 and Figure 1). Several other investigators confirmed, in the forties, this finding (see AGK, 1967, for review); and, in the fifties, WFR, Jr., subsequently a Sollmann Award winner, picked up the matter and developed the paradigm in a somewhat unexpected way as he ascribed the neuromuscular transmission to currents originating in the motor nerve terminal (FFW et al., 1957; FGS and FFW, 1967). Actually, in doing so, she was essentially taking up the position embraced some 20 years earlier by JCE! Ultimately, she published data which seemed to demonstrate that no more ACh is released during the stimulation of the motor nerve than during the rest periods (AHH and FFW, 1963) - hence the quotation listed in Table 1. To add oil to fire, FGS and WFR opined subsequently (1967) that there is no more reason to assign the transmitter role to ACh than "to decamethonium, tetraethylammonium, succinylcholine...", as all these agents exert similar actions at the neuromyal junction! All this sent the pharmacologists scurrying back to the laboratory, as they thought that since the days of Dale, Feldberg and Vogt the stimulation-bound release of ACh was a demonstrated fact (see, AGK, 1967, for the review of the pertinent literature); finally, no lesser investigators than Krnjevic and Straughan (1964) repeated the early work one year after the AHH and WFR publication (1963), only to obtain the same, classical results!

Now, it happens that T.P. Feng realized that his findings may appear to support the notion of electrical transmission at the neuromyal junction. However, upon further studies he came down in full on the side of the chemical, cholinergic nature of the neuromyal transmission (see, for example, Feng and Li, 1941). Notabene, his actual findings were essentially the same as those of FFW some 15 years later.

The interesting link of the work of T.P Feng with that of JCE is as follows. Mostly on the basis of the story as told by Zenon Bacq (1975) it is generally accepted that JCE accomplished his "conversion" because the Austrian-British mathematician and philosopher, Hans Popper, who visited JCE in the early fifties in New Zealand, convinced him of the necessity of scientific flexibility and "openness" for creativity. I think differently, as JCE's method of thinking - in matters of science rather than metaphysics...- is pragmatic rather than philosophical. In fact, when in an argument with JCE re ionic currents I asked her how did she arrive at a certain conclusion (see Figure 2), she said: "you must think like the neuron". On the other hand, in his nineteen sixties lecture on

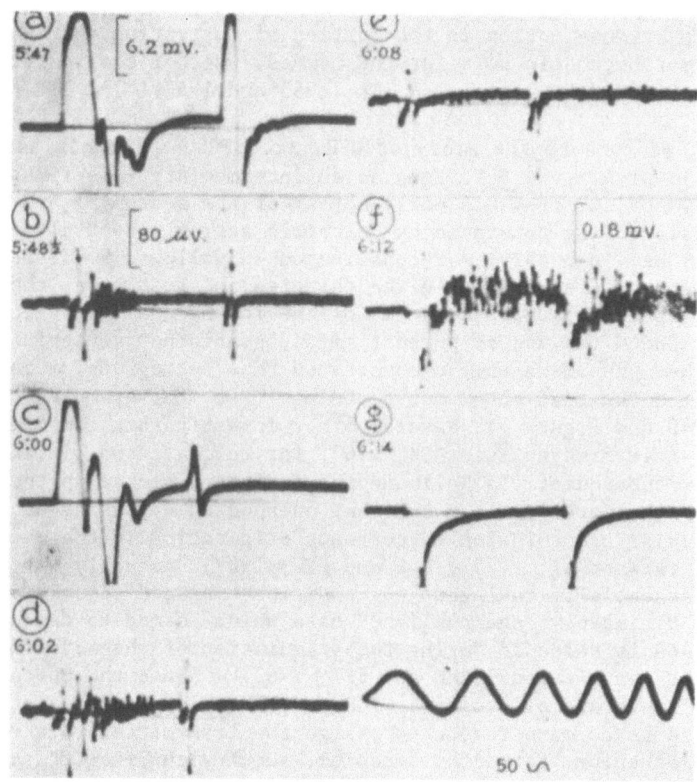

Fig. 1. From Feng and Li, 1941. The figure shows the retrograde discharges
from the motor nerve endings of the soleus and other muscles.
Records **a** and **c**, made with leads in the soleus muscle; all others
made with leads on the ventral root. For **a-e**, stimulus on the
soleus nerve; for **f** and **g**, stimulus on the sciatic nerve. At
5:42½-45, the soleus as well as its neighboring muscles irrigated
with about 8 c.c. 1/100,000 eserine solution; at 5:49-54, further
irrigated with about 14 c.c. 1/10,000 solution. Between **d** and **e**,
at 6:07, the soleus nerve sectioned at its entrance to the muscle.
Between **f** and **g**, at 6:13½, the sciatic nerve sectioned below the
stimulating electrodes. The disappearance of the after-discharges
following sectioning of the nerve in **e** and **g** proves their origin
in the motor nerve endings. See other comments in text.
Amplifier sensitivity same for **a** and **c**, as in **a**; same for **b**, **d** and
e, as in **b**; same for **f** and **g** as in **f**.

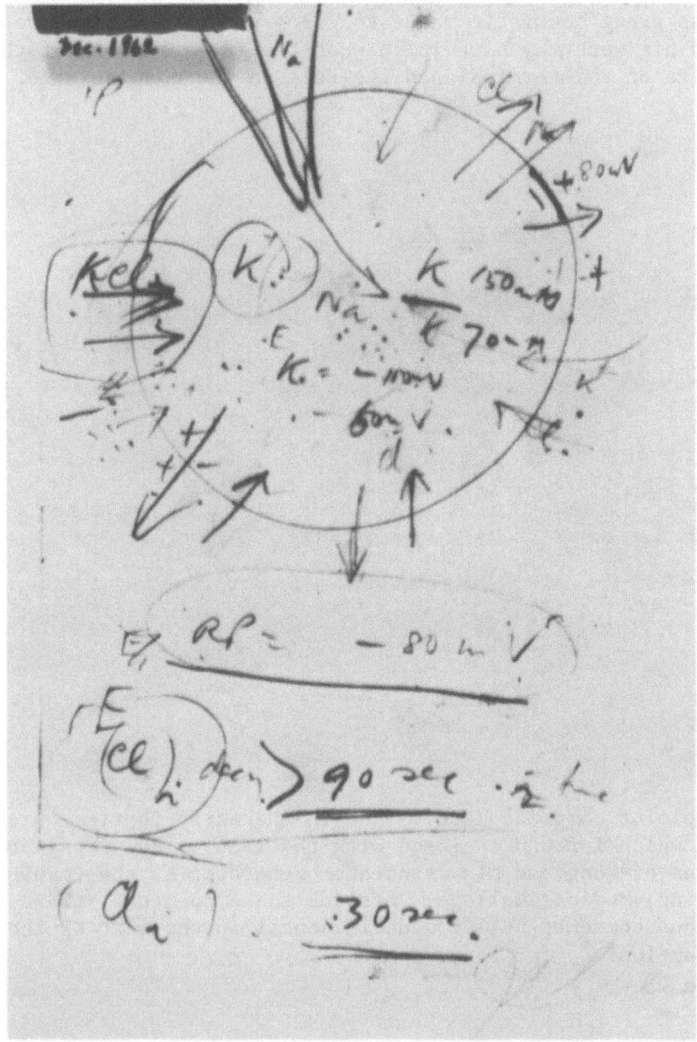

Fig. 2. A diagram drawn by JCE in 1962 to illustrate the concept of ionic movements, particularly those of Cl, underlying the synaptic potentials.

the Strategy of Scientific Discovery, JCE emphasized the importance of a free-wheeling creativity.

What I think is that this latter position of JCE is somewhat disingenuous. I believe that what induced JCE to shift conceptually was the evidence accumulating at the time in support of chemical transmission. And, in particular, he was influenced by the work - and, presumably, conclusions - of T. P. Feng. I have a volume of Feng's reprints which is full of text and full page comments by JCE (Figure 3) all indicative of the alteration in JCE's thinking as she perused the papers in question.

Altogether, Feng's papers helped JCE in his "recovery" (Table 1), but led WFR (and WKR) and FGS into excesses! Having said that, I must add that there was great heuristic value to WFR's stress of the nerve terminal, and his work may have influenced the investigators of the receptor nature of the terminal and its role in the modulation of

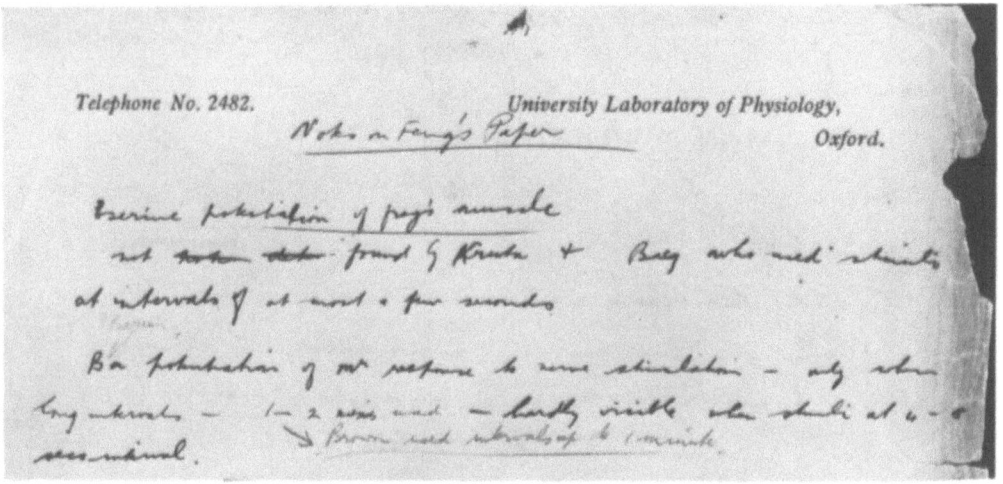

Fig. 3. A sample of comments made by JCE in the early forties on separate pages which I found enclosed with the 1936 - 1941 reprints of the studies of Feng and his associates generated at the Peiping (Beijing) Medical College. Besides these comments, there are numerous comments by JCE made in pencil in the text of the papers in question.

transmission (mostly, via the modulation of ACh release, as in the case of George Koelle's - another Sollmann Award winner - percussive hypothesis of the release of ACh).

MAJOR STEPS AND "FANFARES"

Beyond the "goofs" and recoveries, there is a brilliant succession of giant steps in the cholinergic story - this can be readily illustrated, and this illustration will demonstrate also the richness of the cholinergic field and the variety and increasing sophistication of cholinergic research.

The illustration in question is provided in Tables 3 through 7; these Tables define the steps and associate them with at least some of the notable investigators who helped in making the steps in question. There may be a problem here: as I reach in my enumeration the contemporary times including those of my own generation I will be surely accused of bias and incompleteness by those cholinergikers who are still around and who are not listed here - and colinergikers, including myself, are quite sensitive! Let me just say that generally I limited myself to pioneers and certainly did not intend to list all the current leaders; of course, in my listing I was a bit biased with respect to my friends...

Let me now make a few comments. First, there was a heurististic and, indeed, brilliant XIXth century era of cholinergic research which preceded and laid the grounds for, the actual demonstration of the chemical, cholinergic transmission; the first 3 steps listed in Table 3 concern this era. What was accomplished by the Edinburgh pharmacognosists such as Wadell, Fraser and Christison, and German pharmacologists such as Schmiedeberg and Koppe (see Brazier, 1959; Holmstedt and Liljestrand, 1963; Holmstedt, 1972; Karczmar, 1967 and 1990) was the description of the pharmacodynamics of physostigmine (or, more properly, the extracts of Calabar bean), muscarine, pilocarpine and atropine. The evaluation of the activities of choline derivatives was equally important for the future progress, as was the clarification of the pertinent structures, particularly of carbamate antiChEs and, ultimately, their synthesis which was first carried out by my late Oak Park neighbor and friend, Percy Julian, who may be better known for his synthesis of cortisone.

Establishing the cholinergic transmission at peripheral and central sites followed (Tables 3 and 4). This is a newer and better known story, which includes such notorious episodes as JCE (during his electric transmission phase) challenging Feldberg to find ACh in his sweaty tennis socks... This brief listing cannot give full justice to many aspects of this part of the story - one of such aspects in the case of the neuromyal transmission was the synthesis and structure- activity studies by the Nobel Prizeman, Daniele Bovet (who defined curaremimetcs and depolarizers as pachycurares and leptocurares; Bovet, 1959); this work led to the definition by William Paton, Eleanor Zaimis and Stephen Thesleff (Table 3; for references, see Karczmar, 1967) of the non-depolarizing phase of block induced by the depolarizing neuromyal blockers; this notion culminated in the discovery of the processes of sensitization (Karczmar, 1957 and 1967; Thesleff, 1960) and desensitization (receptor inactivation; Thesleff, 1955 and 1960).

Another interesting general theme which was probably initiated for the first time with respect to the cholinergic system is that of the ontogenesis of transmitter systems and of their comparative pharmacology (Table 3). In the present, cholinergic context these studies were initiated not for their own sake but for an entirely different purpose. Indeed, when DN and Zenon Bacq first approached this subject they had the notion that, if in terms of either development or species comparison they could relate the appearance of ChEs and, better still, AChE to that of function such as motility (they did not have available, in the late thirties, the methodology needed to establish this relationship for the transmitter itself); this would prove the existence of synaptic cholinergic transmission (see Karczmar, 1990).

Another pioneer in the field of comparative and ontogenetic pharmacology must be mentioned here. Theodore Koppanyi (Fig. 4) actually may be referred to in any or all of the rubrics of the Tables 3 - 7. In the leaflet describing the Koppanyi-Georgetown Lecture Series it is stated that Koppanyi could claim discoveries in just about any field of pharmacology generally, and of cholinergic pharmacology specifically, and, if one doubted it, he could pull out a reprint of his which would prove the point! The leaflet is right, and I cite Koppanyi in this

Table 3. Fanfares: Cholinergic Transmission.

Edinburgh and Halle-Dorpat Schools	Peripheral and central actions of Calabar Bean, pilocarpine and "atropia".
Max and Michael Polonowski, Percy Julian	Structure and synthesis of physostigmine and neostigmine.
Reid Hunt, Walter Dixon	Choline derivatives
Walter Dixon, Otto Loewi, John Gaddum, William Feldberg, Geoffrey Brown, A. Kibjakov	Cholinergic parasympathetic and ganglionic transmission.
Henry Dale, William Feldberg, Martha Vogt, Geoffrey Brown, P.T. Feng Bernard Katz, Jose Del Castillo, William Paton, Eleanor Zaimis, Daniele Bovet, Stephen Thesleff, Paul Fatt, Ricardo Milecli, AGK	Cholinergic neuromyal transmission
Henry Dale, Alfred Schweitzer, Samson Wright, John Eccles, Kris Krnjevic, David Curtis, Giancarlo Pepeu	Central cholinergic transmission
David Nachmansohn, Zenon Bacq, K.A. Youngstrom, Theodore Koppanyi, AGK, G.A. Buzikov	Cholinergic ontogenesis and comparative pharmacology
Walter Holbrook Gaskell, A.E. Smirnov, John Newport Langley, J.H. Burn,	Peripheral cholinergic pathway
Pat & Edith McGeer, George Koelle, Lewis Shute, Peter Lewis	Central cholinergic pathways
Thomas Hokfelt, Y.N. and Lilly Jan, Stephen Kuffler, Nae Dun	Release of non-ACh transmitters from cholinergic terminals
Gerhard Schrader, Fritz Lange, Bo Holmstedt, George Koelle, Earl Usdin, Henry Wills, John-Paul Long, AGK	AntiChEs of OP and carbamate type
Henry Dale, Klas-Bertil Augustinsson, F. Plattner, H. Hintner, Harry Rudney, E. Stedman, Bruno Mendel, David Nachmansohn, Werner Kalow	Cholinesterases

particular single context only because he is called the Father of Comparative Pharmacology (see Koppanyi and Sun, 1926), and because he introduced me to the cholinergic lore via the fields of comparative and ontogenetic pharmacology.

Fig. 4. Photograph of Theodore Koppanyi, taken in the late forties in Washington, D.C.

In our fast perusal, we are coming now to the modern - if not quite contemporary - era of cholinergic synaptic research. An important step for our understanding of cholinergic function was the establishing of the cholinergic peripheral and central pathways (Table 3), this latter story occupying us for the last 30 years and extending to the present day (for references, see Karczmar, 1963, 1987 and 1990; Holmstedt and Liljestrand,

1963). Both for the peripheral and central pathways the necessary steps included establishing the cytochemistry of AChE by George Koelle and the immunobiology of choline acetyltransferase (ChAT) by the McGeers and Yamamura. And, the cholinergic synaptology hit a new highpoint when it became established that cholinergic terminals may release more than ACh, i.e, that they may release several transmitters or modulators (Table 3). Parenthetically, this most important late seventies discovery does not disprove the famous Dale's principle concerning transmitter release at neuronal nerve endings; only, instead of stating that the cholinergic neuron releases ACh at all its terminals - the notion which underlies the work of JCE with the Renshaw cell - one must say that a neuron releases the same substance or the same substances at all its terminals (however, as thus phrased, this theorem remains to be proven).

Finally, still within the general area of the progress made with respect to cholinergic transmission, an important aspect of this progress was the research concerning cholinesterases and antiChE agents, including carbamates and organophosphorus drugs (Table 3). Dale suggested already in 1914 that there must be a fast enzyme in the blood which is capable of hydrolyzing ACh, but the actual discovery of cholinesterases dates from the investigations of F. Plattner and H. Hintner; subsequently, the Stedmans and Mendel and Rudney initiated the research concerning the family of cholinesterases (see Augustinsson, 1948, 1963) which today include a number of enzymes and their isozymes, as well as a number of physical forms of AChE (Massoulie, 1986). But, the important aspect of the research in question was establishing AChE as a very fast enzyme, capable of providing hydrolytic kinetics needed to terminate the action of ACh ; this was an important component in the reasoning causing JCE to accept ACh as a transmitter.

In this context, the OP drugs are a most important tool in the study of cholinergic transmission as they work irreversibly and as they exert differential actions at cholinergic synapses. In fact, it was thought for many years that they have only one mechanism of action, a blasphemy to a pharmacologist; fortunately, this proved to be not quite so, a point embraced quite early, although perhaps erroneously by CH (see above, and Table 1), and, in a less extreme form, by Theodore Koppanyi (see Koppanyi and Karczmar, 1951). An interesting point is that this research, as carried out in the forties at Edgewood Arsenal, Maryland, constituted teething grounds for a number of prestigious USA - and, to an extent, also Swedish - investigators, such as George Koelle, Alfred Gilman, Bo Holmstedt, Henry Wilson, Henry Wills, Amadeo Marrazzi, Steve Krop, Bernard McNamara, and others.

Following the demonstration and the description of cholinergic transmission at a number of peripheral and central sites, more intimate, cellular and molecular aspects of this transmission and its processing emerged.

Among the first considerations here was the significance of the ionic size as underlying the generation of the IPSP and the EPSP (JCE, 1964); this, and electrophysiological studies initiated in the fifties by the Canberra team, led to the discovery of cholinergic muscarinic (slow), nicotinic (fast) and, subsequently, additional excitatory potentials described originally - first by means of extracellular recordings, and then intracellularly - for the rabbit sympathetic ganglia by Rosamond Eccles (Table 4; for references, see Karczmar, 1967 and 1987). Her teaming up with Benjamin Libet led to a diagram of multipotential ganglionic transmission and to the well known postulate that the IPSP

Table 4.   Fanfares Molecular and Cellular Aspects of Cholinergic
           Transmission.

Eduardo de Robertis, Victor Whittaker, Paul Fatt, Jose del Castillo, Maurice Israel, Yves Dunant, Brian Collier	Vesicular and cytoplasmic ACh; quantal phenomena elementary events
Earl W. Sutherland, Paul Greengard, L.E. and M. Hokin, Yasutoni Nishizuka	Second messengers and phosphorylations
Henry Dale, Carlos Chagas, Jean-Pierre Changeux, A. Karlin, Jan Lindstrom, Peter Waser, Michael Rafftery, James Patrick, Nigel Birdsall	Cholinergic receptors and their subtypes
John Eccles, Rosamond Eccles, Ben Libet, Kris Krnjevic, Paul Adams, Syogoro Nishi and Kyo Koketsu	Ionics of cholinergic transmission and of synaptic potentials
George Koelle, T.P. Feng, Robert Polak, John Szerb	Regulation of ACh release in the CNS and periphery
Paul Salvaterra, Michael Rafftery, Jean Massoulie, Hermiona Soreq, Jan Lindstrom	Molecular biology of components of cholinergic system

component of these potentials arises via a disynaptic pathway which terminates with the release of dopamine (Eccles and Libet, 1961).

This constituted the beginning of a controversy that lasts now for some 30 years: is the sympathetic ganglionic transmission a disynaptic pathway which includes a chromaffin cell or interneuron, as proposed originally by Rosamond Eccles and Ben Libet, or do we deal here with a monosynaptic pathway? This controversy was broadened when Greengard and his associates (Greengard and Kebabian, 1974) proposed that the disynaptically generated IPSP is, ultimately, the resultant of the activation of a second messenger process which involves the cyclic nucleotide system (see below). Today, it appears that in many species the IPSP is generated monosynaptically, as demonstrated, among others, by my students and associates, Nae Dun and the Gallaghers (for references, see Karczmar and Dun, 1986): furthermore, it is doubtful that Greengard's hypothesis is tenable in so far as the IPSP of the sympathetic ganglion is concerned, although it is provable in the case of certain central sites.

Additional research concerning the molecular and cellular aspects of cholinergic transmission led to our current understanding that ACh is released in quantal packets, these packets being vesicular or cytoplasmic in nature. The phraseology which I employed here diplomatically - "quantal packets... vesicular or cytoplasmic in nature" - glosses over considerable controversy which arose in the sixties and continues to an extent today between Victor Whittaker, the codiscoverer with the late Eduardo De Robertis, of the cholinergic vesicles, on the one hand, and Maurice Israel and others on the other (Table 4).

A major step in the further progress was the development of the concept of the second messengers and of their operations via protein phosphorylations at the effector sites (Table 4). Again, there was here a historical first for the cholinergic system, as the second messenger concept and the role of the second messengers in the evocation of ionic permeability changes in the membrane was identified in the fifties by the Hokins with respect to the muscarinic, cholinergic system (see Hokin and Hokin, 1960); the second messenger system in question was constituted by the phosphotidylinositol cascade. The parallel concept of cyclic nucleotides as second messengers was elaborated by the late Nobel Prizeman Earl Sutherland and particularly related to the cholinergic system by Paul Greengard (Greengard, 1978); additional second messenger systems such as those which are mediated by calmodulin were described subsequently (for references, see Karczmar, 1990). Today, with much elegance, Nishizuka, Greengard, Nahorski and others (see Aquilonius, 1990 and Karczmar, 1990) describe in detail the processes involving specific protein kinases; these processes, generated by the second messenger mechanisms result in site- and receptor-specific effects such as the ionic fluxes at the pre-and postsynaptic membranes, receptor inactivations, and transmitter release.

The question of receptors and receptor subtypes originated with Dale's nicotinic and muscarinic division of cholinergic receptors, or, perhaps better still, with Paul Ehrlich; the division in question was based on the pharmacological differentiation. How exciting it was to hear, much later, about the first attempts - by Carlos Chagas, Sy Ehrenpreis and DN - to obtain receptors as chemical entities. When these first efforts were described at the Rio de Janeiro Meeting of 1960 (Chagas and Paes de Carvalho, 1961), I for one thought that we were dealing here with chimerae: I did not realize that what actually these investigators accomplished was a drastic "paradigm shift" (Kuhn, 1962) in the meaning of the term "receptor" - from an abstract or, at best, pharmacological notion to a chemical definition! Once this semiotic breach was made, several leads were followed, including Peter Waser's (1967) radioautographic visualization of the nicotinic, neuromyal receptors. The subsequent progress was exponential and it led to the identification of the subtypes of the nicotinic and muscarinic receptors, the elucidation of their chemical structure, and the genetics involved in their ontogenesis and regulation. The molecular approaches, particularly, delivered remarkable results concerning the genetics of cholinesterases as obtained in Paris by Jean Massoulie (Massoulie, 1986) and, in Jerusalem, by Hermiona Soreq, as well as the genetics and the evolution of cholinergic receptors as established in this country by Paul Salvaterra and his associates (for references, see Karczmar, 1990; see also Aquilonius, 1990).

Regulation of ACh release may be considered as still another aspect of the cellular mechanisms involved in cholinergic transmission. The story here may begin with the discovery of the responsiveness of the motor nerve terminal to cholinergic drugs (other drugs such as guanidine, etc.) by T. P Feng (see above and Feng et al., 1939); this sensitivity relates to the regulation of ACh release from the nerve terminal via, first of all, ACh itself, as proposed first by George Koelle (1963). Now, we know that these phenomena are present at both the peripheral and central junctions (Table 4); in fact, ACh controls also, via heteroreceptors, the release of non-ACh transmitters at noncholinergic

Table 5. Fanfares: Modulations, Metabolism, etc.

Kyo Koketsu, Erminio Costa, AGK	Pre-and post-synaptic modulations
Frank McIntosh, R. Birks, Brian Anzell, Richard Wurtman, Stanislav Tucek, J. Kazimierz Blusztajn, Erminio Costa, Israel Hanin, Don Jenden, Bo Holmstedt, Lynn Wecker	Measurement, metabolism and turnover of ACh
Fred Schueler, Frank McIntosh, Kazuko and Taksuga Haga	Hemicholinium and uptake of choline

synapses; reciprocally, other than ACh transmitters regulate the release of the latter at cholinergic synapses.

This matter of the regulation of ACh release leads us to the consideration of the processes of modulation that exist both post and presynaptically and which bring about a subtle point-to-point control of synaptic processes, as suggested early by Koketsu (1977) and Karczmar and Nishi (Karczmar et al., 1972; Table 5). Transmitters and non-transmitter modulators (i.e., substances that, per se, do not activate whether inhibitory or excitatory transmission; see Karczmar, 1986 and 1990) are involved in these processes, as are postsynaptic biochemical processes, referred to by Erminio Costa as trans-synaptic phenomena (see Costa et al., 1987).

Modulations must embrace, as well, the metabolism of ACh, as this metabolism is sensitive to environmental and biochemical signals. Before this flexibility of ACh metabolism could be established, its characteristics had to be evaluated and clarified. This evaluation was promoted by the discovery and development, by the late Fred Schuler and John Paul Long, of the choline uptake blockers, the hemicholiniums (originally designed with their possible antiChE potency in mind), and these compounds were used in the pioneering research of Frank McIntosh (Table 5). Soon, the story of ACh metabolism and turnover - a term coined first, presumably, by Erminio Costa - went beyond the uptake of transmission-generated choline, as it was shown, particularly by the late Brian Ansell and Sheila Spanner that this metabolism involves phospholipids; further complications and several pools of this metabolism were described by Richard Wurtman, Kazimierz Blusztajn, Stanislav Tucek, Steve Zeisel, Lynn Wecker, and others (Table 5). This progress could not have been made without advances in the methodology for the measurement of ACh which was developed particularly by Don Jenden, Bo Holmstedt and Israel Hanin.

As the knowledge of the cholinergic system became somewhat settled and certain mainlines of investigation established, new vistas began to emerge. One of those concerns behavior and its close correlate, the EEG. While even the early Edinburgh school referred to the behavioral effects of physostigmine, particularly in the case of the self experimentation carried out by Robert Christison, the corresponding animal studies and, particularly, the use of the operant behavior and conditioning paradigm

Table 6. Fanfares: New Functions.

William Funderburk, Theodore Case, William Hursley Gantt, William Feldberg, Stephen Sherwood, Carl Pfeiffer, Daniele Bovet, Donald Reis, Lembit Allikmets, Luigi Valzelli, Robert Myers, AGK	Behaviors
V. Bonnet and Frederic Bremer, Donald Tower, K.A.C. Eliott, Giuseppe Moruzzi, Horace Magoun, Raul Hernandez-Peon, Peter Andersen, Sven Andersson, F.T. Brucke, Christof Stumpf, Vincenzo Longo, Philip Bradley, Joel Elkes, Michel Jouvet, Herbert Jasper, Robert McCarley, J.A. Hudson, Edward Domino, Chris Gillin, G.H. Vanderwolf, R. Kramis, Frank Duffy, Taddeus Marczynski, AGK	EEG, theta rhythms, and REMS
Victor Hamburger, Paul Weiss, Rita Levi-Montalcini, John Zachary Young, Paul Guth, E.D. Bueker, Ian Hendry, Hans Thoenen, E. Cohen	Retro-and antero-grade trophic factors
Mary Pickford, George Hedge, Charles Sawyer, E.W. Naumenko, I. Bern	Endocrine and immuno-biological effects

by William Funderburg, Carl Pfeiffer (Pfeiffer and Jenney, 1957) and others (see Table 6) led to new understanding of the behavioral effects of the cholinergic drugs and their site of action. One of the interesting suggestions made at that time (in the late forties) by Carl Pfeiffer concerned the role of the cholinergic system as an "antischizoid" factor, a concept which is still under inspection (see Karczmar, 1988). This then led to establishing the role of the cholinergic system in memory and learning processes as well as in just about any behavior which can be distinguished or measured (see next Section). Furthermore, the EEG counterparts of some of these behaviors were established with the pioneering work of V. Bonnet and Frederic Bremer, R. G. Hernandez-Peon, Joel Elkes and Phil Bradley, Enzo Longo, Ed Domino and others (Table 6), and this led then to the understanding of the role of the cholinergic system in cognition and the REM sleep (Table 6; see also below, and Karczmar, 1979, 1990 and 1991).

An entirely new departure was the development of the concept of trophic influences acting upon (retrograde trophic factors) or evoked by (anterograde trophic factors), the cholinergic system. While some vague premonitions existed with respect to this concept already in the thirties (see above), we owe the first demonstration, in the late forties, of the existence of the trophic factors to Victor Hamburger, E.D

Table 7. Fanfares: Models for Diseases with Cholinergic Implications.

J. Simpson, James Patrick and Jan Lindstrom, David Drachman, A.G. Engel	Immunological model for myasthenia gravis
Rita Rudel, J.E. Somers	Chemical model for myotonia congenita
Israel Hanin	Action of AF64A as model for Alzheimer's disease
Volia (W.T.) Liberson, Charles Scudder, AGK	Non-goal behavior as model for affective disorders
Stanley Appel	Immunological model for Amyotrophic lateral sclerosis (ALS)

Bueker, and, particularly, Rita Levi-Montalcini (see Table 7); this demonstration concerned the Nerve Growth Factor (NGF). This discovery and the characterization of the activities of the NGF earned the Nobel Prize for Rita Levi-Montalcini; as she descended, in the course of the acceptance ceremony, the staircase of the Stockholm City Hall on the arm of the King of Sweden she appeared easily as regal as the King (Giacobini, personal communication; see also Levi-Montalcini, 1987). Many additional retrograde as well as anterograde-acting trophic factors have been discovered since.

There is still another instance of the involvement of the cholinergic system with an entirely different, non-transmitter system - in this case, the neuroendocrines and immunobiological regulation (Table 6). While the first, pertinent steps were made in the late forties by Mary Pickford with respect to vasopressin and oxytocin, many additional ramifications of the cholinergic-endocrine interactions have been demonstrated since. These interactions involve the immune system as well, and, while catecholamines are more generally implicated in the processes in question, a cholinergic influence was demonstrated as well (see Ader et al., 1990). Another aspect of the concept of cholinergic-immunobiological interaction is constituted by the possibility that certain diseases that exhibit cholinergic features, may be caused by self-immune processes (see next paragraph).

The final - but, I am sure, not the last! - step in cholinergic progress concerns the development of animal models, with cholinergic implications, of disease states, the earliest -and dramatic! - instance being the immunobiological model of myasthenia gravis (Table 8), presented originally by J. Simpson, J. Patrick and Jan Lindstrom, and David Drachman. Additional models have become recently available as well (Table 7).

Table 8. CNS (organic) effects of cholinergic drugs.

I. Motor Behavior and Related Neurological
    Syndromes

    A. Catalepsy
    B. Locomotor and Related Actions: gnawing,
        self-biting, head motion, sniffing;
        compulsive circling; hypokinesia
    C. Tremor
    D. Convulsions

II. Respiration

III. Appetitive Behavior

    A. Hunger and Feeding: effect dependent on
        brain site and species
    B. Thirst and Drinking: effect dependent on
        brain site and species

IV. Thermocontrol

    A. Heat Production
    B. Hearing Loss

V. CV Control

VI. Endocrine and immune function

VII. EEG and Brain Excitability

VIII. Sleep

    A. REMS (rapid eye movement sleep)
    B. SWS (slow wave sleep)

IX. Chronobiology

    A. Diurnal Rhythms
        B. Hibernation, Seasonal Changes
    C. Aging

X. Sensorium
    A. Nociception
    B. Audition
    C. Vision

XI. Sexual Activity

I cannot avoid the temptation to end this presentation with some allusions as to the functional and behavioral implications of the central cholinergic system. These few remarks are pertinent, as this Symposium concerns certain types of behavior, namely the cognitive and learning functions, and it is well to remember that these functions constitute just a small part of the spectrum of the behavioral role of the cholinergic system.

I wish to emphasize here a special point. The major steps in the cholinergic progress which I have summarized above illustrate well the rich inventory of cholinergic phenomena and their bearing on a multitude of biological processes. But, as with the Mandelbrot set, the richness of this pattern is repeated as one focuses on any of the steps in question, and as the same rich spectrum and wide range of phenomena spreads then before one's eyes, and this process can be repeated, over and over again, ad infinitum! Let us then focus on the step concerning the behavioral implications of the cholinergic system.

What I am showing now is two lists: one concerns what I refer to as organic central functions (Table 8); the other list concerns what can be called behavioral or mental phenomena (Table 9). Some of the activities listed may defy this organization. For instance, the EEG and REM sleep phenomena which are listed in Table 8 as organic in nature present, of course, behavioral and mental aspects as well. Altogether, one cannot help but be impressed with the significance of the cholinergic system for a multitude of organic functions which include respiration and cardiovascular and motor phenomena as well as hypothalamically regulated homeostasis of energy processes which involve thermocontrol, feeding and drinking, and its significance for a wide behavioral spectrum, ranging from aggression to addiction to schizophrenia (for references, see Karczmar, 1976, 1984 and 1988).

There is one matter that is particularly pertinent to the focus of this Symposium on cognitive function. Indeed, the cholinergic system is crucial for cognition, particularly as it concerns the organism-environment interaction. This particular regulation results in what I termed Alert Cholinergic Nonmobile Behavior (CANMB, Table 10; see Karczmar, 1979, 1990 and 1991). The CANMB partakes of cognitive processes, REM sleep and EEG phenomena (slow theta rhythms; Vanderwolf, 1975; and Karczmar, 1979), and "anti-schizoid" modulation, and it may constitute the most important aspect of the cholinergic regulation of mental behavior.

ENVOI

As I indicated in my introductory remarks, the notion of the importance of the cholinergic system constitutes a truism which really does not require proof. But, perhaps, this notion needs an occasional reminder. This chapter represents this reminder, and also, in telling the cholinergic story, it alludes, as incompletely as it undubitively does, to these brilliant and unique researchers who made this story so rich and so heuristic. But, besides constituting a reminder of the past, this chapter, as it illustrates the continuity of the progress of cholinergic research which extends into the contemporary areas ranging from molecular biology to cognition, affords the prediction that the future of cholinergic research will be no less brilliant than its past.

Table 9. Behavioral, "psychological," and "mental" functions
In animals with cholinergic connotations.

I.   Aggression

     A.   Emotional (Affective)
     B.   Predatory
     C.   Irritable

II. Learning and Related Phenomena

     A.   Conditioning
     B.   Memory (Short and Long Term)
     C.   Habituation
     D.   Retrieval

III. Imprinting

IV. Emotional Behavior and Fear

V.   Addiction, Dependence, Withdrawal
     Syndrome

     A.   Opiate Addiction
     B.   Ethanol Addiction

VI. "Schizoid" Behavior

VII. Organism - Environment Interaction (OEI)

     A.   Cognitive Functions
     B.   Cholinergic Alert Non-Mobile Behavior

REFERENCES

Ader, R., Felten, D., and Cohen, N., 1990, Interactions between the brain and the immune system, Ann. Rev. Pharmacol. Toxicol., 30:561.

Aquilonius, S.-M., and P.-G. Gillberg, ed.: Cholinergic. Neurotransmission: Functional and Clinical Aspects. Nobel Symposia, Stockholm, 1990.

Augustinsson, K.B., 1948, A study in comparative enzymology, Acta Physiol. Scand. 15, Suppl. 52, 1.

Augustinsson, K.B., 1963, Classification and comparative enzymology of cholinesterases, and methods for their determination, in: "Cholinesterases and Anticholinesterase Agents", Hdbch. d. Exptl. Pharmakol., Ergänzungswerk, G.B. Koelle, ed., Springer, Berlin, 15(4):89.

Bacq, Z.M., 1975, "Chemical Transmission of Nerve Impulses". Pergamon Press, Oxford.

Baghdoyan, H.A., Rodrigo-Angulo, M.L., McCarley, R.W., and Hobson, J.A., 1987, A neuroanatomical gradient in the pontine tegmentum for the cholinoceptive induction of desynchronized-sleep signs, Brain Res., 414:245.

Bovet, D., 1959, Rapports entre constitution chimique et activite pharmacodynamique dans quelques series de curares de synthese, in: "Curare and Curare-like Agents", D. Bovet, F. Bovet-Nitti and G.B. Martini-Bettolo, eds., Elsevier Publishing Co., Amsterdam, p. 252.

Brazier, M.A.B., 1959, The historical development of neurophysiology, in: "Neurophysiology", H.W. Magoun, ed., Handbk. Physiol. 1:1, Am. Physiol. Soc., Washington, D.C.

Chagas, C. and Paes de Carvalho, A., eds., 1961, "Bioelectrogenesis. Comparative Survey of its Mechanisms, with Special Emphasis on Electric Fishes", Elsevier Publishing Company, Amsterdam.

Costa, E., Hanbauer, I., and Guidotti, A., 1987, Receptor-receptor interactions in the modulation of nicotinic receptors in adrenal medulla, in: "Neurobiology of Acetylcholine", Symposium Held in Honor of Alexander G. Karczmar, N.J. Dun and R.L. Perlman, eds., Plenum Press, New York, p. 355.

Dun, F.T., and Feng, T.P., 1940, Studies on the neuromuscular junction, XX. The site of origin of the junctional after-discharge in muscles treated with guanidine, barium or eserine, Chinese J. Physiol. 15:433.

Eccles, J.C., 1987, The story of the Renshaw Cell, in: "Neurobiology of Acetylcholine", Symposium Held in Honor of Alexander G. Karczmar, N.J. Dun and R.L. Perlman, eds., Plenum Press, New York, pp. 189.

Eccles, J.C., Fatt, P., and Koketsu, K., 1953, Cholinergic and inhibitory synapses in a central nervous pathway, The Australian J. Sci. 16:50.

Eccles, J.C., Fatt, P., and Koketsu, K., 1954, Cholinergic and inhibitory synapses in a pathway from motor-axon collaterals to motoneurons, J. Physiol. (Lond.) 126:524.

Eccles, R.M., and Libet, B., 1961, Origin and blockade of the synaptic responses of curarized sympathetic ganglia. J. Physiol. (Lond), 157:484.

Feng, T.P., Li, T.H. and Ting, Y.C. 1939, Studies on the neuromuscular junction, XII. Repetitive discharges and inhibitory after-effect in post-tetanically facilitated responses of cat muscles to single nerve volleys, Chinese J. Physiol., 14:55.

Feng, T.P., and Li, T.H. 1941, Studies on the neuromuscular junction, XIII. A new aspect of the phenomenon of eserine potentiation and post-tetanic facilitation in mammalian muscles, Chinese J. Physiol., 16:37.

Greengard, P., 1978, "Cyclic Nucleotides, Phosphorylated Proteins and Neuronal Function. Raven Press, New York.

Greengard, P., and Kebabian, J.W., 1974, Role of cyclic AMP in synaptic transmission in the mammalian peripheral nervous system. Fed. Proc. Fed. Am. Soc. Exp. Biol. 33:1059.

Hokin, M.R., and Hokin, L.E., 1960, The role of phosphatidic acid and phosphoionositide in transmembrane transport elicited by acetylcholine and other humoral agents. Int. Rev. Neurobiol. 2:99.

Holmstedt, B., 1972, The ordeal bean of old calabar: The pageant of physostigma venenosum in medicine. in: "Plants in the Development of Modern Medicine", T. Swain, ed., pp. 303-360, Harvard U. Press, Cambridge, Mass.

Holmstedt, B., and Liljestrand, G., 1963, "Readings in Pharmacology", Pergamon Press, Oxford.

Karczmar, A.G., 1957, Antagonism between a bis-quaternary oxamide, WIN 8078, and depolarizing and competitive blocking agents, J. Pharmacol. Exp. Therap. 119:39.

Karczmar, A.G., 1967, Pharmacologic, toxicologic and therapeutic properties of anticholinesterase agents, in: "Physiological Pharmacology", W.S. Root, and F.G. Hofman, eds., 3:163.

Karczmar, A.G., 1976, Central actions of acetylcholine, cholinomimetics and related drugs, in: "Biology of Cholinergic Function", A.M. Goldberg, and I. Hanin, eds., Raven Press, New York, p. 395.

Karczmar, A.G., 1979, Brain acetylcholine and animal electrophysiology, in "Brain Acetylcholine and Neuropsychiatric Disease", K.L. Davis and P.A. Berger, eds., Plenum Press, New York, p. 265.

Karczmar, A.G., 1980, Drugs, transmitters and hormones and mating behavior, in: "Modern Problems in Pharmacopsychiatry", T. Ban, ed., Karger, Basel, Vol. 5, p. 1.

Karczmar, A.G., 1984, Acute and long lasting central actions of organophosphorus agents. Fund. Appl. Toxicol. 2:S1.

Karczmar, A.G., 1987, Concluding remarks: past, present and future of cholinergic research, in: "Neurobiology of Acetylcholine", Symposium Held in Honor of Alexander G. Karczmar, N.J. Dun and R.L. Perlman, eds., Plenum Press, New York, p. 561.

Karczmar, A.G., 1988, Schizophrenia and cholinergic system, in: "Receptor and Ligands in Psychiatry" A.K. Sen, and T. Lee, eds., Cambridge University Press, Cambridge, p. 29.

Karczmar, A.G., 1990, Physiological cholinergic functions in the CNS, in: "Cholinergic Neurotransmission: Functional and Clinical Aspects", S.-M. Aquilonius, and P.-G. Gillberg, eds., Nobel Symposium, Stockholm, p. 437.

Karczmar, A.G., 1991, "The Cholinergic System, its Pharmacology and Function", Plenum Press, New York (in preparation).

Karczmar, A.G., and Dun, N.J., 1986, Pharmacology of synaptic ganglionic transmission and second messengers, in: "Autonomic and Enteric Ganglia", A.G. Karczmar, Koketsu, and S. Nishi, eds., Plenum Press, New York, p. 297.

Karczmar, A.G., and Dun, N.J., 1988, Effects of anticholinesterases pertinent for SDAT treatment but not necessarily underlying their clinical effectiveness, in: "Current Research in Alzheimer Therapy: Cholinesterase Inhibitors", E. Giacobini, and R. Becker, eds., Taylor and Francis, New York, p. 15.

Karczmar, A.G., Nishi, S., and Blaber, L.C., 1972, Synaptic modulations, in: "Brain and Human Behavior", A.G. Karczmar, and J.C. Eccles, eds., Springer-Verlag, Berlin, p. 63.

Koelle, G.B., 1963, Cytological distributions and physiological functions of cholinesterases, in: "Cholinesterases and Anticholinesterase Agents" G.B. Koelle, ed., Handbch. d. exper. Pharmakol., Ergänzungswerk, Springer-Verlag, Berlin, 15: 187.

Koketsu, K., 1977, Neurohumoral controls of neurone activities, in: "Neurohumoral Correlates of Behavior", S. Subrahmayan, ed., Thomson Press, Faridabad, p. 21.

Koppanyi, T., and Sun, K.H., 1926, Comparative studies on pupillary reaction in tetrapods. II. The effect of pilocarpine and other drugs on the pupil of the rat, Amer. J. Physiol., 75:355.

Koppanyi, T., and Karczmar, A.G., 1951, Contribution to the study of the mechanism of action of cholinesterase inhibitors, J. Pharm. Expt. Therap. 101:327.

Krnjevic, K., and Straughan, D.W., 1964, The release of acetylcholine from the denervated rat diaphragm, J. Physiol. (London) 170:371.

Kuhn, T., 1962, "Structure of Scientific Revolution", Chicago Press, Chicago, First Edition.

Lindgren, S., Lijestrand, G., and Zotterman, Y., 1952, The effect of certain autonomic drugs on the action potentials of the sinus nerve. Acta Physiol. Scandinav., 26:264.

Levi-Montalcini, R., 1987, "Elogio dell' Imperfezione", Garzanti Editore, Milano.

Massoulie, J., 1986, Polymorphism of cholinesterase: possible insertion of the various molecular forms in cellular structures, in: "Dynamics of Cholinergic Function", I. Hanin, ed., Plenum Press, New York, p. 727.

Pfeiffer, C.C., and Jenney, E.H., 1957, The inhibition of the conditioned response and the counteraction of schizophrenia by muscarinic stimulation of the brain, Ann. N.Y. Acad. Sci. 66:753.

Thesleff, S., 1955, The mode of neuromuscular block caused by acetylcholine, nicotine, decamethonium and succinylcholine, Acta Physiol. Scand. 34:218.

Thesleff, S., 1960, Supersensitivity of skeletal muscle by botulinum toxin, J. Physiol. (London) 151:598.

Vanderwolf, C.H., 1975, Neocortical and hippocampal activation in relation to behavior: Effects of atropine, eserine, phenothiazines, and amphetamine, J. Comp. Physiol. Psychol. 88(1):300.

Waser, P.G., Receptor localization by autoradiographic techniques, Acad. Sci., 144:737.

REFERENCES FOR THE QUIZ

AGK, 1946, The role of amputation and nerve resection in the regressing limbs of urodele larvae, J. Exp. Zool. 103:401.

AGK, 1963, Ontogenetic effects of anticholinesterase agents, in: "Cholinesterases and Anticholinesterase Agents", G.B. Koelle, ed., Handbch. d. Exper. Pharmakol., Ergänzungswerk., Springer-Verlag, Berlin, 15:179.

AGK, 1963, Ontogenesis of cholinesterases, in: "Cholinesterases and Anticholinesterase Agents", G.B. Koelle, ed., Handbch. d. Exper. Pharmakol., Ergänzungswerk., Springer-Verlag, Berlin, 15:179.

AGK, 1967, Pharmacologic, toxicologic and therapeutic properties of anticholinesterase agents, in: "Physiological Pharmacology", W.S. Root and F.G. Hofman, eds., Academic Press, Inc., 3:163.

AGK, 1969, Is the central cholinergic nervous system overexploited? in: "Symposium on Central Cholinergic Transmission and its Behavioral Aspects", Fed. Proc., 28:147.

AGK, 1972, History of the research with anticholinesterase agents, in:

"Anticholinesterase Agents" AGK, ed., International Encyclopedia of Pharmacology and Therapeutics, 1(13):1.

AGK, 1991, "The Cholinergic System, its Pharmacology, Physiology and Function", Plenum Press, New York (in preparation).

AHH, and WFR, 1963, Acetylcholine release at the neuromuscular junction, J. Pharmacol. Exp. Therap. 142, 200.

CH, 1955, Action of drugs on carotid body and sinus, Pharmacol Rev. 7:119.

CH, 1950, Les antidotes du di-isopropylfluorophosphonate (DFP), Arch. Int. Pharmacodyn., 81:230.

DK, MRR, and ELB, 1966, Environmental impoverishment, social isolation and changes in brain chemistry and anatomy, Physiol. Behav. 1:99.

DN, 1961, Chemical factors controlling nerve activity, Science, 134: 1962.

DN, 1963, The chemical basis of Claude Bernard's observations on curare, Biochem. Zeitsch. 338:454.

DN, 1963a, Choline acetylase, in: "Cholinesterases and Anticholinesterase Agents", G.B. Koelle, ed., Handbch. d. Exper. Pharmakol., Ergänzungswerk, Springer-Verlag, Berlin, 15: 40.

FFW, 1968, Cholinergic mechanisms in recurrent inhibition of motoneurons, in: "Psychopharmacology", Proc. Amer. College of Neuropsychopharmacology, Government Printing Office, Public Health Service Publication No. 1836, M.A. Lipton, ed., p. 69.

FGS and WFR, 1967, The consequences of cholinergic drug actions on motor nerve terminals, Ann. N.Y. Acad. Sci. 144:517.JCE, 1964.

JCE, FD, and KK, 1953, Cholinergic and inhibitory synapses in a central nervous pathway, The Australian J. Sci., 16:50.

JCE, PF, and KK, 1954, Cholinergic and inhibitory synapses in a pathway from motor-axon collaterals to motoneurons, J. Physiol. (Lond.) 126:524.

JCE, 1964, "The Physiology of Synapses", Springer-Verlag, New York, Inc. JCE, 1987, The story of the Renshaw cell, in:" Neurobiology of Acetylcholine, a Symposium Held in Honor of Alexander G. Karczmar", N.J. Dun, and R.L. Perlman, eds., Plenum Press, New York, p. 189.

MJ, 1967, Regulation neuro-humorale des etats de sommeil, in: "Aspects Biologiques et Cliniques du Systéme Nerveux Central", Symposium de Sandoz SA, Bale, pp. 103.

MJ, 1972, Some monoaminergic mechanisms controlling sleep and waking, in: "Brain and Human Behavior", A.G. Karczmar and J.C. Eccles, eds., Springer-Verlag, Berlin, p. 131.

MRR, DK, and ELB, 1958, in: " Biological and Biochemical Bases of Behavior", H.F. Marlow and C.N. Woolsey, eds., Madison Univ. Wisconsin Press, p. 367.

MRR, ELB, and MCD, 1967, in: "Psychopathology of Mental Development", J. Zubin and G. Jervis, eds., New York, Grune & Stratton, p. 45.

WFR and FGS, 1957, The motor nerve terminal as the primary focus for drug-induced facilitation of neuromuscular transmission. J. Pharmacol. Exp. Therap. 121:286.

Editor's Note: For those of you who would enjoy knowing the names of the individuals in Dr. Karczmar's quiz, please refer to the table provided on the following page.

Names of Individuals Whose Initials are Used in the Text of
"Historical Aspects of the Cholinergic Transmission"
by Alexander G. Karczmar.

INITIALS	NAMES
AGK	Alexander G. Karczmar
AHH	Arthur Hull Hayes
CH	C. Heymans
DK	David Krech
DN	David Nachmansohn
ELB	Edward L. Bennett
FD	Frank Duffy
FFW	F.F. Weight
FGS	Frank G. Standaert
JCE	John C. Eccles
JR	Jay Roberts
KK	K. Koketsuo
MCD	Marian C. Diamond
MJ	M. Jouvet
MRR	Mark R. Rosenzweig
WFR	Walter F. Riker

Index